17th International Scientific Conference on Current Problems of the Corporate Sector 2020 (ICRTEG 2020)

SHS Web of Conferences Volume 83 (2020)

Bratislava, Slovakia
13 - 14 May 2020

Editors:

Dana Hrusovska
Maria Kmety Bartekova
Maria Kozakova

ISBN: 978-1-7138-1977-6

TABLE OF CONTENTS

UNIVERSITY OF ECONOMICS IN BRATISLAVA

Faculty of Business Management

Department of Business Economy

CURRENT PROBLEMS OF THE CORPORATE SECTOR 2020

17th International Scientific Conference

13th – 14th May 2020

Bratislava, Slovak Republic

ORGANIZED BY

FOUNDATION MANAGER
UNIVERSITY OF ECONOMICS IN BRATISLAVA, Faculty of Business Management, Department of Business Economy

GUARANTEED BY

prof. Ing. Helena Majdúchova, CSc.

prof. Ing. Štefan Majtán, CSc.

doc. Ing. Alena Tršťanská, CSc.

EDITED BY

Ing. Dana Hrušovská, PhD.
PhDr. Mária Kmety Barteková, PhD.
Ing. Monika Raková, PhD.
Ing. Mária Kozáková, PhD.

PROGRAMME COMMITTEE

prof. Ing. Peter Markovič, PhD.	University of Economics in Bratislava, Slovakia
doc. Ing. Michaela Krechovská, PhD.	University of West Bohemia, Czech Republic
dr hab. Grzegorz Głód, prof. UE	University of Economics in Katowice, Poland
prof. Mgr. Peter Štarchoň, PhD.	Comenius University in Bratislava, Slovakia
doc. Ing. Mgr. Gabriela Dubcová, PhD.	University of Economics in Bratislava, Slovakia
Ing. Mgr. Jakub Procházka, PhD.	Masaryk University, Czech Republic
prof. Ing. Rastislav Rajnoha, PhD.	Pan European Uninversity, Slovakia
doc. Ing. Jindra Peterková, PhD.	Technical University of Ostrava, Czech Republic
prof. Ing. Zdeněk Mikoláš, CSc.	The College of Entrepreneurship and Law, Czech Republic
prof. Ing. Eleonora Fendeková, PhD.	University of Economics in Bratislava, Slovakia
prof. Ing. Ladislav Blažek, CSc.	Masaryk University, Czech Republic
prof. Ing. Lilia Dvořáková, CSc.	University of West Bohemia, Czech Republic
doc. Ing. et Ing. Renáta Myšková, PhD.	University of Pardubice, Czech Republic
doc. RNDr. Ing. Hana Scholleová, PhD.	Czech Technical University in Prague, Czech Republic

ORGANIZING COMMITTEE

CURRENT PROBLEMS OF THE CORPORATE SECTOR is an international scientific conference, thematically focused on the various current themes of the corporate sector and the topics linked to the business environment. Its main aim is the presentation of current business issues that concern managers and entrepreneurs operating in CEE markets, presentation of scientific research outcomes by the conference participants and supporting the knowledge transfer by finding the links between scientific research and business practice and the cooperation with domestic and foreign educational and research institutions.

The condition of publishing the paper in the proceeding is its originality, its contribution, presentation of empirical knowledge both at home and abroad. The 17th volume of the conference focuses on following fundamental issues:

1. Business economics and finance
2. Marketing research and strategy
3. Business Information Systems
4. E-business
5. Corporate Social Responsibility
6. Supply Change Management
7. Public Administration and Small Business Entrepreneurship
8. Entrepreneurship in small and medium-sized enterprises
9. Economic development of regions and international studies
10. Corporate finance and governance

Statement of Peer review

In submitting conference proceedings to *Web of Conferences*, the editors of the proceedings certify to the Publisher that

1. They adhere to its **Policy on Publishing Integrity** in order to safeguard good scientific practice in publishing.
2. All articles have been subjected to peer review administered by the proceedings editors.
3. Reviews have been conducted by expert referees, who have been requested to provide unbiased and constructive comments aimed, whenever possible, at improving the work.
4. Proceedings editors have taken all reasonable steps to ensure the quality of the materials they publish and their decision to accept or reject a paper for publication has been based only on the merits of the work and the relevance to the journal.

Title, date and place of the conference

Current Problems of the Corporate Sector 2020
13-14 May 2020, Bratislava (Slovakia)

Proceedings editor(s):

Dana Hrušovská
Maria Kmety Barteková
Maria Kozáková

Date and editor's signature

Using Price Gaps for Hedging Prices and Making Financial Decisions

Peter Badura[1,*]

[1]University of Economics, Faculty of Business Management, Department of Business Finance, Dolnozemska cesta 1, 851 04 Bratislava, Slovakia

Abstract. Price gaps in assets pricing are relatively rare. Gaps arise at the moment when the open price of a new period opens significantly lower or higher than the close price of the previous period. The aim of this paper is to find out how often gaps are created in the prices of a selected underlying asset and how they can be used for improving the corporate financial situation. The object of our examination was a soybeans oil commodity traded on the e-CBOT futures market while the subject of the research were the price gaps themselves, the frequency of their occurrence and the likelihood of their closing. Data were analyzed over a period of 30 years. The fact that it is more likely than unlikely that the price will return and close the gap has been confirmed. The larger the price gap was, the longer it was necessary to wait for it to close. However, as for trading, it was also possible to take advantage of the low probability that the price gap would be closed - to set up a suitable stop loss order.

1 Introduction

Gaps are relatively rare phenomena in the financial assets valuation in financial markets. The more rarely they occur, the more reliable signals they can generate (for example, they are one of the ways how choose trades with a profit probability of more than 70 %). That is the reason why price gaps could also be used by enterprises to improve their financial situation and stability. This applies in particular to companies that have access to financial markets and actively trade in them (e.g. agricultural companies that use futures markets).

Gaps arise when the opening price of a new period opens significantly lower or higher than the closing price of the previous period. There is a kind of free space between these prices, which is not filled. This "vacuum" is an illustration of the range of prices that were skipped and for which no trades were concluded. More precisely - it should be noted that this is the blank space between the highest and lowest price of two consecutive periods. As stated by C. M. Corcoran: "Discontinuities in price behavior or “gaps” are one of the most intimidating characteristics of financial markets. Gaps or discontinuities arise because price does not follow a trajectory in the same way that physical objects move continuously through spatial dimensions. Although price can be mapped into two or even three dimensional

* Corresponding author: peter.badura@euba.sk

graphical surfaces to depict the movement of price "through" time, there will be discontinuities in the path of movement." [1]

2 The aim of the paper and methodology

The aim of this paper is to find out how often gaps are being created in the prices of a selected underlying asset and how they can be used for hedging prices or financial decisions of enterprises. Especially companies dealing with agricultural products can actively trade e.g. futures contracts and take advantage of the price gaps once created.

We are using MS Excel and its functions, in which the underlying prices of the asset are being analyzed. Soybean oil commodity traded on the e-CBOT futures market (a member of the CME group) has been chosen as the object of our examination. The subject of the examination were price gaps themselves, the frequency of their occurrence and the likelihood of their closing. Data were analyzed over a period of thirty years.

3 Price gaps theory

The reason for creating price gaps is that new information has entered the market, usually at a time when it was closed for trading. The surprising information subsequently had an impact on extreme price moves after the market opening (it could be e.g. bigger than expected changes in interest rates, natural disasters, unexpected news, etc.). Author T. Knight states: "Many things can cause a price gap, such as surprisingly good earnings, surprisingly bad earnings, FDA approval or disapproval of an important drug at a pharmaceutical company, a takeover of a company, and so on. Gaps are the market's way of signaling a big change, but knowing what to do with that change separates profitable trades from losing ones." [2] D. H. Weis & A. Elder concluded: "The market we are watching can gap sharply higher or lower because of unexpected news events. These are the extremes of the probability curve but they must be accepted by anyone who ventures into speculation as part of the territory." [3] More information on the topic of unexpected news can be found in the paper by K. P. Evans: "Intraday jumps and US macroeconomic news announcements". [4] Alike, J. Bernstein on the topic of price gaps: "Gap days are very important because they often provide very reliable day trading and short-term trading opportunities." [5] Author C. A. Kase confirms: "The placement of gaps before or after key patterns, or during trends, can give clues as to the market's future action." [6]

Gaps are usually created on market openings. Due to the information received during a break or a market close, an enormous amount of market orders (buy or sell) has been accumulated, which means that traders are willing to sell or buy significantly cheaper or more expensive than they were at the closing price of the previous period. The pressure to move in one direction is so huge that the market will open up at a significantly higher or lower price level.

T. N. Bulkowski states in his book: "Traders define an opening gap as one in which the opening price gaps away from the prior day's close, leaving a hole on the chart - at least temporarily. Some day traders fade the opening gap and others trade with the trend. Fade means to go against the prevailing trend, like trying to spit into the wind." [7] Alike, E. Ponsi states: "Gaps commonly occur when a market opens. Illiquid trading vehicles have a higher tendency to gap than liquid trading vehicles. Some investments will gap in the middle of the trading session. These intraday gaps are particularly prone to occur during lunch hours, when many participants are away from their desks." [8]

However, there are many authors dealing with the price gaps and the basic groups they can be put into, e.g. T. N. Bulkovski in his other books about technical analysis and the price gaps [9, 10], A. Plastun et al. [11], J. Stubinger & L. Schneider, [12], M. C. Thomsett [13], G. M. Caporale & A. Plastun [14], GF. Gu et al. [15]. Most authors agree that price gaps can be divided into four basic groups:

1. Common gaps (also Trading gaps) – are the most common gaps and do not necessarily increase the volume of trades of the underlying asset. They arise coincidentally, when multiple market orders accumulate in one direction. These orders are supported by most market players after the market opens. However, such gaps are closed within the next few days.

2. Breakaway gaps – are the ones that follow the breakthrough of a strong technical formation (usually triangles or levels of support and resistance). Above and below this range, in which the market had moved for days, traders have meanwhile set up a number of buy and sell orders. But once strong fundamental changes and reports reach the market, a price gap is created immediately after the opening. This price jump is later supported by a number of additional sell bellow / buy above orders which were activated. Such price gaps are also often the beginning of new strong trends, especially if these gaps have not been closed. Breakaway gaps are accompanied by a sharp increase in trading volume.

3. Runaway gaps – are created during strong trends, when the price gap is caused by a further increase in interest in the given asset, supporting the current price trend even more. In other words, in the current constant price move in one direction, such price gap just creates a kind of sudden price jump, in the same direction. This kind of price gap is not usually closed.

4. Exhaustion gaps – are usually created during strong trends, but unlike Runaway gaps, the price use to return and the gap is filled. The fact that the price has reversed and filled the gap signifies that the price jump was not strong enough and is no longer supported by most subjects. Exhaustion gaps signalize the end of trend and its change to the opposite one.

Despite the fact that the price gap theory speaks clearly, the practical side of their trading is partly different. In fact, at the time of formation of a new price gap, the company and the financial manager could hardly clearly determine whether it is really e.g. the Runaway gap, or the price reverses later and forms the Exhaustion gap. All price gaps are also usually accompanied by a sharp increase in trading volume (with the exception of the Common gaps), which also does not make the situation easier. In addition to the abovementioned facts, the gaps could also be closed later just partially, which can be confusing even more. The most reliable ones are probably the Breakaway gaps, where movement in one direction is supported by a number of pre-placed orders from other entities. An example of a perfect closing of the price gap within six days is shown in Figure 1.

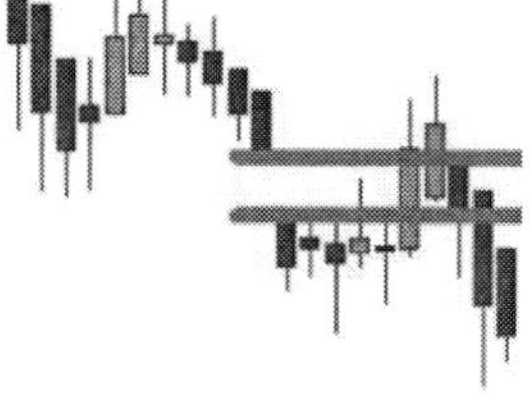

Figure 1. Price gap closed.

An example of a price gap that keeps being open is shown in Figure 2. Not even the next 10 trading days were enough to close the gap.

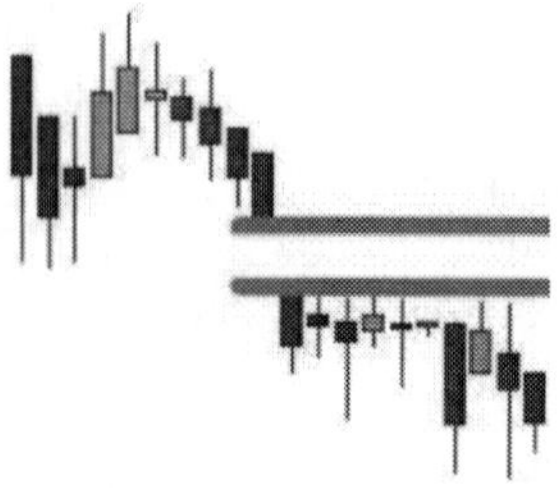

Figure 2. Price gap was not closed.

A common way of trading price gaps is to rely on reverse price movements and thereby on closing the gap. This is especially true for Common gaps as well as Exhaustion gaps. On the other hand, if traders would rely solely on the reverse movement at Runaway or Breakaway gaps, they could pay significant financial costs later. The theoretical dividing into four basic groups is more of an informative nature on the nature of price gaps rather than of practical commercial importance. It is not always possible to tell what kind of price gap we see and how it can be used for a profitable trade right after the gap has been formed (we do not know whether it will be closed or remain open for many days). Anyway, what the financial managers can do is to use the long-term test results, which can tell them what the probability of closing a gap for their selected asset will be.

4 Research results and discussion

Table 1 shows the results of the tests we have performed. The tests were conducted on a soybean oil commodity (e-CBOT prices) for a 30-year period (1979-2009).

Table 1. Price gaps long-term trading tests (soybean oil)

GAP	Incidence	Period	Gap filled	GAP	Incidence	Period	Gap filled
$60	5,19%	5 days	42,49%	$210	1,63%	5 days	30,08%
		10 days	55,47%			10 days	43,09%
		15 days	62,09%			15 days	49,59%
		20 days	66,41%			20 days	54,47%
		25 days	68,45%			25 days	56,91%
		30 days	70,74%			30 days	58,54%
$90	3,93%	5 days	38,38%	$240	1,48%	5 days	29,46%
		10 days	51,85%			10 days	39,29%
		15 days	58,92%			15 days	46,43%
		20 days	63,97%			20 days	51,79%
		25 days	65,99%			25 days	54,46%
		30 days	68,69%			30 days	56,25%

GAP	Incidence	Period	Gap filled	GAP	Incidence	Period	Gap filled
$120	3,07%	5 days	36,64%	$270	1,20%	5 days	23,08%
		10 days	48,28%			10 days	34,07%
		15 days	54,74%			15 days	42,86%
		20 days	59,91%			20 days	48,35%
		25 days	62,07%			25 days	51,65%
		30 days	65,09%			30 days	53,85%
$150	2,45%	5 days	33,51%	$300	1,00%	5 days	22,37%
		10 days	47,03%			10 days	32,89%
		15 days	52,97%			15 days	39,47%
		20 days	58,38%			20 days	44,74%
		25 days	60,54%			25 days	48,68%
		30 days	63,24%			30 days	51,32%
$180	1,97%	5 days	31,54%	$330	0,93%	5 days	22,86%
		10 days	44,97%			10 days	32,86%
		15 days	51,68%			15 days	40,00%
		20 days	57,05%			20 days	45,71%
		25 days	59,73%			25 days	50,00%
		30 days	61,07%			30 days	52,86%

Note: own calculations

The first column shows the minimum level of the price gap that had to be created to be included in the calculations. The second column shows the percentage of all days that saw an opening at or above this price gap. The third column shows the number of days that were included in the tests to see whether the price gap has been closed. The last, fourth column shows the percentage of price gaps that have returned to the initial price over a given time and caused closing a gap.

As can be seen in the Table 1, with a price gap of $ 60, the incidence was more than 55 % that the gap was closed within next 10 days. With an incidence rate of more than 70 % the gap was closed no later than 30 days from its arising. However, if $ 330 price gaps were traded, it would be necessary to wait at least 25 days or more so that financial managers could rely on a more than 50% probability of closing the gap.

In general, the smaller the price gap which was created, the more likely it was that it will close quickly. Vice versa, the larger the price gap which was created, the longer it was necessary to wait until it was closed. All tested price gaps and time periods confirmed that it was more likely that the price would return and close the gap. We can also state that the larger the price gap was, the longer it was necessary to wait for it to close. Of all the price gaps included in the tests, more than 50 % of them were closed within next 30 days.

If a financial manager is interested in using price gaps to hedge prices or even improve the financial situation of his company and wants to reach high probability of profit, he should focus on small price gaps. Small gaps were closed with the incidence of more than 70 % of all cases and were also closed very quickly. The opposite applies to big price gaps.

In addition, our tests have shown that it is not only possible to take advantage of the high probability of closing the gap, but also vice versa - the very low probability that the price gap will be closed. How? This finding can be used for the correct setting of a stop loss order. This can be illustrated by Table 1 - when the market created a price gap of $ 300, there was only a 22 % probability that such a gap would be closed within 5 days, or 33 % that it would be closed within 10 days. Stop loss can therefore be set at the initial price of this price gap - of course, this only applies if the financial manager wants to enter the market in the direction of the prevailing trend. Such a time period was sufficient to allow the market to begin moving in the expected direction.

5 Conclusion

The size of the gaps and their incidence depends on the market liquidity. In general, the more liquid the market, the less often price gaps are being created. If they are created, they are not too big. Market where price gaps do occur rarely is e.g. the foreign exchange market (forex). Not only is the forex liquidity the highest of all financial markets, but it does non-stop trading, so any information that reaches the market is immediately taken into account, leaving no room for excessive panic and accumulation of one-way orders before market opening.

If some enterprises or financial managers are interested in using the price gaps to hedge prices or improve the financial situation and stability, they should first make the long-term tests of gaps incidence in asset prices. They could find out the best moments to enter the financial market or where exactly to place the stop loss order. On the contrary, the main risk of price gaps, especially for enterprises that remain in the market overnight and do not trade the price gaps at all, is that the price could skip their set stop loss - thus, their loss could be much higher than originally intended.

Acknowledgement

The presented paper is the output of the scientific grant: VEGA No. 1/0007/19 "Asset allocation in a low interest rate environment in the financial and non-financial companies in the Slovak republic" (100 %).

References

1. C. M. Corcoran, *Long/Short Market Dynamics: Trading Strategies for Today's Markets* (Wiley Online Library, DOI:10.1002/9781119208914.ch7, 2012).

2. T. Knight, *High - Probability Trade Setups: Chartist's Guide to Real - Time Trading* (Wiley Online Library, DOI:10.1002/9781119202431.ch13, 2012).

3. D. H. Weis & A. Elder, *Trades About to Happen: A Modern Adaptation of the Wyckoff Method.* (Wiley Online Library, DOI:10.1002/9781118656389, 2013).

4. K. P. Evans, Intraday jumps and US macroeconomic news announcements, *Journal of Banking & Finance. Vol. 35, Issue 10.* Accession Number: WOS:000294037800001. (2011).

5. J. Bernstein, *30 Days to Market Mastery: A Step by Step Guide to Profitable Trading.* (Wiley Online Library, DOI:10.1002/9781119198147, ISBN: 978-1-119-19814-7, 2015).

6. C. A. Kase, *Kase on Technical Analysis Workbook: Trading and Forecasting.* Wiley Online Library, DOI:10.1002/9781119209737, ISBN: 9781119209737, 2015).

7. T. N. Bulkowski, *Fundamental Analysis and Position Trading: Evolution of a Trader.* John Wiley & Sons Inc., Hoboken, USA, Accession Number: WOS:000429855400025, (2013).

8. E. Ponsi, *Technical Analysis and Chart Interpretations* (Wiley Online Library, DOI:10.1002/9781119204800.ch9, 2016).

9. T. N. Bulkowski, *Encyclopedia of Candlestick Charts* (Wiley Online Library, DOI:10.1002/9781119202288, 2012).

10. T. N. Bulkowski, Gaps. *Visual Guide to Chart Patterns.* Blackwell Science Publ, Oxford, Accession Number: WOS:000416847300006, (2013).

11. A. Plastun, X. Sibande, R. Gupta, M.E. Wohar, Price gap anomaly in the US stock market: The whole story. *North American Journal of Economics and Finance 52*. Accession Number: WOS:000526121400047, (2020).

12. J. Stubinger, L. Schneider, Statistical Arbitrage with Mean-Reverting Overnight Price Gaps on High-Frequency Data of the S&P 500. *Journal of Risk and Financial Management, Vol 12, Issue 2.* Accession Number: WOS:000475294000002, (2019).

13. M.C. Thomsett, Mind the Gap: When Price Jumps Signal Change. *Practical Trend Analysis: Applying Signals and Indicators to Improve Trade Timing,* Walter De Gruyter, Berlin, Accession Number: WOS:000489944600011, (2019).

14. G.M. Caporale, A. Plastun, Price gaps: Another market anomaly? *Investment Analysis Journal, Vol 46, Issue 4*. Accession Number: WOS:000413708400003 (2017).

15. GF. Gu, X. Xiong, YJ. Zhang, W. Chen, W. Zhang, WX. Zhou, Stylized facts of price gaps in limit order books, *Chaos Solitons & Fractals. Vol. 88.* Accession Number: WOS:000377731100006. (2016).

Belbin team roles in a start-up team

Richard Bednár[1,*] and *Ivana* Ljudvigová[1]

[1]University of Economics in Bratislava, Faculty of business management, Dolnozemská cesta 1, 852 35 Bratislava, Slovakia

Abstract. Leadership in start-ups has its specific features, which are connected to a high uncertainty, moving of current boundaries, outstanding adaptability and perseverance of the founder. The main goal of this research paper was to identify the role of founders in successful start-ups. The purpose of this research is to determine whether there are certain common characteristics of successful entrepreneurs- start-up founders. In our analysis, we used the questionnaire- Belbin Full Individual Report. This questionnaire analyses and compares self-evaluation of an examined individual with 360-degree analysis of four members of his team. As a successful start-up we defined a company which fulfilled all of the following conditions: creates a unique scalable product/service, exists on the market 4 to 6 years and is achieving revenues in the amount of the third quartile (Q3) in its industry. To make the analysis comparable, analysed figures are only men. In our research we analysed nine founders and CEOs of start-ups base on a 360-degree evaluation- self-evaluation and evaluation of four members of the team. To view cumulative results, we have used the function boxplot.

1 Introduction

Establishment and development of start-ups is inevitably associated with leadership, because applied leadership has a decisive impact on the success or failure of a start-up [1]. For a start-up to work like a team it is necessary to unite in order to achieve a common goal and that is precisely the sense of leadership. Where teamwork is important, leadership is also important and vice versa. Team work and leadership are necessary for the success of any start-up [2]. On the contrary the absence of leadership can have a range of negative impacts for start-up business. Start-up without a leader, who creates and communicates a vision, inspires and coaches and interferes with processes and business activities, when it is necessary, operates on the market aimlessly [1]. Nevertheless, there is relatively little empiric research done, that would deal with the topic of leadership in start-ups. Many of the finding in this area are still based on theoretical inquiry. The main goal of this contribution therefore is to examine the role of a leader in start-up teams.

2 Theoretical basis of leadership in start-ups

* Corresponding author: richard.bednar@euba.sk

The uncertainty, or the uncertain future represents the next important factor when characterizing the concept of start-up. Blank and Dorf [3] argue, that "start-up goes from failure to failure in order to learn from each failure and find out what works in the process of finding repeating business model with high growth". The uncertainty inside of a novice business is caused by short business tradition, lacking experience and low level of developing routines and processes. The uncertainty is however also linked to external environment in which start-up operates and is given by in particular specific conditions of the environment, complexity and dynamics on the market. The third crucial parameter in the defining of a start-up is the criteria of growth. The value of a start-up lies exclusively in its future growth potential. Aspects of growth are stressed in Graham´s definition [4], who refers to start-ups as "an enterprise, which is created for rapid growth".

The personality of a founder, his business history, engagement, impression, confidence and affinity, which are for investors and partners decisive criteria when it comes to obtaining necessary investment resources. It is confirmed by Kets DeVries and Miller [5], according to which individual characteristics and behaviour of leaders influences the development and long-term sustainability of start-ups As well as according to Hambrick and Mason [6] start-ups are often seen as a reflection of its management. The demonstration of perseverance and resilience, reliability and honesty [7] are the most important characteristics of a novice entrepreneur, which is appreciated by the potential investor. Butler [8] indicates, that business leaders are successful in the environment of uncertainty, they are motivated by the risk, they are characterised by an extraordinary ability to convince and keen curiosity towards external environment. Similarly, Bussgang [9] ranks the ability to deal with uncertainty and pushing the boundaries of traditional business as the most important characteristics of leaders in start-ups. Many empiric studies have shown a positive relationship between the leaders behaviour and different indicators of the performance of start-upists [10, 11, 12].

Most investors, when deciding on investing into a particular business idea look for competent, highly biased and credible management team with relevant experience in the relevant field. The quality of a start-up team is according to Timmons and Spinelli [7] judged based on these criteria: experience, motivation, perseverance, creativity, responsibility, courage, adaptability, tolerance of risk and uncertainty. It is confirmed by Blank and Dorf [3], according to whom a start-up team has to be used to change, chaos, learning from mistakes, risk-taking and unstable situations without instructions on the solution, must be curious, searching, creative and eager in search of repeatable and scalable business model.

One of the experts, who are intensely devoted to the issue of team and team roles is Meredith Belbin [13, 14]. the object of his interest was to find out the reasons, why some teams achieve great results and others face failure. In 1970 he, with a group of experts from Henley Management College, decided to conduct a study, which would bring answers to this question. He intensely watched the behaviour of managers using a series of different psych-evaluation for the duration of ten years. On the basis of the results of the evaluations was the sample of managers divided into seven categories. Subsequently, these managers were divided into teams, in which they carried out different managerial tasks. During the observation of task-solving, certain patterns of behaviour of this research sample were identified. On the basis of these patterns the researchers identified the basic team roles. Initial research team defined eight team roles, but over the time added a ninth role. These nine team roles were divided into three groups based on their orientation - roles oriented in dealing with people (coordinator, team worker and spotter of sources), roles oriented in action (rectifier, producer, finisher) and roles oriented in thinking (Specialist, Plant, Monitor evaluator). A short description of team roles according to their strengths and weaknesses is illustrated in Table 1 [14].

Table 1. Typology of team roles according to Belbin.

	Benefits	**The negatives**
Shaper (SH)	Call for performance, is dynamic, performs well under pressure. Is enterprising and daring when it comes to overcoming obstacles.	Is inclined to provoke. Offends the feelings of others.
Co-ordinator (CO)	It is a matured and self-confident, is able to recognize talent. Clarifies objectives. Effectively delegates.	He may seem manipulative. He exonerates his share of the work.
Plant (PL)	Creative, has imagination, free-minded. Generates ideas and solves difficult problems.	Ignores details. He is fully preoccupied with his own thoughts to the detriment of effective communication.
Teamworker (TW)	Works, is responsive, and diplomatic. Listens and to prevent disagreements between people.	Is indecisive in critical situations. Avoids sharp conflicts.
Completer finisher (CF)	Is attentive, conscientious, diligent. Searches for errors on the way to perfection.	Is inclined to excessive worrying. Reluctant to delegate.
Resource investigator (RI)	Is sociable, enthusiastic and communicative. Detects opportunities and develops contacts.	Is excessively optimistic. May lose interest after the initial enthusiasm.
Specialist (SP)	Is focused on one single thing, proactive and focused on his own area. Provides rare knowledge and skills.	Contributes only in the demarcated area. Deals with professional matters.
Implementer (IMP)	Is practical, reliable, powerful. Transforms ideas into actions and organises work that needs to be done.	Is a little inflexible. Slowly responds to new opportunities.
Monitor evaluator (ME)	He is factual, strategist and smart. He sees all possibilities. He has a precise judgement.	He may lack the drive and ability to inspire others. May be over-critical.

3 The objectives and Methods

The result of Belbin's study was the finding that most successful were those teams that were composed of a variety of people (different team roles). It is necessary to mention that some members of the team are able to uphold multiple roles at the same time. Businesses, world advice organisations and the leaders themselves constantly use this typology as a guide to create a team. So far however the Belbin typology of team roles has not been applied to start-ups and their founders. The main aim of the implemented research team was to identify the role/roles of the founders in start-up teams and based on the that examine the importance of leaders in start-ups. The purpose of this mentioned research is to determine whether there as certain common characteristics of successful entrepreneurs (start-up founders). A prerequisite for the fulfilment of the objective is to verify the following hypothesis:

1. Every leader upholds several team roles, of which as least four proceed the boarder of occurrence by 50 percent.
2. The combination of team roles in leaders of successful start-ups will be the same.
3. At least in one team role it will be the same or similar tendency (high or low representation).

The output of the conducted survey may serve as a role model for people who are about to establish their own start-up and do not know what team role they should as the founder hold and similarly for leaders of already existing star-ups who face the problems caused by unclear roles and failure of teamwork. The results of the survey may also be a tool for investors, business angels and institutions supporting a developing start-up (whether they are private or state) when deciding which criteria, they should consider when assessing a new potential start-up business.

The main research method was certified survey questionnaire Belbin. Belbin analysis creates a personalized report on the individual behaviour. The starting point is the individual report generated directly by Belbin, who identified which combination of the nine is represented in the studied personality, in what proportions, defines the strengths and weaknesses of the personality and contains a proposal to self-improvement personality, in order to achieve higher efficiency.

In our analysis, we used the questionnaire- Belbin Full Individual Report. This questionnaire analyses and compares self-evaluation of an examined individual with 360-degree analysis of four members of his team. We have defined a successful start-up as a business that fulfils all of the following conditions:
- creates a unique, scalable product/service
- is on the market 4 to 6 years,
- achieved sales in the amount of the third quartile (Q3) in its industry.

To make the analysis comparable, analysed figures are only men. In our research we analysed nine founders and CEOs of start-ups based on a 360-degree evaluation - self-evaluation and evaluation of four members of the team. To view the cumulative results, we used the function of boxplot which is rather descriptive but also allows the assess the variability and symmetry of the data, further serves to detect outliers and extreme observations. The latter consists of a rectangle (i.e. a box) and whiskers (i.e. an antenna). The rectangle takes about half the data set, its lower part is determined by the lower quartile, the upper edge defining the upper quartile. The height of the box is therefore given by so called Interquartile Range which we obtain from the relation $IQR = x0{,}25 - x0{,}75$ where $x0{,}25$ indicates the lower quartile and $x0{,}75$ the upper quartile. The box is divided into two parts of the median, which represents the skew. We obtain the lower part in the relation $x0{,}25 - 1{,}5 \cdot IQR$, the upper then similarly $0{,}75 + 1{,}5 \cdot IQR$. The values, that do not belong into the interval bounded by walls are considered suspicious and are shown as a ring.

5 Results and discussion

In the results of the research we show brief characteristic of the examined start-ups and their founders including the resulting table of the representants of team roles of each leader. Analysis of each founder is completed by brief pointing his strengths and weaknesses and from that resulting space for improvement in the role of the team leader:

Table 2. Business model and results of Belbin analysis.

	Company Description	**SH**	**CO**	**PL**	**TW**	**CF**	**RI**	**SP**	**IMP**	**ME**
Powerogy CEO - Dušan	Online shop of of improved products such as organic chocolate, coffee, special coconut oil, almond butter and others.	78%	6%	100%	0%	0%	98%	84%	11%	0%
Zelená pošta CEO - Adrián	Sending and receiving mail via the online system.	48%	57%	28%	44%	67%	21%	19%	58%	68%
Octago CEO - Dušan	Building and selling customers customized fields for street workout.	38%	24%	20%	24%	55%	98%	38%	68%	92%

Venzeo CEO - Jozef	Helping companies document manual activities and problems in the field.	78%	30%	77%	24%	60%	66%	50%	39%	56%
sli.do CEO - Peter	Online platform for conferences: Q&A sessions, live polls, live quizzes, brainstorm ideas	50%	22%	76%	60%	26%	84%	97%	20%	49%
MeetNlearn CEO - Juraj	Online platform for mediation of tutoring between the tutors and the students.	3%	2%	94%	25%	60%	33%	92%	10%	93%
StartitUp CEO - Marek	An information web which would connect all start-up enthusiasts. Startitup Group has a portfolio of eight websites.	72%	76%	98%	20%	36%	88%	0%	0%	19%
Zaraguza CEO - Michal	Ranked by Facebook as one of the TOP 10 most creative agencies in the world.	90%	71%	100%	25%	36%	67%	0%	0%	0%
Aktivizer CEO - Ján	Application which offers various sports activities to choose from in one's leisure-time.	82%	92%	62%	50%	14%	66%	20%	30%	38%

Based on comparing the individual types of roles demonstrated by founders of the examined start-ups (Figure 1) we can observe that each of the analysed leaders displays a different combination of roles carried out in his team (hypotheses no. 1). What this also means is that there is no role or a combination of roles which would be sure to guarantee a start-upist's success (hypotheses no. 2).

Figure 1. Boxplot.

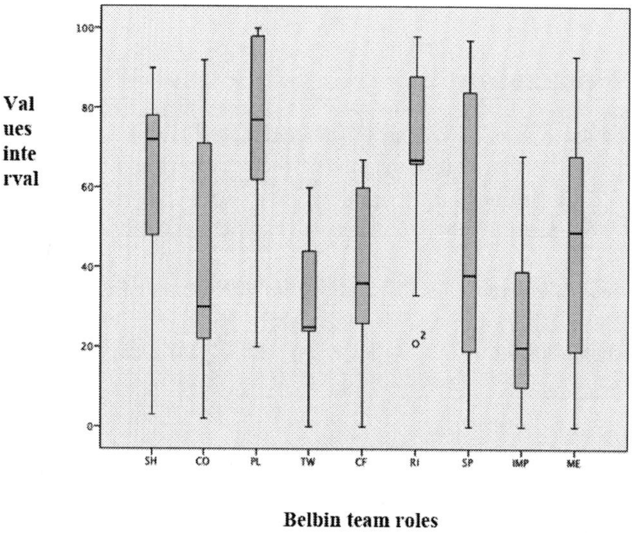

Regards to two particular types of roles, the founders' tendency to demonstrate them is rather low (Table 10). We're speaking of the role of a Teamworker (an average value of 30.22) and an Implementer (an average value of 26.22). Based on the general characteristics of a Teamworker – this person is perceptive, diplomatic and easy to cooperate with. He listens, pays attention and avoids disagreements; he builds relationships and diverts conflicts. On the other hand, there is his indecisiveness in crucial situations

(Belbin, 2010). Start-up leaders are in their nature very competitive, which also shows in a lowered ability to cooperate with others. Start-up founders tend to pay more attention to goals and tasks rather than to interpersonal relationships at their workplace, which may be due to their age, considering these businesses are often pretty young. However, a leader's lack of ability to cooperate and solve conflicts may have a negative impact on his future. Research proves that start-ups lead by teams of several people are in the long-run more successful than those lead by an individual. Team (collective) leadership, in start-ups especially, undeniably dominates at the expense of vertical leadership (based on a formal leader). Moreover, unresolved conflicts between the members of start-up teams are one of the most common reasons why people leave them, which eventually causes these start-up businesses to perish.

The team role of an Implementer was least demonstrated by the founders of examined start-ups (26.22). According to our theoretical starting point, an Implementer is someone practical, reliable and efficient. He transforms ideas into action and arranges any task there is to be done. He implements solutions which his team has agreed on. He's disciplined, dependable, conservative in his habits, yet always able to act and take practical steps. His downside lies in his rigidity, best shown in his slow reactions to new possibilities (Belbin, 2010). It is indeed for this reason that the role of an implementer is much less demonstrated by the examined start-ups' founders. A start-up business is characterized by its bumpy ride to success, full of unsuccessful projects and having to adjust its business model to the needs of its customers. Flexibility, quick reactions to changes and proactivity are therefore the typical traits of a successful start-up-team leader.

The samples we've decided to analyse point out one team role which is relatively highly demonstrated by all leaders (hypotheses no. 3). It was the role of a Shaper, a person who's able to deal with difficult problems and comes up with solutions. His personality could be defined as individualistic, creative, innovative and unorthodox. His best quality is imagination, whereas his weaknesses encompass his negligence towards being subsidiary and a preoccupation with his own thoughts at the expense of effective communication [14]. After its realization, the research project discovered that founders of successful start-ups are sensible, open, energetic and creative. Based on these characteristics, leaders create visions for their companies, as it turns out that innovativeness, curiosity and creativity are indeed some of the most important characteristics leaders of successful start-ups can possess. At the same time, it is important for the long-term future of a start-up business to encourage and develop creativity of one's co-workers.

Table 10. A comparison individual team roles demonstrated by founders of examined start-ups.

	Minimum	Maximum	Mean	Std. Deviation
Shaper	3	90	59.89	27.733
Co-ordinator	2	92	42.22	32.568
Plant	20	100	72.78	30.626
Teamworker	0	60	30.22	18.089
Completer finisher	0	67	39.33	23.06
Resource investigator	21	98	69.00	27.134
Specialist	0	97	44.44	38.497
Implementer	0	68	26.22	24.601
Monitor evaluator	0	93	46.11	35.254

The research project has shown that leaders of start-ups differ from leaders of traditional businesses. Several examinations concerning team roles, which are executed in standard teams inside regular organizations, have indicated that the "ideal" leadership roles are those of a coordinator, source seeker and a team operative [15, 16, 17]. Furthermore, the team roles of a coordinator and a team operator have reached a below-average presence in the examined start-up leaders, meaning that whatever helps a leader to manage his team in a traditional business is not the same thing as that which stands behind a start-upist's success.

There are many well educated and skilful people on the Slovak market, but the question is whether they are the right people to work on a start-up. Not everyone is inclined to dynamism, which is a typical feature of innovative businesses. Working on a start-up requires a creative, yet demanding activity without any actual limits. There is no precise prescription of work activities for individual positions which means that especially in their first years of existence, innovative businesses have to engage everyone in everything. The results of aforementioned research project prove this to be true as they show that leaders of successful start-ups tend toward several different team roles in various combinations. A leader's current choice of a team role mainly depends on the situation in which he finds himself. It turns out important for the success of a startup that he knows how to use the strengths of each team role to the best of his ability. Here, the analysis carried out according to Belbin's questionnaire could prove to be a suitable and explicit additional tool, used by institutions supporting the development of start-ups to evaluate these businesses and their founders. The biggest perk of this tool lies in its simplicity, not being time demanding and in the credibility of its conclusions, confirmed by decades of experience with businesses and consulting firms all over the world.

6 Conslusion

Every entrepreneur wants to enter the market with an innovative idea, put it into practice in the form of a start-up, impress potential customers and become successful. While it is true that an idea is truly an important stepping stone to launching a start-up, identifying the right people for such innovative business is one of the hardest tasks. Here, Belbin's analysis, identifying the team role displayed by each given start-up team's founder, becomes a great aid at determining the suitability of given leadership traits for leading and managing a start-up. Being put into practice and examining nine successful start-ups and their founders, the research project has shown that those who take the lead in start-up businesses display various team roles. It turns out to be very important, in start-up businesses especially, for a leader to demonstrate multiple team roles in order to strengthen his versatility. In regards to start-ups, the most important roles seem to be those of a Plant, source seeker and a director as opposed to the roles of a coordinator and a team operator, crucial when leading teams in traditional business.

Acknowledgement

The paper is a part of solving the task of VEGA no. no. 1/0631/19. Title: Metamorphoses of startups on the road to business success.

References

1. Zäch, S., Baldegger, U. (2017). Leadership in start-ups. *International Small Business Journal: Researching Entrepreneurship*, 2017, 35, pp. 157-177.

2. Cogliser, C. C., Brigham, K. H. 2004. The intersection of leadership and entrepreneurship: Mutual lessons to be learned. *The Leadership Quarterly*, 15 (6), pp. 771-799.
3. Blank, S., Dorf, B. 2012. *The start-up owner's manual. The step-by-step guide for building a great company*. K&S Ranch Publishing Division. ISBN-10: 09849993032.
4. Graham, P. 2013. Retrieved from http://www.forbes.com/sites/natalierobehmed/2013/12/16/what-is-a-startup/.
5. Kets de Vries, M. R., Miller, D. 1986. Personality, culture, and organization. *Academy of Management Review*, 11(2), pp. 266-279.
6. Hambrick, D. C., Mason, P. A. 1984. Upper echelons: The organization as a reflection of its top managers. *Academy of Management Review*, 9(2), pp. 193-206.
7. Timmons, J. A., Spinelli, S. 2008. New Venture Creation: Entrepreneurship for the 21st Century. 8. ed. London: McGraw-Hill. ISBN 978-0-0733-8155-8.
8. Butler, T. 2017. Hiring an entrepreneurial leader. *Harvard Business Review*, 2 (March-April 2017), pp. 85-93.
9. Bussgang, J. 2017. Are you suited for a start-up? *Harvard Business Review*, 6 (November-December 2017), pp. 150-153.
10. Gooty, J., Gavin, M., Johnson, P. D., Frazier, M.L, & Snow, D. B. 2009. In the eyes of the beholder: Transformational leadership, positive psychological capital, and performance. *Journal of Leadership & Organizational Studies*, 15 (4), pp. 353-367.
11. Gumuslouglu, L., Ilsev, A. 2007. Transformational leadership, creativity, and organizational innovation. *Journal of Business Research*, 62 (4), pp. 461-473.
12. Wang, H., Tsui A. & Xin, K. 2011. CEO leadership behaviors, organizational performance, and employees' attitudes. *The Leadership Quarterly*, 22 (1), pp. 92-105.
13. Belbin, R. M. 1993. *Team Roles at Work*. Butterworth-Heinemann, Oxford.
14. Belbin, R. M. 2010. *Management Teams: Why They Succeed or Fail*. 3rd ed., Butterworth- Heinemann, Oxford.
15. Fisher, S. G., Hunter, T. A., Macrosson, W. D. K. 2000. The distribution of Belbin team roles among UK managers, *Personnel Review*, Vol. 29 No. 2, pp. 124-140.
16. Dulewicz, V., Higgs, M. J. 2003. Design of a new instrument to assess leadership dimensions and styles, *Henley Working Paper Series* No. HWP 0311, Henley Management College, Henley-on-Thames.
17. Batenburg, R., Walbeek, W. Van, Maur, W. 2013. Belbin role diversity and team performance: is there a relationship? *Journal of Management Development*, 2013, 32(8), pp. 901-913.

Human resources policy in relation to cybernetic security in Slovak medium and small companies

Benita Beláňová[1*]

[1]University of Economics in Bratislava, Bratislava, Slovak Republic

Abstract. Due to a mass penetration of information technologies into all business processes, researches in area of IS/IT field within medium and small companies has gained on significance. With the aim to acknowledge the status of information safety in the environment of Slovak medium and small companies a research has been carried out, which focused on identifying: threats on information systems; methods for treating information safety; and major obstacles in implementation of secure and safe policy. Within this research a special part was dedicated to companies' human resources policy in relation to cybernetic security. The research separately focused on staff provisioning in area of IS/IT. It surveyed who and on what position is responsible for this field, what requirements should be fulfilled by such people, which knowledge is the most valued and which they miss. The second area we focused on, was the organization of the security policy in relation to all employees, but mainly to users of information systems. The research was performed in the years 2008, 2012, 2015 and 2017 and the results about human resources are presented in this article. The respondents were company owners, managers and specialists form IS/IT field.

1 Introduction

Steering of medium and small companies has become a subject of economic investigation in the recent years. Regarding the influence of big organizations on national economy, the medium and small companies seemed to be not much attractive object of research and base for scientific knowledge implementation. The change in the mind-set was caused by new trends based on the acknowledgement of globalization and informatization in all areas of the private and corporate sphere [1-4].

Together with informatization there is a new field, which should be systematically managed and that is the security of information systems in the company [5]. The significance, which is attributed to the information security, is justified by an increased interest of companies about drafting, acknowledging and implementation of security policy. The information systems consist of information technologies, processes and people, and that's why human resource form an inseparable part of this policy [6,7].

[*] Corresponding author: benita.belanova@euba.sk

The employees of the company are the one, who represent one of the most significant threats in area of information security [8-11]. In generally, these employees can be split into two groups. The first group are the employees, who are responsible for administration and security of information systems and technologies in the company. In the second group there are the so called "passive users" of IS/IT, while mainly this group are the weakest element in the security of information in a company. Not only because of the fact, that the users generally don't read guidelines and so they don't follow any rules, but also because they like to experiment by breaking essential principles of security. These two groups of employees should be assessed separately within human resources policy in relation to information security of the company, as the expectations and requirements on their knowledge and skills diametrically differ. Furthermore, their place in the hierarchy of the company is diverse. While the first group can be easily identified a localized in the company's structure, the passive users of IS/IT range from managers to the lowest level of executive employees.

2 Results of the research

Presented results are partial outputs of a larger project mapping the development of information security in small and medium-sized companies in Slovakia. The first part of this research was passed out in 2006, followed by a much wider range in years 2008, 2012, 2015 and 2017. Respondents responded to 105 questions, divided into 15 thematic units. One area was also personal staffing.

The results of the research are presented in the following structure:
- Methodology and Research Methods
- Features of the investigated sample
- Results of investigation in area of human resources provisioning of IS/IT security
- Results of security policy investigation in relation to people – employees of the company

2.1 Methodology and Research Methods

The method of analysis, which allows for the breakdown of the whole into its individual parts and elements in order to know the sub-system as much as possible, has been used to analyse the current state of affairs regarding this issue at home and abroad, and we have combined the knowledge gained into a new whole by using the method of synthesis. Using the method of comparison the theoretical part was compared with opinions on the given issue and the terminological apparatus was subsequently specified by method of abstraction. The descriptive method is used to describe the principles of security information system management.

In the testing process, we investigated the hypotheses using one of the exploration methods - the questionnaire method. When applying this investigative method, the information obtained from the investigated subjects is gathered by querying, i.e., by deliberately targeted questions. This method makes it possible to obtain more information about the individual respondent at the same time. This information may relate to the extent and depth of the respondent's knowledge, preferences, past, present, or intended behaviour, and the characteristics of the respondent. Detection methods can be used to examine any phenomena, that is, companies of different sizes, different industry structures, ownership, etc. The application of detection methods can be adapted to both descriptive and causal research. For proper analysis and subsequent comparison of phenomena, it is essential that the analysed phenomena occur under the same conditions. Otherwise, erroneous conclusions about the causes and consequences of the phenomenon due to the complexity of the conditions of economic life may occur.

Therefore, in order to obtain relevant and quantifiable information on the state of the security of information systems in small and medium-sized enterprises in Slovakia, the enterprises were sized according to the European Commission's directive no. 96/280 / EC, according to the industry classification of economic activities of SK NACE Rev. 2, then according to the legal form and ownership of the enterprise. The questionnaire was processed in electronic form. The MS Access database and the MS Excel spreadsheet were used for processing. Mathematical and statistical methods were used to evaluate the questionnaire. The results were transformed into tables and graphs.

By comparing the phenomena in time and space, we have identified, on the basis of the questionnaire survey, the common and different aspects of the behaviour of individual types of enterprises. The induction method allowed us to explore all available phenomena and their parts to generalize the results obtained.

2.2 Methodology and Research Methods

The research has been performed on a sample of randomly chosen medium and small companies in the Slovak Republic. A questionnaire was created in line with a template on the information security survey in the entrepreneur environment in Slovakia, which was carried out by the National Security Authority in cooperation with the magazine DSM – Data Security Management and company Ernst & Young in the years 2006 [12] and 2008 [13] and the Guideline on performing security audit of IS [14].

The research in the form of moment observation and random selection has in comparison with intentional interviewing of an earmarked group of companies selected in advance certain deficiencies. However, the objectivity of the results is justified by the high number of companies and their considerable sectorial variety. 247 companies took part in the research in the year 2008, 129 in 2012, 209 in 2015 and 219 companies in 2017. According to the legal form, the highest participation was represented by limited companies. According to the structure of the ownership, companies with exclusively local ownership or mainly local ownership prevailed.

2.3 Provisioning of human resources in area of information security

Qualitative provisioning of human resources is one of the basic and inevitable conditions for the execution of security policy of IS in the company, therefore it is necessary to give special emphasis on this factor [15]. More specifically on the definition, creation and staff occupation of roles, which ensure a continuous development of information security in the company that is in line with new legislative elements, organizational changes or evolution of the society, IS changes and finally in line with the security documentation itself [16].

Key roles, that in practice present the primary element of each steering system of information security should respect the existing elements of the company's management (e.g. linear or process steering, expert competence etc.). The creation of these roles doesn't necessarily require creation and occupation of new positions, as often a new role is allocated to existing employees. It is however important, that the respective roles ensure the execution of rules and principles specified in the security policy and their practical application and respect [17].

Within our research we have been investigating how is the field of information technologies included into the organizational structure and on which hierarchical level is the information security. Nonetheless, the expectations, which are required from the employees involved in IS/IT and information security are significant.

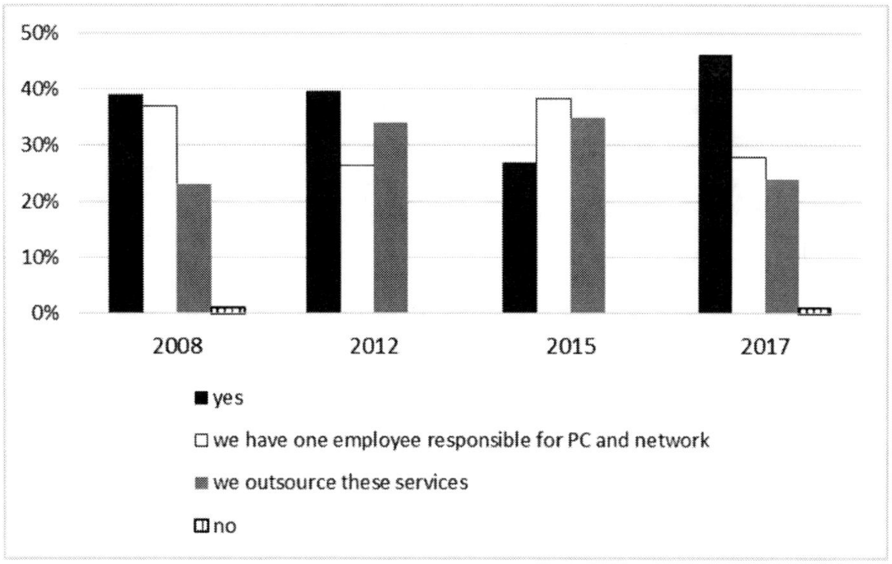

Fig. 1. Does your company have a separate business area responsible for PC services?
Source: own elaboration

Usually a business area specifically responsible for information technologies and systems in general is present in medium sized companies. In small companies this share is much lower. Small companies usually dispose with one person, who is in the position of network administrator and takes care about the operation of the information systems in the company. There are almost 10% of companies, mainly medium sized which dispose with employees, who are specifically involved into information security as their main work objective. The usual practice is to combine the information security with the IS/IT department. Companies, which don't have such department or administrator, get these services outsourced. In such cases, mostly in small companies, which get these services outsourced, no one is responsible for the security of IS itself (no department). Companies using the network administrator, usually pass the responsibility onto the economic and financial department.

The research has revealed the fact, that the information security is not integrated into business processes of the companies. In case of a separate solution regarding information security and risk management, an inadequate usage of company's resources just like a weak coverage of business risks, which come hand in hand with security risks on the level of information systems, may happen [12].

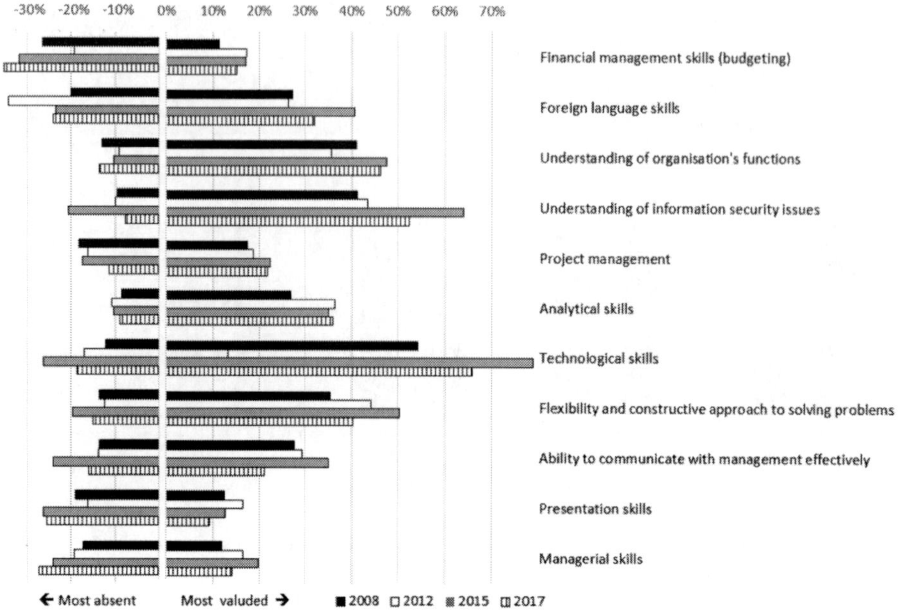

Fig. 2. The most absent and most valued knowledge and skills of information security staff
Source: own elaboration

We have found out a notable shortage of knowledge and skills of employees from the field of company's financial management, what we don't consider as a that much significant factor in the area of IS/IT, when comparing the most valued and on the contrary the most absent knowledge and skills of employees from the field of securing information systems. It is much worse in the field of technological skills of IS/IT and presentational and managerial skills, which belong to the key skills of IS/IT employees. The knowledge of foreign languages of the employees working in the field of information security has in comparison with previous periods improved, however still missing in 23% of companies. The respondents of the research considered technological knowledge of IS/IT, factual knowledge of the information security problem and flexibility and constructive approach in problem solving as the most valued knowledge and skills. The skill of an effective communication with the management of the company recorded a significant drop, when comparing to previous research. This is probably connected with the growing skills of employees in this field, hence their deficiency has proportionally decreased at the same time.

With regard the representation of IS/IT employees in the company, 1/4 of the companies considers the existing state as insufficient. A significant fluctuation of these employees was detected by 10% of companies. The compliance between the expertise and requirements is present at 2/3 of companies. The rest of 1/3 companies consider knowledge and skills as insufficient, or only partially compliant with the set requirements.

One of the reasons of this situation may be the financial compensation of these employees. The salaries of employees, who are responsible for IS/IT area in the company copies the development of salaries in Slovakia. In terms of time, there is an obvious shift from lower salary grades of employees responsible for information security into higher grades. However, we cannot claim that these salaries would reach a higher standard level, as it is in other matured economies, what leads to continuous deficiency of this qualified workforce, respectively to its fluctuation. Income over 2000€ were recorded in the segment of Business and Transportation.

Table 1. Gross average monthly remuneration of employees responsible for information security.

Salary	2008	2012	2015	2017
less than 500 €	42%	24%	16%	21%
501 – 1 000 €	37%	35%	34%	23%
1 001 – 1 500 €	11%	22%	31%	31%
1 501 – 2 000 €	6%	12%	14%	24%
more than 2000 €	4%	7%	5%	1%

Source: own elaboration

2.4 Security policy of the company in relation to other employees

Basic criteria were chosen to evaluate the security policy of the company in relation to "passive users". These criteria are considered from the IS/IT security perspective as main assumptions for security policy operation in medium and small companies. In the case of employees from the IS/IT field responsible for security, the fulfilment of these evaluation criteria is taken for granted.

The first evaluation criteria are trainings. The basic objective of the training in general is to increase the awareness of the employees about possible risks and instruct them, how to avoid these risks, or how to act in case a certain situation/state occurs. Regarding the security of IS/IT we can talk about rules on safe HW and SW usage, on safe data manipulation in the information system, their protection, policy of passwords and others. Employees should attend such trainings at least as newcomers. It is however desirable to organize such trainings periodically; the optimal frequency is one year. According to our research (2017) 39% of small and 56% of medium companies organize trainings on protection of data and usage of passwords.

In connection with the security policy, it is necessary to treat the leaving employees, where all access, manipulation and entering authorizations and means should be retained, or access passwords which the employee dispose with should be changed. As demonstrated on the Fig. 3, there is a percentage of companies, on average about 15% higher than in trainings, which follow this procedure.

The situation is worse in keeping written documentation about granted entering and access authorizations of employees to respective applications. When commenting these results, it should be stated, that in all monitored parameters in security area there is a decrease in 2012 in comparison with 2008. We link this phenomenon to the then ongoing economic crisis, when the companies had to cut their costs and the trend was to cut the costs mainly in security.

The access password is the main significance of the company's and its employees approach to information security. It is the simplest, cheapest and most effective security element which protects unauthorized penetration into the system. However, the practice shows, that companies don't focus sufficiently on password creation and definition of access rights within their security policy, as they should have. The rules of password creation are an internal matter of each company, just like the principles that should be followed when creating a strong, memorable password. These should be defined in the internal guideline of the company, with which should be all employees familiar (at least when joining the company). It is very usual that the employees have within applications the same access rights or that these rights are not specified in writing within their function. From this reason, we have specifically looked to the problemacy of password creation, their form and manipulation with them.

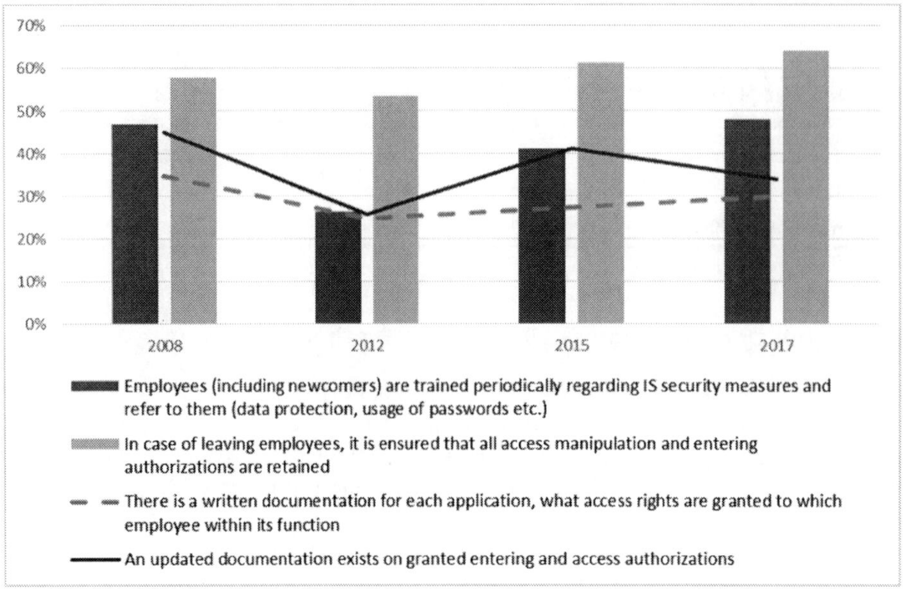

Fig. 3. Security trainings and documentation about access rights and data protection
Source: own elaboration

The outcome of the research presented, that more than 80% of users have an access password. However, the results confirmed that at least a half of them use the same access password to more applications, thus there is an increased risk from violation of more programs with the breach of one password. Furthermore, not all applications require to enter an access password. The main problem in password creation is, that they don't contain sufficient number of symbols and proper combination of symbols, which would prevent them from password breach.

Table 2. The number of symbols in the access passwords of employees

Number of symbols	2008	2012	2015	2017
More than 8 symbols	5%	6%	13%	30%
8 symbols	**22%**	**40%**	**42%**	**57%**
Less than 8 symbols	64%	54%	32%	14%

Source: own elaboration

In 2008 the most frequently used password was with the length of 5 and 6 symbols. In 2017 a more than a half of users have a password with 8 symbols, what is taken as a positive trend. The entrepreneur practice showed, that the security standard of an unbreakable password in all business applications is a predefined password format for server operation systems Windows from Microsoft, which requires a minimum of 8 symbols, 1 small letter, 1 capital letter and one digit. We haven't investigated on the password composition, but as we can see, already 30% of companies use a password with the length of more than 8 symbols. According to experts on information security, by adding 2 symbols you can substitute the missing capital letters, digits or random signs in the password.

A safe password is a password which change in regular interval. It is recommended that this interval takes 2 months. However, the entrepreneur practice showed that this safety rule is not kept in the companies. In 2/3 of the addressed companies, the passwords are not changed during the overall operation of the systems and it is not usual to monitor and protocol any attempts of unauthorised access or breach of authorization.

Even the most perfect password is useless, when the employees don't use it in line with the purpose, for which it was created. Usually, the employees sign off and turn off their PC at the end of the working time, but they automatically don't, when they leave their working place, e.g. for lunch. In the years 2008-2015 we have encountered a group of employees, who didn't turn off their PC at all. This trend was however decreasing and in the year 2017 it didn't appear at all. Despite of this, there is still a big group of employees, who don't sign off their workstation during their absence, creating thus suitable conditions for data manipulation and undesired information outflow inside the company. According to our findings, 65% of companies recorded internal unauthorized access.

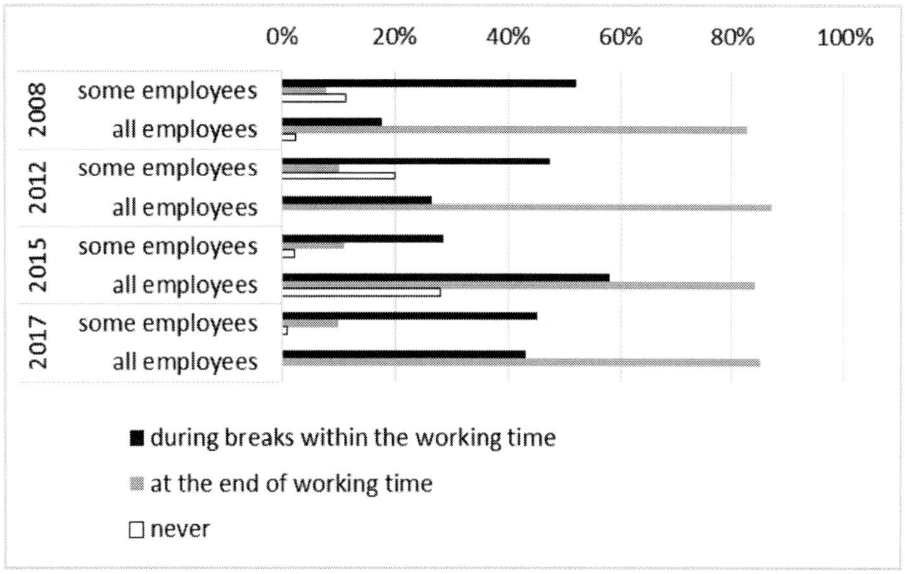

Fig. 4. Employees turning off/signing off PC
Source: own elaboration

3 Conclusion

The risks of the information security are caused by various threats, which influence vulnerable points of the company's information and communication systems. In the article we have focused on one of these – the human factor. Especially we were interested in investigating how the companies treat their cyber security in relation to their own employees. The conducted research brings more knowledge and outcomes for the area of information security management focusing on human resources. Resulting from the problem as it stands at present, we can confirm that 41% of companies create only a basic information security management structure, that is reflected also in the recruitment process of employees for this area and total costs incurred in connection with security awareness and employees acting safely. 13% of employees doesn't develop any kind of activity in this area.

If the companies want to reach an adequate level of their information assets security, they need to activate and effectively link all management processes, including management of human resources. In order that the company would fulfil the new requirements, it should change the basic awareness of its functioning. The organizational structure should be taken as a basis, as a firmly defined distribution of activities, relations and competencies, resulting in present responsibilities and financial compensations.

With regards the employees in general, just like in other managerial disciplines, it is not easy to make a universal guidance on managing people and forming security awareness, that should be followed to achieve a guaranteed success. It is influenced by several factors, such as the size of the company, field of operation, structure of employees and other. According to our view, we have chosen the basic aspects and principles that should be followed by all types of companies. As can be seen from the results, the companies have significant deficiencies in effective application of security policy in relation to employees.

As we know, the research of information security in small and medium-sized companies in Slovakia has not been implemented to such an extent and such a focus by any other organization and we have no knowledges that a similar research was realized in any other country. Therefore, we can not compare our results with another similar research.

The gained results don't mean the termination of research activity in the field of information security management, but they provide further knowledge on the development of this problem and consequent assessment in the following periods.

The article was elaborated within the VEGA project no. 1/0309/18 Social Networks in Human Resources Management.

References

1. V. Bolek, A. Romanova, P. Richnak, K. Porubanova, The use of pervasive technologies in business processes. *Ad Alta-Journal Of Interdisciplinary Research,* **9**, 2, 293-298 (2019)
2. A. Romanova, P. Richnak, K. Porubanova, V. Bolek, Application of modern information technology in innovation of business logistics processes. *Ad Alta-Journal Of Interdisciplinary Research,* **9**, 1, 245-248 (2019)
3. M. Kokles, A. Romanova, A. Hamranova, Information Systems in the post-transition period in enterprises in Slovakia. *Journal of Global Information Technology Management*, **18**, 2, 110-126 (2015)
4. A. Hamranová, Š. Marsina, P. Molnár, F. Okruhlica, Development of Information and Communication Systems within the Building of Project-Oriented Manufacturing Organization. *Advances in production management systems: innovative and knowledge-based production management in a global-local world, pt 1 Book Series: IFIP Advances in Information and Communication Technology*, **438**, 241-+ (2014)
5. F. Korcek, V. Bolek, A. Romanova, P. Richnak, Practicing information security management system in e-commerce. *Ad Alta-Journal Of Interdisciplinary Research,* **8**, 1, 207-212 (2018)
6. D. G. Moody, M. Siponen, S. Pahnila, Toward a unified model of information security policy compliance. *MIS quarterly*, **42**, 1, 285-+ (2018)
7. Y. Zhang, Y. L. Zhang, J. Zhou, L. Liu, F. Chen, X. He, A review of compressive sensing in information security field. *IEEE access*, **4**, 2507-2519 (2016)
8. J. D'Arcy, P. B. Lowry, Cognitive-affective drivers of employees' daily compliance with information security policies: A multilevel, longitudinal study. *Information Systems Journal*, **29**, 1, 43-69 (2019)
9. V. Bolek, A. Látečková, A. Romanová, F. KORČEK, Factors affecting information security focused on SME and agricultural enterprises. *Agris on-line Papers in Economics and Informatics*, **8**, 4, 37-50 (2016)

10. P. Balozian, D. Leidner, M. Warkentin, Managers' and employees' differing responses to security approaches. *Journal of Computer Information Systems*, **59**, 3, 197-210 (2019)
11. P. Ifinedo, Roles of organizational climate, social bonds, and perceptions of security threats on IS security policy compliance intentions. *Information Resources Management Journal*, **31**, 1, 53-82 (2018)
12. PSIB SR ´06, Ernst & Young, NBÚ SR, DSM – data security management, TATE International Slovakia. http://www.dsm.tate.cz/cz/psib-sr-2006/ last accessed 2007/09/11
13. PSIB SR ´08, Ernst & Young, NBÚ SR, DSM – data security management, TATE International Slovakia. http://www.dsm.tate.cz/cz/psib-sr-2008/ last accessed 2009/05/06
14. Guideline on performing security audit of IS FM SR, 331/2003-22 (2003)
15. M. Szarková, M. Andrejčák, Personnel audit in financial institutions in Slovac Republic. *International scientific conference Financial management of Firms and Financial Institutions,* Part III, 899-902 (2013)
16. N. Matkovčíková, Health and safety management in companies of all size. *International scientific conference Hradec Economic Days 2017*, 597-603 (2017)
17. J. Koubek, *Řízení lidských zdrojů* (Management Press, Praha, 2015)

Circular economy as a way of sustainable production and consumption

Brigita Boorová[1,*]

[1]Department of Production Management and Logistics, Faculty of Business Management, University of Economics in Bratislava, Dolnozemska cesta 1, 852 35 Bratislava, Slovakia

Abstract. The paper deals with a circular economy as a way of sustainable production and consumption. The introductory part describes the basic features of circular and linear economics as well as their differences. The linear economy is based on the principle of "exploit - produce – discard". The aim of the circular economy is to preserve the value of products and materials for as long as possible, to minimize waste and to conserve natural resources. The paper presents examples of good practice of business entities and self-governments whose activities are in accordance with the principles of the circular economy.

1 Introduction

Our lifestyle is focused on consuming as much as possible, and we can say that we have been living in a linear economy since the industrial revolution. In a linear economy, predominantly single-use products are produced, resulting in the disproportionate depletion of non-renewable natural resources, the production of huge amounts of waste, and the associated degradation and pollution that results in climate change. This one-way economic model applies in production to the extraction of natural resources, which serve as production inputs, and are then used in the material flow in the mass production of products that are pushed to consumer markets. For consumption, most of these mass-produced products are for single use and are usually disposed of after use. This one-way mass production and consumption model is unsustainable and we are beginning to feel its impact. The transition to a circular economy is a way of sustainable production and consumption.

2 Characteristics of the circular economy

The circular economy is a useful and regenerative design of an economic model, which means the constant maintenance of raw materials, materials, and products in their useful position, within the life cycle. Common sources of raw materials are being depleted all over the world. This model uses waste to create new sources of raw materials and increases the circular economy. In the traditional economy, everything is direct, resp. in the line, from the

* Corresponding author: brigita.boorova@euba.sk

extraction of raw materials through production and consumption to waste. The circular economy is more environmentally friendly, making it more suitable for our country. [15]
The currently popularized concept of the circular economy expands the possibilities of reuse and recycling of conventional waste and by-products, by emphasizing the use of value built into waste materials. [1, 9, 15, 2, 17] The new concept of the circular economy, promoted by the business community, is assessed in terms of the concept and scientific research of sustainable development. In particular, sustainability science. [3, 11]
The model of circular economy is an in-depth analysis of a case study over a 30-year period, highlighting the successes, difficulties, and conflicts of adopting a circular economy model. The case study is based on interviews, key documents and information from customers. The findings demonstrate how sustained circular economy business practices can deliver significant new revenues, resource productivity, and business continuity benefits, but also require managers and practitioners to develop competencies and capabilities, such as balancing linear and circular systems, to address complex and highly dynamic factors, including rapid technological shifts and market volatility. [6]

2.1 Circular economics versus linear economics

Circular economics
According to the Institute of Circular Economics SR [7], the circular economy is a sustainable development strategy that creates functional and healthy relationships between nature and human society. By perfectly closing the flows of raw materials and materials in the long cycles shown in figure 1, it opposes the current linear system, where the raw materials are transformed into products, sold and burned or landfilled at the end of their life.

Fig. 1. Circular economics. [4]

Its basic principles are based on the idea that all raw material, material and product flows can be re-integrated into the life cycle after their use, where they become a source of resources for new products and services. While the replacement of primary materials by secondary ones may offer part of the solution, recycling is not a final and at the same time attractive solution, as its processes are energy-intensive and generally involve the degradation of materials. All this leads to an increase in demand for original materials. The circular economy goes beyond recycling, as it is based on a renewable industrial system leading to the elimination of waste. Recycling can be understood as the outer packaging of the entire circular economy, requiring greater energy consumption than the inner packaging of the circular economy, which means, above all, repair/modification, reuse, or processing. The aim is not only to improve the life cycle and use of the product itself but also to minimize energy consumption.

Linear economics
But companies usually operate on the principle of a linear economy. This means a one-way economy based on consumption, the sequence of which is shown in figure 2. In principle, the initial minerals are extracted, then enter the transformation process, where they are mainly mass production into a product that travels through distribution channels to B2B or B2C. If it is B2B, it means from business to business for further processing, such as semi-finished products, unfinished products, material, etc. Within B2C, the product travels to the end customer for final consumption. Products in a linear economy are either disposable or short-lived. At the end of their service life, they end up as waste, in the conditions of the Slovak Republic mostly in landfills.

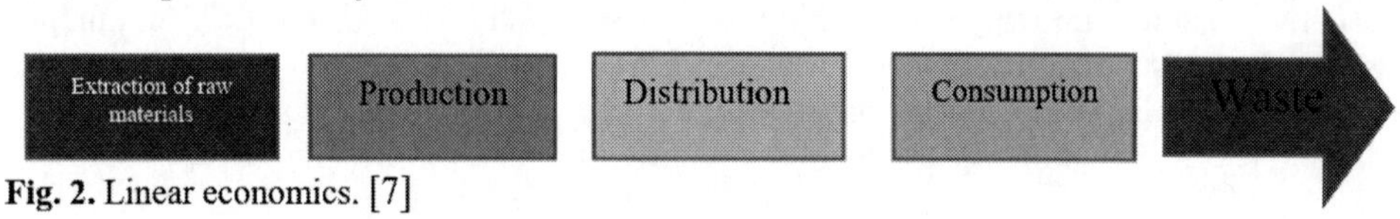

Fig. 2. Linear economics. [7]

This type of economy does not follow the principles of sustainability, it is based on the disproportionate extraction of minerals, on massive production, of which the products are mostly of poor quality and do not have a long life. Massive (almost 24 hours) production, as well as distribution associated with logistics and massive consumption, resulting in the depletion of minerals, consumption of huge amounts of energy, waste of resources as well as the waste of products and especially the generation of large amounts of waste of all categories throughout the process.
Monitoring the progress of the goals in the area of the transition to a circular economy is necessary but also a challenging task. It is also challenging because this transition is a systemic change that affects the whole economy and includes almost all products and services. The monitoring framework aims to measure progress towards a circular economy so as to cover individual aspects at all stages of the life cycle of resources, products, and services. [5]

3 Research design

The main goal of the paper is to describe the basic characteristics of the circular economy, to describe its difference from the linear economy, and to give examples of good practice used by the circular economy in Slovakia. The introductory part of the paper describes the basic characteristics of concepts that are related to the issue and relate to the circular economy. Subsequently, we gave examples of good practice of business entities and local governments, whose activities are in accordance with the principles of the circular economy. We used classical methods of analysis, synthesis, induction, and deduction to process information and knowledge in the paper.

4 Results and Discussion

It is clear that the current linear model of economic growth no longer meets the needs of today's society and cannot sufficiently reflect on the problems associated with it. To ensure sustainable growth on a global scale as well as at the EU level, resources need to be used in a smarter, more sustainable way. One of the efforts of the world community to address these issues in 2015 was the adoption of the 2030 Agenda for Sustainable Development. This document, adopted by the United Nations (UN), defines a total of 17 goals with a priority on Goal 12 to ensure sustainable consumption and production patterns, which defines the needs of the communities and economies of the countries in order to mitigate, respectively, eliminate existing and projected problems with the availability of natural resources on a global scale. Other selected goals also deal with the change of production models, the development of new technologies, as well as other forms of innovation. In 2015, the European Commission adopted an ambitious package of rules on the circular economy. The focus is on the EU Circular Economy Action Plan, with measures for the entire product life cycle: from design, resources, production, and consumption to waste management and the secondary raw materials market. The aim of the circular economy is to preserve the value of products and materials for as long as possible in order to minimize waste and the use of new resources. The transition to a circular economy supports the fulfilment of the main objective of the Resource Efficiency Program under the Europe 2020 Strategy for Smart, Sustainable, and Inclusive Growth. The shift towards a circular economy means, in addition to reducing the negative impact on the environment, also increasing the security of supply of raw materials, strengthening competitiveness and innovation, and thus higher growth and job creation. Monitoring the progress of the objectives in the area of the transition to a circular economy is a necessary but also challenging task. It is also challenging because this transition is a systemic change that affects the whole economy and includes almost all products and services. The monitoring framework aims to measure progress towards the circular economy so as to cover individual aspects at all stages of the life cycle of resources, products, and services. In January 2018, the European Commission proposed a list of indicators to measure progress. The list consists of a set of indicators grouped into four phases and aspects of the circular economy: 1) production and consumption, 2) waste management, 3) secondary raw materials, 4) competitiveness and innovation. In general, this copies the logic and structure of the EU Circular Economy Action Plan. [5]

In the article, we will then focus on the phase and aspect of the circular economy no. 1) Production and consumption.

Changes in production and consumption (processed according to [5]) are a necessity in the introduction of the circular economy. Emphasis must be placed on the circulation of products when they are created. At the same time, the sectors of the economy that produce and consume materials and products, as well as households, need to reduce the amount of waste they generate. For production and consumption, progress is monitored in the areas of:

- supply of critical raw materials,
- voluntary environmental tools,
- waste generation,
- food waste.

4.1 Examples of good practice of selected business entities and local governments, whose activities are in accordance with the principles of circular economy

Not only is the business community aware of the need for sustainable consumption and production, as well as maintaining a suitable environment for future generations. The following section provides examples of good practice.

Green County

Green County in the Trenčín Self-Governing Region aims to reduce its ecological footprint and increase the environmental awareness of the region's inhabitants. Green County is so far the only project of its kind in Slovakia at the regional level at all. The project is divided into the following areas:

- energy saving,
- electromobility support,

- environmental education,
- electronic county,
- bicycle paths and cycle paths. [16]

Circulation Slovakia platform
Seven partners from the public, private, and non-governmental sectors established the Circulation Slovakia platform in 2019. In December, 2019, they signed a memorandum of cooperation on the premises of the Ministry of the Environment of the Slovak Republic. The platform aims to contribute to the transition to a circular economy that conserves natural resources, minimizes waste, reduces dependence on imports of primary raw materials, and provides new business opportunities. The founding members of the Circulation Slovakia platform are the Ministry of the Environment of the Slovak Republic, the Embassy of the Kingdom of the Netherlands, Slovak Business Agency (SBA), PricewaterhouseCoopers Slovakia, the Institute of Circular Economics (INCIEN), the Slovak Environment Agency and the Dutch Chamber of Commerce in Slovakia. [8]

Voluntary tools of the environmental policy used in selected enterprises
The use (application) of voluntary tolls of the environmental policy is one of the principles of the circular economy, as they are of great importance in influencing consumption and production and at the same time can be its driving force. The most frequently used in Slovakia are EMS (Environmental Management System) according to ISO 14001 and EMAS (Eco-Management and Audit Scheme), Environmental Labelling of products and services (Ecolabelling). The Slovak Environment Agency, the Department of Environmental Management and the Basel Convention in Bratislava, which also acts as the competent authority for the EMAS scheme, participate in their implementation.

Ecomanagement and Audit Scheme (EMAS) is a voluntary environmental management tool for companies that want to enhance and improve their environmental performance. By implementing the scheme, companies declare compliance with environmental legislation, local responsibility, and active involvement of employees, reliability, and credibility of published environmental information. The aim of the EMAS scheme for businesses is the continuous improvement of environmental behaviour, which is monitored in six areas of the environment, namely energy efficiency, material efficiency, water, waste, biodiversity, and emissions. They are required to publish to the general public and other interested parties in the Environmental Statement throughout the registration period. EMAS is a reliable and effective management tool on the market for companies that want to improve their environmental performance through added value compared to the requirements of environmental management systems according to the international standard EN ISO 14001, in particular:

- in accordance with environmental legislation, the implementation of which is guaranteed by the state,
- mandatory information to the public through the organization's environmental statement,
- increased employee engagement. [12]

The following business entities are currently registered in the EMAS scheme with their registered office in the Slovak Republic:

- SEWA, a. s., Bratislava (engineering services in the field of collection, treatment, and recycling of electrical waste and waste),
- ŽOS – EKO, s. r. o., Vrútky (social services in the field of waste management),
- EUROVIA SK, a. s., Košice (construction sector),
- PROSPECT, s. r o., Nové Zámky (construction sector),

- STRABAG, s. r.o., Bratislava (construction sector),
- VÁHOSTAV – SK, a.s., Bratislava (construction sector),
- PORR, s.r.o., Bratislava (construction sector),
- SKANSKA SK, a.s., Bratislava (construction sector),
- CED Consulting, s.r.o., Bratislava (construction sector),
- AVA – stav, s.r.o., Galanta (construction sector),
- FERRMONT, a.s., Púchov (water and air technologies). [10]

The National Eco-friendly Product label scheme has been in place in our country since 1997. It refers to products or services based on an assessment of their entire life cycle that is less harmful to the environment than their alternatives. After Slovakia's accession to the EU, the EU Ecolabel (formerly the European Flower) is awarded through the European scheme. For companies with an environmental mindset, Ecolabelling is a suitable marketing tool that can strengthen their market position and gain a competitive advantage. The schemes are intended for promotion and represent an important information tool for environmental and consumer policy.
Under the national scheme, the following companies are currently granted the right to use the national label "Environmentally Friendly Product":

- Johan ENVIRO s.r.o. Bratislava: for products belonging to the product group "Sorption materials",
- CRH (Slovensko), a. s. Turňa nad Bodvou: for products belonging to the product group "Cement",
- Maccaferri Manufacturing Europe s. r. o. Senica: for products belonging to the product group "Gabion, Terramesh, Reno Mattresses". [13]

Under the European scheme, the following company is currently granted the right to use the "EU Ecolabel" in the Slovak Republic:

- SLOVENSKÁ GRAFIA, a.s. Bratislava: for products belonging to the product group "Printed paper". [14]

5 Conclusion

The paper deals with the circular economy as a way of sustainable production and consumption. In the introductory part, we characterized the analysed issues and described the basic features of circular and linear economics, as well as their differences. Subsequently, we gave examples of good practice of business entities and local governments, whose activities are in accordance with the principles of the circular economy.
The linear economy is based on the principle of "extract - produce - discard". This type of economy does not follow the principles of sustainability, it is based on the disproportionate extraction of minerals and massive production, from which the products do not have a long life and after their end they become waste.
The aim of the circular economy is to preserve the value of products and materials for as long as possible in order to minimize waste and the use of new resources. Its basic principles are based on the reintegration of material and product flows into the life cycle after their use, where they become sources (raw materials) for new products and services, i.e. waste generation and environmental pollution are partially or completely eliminated.
Good examples of the practice of selected business entities that eliminate their impact on the environment and adhere to the principles of the circular economy are companies that use voluntary instruments of environmental policy, such as. EMAS and Ecolabelling scheme. The use of voluntary environmental policy instruments in enterprises is linked to the transition to a circular economy, as these instruments are linked to its principles. These

tools are of great importance in influencing consumption and production and guarantee their sustainability.

Acknowledgement

This paper is an outcome of research project: „Socio-economic determinants of sustainable consumption and production in terms of impact on performance competitiveness of enterprises", VEGA no. 1/0708/20 - project share is 100%; led by doc. Ing. Daniela Rybárová, PhD.

References

1. F. M. Asif, M. Lieder, A. Rashid, Multi-method simulation based tool to evaluate economic and environmental performance of circular product systems. *Journal of Cleaner Production,* **139,** 1261-1281 (2016)
2. B. CH. Barman, Role of Green Economy in the Context of Indian Economy. *Published in International Journal of Trend in Scientific Research and Development (ijtsrd), ISSN: 2456-6470,* Volume-**2,** Issue-1, (2017)
3. G. Broman, K. Robèrt, T. J. Collins, G. Basile, R. J. Baumgartner, T. Larsson, D. Huisingh, Science in support of systematic leadership towards sustainability. *Journal of Cleaner Production,* **140,** 1-9. (2017)
4. European Commission, *https://ec.europa.eu/,* (2020)
5. T. Guštafíková, A. Kostúriková, Z. Lieskovská, Obehové hospodárstvo – budúcnosť rozvoja Slovenska. *Ministerstvo životného prostredia Slovenskej republiky, Slovenská agentúra životného prostredia.* 12 – 28. (2019)
6. P. Hopkinson, M. Zils, P. Hawkins, S. Roper, Managing a Complex Global Circular Economy Business Model: O PPORTUNITIES AND C HALLENGES. *California Management Review 2018, © The Regents of the University of California,* **60** (3) 71 – 94 (2018)
7. Inštitút cirkulárnej ekonomiky (ICE), Cirkulárna ekonomika. *ICE: https://www.incien.sk/cirkularna-ekonomika/,* (2020)
8. Ministerstvo životného prostredia SR (MŽP SR), Obehové Slovensko. *(MŽP SR): https://www.minzp.sk/spravy/2019/november/platforma-obehove-slovensko-chce-prispiet-k-zmene-nasho-ekonomickeho-modelu.html,* (2019)
9. A. Rashid, M. A. Farazee, P. Krajnik, C. Nicolescu, C. Resource conservative manufacturing. *Journal of Cleaner Production,* **57**, 166–177. (2013)
10. Register EMAS v SR, *https://www.emas.sk/register-emas-v-sr/,* (2020)
11. K. H. Robért, G. Broman, G. Basile, Analyzing the concept of planetary boundaries from a strategic sustainability perspective: how does humanity avoid tipping the planet? *Ecol. Soc.* **18** (2):5. (2013)
12. Slovenská agentúra životného prostredia (SAŽP), Čo je EMAS? *SAŽP: https://www.sazp.sk/zivotne-prostredie/environmentalne-manazerstvo/schema-pre-environmentalne-manazerstvo-a-audit-emas/co-je-emas.html,* (2020)

13. Slovenská agentúra životného prostredia (SAŽP), Environmentálne vhodný produkt "EVP". *SAŽP:https://www.sazp.sk/zivotne-prostredie/environmentalne-manazerstvo/environmentalne-oznacovanie-produktov/narodna-znacka-environmentalne-vhodny-produkt.html*, (2020)
14. Slovenská agentúra životného prostredia (SAŽP). Environmentálna značka EÚ. *SAŽP:*https://www.sazp.sk/zivotne-prostredie/environmentalne-manazerstvo/environmentalne-oznacovanie-produktov/environmentalna-znacka-eu.html, (2020)
15. K. M. Srisruthi, Circular Economy. *In International Journal of Trend in Scientific Research and Development (ijtsrd),* Volume **1,** Issue-6, (2017)
16. Trenčiansky samosprávny kraj (TSK), Zelená župa. *TSK: https://www.tsk.sk/regionalny-rozvoj/zelena-zupa.html?page_id=323084,* (2017)
17. V. S. C. Tunn, N. M. P. Bocken, E. A. van den Hende, J. P. L. Schoormans, Business models for sustainable consumption in the circular economy: An expert study. *Journal of Cleaner Production,* **212**, 324-333. (2019)

Sustainable Development and its Impact on the Quality of Human Capital

Sylvia Bukovová[1], *Ľubica* Foltínová[1*]
[1] University of Ecnomics in Bratislave, Dolnozemská 1, 852 35 Bratislava

Abstract. The quality of human capital depends on the conditions in which people live, work and learn. Therefore, the issue of sustainable development as a prerequisite for ensuring the existence and development of human capital is at the center of attention of the governments of each country. The article describes sustainable development, presents its goals and strategies to be achieved at European and national level. The aim of the research is to assess the current state and development of selected indicators of sustainable development related to the quality of human capital in Slovakia and other V4 countries in the period from 2014 to 2018.

1 Introduction

Use Labor as a factor of production is becoming increasingly important in the era of globalisation. The efficiency of business processes, as well as the quality of their outputs depends on this factor, therefore it plays a crucial role in securing the competitiveness of the corporation, economy as well as economic space. The quality of this productive factor represented by people with their abilities, knowledge and skills to produce goods and offer services, is dependent on the conditions in which people live, work and learn. Therefore, the issue of sustainable development as a prerequisite for ensuring the existence and development of human capital is at the center of attention of the governments of each country[17].

1.1 Sustainable development

Sustainable development, which means meeting the needs of present generations without jeopardizing the needs of future generations, has been the leading motive behind mankind political, economic and environmental activities for many years. It is a comprehensive approach that integrates the economic, social and environmental aspects in a synergistic way, linking thus the present with the future[10]:.

The concept of sustainable development includes according to the IHK the following aspects [3]:

[*] Corresponding author: lubica.foltinova@euba.sk.

- it strives to balance economic, social and environmental goals
- it represents mutually interconnected processes of searching, understanding, learning and empathy
- it is a decisive strategic competitive factor for the economy
- it represents a creative process which requires a new understanding of politics

The definition of the term sustainable development has a strong interdisciplinary character. So far, however, there is no uniformly accepted definition of it. It is generally understood as a form of ecological and economic activity designed to provide comparable or better living conditions for present and future generations. The focus is on the environment, the economic and social aspects of society.

The European Parliament has defined sustainable development as the improvement of the standard of living and well-being of people within the limits of ecosystem capacity while preserving natural values and biodiversity for present and future generations.

Several definitions focus on the following aspects:

- sustainability is always focused on the present and the future containing thus a temporal dimension
- resources - material and intangible goods, economic / ecological units and p. should be protected, especially when they are not renewable
- the continuity of the particular object should be sustained not only in short term but also in the long term.

As a rule, the concept of sustainable development is attributed to three different problem dimensions that need to be taken into account while respecting the political objectives of sustainable development. According to BPB (2008), these are the economic, ecological and social dimensions of sustainability, which, however, take on different value in practical implementation [1,5]:

- The economic dimension of sustainability focuses on preserving capital on long-term returns, based on efficient and long-term use of existing resources, thus differentiating from the logic of short-term gains of continuous economic growth, while continuing to support the gradual quantitative growth that is needed. to combat chronic shortages in the poorest countries in terms of catching up in rich developed countries.
- On the other hand, the environmental dimension of sustainability highlights the empirically hardly measurable intrinsic value of nature itself and the definitive nature of natural resources, which implies the economic necessity of preserving existing natural capital and preserving the ecological conditions of human survival. In terms of a qualitative understanding of economic development, ecological costs of production and consumption must therefore be taken into account in the global economy.
- The focus of social dimension is the issue of distributive justice in distribution. This concerns the access to opportunities and resources not only within individual countries and societies, but also in the conflict of the global division of the world between rich industrialized countries and poor, highly indebted developing countries. In addition to meeting the basic needs of present and future generations, the social dimension explicitly addresses gender issues in terms of creating a fairer environment for women and men.

The following figure shows the dimensions and sustainability strategies in their mutual interconnection:

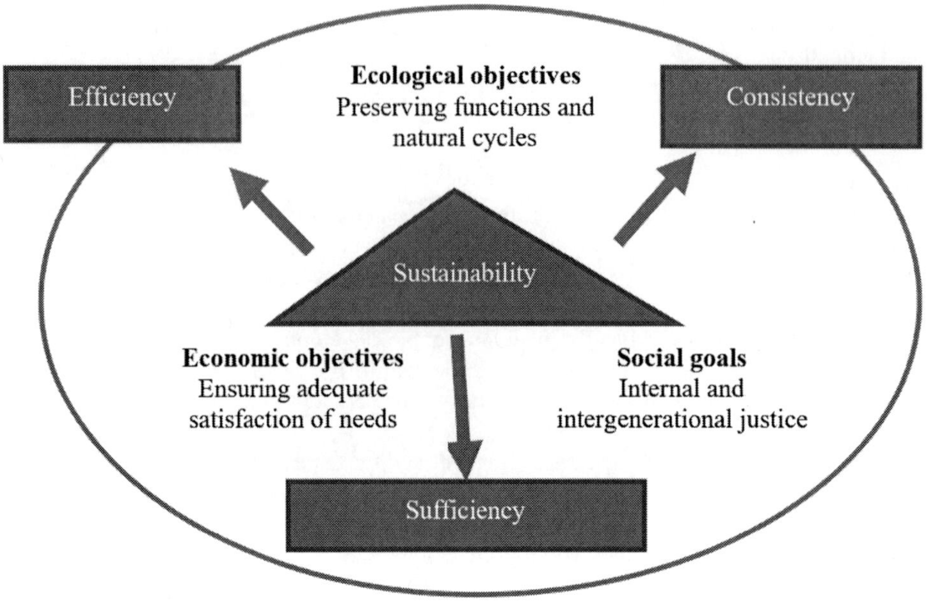

Fig. 1. Sustainability dimensions and strategies Below the figure [1].

Act no. 17/1992 Coll. on the Environment (§ 6) defines sustainable development as a development that preserves the possibility for current and future generations to meet their basic living needs, while not reducing the diversity of nature and preserving the natural functions of ecosystems

The Slovak National Sustainable Development Strategy & Agenda 2030 [7,8] describe it as a targeted, long-term, comprehensive and synergistic process, affecting conditions and all aspects of life , at all levels and towards a functional model of a community that satisfies biological, material, spiritual and social needs and interests of people, while eliminating or significantly reducing interventions threatening, damaging or destroying conditions and life forms, does not burden the country beyond the bearable extent, rationally uses its resources and protects cultural and natural heritage [11,19].

1.2 Agenda 2030 and Sustainable Development Goals

At the historic United Nations Summit in New York in September 2015 leaders around the world committed themselves to ending poverty, fighting climate change and injustice. The United Nations Program, entitled We are transforming our world: The Sustainable Development Program by 2030 - Agenda 2030 was adopted, setting a set of objectives for sustainable development in order to halt poverty, protect the planet, ensure the protection of human rights and guarantee prosperity for all.

The adoption of this program is a historic milestone on the road to a new model of the world because it addresses economic, social and environmental disparities in a global and integrated way. The European values of social justice, democratic governance and the social market economy, as well as environmental protection, are fully reflected in this process.

The 2030 Agenda for Sustainable Development offers a better future not only for billions of people around the world, but also for the planet itself. 193 countries have unanimously adopted the following sustainable development goals [9]:

1. No poverty - to end and eradicate poverty everywhere and in all its forms, anywhere in the world
2. No hunger - to eliminate hunger and malnutrition, ensure food security and better nutrition, and promote sustainable agriculture
3. Quality of health and life - to ensure a healthy life and promote well-being for all and at every age
4. Quality education - to ensure inclusive, high quality and fair education of equal quality for all, to support lifelong learning
5. Gender equality - to achieve gender equality and to empower the position and women and girls in society
6. Clean water and sanitation - to ensure water availability, sustainable water management and sanitation for all
7. Clean energy - to ensure access to affordable, sustainable and modern energy for all
8. Decent work and economic growth - to support sustainable, inclusive and sustainable economic growth, full and productive employment and decent work for all
9. Infrastructure - to build resilient infrastructure, to support inclusive and sustainable industrialization, to foster innovation

10, Reduction of inequalities - to reduce inequalities within and between countries

11. Sustainable cities and communities - to create safe and resilient cities and human settlements
12. Responsible consumption and production - to ensure sustainable consumption and production patterns
13. Climate protection - to take immediate steps to combat climate change and its consequences
14. Life under water - to preserve and sustainably use the oceans and seas and sea resources for sustainable development
15. Life on the mainland - to protect land ecosystems, to support sustainable forest management, to fight desertification and to stop land degradation and biodiversity loss
16. Peace, justice and powerful institutions - to promote a peaceful, inclusive society for sustainable development, to ensure access to justice for all, to create effective, transparent and accountable institutions at all levels
17. Partnership for objectives - to strengthen the means of implementation and revitalize the Global Partnership for Sustainable Development

These sustainable development objectives ensure a balance between the three dimensions of sustainable development: economic, social and environmental. They represent specific goals for the next 15 years and focus on areas such as [2]:

- human dignity,
- regional and global stability,
- a healthy planet,
- fair and resilient societies,
- prosperous economy.

This universal and indivisible Agenda challenges all - developed and developing countries as well as people - to achieve the goals of ending poverty, fighting inequality and combating climate changes by 2030 EC [12].

1.3 The EU and sustainable development

A number of strategies, programs and plans have been developed at European level since the United Nations Conference on Sustainable Development in Rio de Janeiro in

1992. As early as 2001, the EU adopted a long-term Strategy for Sustainable Development, which contains a number of specific objectives and proposals for improving European policy. These objectives are pursued in individual European countries through national sustainability strategies.

1.3.1 Related strategies

According to LS (2020), the strategies are as follows [4]:
2005 - Strategy for the sustainable use of natural resources
2006 - A renewed strategy for sustainable development
2009 - Report on the implementation of the Community Lisbon Program in. 2008 -2010
2010 - European Sustainability Strategy 2010 - Europe 2020 Strategy Represents a growth strategy to deliver smart, sustainable and inclusive growth across the EU.
2011 - The Roadmap to a resource-efficient Europe - Strategy Europe 2020 flagship initiative
2011 - White Paper on competitive and resource efficient transport by 2050
2011 - EU Strategy (2011-2014) on Corporate Social Responsibility
2013 – 7. EU Environmental Action Program 2020
2014 - Horizon 2020 - EU Research Framework Program - 2014-2020
2014 - European Union Alpine Space Strategy - Action Plan
2016 - Towards a sustainable future - European sustainability policy
2017 - EU response to Agenda 2030: Council conclusions of 20.06.2017
Furthermore, mitigation and adaptation measures have been developed - Klimaroadmap 2050 and Energieroadmap 2050.

In November 2016, the European Commission presented its strategic approach to the implementation of the 2030 Sustainable Development Program, including the Sustainable Development Goals. The EU has a strong starting position in the field of sustainable development and, together with its member countries, is committed to leading the implementation of the UN agenda by 2030. The Agenda promotes convergence between EU countries, within societies and with the rest of the world, setting a new universal standard for sustainable development for the benefit of all the people of the planet. Sustainable development goals are part of all10 priorities of the European Commission .

The key measures for implementing the 2030 program are [2]:

- to mainstream sustainable development objectives into all policies and initiatives, with sustainable development becoming the guiding principle that will guide all Commission policies,
- regular reporting on EU progress (first report in 2017),
- ensuring the successful implementation of the 2030 program together with EU Member State governments, the European Parliament, other European institutions, international organizations, civil society organizations, citizens and other stakeholders,
- launching a high-level multi-stakeholder platform, to promote the exchange of best practice in the implementation of the program in different sectors at national and EU level;
- developing a longer-term vision for the post-2020 period.

In order to promote sustainable development in the world, the EU will continue to work with external partners, making use of all the instruments at its disposal in the framework of external policy and, in particular, supporting efforts in developing countries.

1.3.2 The EC has set 10 priorities for sustainable development

1. Jobs, Growth and Investment

The aim is to stimulate investment and create job opportunities. Since the global economic and financial crisis, the EU has suffered from low investment rates. Coordinated efforts at European level are needed to put Europe on the road to economic recovery. Investment plan for Europe, so called Juncker plan aims to create jobs and promote growth by making more intelligent use of financial resources, removing barriers to investment, raising the profile of investment projects and providing technical assistance.

2. Digital Single Market
The Internet and digital technologies are transforming the world. Existing online barriers, however, prevent citizens from accessing goods and services, internet companies and start-ups are limited in scope, and businesses and governments cannot make full use of digital tools. It is therefore essential to prepare the EU single market for the digital age - by removing regulatory barriers and moving from national markets to a single market. It could bring the economy € 415 billion a year and hundreds of thousands of new jobs

3. Energy Union and Climate
The European Energy Union will ensure that Europe has secure and affordable climate-friendly energy. Smarter use of energy, while fighting climate change, is a stimulus to create new jobs and to invest in Europe's future. The state of the Energy Union has shown progress since the adoption of the Energy Union Framework Strategy for the transition to a low-carbon, secure and competitive economy.

4. Internal Market
The single market is one of Europe's greatest achievements as well as its greatest asset in times of increasing globalization. It is the engine for building a stronger and fairer EU economy. The freedom of movement of people, goods, services and capital brought by the single market offers new opportunities for citizens, workers, entrepreneurs and consumers, thus contributing to job creation and long-term growth. Integrated capital markets will provide more funding to companies, in particular SMEs, and to infrastructure projects. Improved labor mobility will allow people to move more freely where their skills are needed. Combating tax evasion and tax fraud will ensure that everyone contributes a fair share.

5. Deeper and Fairer Economic and Monetary Union
- Within the Economic and Monetary Union (EMU), the economic policies of EU countries are coordinated with an aim to:
- support investment and increase competitiveness
- create more jobs and better living conditions.

Currently, the reinforcement of the EMU structure is being addressed, which will enable it to respond to challenges more quickly and decisively in the future.

6. A Balanced and Progressive Trade Policy to Exploit Globalization
In the current world economy, trade is a prerequisite for growth, employment and competitiveness. The EU is committed to maintaining an open trading system. EU trade policy helps to create new jobs and new business and investment opportunities for large and small corporations. For consumers, trade agreements bring lower prices and more choice, while maintaining stringent EU consumer protection standards, social rights and environmental rules. At the same time, trade policy promotes Europe's influence in the world, spreads European values and attracts new investment. And the trade agreements help to achieve all these goals.

7. Justice and Fundamental Rights
The purpose is to promote cooperation between the different EU judicial systems and to protect the rule of law Every EU citizen has the same fundamental rights as a result of the values of equality, non-discrimination, inclusion, human dignity, freedom and democracy. These values are strengthened and protected by the rule of law enshrined in the EU Treaties and the Charter of Fundamental Rights.. The Commission has been working to enable EU citizens to cross borders without difficulty, whether to live, work, study or marry in another EU country, while protecting their personal data and improving their position as consumers.
Peace is one of the fundamental principles of the EU and requires effective and appropriate access to terrorism, trafficking in human beings and cybercrime.

8. Migration
The fate of the thousands of migrants who have risked their lives to reach the EU is shocking. Obviously, no EU country can face enormous migration flows on its own and should not be left alone to tackle them. To tackle this problem The European Commission's Migration Program provides for the integration of internal and external policies, the use of EU agencies and instruments, and the involvement of all actors: EU countries and institutions, international organizations, civil society, local authorities and national partners outside the EU.

9. A Stronger Global Actor - Pooling External Politics Instruments
The EU needs a strong common foreign policy to:
- respond effectively to global challenges, including crises in its neighborhood,
- spread its values,
- reject protectionism and maintain EU business standards; and
- contribute to peace and prosperity in the world.

10. Democratic Change
In 2014, for the first time, EU countries had to take into account the results of the elections when proposing a candidate for the position of the President of the European Commission. Although this was an important step, it is only the first of many on the road to a more democratic European Union and its getting closer to citizens. Europeans have the right to know who the Commissioners, Commission staff, MEPs and Council representatives are meeting in the legislative process. The Commission is committed to giving fresh impetus to its relationship with the European Parliament, while working more closely with national parliaments.

1.4 National Strategy Agenda 2030

The Slovak Republic signed up for the implementation of the 2030 Agenda in the document "Background to the implementation of the 2030 Agenda for Sustainable Development" approved by Government Resolution no. 95/2016. Sustainable development can, however, only be ensured by integrating it into all public policies.

The process of implementing the objectives of the 2030 Agenda on a national level is based on the fact that the topic of sustainable development is not new in the country. It makes it possible to build on existing activities and focus on reviving efforts to implement previously elaborated, more or less completed sectoral strategies and activities, in particular the National Sustainable Development Strategy adopted by the Slovak Government in 2001.

Achievement of Sustainable Development Goals in Slovakia

Agenda 2030 represents an opportunity and an instrument o identify long-term priorities for the development of Slovakia. It will be reflected in a narrower set of six priority areas which take into account the specific features of Slovakia and which will form the basis for further strategic and conceptual work. Six priorities for sustainable development have been set for Slovakia's development by 2030:

Directing towards a knowledge-based, environmentally sustainable and circular economy in the face of demographic change and a changing global environment:

- Education for a dignified life,
- Good health
- Sustainable settlements, regions and countries in the context of climate change,
- Poverty reduction and social inclusion,
- The rule of law, democracy and security.

The priorities are based on the 17 Sustainable Development Goals contained in the UN Agenda 2030 Strategy Paper for Sustainable Development. Slovakia has chosen to combine several goals into a smaller number of national priorities relevant to our country. Representatives of the non-governmental sphere, such as representatives of non-profit organizations, the academic community, local authorities or interest groups (stakeholders), helped in creating the content of the six priorities.

At the same time, it is also necessary to create an environment that supports citizens' initiatives and innovations in the field of sustainable development. Therefore, among other instruments, a new microgrant scheme is being developed in cooperation with non-governmental partners. This scheme will support the best initiatives and local micro-projects focused on the implementation of the 2030 Agenda for Sustainable Development.

1.5 The role of businesses in meeting sustainability objectives

All businesses should pursue the objectives and principles of the 2030 Agenda, not only because the 2016 Paris Agreement already includes an obligation for businesses to comply with these principles, but also because of the sustainability of the enterprise itself. At present, the competitiveness and further development of each company depends mainly on its social responsibility.

Responsible behavior is the company's commitment to developing its economic activities efficiently and responsibly towards society and the environment, taking into account the interests of all stakeholders as part of the company's strategic and risk management. Responsible behavior of companies not only towards their employees, but also in relation to their surroundings should represent a continuous activity - responsible approach of the company to the community, to solving some social problem or the need of the community.

Socially responsible behavior towards the community, or community support is part of the so-called external social pillar of corporate social responsibility. A company or organization operates within the community, it is part of it. Socially responsible companies are, however, not part of the community only as market player, but also as actors in community life, seeking to create, maintain and develop good "neighborhood relations" that is beneficial for both the community and the companies themselves.

According to Ľapinová, socially responsible corporate behavior towards the community includes two groups of activities in terms of form, intensity, time and systematic nature [6]:

- The most common focus is on environmental protection, support for education and assistance, and support for socially weak, young, talented or long-term unemployed citizens.

- To a lesser extent, they are also involved in building community capacities (infrastructure, services) or, together with the non-profit and public sectors, participate in the preparation of strategic development documents in the community territory.

In a broader sense, socially responsible behavior includes also socially responsible behavior of the companies towards employees, suppliers, buyers who also live or operate within the local community, and thus are linked back with multiple relations to these businesses in the local community. Companies are also involved in tackling societal challenges in areas such as social integration (tolerance, social cohesion), health care, education, quality of life (sport, culture), economic recovery and development, local infrastructure, safety, human capital and human rights.

2 Results and discussion

The aim of the article is to assess the current state and development of selected indicators of sustainable development related to the quality of human capital in Slovakia and other V4 countries in the period since. 2014-2018 and compare them with the EU-28 average.

The research is based on data from the Eurostat data base from 2014 to 2018.

To process the data, we use general scientific methods of analysis, synthesis, abstraction, induction, deduction, comparison and standard mathematical-statistical methods.

Selected indicators represent indicators of poverty and monitor compliance with the set objectives: no poverty, no hunger and quality of health and life in the context of sustainable development. In this context, we monitored the state and development of the V4 countries over the period (2014 to 2018) and compared it with the EU-28 average on the following indicators [13,14,15,16]:

- Persons at risk of poverty or social exclusion in% of the population –Table 1
- Persons suffering from material deprivation in % of the population - Table 2 (Indicator expresses the living conditions of people suffering from material deprivation, which are severely constrained due to lack of funds and are unable to cover at least six of the following nine items of expenditure: rent and energy, adequate home heating, unexpected expenditure , meat, fish or equivalent protein every other day, weekly vacation elsewhere, car, washing machine, color TV or phone
- Poverty risk rate by sex in% of employed women over 18 -Table 3
- Inhabitants living in substandard conditions in% of the total population - Table 4 (This indicator shows the population at risk of poverty living in a flat with leaking roof, moisture in walls, floors, foundations or rot in windows and floors)
- Population with unfulfilled need for medical treatment in% of total population over 16Table 5 (Indicator shows the number of people who claim that their needs for medical treatment were not fulfilled due to of high costs, long distance or waiting list)
- Changes in contracts for the indefinite period- average over three years in% of fixed-term workers -Table 6 (Indicator reflects changes in employment relationships from fixed-term employment contracts to employment contracts for the indefinite period)
- Number of 2-adult households with two children under the age of 14 living at risk of poverty (60% median income)- Table 7.

Table 1. Persons at risk of poverty or social exclusion in% of the population [20]

	2014	2015	2016	2017	2018	Index

						2014/2018
EU	24.4	23.8	23.5	22.4	21.9	-2.5
Czech Republic	14.8	14.0	13.3	12.2	12.2	-2.6
Hungary	31.8	28.2	26.3	25.6	19.6	-12.2
Poland	24.7	23.4	21.9	19.5	18.9	-5.8
Slovakia	18.4	18.1	18.1	16.3	16.3	-2.1

If we compare the percentage of people at risk of poverty in the V4 countries with the EU average, we can say that in 2018 all V4 countries have a lower proportion of these people than the EU average. The Czech Republic and Slovakia are among the countries with the lowest shares, Hungary has this indicator as the highest. All countries in the period under review (including the EU-28 average) recorded a favorable development of this indicator. The most favorable development - the highest decrease was recorded in Hungary; the lowest decrease was recorded in Slovakia.

Table 2. Persons suffering from significant material deprivation in% of the population [20]

	2014	2015	2016	2017	2018	Index 2014/2018
EU	8.9	8.1	7.5	6.6	5.9	-3.0
Czech Republic	6.7	5.6	4.8	3.7	2.8	-3.9
Hungary	20.0	19.4	16.2	14.5	10.1	-9.9
Poland	10.4	8.1	6.7	5.9	4.7	-5.7
Slovakia	9.9	9.0	8.2	7.0	7.0	-2.9

The share of persons suffering from material deprivation compared to the EU28 is the most favorable in the Czech Republic and in 2014 in Slovakia, since. 2015 Poland has been ahead. The worst situation is in Hungary in the monitored period. The favorable trend in the monitored period shows an analogous trend to the previous indicator. Slovakia has again the lowest decrease - 2.9%, Hungary the highest - 9.9%

Table 3. Poverty risk rate by sex in% of employed women over 18 [20]

	2014	2015	2016	2017	2018	Index 2014/2018
EU	9.1	8.7	9.1	9.0	9.1	0.0
Czech Republic	3.9	4.5	3.9	3.7	3.7	-0.2
Hungary	5.9	8.7	9.9	10.6	8.9	+3.0
Poland	9.3	9.8	9.5	9.1	8.2	-1.1
Slovakia	5.3	5.5	6.0	5.8	5.8	+0.3

Comparing the development of this indicator in the monitored period shows that the persistently worst situation in this area is in Poland - the first four years the indicator slightly exceeds the EU 28 average, in 2018 is lower, the indicator is worst in Hungary- the risk of poverty among employed women increased by 3%. The Czech Republic has the lowest percentage with favorable development. In Slovakia, however, where the situation is significantly better than the EU-28 average, this indicator worsens in 2018 compared to 2014.

Table 4. Inhabitants living in unsatisfactory conditions in% of the total population [20]

	2014	2015	2016	2017	2018	Index 2014/2018
EU	15.7	15.2	15.4	13.3	13.9	-1.8
Czech Republic	9.2	8.9	8.2	8.0	7.7	-1.5
Hungary	26.9	25.4	26.7	24.8	22.5	-4.4
Poland	9.2	11.9	11.6	11.9	11.6	+2.4
Slovakia	7.0	6.3	6.2	6.7	5.1	-1.9

In contrast to the previous comparisons, the situation in the percentage of the population living in substandard conditions in% of the total population is completely different. Slovakia is on the first place from all V4 countries and it also exceeds the EU 28 average. The fact that the percentage fell by 1.9% in the monitored period can be considered as positive. The worst situation is Hungary, which exceeds the EU-28 average by approximately 10% in the monitored period. It shows, however, a positive trend - a decrease of 4.4%

Table 5. Residents with unfulfilled medical need in% of total population over 16 years [20]

	2014	2015	2016	2017	2018	Index 2014/2018
EU	3.6	3.3	2.6	1.7	2.0	-1.6
Czech Republic	1.1	0.8	0.7	0.5	0.3	-0.8
Hungary	2.5	2.6	1.3	1.0	0.8	-1.7
Poland	7.8	7.3	6.6	3.3	4.2	-3.6
Slovakia	2.1	2.1	2.3	2.4	2.6	+0.5

The share of people experiencing inadequate medical care in the monitored period is the lowest in the Czech Republic, followed by alternating developments in Hungary and Slovakia. While the situation in Hungary improved, it worsened in Slovakia as the only of V4 countries. The most dissatisfied inhabitants are in Poland, but the country shows a significant improvement - almost by 4%.

Table 6. Changes in contracts for indefinite period- average for three years in% of fixed-term employees [20]

	2014	2015	2016	2017	2018	Index 2014/2018
EU	23.6	23.2	24.1	24.7	24.5	+0.9
Czech Republic	33.5	35.1	36.5	39.6	42.9	+9.4
Hungary	38.0	38.5	37.2	32.6	31.2	-6.8
Poland	18.4	18.8	22.0	25.7	30.0	+11.6
Slovakia	39.4	35.7	36.0	x	x	x

In terms of the share of changes in contracts of indefinite period in Slovakia, Hungary and the Czech Republic, this indicator is significantly more favorable than the EU average, unlike in Poland, which lags behind in this area in the first 3 years. While this trend of development is unfavorable in Hungary- the decline is almost 7%, in Poland it is increasing considerably - by almost 12%. Data for Slovakia for the last two years are not available.

Table 7. Number of households at the poverty line with 60% median income [20]

	2014	2015	2016	2017	2018	Index 2014/2018
Czech Republic	9,604	9,353	9,876	10,435	11,451	+19,2%
Hungary	5,686	5,741	6,007	6,285	6,834	+20,2%
Poland	6,724	7,000	7,413	7,491	8,283	+23,2%
Slovakia	8,580	8,732	8,758	9,051	9,402	+9,6%

When comparing the number of households at the poverty line, the highest number of households at risk is in the Czech Republic, followed by Slovakia and Poland. The lowest number of households at risk of poverty is in Hungary. In terms of development trend, growth is the highest in Poland and the lowest in Slovakia.

The results of the survey, which represent a comparison of the development of selected sustainability indicators in the V4 countries between 2014 and 2018, confirm that Slovakia is not characterized by a significant improvement if compared with its initial position. There is a stagnation in this area rather, which in the long term poses a risk of deterioration in the quality of human capital in Slovakia and thus its potential as a source of innovative development of the country.

Conclusion

Ensuring the fulfillment of sustainable development goals, defined and adopted at the UN and EC levels and transformed into the national strategy of the Slovak Republic, as well as their subsequent consistent application in everyday economic practice, is a prerequisite for further development of the knowledge economy of the Slovak Republic. This requires, however, strategic, targeted state measures to facilitate the creation of optimal conditions for all economic actors to fully exploit available human capital.

Acknowledgement

This paper is the outcome of projects: VEGA No. 1/0569/18 (50%) & VEGA No. 1/0309/18 (50%) - funded by the University of Economics in Bratislava; with the project membership of both authors.

References

1. BPB 2008. Leitbild der Nachhaltigen Entwicklung. Available online: https://www.bpb.de/izpb/8983/leitbild-der-nachhaltigen-entwicklung?p=all. [accessed 28.03.2020].
2. EK 2020. Prístup EÚ k trvalo udržateľnému rozvoju. Available online: https://ec.europa.eu/info/strategy/international-strategies/sustainable-development-goals/eu-approach-sustainable-development_sk. [accessed 28.03.2020].
3. IHK 2020. Aspekte der Nachhaltigkeit. Available online: https://www.nachhaltigkeit.info/artikel/politik_1976.htm. [accessed 28.03.2020].
4. LS 2020. Strategien und Programme der Europäischen Union. Available online: https://ec.europa.eu/environment/sustainable-development/SDGs/index_en.htm. [accessed 28. 03. 2020].

5. Nachhaltigkeit. Available online: https://www.nachhaltigkeit.steiermark.at/cms/ziel/139358118/DE/. [accessed 28.03.2020].
6. Ľapinová, E. 2013. Spoločenská zodpovednosť firiem a jej význam pre miestny a regionálny rozvoj. Available online: http://www.ef.umb.sk/dsr_2013/pdf/%C4%BDapinov%C3%A1.pdf. [accessed 28.03.2020].
7. Národná stratégia trvalo udržateľného rozvoja 2001. Available online: https://hpur.vlada.gov.sk/data/files/5636_narodna-strategia-trvalo-udrzatelneho-rozvoja.pdf. [accessed 28.03.2020].
8. Agenda 2030 pre udržateľný rozvoj. 2018. Available online: https://www.minzp.sk/agenda-2030/. [accessed 28.03.2020].
9. UNIS. 2020. Ciele udržateľného rozvoja. Available online: http://www.unis.unvienna.org/unis/sk/topics/ sustainable_development_goals.html#.[accessed 28.03.2020].
10. Z. Kapsdorferová, M. Kadlečíková, E. Svitová, Utilization of quality managerial systems in business entities in the Slovak Republic. International Journal for Quality Research 9, 2, 197-208 (2015).
11. A. Pilkova, M. Holienka, Entrepreneurship Development in Slovakia. Entrepreneurship in Transition Economies: Diversity, Trends, And Perspectives, Book Series: Societies and Political Orders in Transition, Pages: 225-241 (2017).
12. J. Blštáková, Z. Joniaková, Z. Skorková, I. Némethová, R. Bednár, Causes and Implications of the Applications of the Individualisation Principle in Human Resources Management. AD ALTA journal of interdisciplinary research. MAGNANIMITAS 9, 2, 323-327 (2019).
13. M. Szarková, Reasons of Employees´ Fluctuations from the Perspective of Employers in the Slovak Republic. Hradec Economic Days 2019. International Scientific Conference. Hradec Economic Days 2, 404-413 (2019).
14. T. Bobenič, A. Bobenič Hintošová, L. Hliboká, I. Vasková, Effects of changes in steel industry concentration. Croatian Metallurgical Society 54, 3, 571-574 (2015).
15. A. Bobenič Hintošová, Z. Kubíková, The Effect of the degree of foreign ownership on firms' performance. Národohospodářský obzor 16, 1, 29-43 (2016).
16. S. Goda, Slovakia and international crisis management: Focus Ukraine. Medzinárodné vzťahy 2015. Aktuálne otázky svetovej ekonomiky a politiky. Medzinárodná vedecká konferencia. Medzinárodné vzťahy 16, 293-297 (2015).
17. A. Remišová, A. Lašáková, Z. Kirchmayer, Infuence of Formal Ethics Program Components on Managerial Ethical Behavior. Journal of Business Ethics 160, 151 - 166 (2019).
18. K. Pavlyk, J. Péliová, Social enterprises in Slovakia and Ukraine: focus on finances. Finance and risk 2015. International scientific conference. Finance and risk 17, 233-241 (2015).
19. E. Pastoráková, Z. Brokešová, Risks covered by pension systems and the possibility of individual saving for old age. Managing and modelling of financial risks. International scientific conference. Managing and modelling of financial risks 8, 783-790 (2016).
20. Eurostat 2020. Available online: https://ec.europa.eu/info/statistics/search-eurostat-statistics_en. [accessed 28.03.2020].

Modern trends and emerging practices applied by organisations operating in Slovakia

Andrea Cambalikova, *Luboslav* Szabo*

Abstract. The interest in the modern trends in management comes both from the academic field as well as from the practice. Based on the results of our research and literature review we compiled a theoretical basis related to the modern managerial methods and tools used in the organization. We researched the traditional factors of management, hard factors and soft factors, more specifically what importance in managing of the organization is assigned to each of them. Our research has shown that the soft factors of management consider enterprises in Slovakia as those more important. A drawback is a fact that respondents rated corporate culture as the least important factor, while the successful implementation of the most modern management trends requires a change in corporate culture therefore it should be an important part worth paying attention.

1 Introduction

There is a couple of new challenges enterprises are facing to, such as changing economic conditions, coupled with decreasing trust in markets, increasing pressures on natural resources and significant changes in customer and employee demographics. Due to globalization there is a turbulent growth of competition. Globalization of markets and businesses brings equal conditions for access to information, sources and technologies. [1] On account of that enterprises will need to react to meet these new challenges in order to be successful. It becomes necessary to adapt their traditional methods of business to changing circumstances and new opportunities in the market. The role of the managers of enterprises is to monitor these changes continuously, evaluate them and react accordingly. The changed environment has resulted into the creation of new management methods, techniques, tools and practices that enable organizations to adapt to the current trends in the market. Managers of the organizations apply several methods and tools and the final selection depends on

* prof. Ing. Luboslav Szabo, CSc., University of Economics in Bratislava, Faculty of Business Management, Department of Management. Dolnozemska cesta 1/b, 852 35 Bratislava, Slovakia, e-mail: luboslav.szabo@euba.sk, co-authors share 50%.
Ing. Andrea Cambalikova, PhD., University of Economics in Bratislava, Faculty of Business Management, Department of Management. Dolnozemska cesta 1/b, 852 35 Bratislava, Slovakia, e-mail: andrea.cambalikova@euba.sk, co-authors share 50%.
The contribution is a part of the solution task the Ministry of Education VEGA no. 1/0375/20 in the range 100%.

knowledge, skills and personal attributes of the manager as well as on his disposal to run a business and a working team. There were times when modern approaches to organization management were applied only by the largest companies and those that led the changes. Gradually, their knowledge and using are one of the market survival requirements. In recent years, the management methods and tools have been more and more becoming an integral part of the manager´s work. Whether it is revenue-raising effort, innovation, improving quality and efficiency, as well as planning for the future, managers continually seek for helpful tools. In our fast-moving world, the winners will be companies that can sense change and respond accordingly.

2 Methods

The subject of the paper has been processed using the analysis of secondary data, the available foreign scientific literature related to the research issue, which is listed in the references. We have used a wide range of articles from scientific journals related to the processing issue and electronic resources. We have also worked with the results of the consulting companies′ surveys, mainly with the results of an international survey conducted by Bain & Company entitled "Management Tools & Trends", which served us for the needs of comparison of the state in modern trends in management in the world and in the Slovakia. The object of the survey was represented by a sufficiently large and representative sample of enterprises operating in Slovakia for the purpose of realization a primary survey. The questionnaire survey was aimed at determining the extent of application of individual modern methods and tools used during realization of managerial functions, in different types of organizations operating in different sectors. When processing the research topic, primary data were obtained by the realization of the survey using standardized questionnaires and by the realization of a qualitative survey through interviews with the executives. Respondents, in view of the questionnaire focus, were mainly executives of companies, managing directors and managers at the least intermediate level of management. The survey was conducted in the year 2017 with a total sample of 203 enterprises.

3 Results and Discussion

In the first part of this chapter we provide a theoretical overview of modern management methods and procedures that we have identified based on a literature review. In this paper we deal with and evaluate thirteen such management methods and tools.

The most adequate performance measurement method used to identify and illustrate the main drivers of the business, and which offers also an important view on organization's strategy is the ***Balanced Scorecard*** model that aligns and focuses every part of the organisation on improving and implementing the strategy.

Benchmarking is the process of comparing one's business processes and performance metrics to industry bests and best practices from other companies. Dimensions typically measured are quality, time and cost.

Big data analytics (BDA) is defined as a holistic approach to managing, processing and analysing the 5 V data-related dimensions (i.e., volume, variety, velocity, veracity and value) to create actionable ideas for delivering sustained value, measuring performance and establishing competitive advantages. [2] The survey showed that the most powerful businesses use BDA tools up to five times more often than their less efficient competitors. This survey also confirmed that these tools bring added value to businesses. [3] Gartner listed the top 10 technology trends for the next five years. In their view, big data will be a revolution in many areas, including business, scientific research, public administration, etc. [4] The strong interest in big data analytics comes mainly from the financial services, healthcare and industrial sectors. [5]

Over the past several years have hundreds of managers talked about the negative impact of complexity on both productivity and workplace morale. This message has been reinforced by the findings of major CEO surveys conducted by IBM and KPMG, both of which identified complexity as a key business challenge. [6] For years, companies have been compelled to reduce complexity, by tool ***complexity reduction***, in products, services, internal operations and assets, all to eliminate unnecessary variability while improving performance. [7]

Customer relationship management (CRM) refers to building one-to-one relationships with customers that can drive value for the firm. This manuscript discusses proven CRM strategies that companies can use as means of effective customer management (CM). [8] In the current period can be observed more intensive use of information technologies for collecting and storing information about clients. This trend is also reflected in the use of innovative approaches to relationship marketing, which gave rise to new types of customer relationship management (CRM). These are primarily electronic (eCRM) and social (SCRM) CRM. [9]

Digital transformation (DT) means the use of technology to radically improve performance or reach of enterprises. It is becoming a hot topic for companies across the globe. [10] DT opens new possibilities to improve operating efficiency, develop more accurate and agile planning, heighten vendor awareness and collaborate with business partners throughout the value chain. [11]

Employee engagement surveys examine the employee's connection to the company ´s mission. Based on 7,939 business units in 36 companies, study [12] used meta-analysis to examine the relationship at the business-unit level between employee satisfaction-engagement and the business-unit outcomes of customer satisfaction, productivity, profit, employee turnover, and accidents.

Change management refers to a set of basic tools or structures intended to keep any change effort under control. The goal is often to minimize the distractions and impacts of the change. [13] Top managers are not only important to the macro level of the organization, but they also have a key role in supporting the implementation of change management and direct transmission of information to all employees about the importance of change. [14]

Kotler and Armstrong [15] stated that customer satisfaction is the level of one's feelings after comparing products performance that gets along with his expectations. Customer satisfaction is a term frequently used in marketing. It means a measure of how products and services supplied by a company meet or surpass customer expectation. Customer satisfaction is defined as the number of customers, or percentage of total customers, whose reported experience with a firm, its products, or its services (ratings) exceeds specified satisfaction goals. This reffers to ***satisfaction and loyalty management***.

Social media programs allow individuals and organizations to interact with their employees, customers and partners electronically across a range of devices. [16] Social media are Internet-based applications that build on the ideological and technological foundations of Web 2.0, allowing the creation and exchange of user generated contents. [17]

Strategic planning is an organization's process of defining its strategy, or direction, and making decisions on allocating its resources to pursue this strategy. The leader tries by strategic or long-range planning to assist the company in establishing priorities and to better serve the needs of the membership. [18]

Talent management (TM) is the systematic attraction, identification, development, engagement, retention and deployment of those individuals who are of particular value to an organisation, either in view of their 'high potential' for the future or because they are fulfilling business/operation-critical roles. [19] TM is management of high performing staff regardless of their position in the organisation. [20]

Time management is the act or process of planning and exercising conscious control over the amount of time spent on specific activities, especially to increase effectiveness, efficiency or productivity. Senior teams can create time budgets and formal processes for allocating their time. Leaders can pay more attention to time when they address organizational-design matters such as spans of control, roles, and decision rights. Companies can ensure that individual leaders have the tools and incentives to manage their time effectively. [21]

Currently, innovations represent factor influencing not only competitive ability of business, but also population quality of life, environmental behaviour of citizens and businesses, social sphere, informational literacy of business. Innovations are considered to be result of utilisation of different stimulus, tools, techniques, which in practical application represent, apart from satisfaction of basic needs, something more for consumer „over added value". [22] Businesses should be opened and flexible towards new market options, global production trends, modern technologies, innovations, research and development, quality working labor and creation of new organisational forms. [23]

The submitted paper is a part of more comprehensive research, wherein we have evaluated the application of modern trends in business management in Slovakia, in terms of methods and tools used by the realization of the management functions, namely: planning, organizing, staffing, coordinating, and controlling. In this part of the chapter we provide an analysis of achieved results compared with an international survey.

Before we proceed to evaluate modern management trends, methods and tools, we would like to point out the traditional approaches that the managers use. "Hard factors" in organizational management are the elements that can be formally defined and implemented in the enterprise and affect the functioning of the organization and its processes. They are the result of rational thinking and have a factual orientation. In each business, certain hard factors are present, forming a visible part of the environment. Respondents were required to rate the selected hard factors by attribution of importance (5- point scale, 1=very important, 5= totally unimportant). The results are shown in the Fig. 1, where most of the companies consider these factors generally important (score very important and rather important). Most of the respondents consider a very important business strategy, 106 of them (52%). The company's business planning system has also been very positively valued, which implies the importance (very and rather) of 169 businesses, what is 84% of the research sample. The third most important element is the control system, what do think 70% of the respondents.

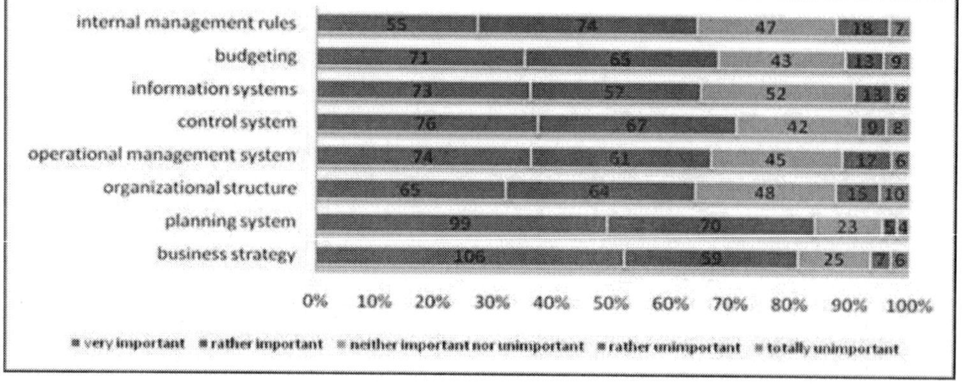

Fig. 1. Hard factors in management.

Hard factors are supplemented by "soft factors" that exist due to the presence of people in enterprise, i.e. they relate to human factor and internal social relations within the enterprise. Soft factors cannot be formally defined and ordered. Based on the results from Fig. 2, the greatest importance (rating very and rather important) was placed by businesses on communication (88%), and teamwork (83%). On the other hand, the lack of importance (assessed by the options totally unimportant and rather unimportant) was attributed to the corporate culture (22 enterprises, i.e. 11% of enterprises) and to the qualification and skills improving (19 enterprises, i.e. 9%). In the past, the trend was putting emphasis on hard factors and the soft ones were overlooked, what remains in many businesses to this day. At the present, the importance of the balance between hard and soft factors is growing. Their mutual balance and conditionality are key success factors in the market.

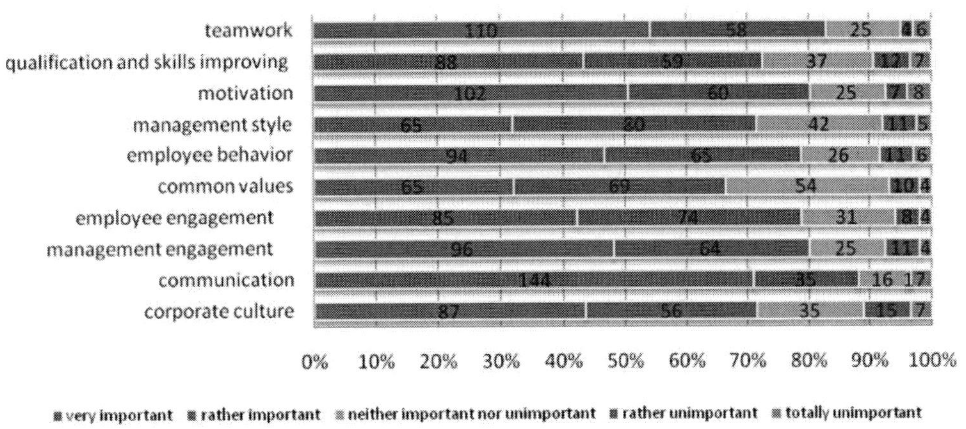

Fig. 2. Soft factors in management.

A drawback is a fact that respondents rated corporate culture as the least important factor, while the successful implementation of the most modern management trends requires a change in corporate culture therefore it should be an important part worth paying attention.

Table no. 1 shows a summary of thirteen researched modern trends in management and assessment their knowledge and use rates in absolute and relative frequencies.

Table 1. Knowledge and use of selected management trends.

	NOT KNOWN		KNOWN, NOT USED		USED	
	Absolute	Relative	Absolute	Relative	Absolute	Relative
Balanced Scorecard	106	52,20%	45	22,20%	52	25,60%
Benchmarking	45	22,20%	46	22,70%	112	55,10%
Big Data Analytics	75	36,90%	66	32,50%	62	30,60%
Complexity Reduction	100	49,30%	47	23,20%	56	27,50%
Customer Relationship Management	29	14,30%	32	15,80%	142	69,90%
Digital Transformation	90	44,30%	45	22,20%	68	33,50%
Employee Engagement Surveys	62	30,50%	57	28,10%	84	41,40%
Change Management	54	26,60%	53	26,10%	96	47,30%
Satisfaction and Loyalty Management	46	22,70%	56	27,60%	101	49,70%
Social Media Programs	48	23,60%	59	29,10%	96	47,30%
Strategic Planning	27	13,30%	31	15,30%	145	71,40%
Talent Management	66	32,50%	55	27,10%	82	40,40%
Time Management	36	17,70%	45	22,20%	122	60,10%

Respondents from the sample use strategic planning tool the most, i.e. 145 enterprises (71%). For 27 respondents only (13%) is this tool unknown. Slightly fewer enterprises use customer relationship management (CRM), 142 enterprises (70%), while CRM do not know 29 respondents from the sample (14%). The third most widely used management tool is time management, which is applied by 122 enterprises (60%). As unknown was the tool marked by 32 enterprises (18%). The least used tools by enterprises were a balanced scorecard (26%), complexity reduction (27.5%), and big data analytics (31%). As we consider the survey of Bain & Company [24], as the most comprehensive and relevant survey on modern trends and tools in management, we compared the results of our research with mentioned survey. In top 5 most used tools have strategic planning, CRM and benchmarking been in common identified in both surveys. The most applied managerial tool according to the international survey was CRM (used by 46% of companies). In our research, CRM was the second most widely used tool with the rate 70% in usage. In our research, the most widely used method was strategic planning (used by 71% of companies), what was the fourth most widely used method in the international survey, with the use of 44% of enterprises. The same percentage of enterprises also applies benchmarking, in our sample of enterprises it was up to 55%. As we can see, in our research the rates of application of the most widely used tools are much higher than in the global survey. By contrast, for example, balanced scorecard and employee engagement surveys have been used less by companies operating in Slovakia than those involved in the global survey.

References

1. P. Richnak, K. Porubanova, *Modern Trends in Business Logistics: Corporate Logistics in the 21st Century*. Prague: Wolters Kluwer, 978-80-7598-183-7 (2018).
2. S. F. Wamba, A. Gunasekaran, S. Akter, S. J. F. Ren, R. Dubey, S. J. Childe, Big data analytics and firm performance: Effects of dynamic capabilities. *Journal of Business Research* **70**, 356-365 (2017)
3. S. La Valle, Big data, analytics and the path from insights to value. *MIT Sloan Management Review* **52**, p. 21-31, 1532-9194 (2011)

4. P. Chen, Ch. Y. Zhang, Data-intensive applications, challenges, techniques and technologies: A survey on Big Data. *Information Sciences*, p. 314-347, 0020-0255 (2014)
5. D. Niknamian, *Hadoop practitioners alike should rejoice in the rise of spark*. [online]. Available: <http://dataconomy.com/2016/05/hadoop-practitioners-alike-rejoice-rise-spark-interview-altiscales-mike-maciag/> (2016)
6. R. Ashkenas, *How To Reduce Complexity In Seven Simple Steps*. [online]. Available: <https://www.forbes.com/sites/ronashkenas/2014/03/24/how-to-reduce-complexity-in-seven-simple-steps/#28cbe091e43a> (2014)
7. R. Kozole, *Reuse, recycle, and reduce complexity*. [online]. Available: <https://www.middleeast.atkearney.com/documents/20152/435359/Reuse_Recycle_Reduce_Complexity.pdf/c939734c-bf61-c1d8-e8e9-cadf3c8fa364?t=1493942411834> (2009)
8. V. Kumar, *Customer relationship management*, [online]. Available: < http://onlinelibrary.wiley.com/doi/10.1002/9781444316568.wiem01015/abstract?userIsAuthenticated=false&deniedAccessCustomisedMessage=> (2010)
9. M. Kubina, V. Lendel, Successful Application of Social CRM in The Company. *Procedia Economics and Finance* **23**, p. 1190-1194, 2212-5671 (2015)
10. Capgemini Consulting, *Digital transformation: a roadmap for billion-dollar organizations*, [online]. Available: <https://www.capgemini.com/resources/digital-transformation-a-roadmap-for-billiondollar-organizations/> (2011)
11. C. Sganzerla, C. Seixas, A. Conti, Disruptive innovation in digital mining. *Procedia Engineering* **138**, p. 64-71, 1877-7058 (2016)
12. J. Harter, F. L. Schmidt, T. L. Hayes, Business-unit-level relationship between employee satisfaction, employee engagement, and business outcomes: a meta-analysis. *Journal of Applied Psychology* **87,** p. 268-279, 268 –279 (2002)
13. J. Kotter, *Change management vs. change leadership- what's the difference?* [online]. Available: <https://www.forbes.com/sites/johnkotter/2011/07/12/change-management-vs-change-leadership-whats-the-difference/#77d548e14cc6> (2011)
14. E. Ionescu, A. Meruta, R. Dragomiroiu, Role of managers in management of change. *Procedia Economics and Finance* **16**, p. 293-298, 2212-5671 (2014)
15. P. Kotler, G. Amstrong, *Principel of marketing.* 15 th edition. Global edition, Person. ISBN 978-0133084047 (2012)
16. Bain & Company, *Social media programs* [online]. Available: <https://www.bain.com/insights/management-tools-social-media-programs/> (2013)
17. A. M. Kaplan, M. Haenlein, Users of the world, unite! The challenges and opportunities of Social Media. *Business Horizons* **53,** p. 59-68, 0007-6813. (2010)
18. R. Bouhali, Y. Mekdad, H. Lebsir, L. Ferhka, Leader Roles for Innovation: Strategic Thinking and Planning. *Procedia Social and Behavioral Sciences* **181**, p. 72-78, 1877-0428 (2015)
19. A. Weeks, *Talent management* [online]. Available: <https://www.cipd.co.uk/knowledge/strategy/resourcing/talent-factsheet> (2020)
20. A. McDonnell, D. G. Collings, K. Mellahi, R. Schuler, Talent management: a systematic review and future prospects. *European J. International Management* **11**, 1751-6757 (2017)

21. F. Bevins, A. Smet, *Making time management the organization's priority* [online]. Available: <https://www.mckinsey.com/business-functions/organization/our-insights/making-time-management-the-organizations-priority> (2013)
22. P. Richnak, K. Porubanova, *Innovations in Logistics: Digitalization Challenges in Logistics.* Prague: Wolters Kluwer, 978-80-7598-627-6 (2019)
23. K. Porubanova, P. Richnak, Usability of Intangible Assets in the Terms of Manufacturing Companies. *Ad Alta: Journal Of Interdisciplinary Research* **7**, p.177-180, 2464-6733 (2017)
24. D. Rigby, B. Bilodeau, *Management Tools & Trends 2015* [online]. Available: <https://www.bain.com/insights/management-tools-and-trends-2015/> (2015)

Creative economy as a tool of competitiveness

Jana Centárová[1]

[1]University of Economics in Bratislava, Faculty of Business Management, Department of Business Economy, Dolnozemská cesta1, 852 35 Bratislava, Slovak republic

Abstract Creativity is a fundamental manifestation of human existence and is present in every human being in different forms. The creative economy is a natural environment for innovative ideas, development and exploitation of creative potential. Over the last two decades, it has been observed that creativity is gradually becoming a driving force for economies, and as a result the importance of creative industries is also increasing. The creative industry is increasingly becoming part of EU documents and policies. The governments of the creative countries gradually introduced the concept of the creative industry, the creative class and the creative city and their importance. The creative industry refers to those parts of the economy that create economic value on the basis of individual creative input or artistic talent. It is a sector based on the exploitation of intellectual property. The aim of this work is to briefly map the concept of creative economy and its measurement in the work of foreign and Slovak authors and describe the possible potential of this industry.

Introduction

Creativity is an opportunity to create, develop, see things from different perspectives. Every person has a certain creativity, which he can develop to a certain extent. A creative person can discover original and new solutions and make meaningful use of them. Creativity as the ability to create new solutions and problem-solving becomes a competitive advantage positively influencing economic and social development derived from talented people forming the creative class. Creative people only come up with new ideas or technologies if the right conditions are created in the form of a creative and open environment. If we want to support the development of a creative economy, it is necessary to ensure a certain degree of maturity of the economic, technological, social and cultural environment.

1 Current state of the issue

Many authors deal with the topic of creative economy. John Howkins and Richard Florida were among the first to address this issue. John Howkins is the author of "The Creative Economy" book, which describes the advent of a new creative workforce and creative industry. It describes the creative industry and the creative economy as a summary of four types of intellectual property:

- copyright industry,
- patent industry,
- trademark industry
- design industry.

They argue that the creative industry should be limited to the arts and culture without the addition of science, research or the patent industry. [1]

Richard Florida is the author of"The Rise of Creative Class" book. He sees the present as the "age of creativity". He believes that creativity represents an infinite potential, the release of which is the key to economic growth. He published research results confirming the fact that the presence of creative people living and working in the city affects the attractiveness, economic performance and competitiveness of the city. Cities are overtaking each other in attracting talent, because only those cities that are creative and supportive of interactions, knowledge exchange, generation of innovation clusters are creative. [2]

The author Gašparík and the collective define the creative economy as an industry that deals with the connection between economy, culture, technology and asocial aspects. [3]

Sava and Badulescu are currently considered to be one of the most dynamically developing sectors of the economy. They consider the emergence of a creative economy as a shift towards a global change towards a tertiary economy, where creativity, cultural and scientific contributions by employees are its most important features. [4]

The authors of KEA European Affairs organisation have defined the creative industry on the basis of the key elements of the creative economy, based on:

- artistic creativity, which is the creation of new thoughts and ideas expressed in music, lyrics, works of art. In today's economy we can perceive it as creativity in science, economic creativity, technological creativity and cultural creativity.
- scientific creativity, which includes experiments and research to solve problems
- economic creativity, where creativity is a dynamic process and leads to technological, business, marketing innovation. [5]

The world organisation UNESCO has also enshrined the definition of the creative industry in its documents. It defines the cultural and creative industries as sectors of organized activity whose main purpose is the production or reproduction, promotion, distribution of goods, services and activities of a cultural, artistic or natural nature. This approach emphasizes more than just industrially produced products of human creativity, highlights the importance of the entire product chain as well as the specific functions of each sector involved in bringing these products to the public. [6]The UNESCO Report in 2009 sets out the characteristics of four models of the creative industry

- Model of the UK Department of Culture (UK DCMS)
- Symbolic text model
- Model of concentric circles,
- WIPO copyright model. [7]

Based on four models, UNCTAD has created its own definition of the creative industry, which we can see in more detail in the following picture 1.

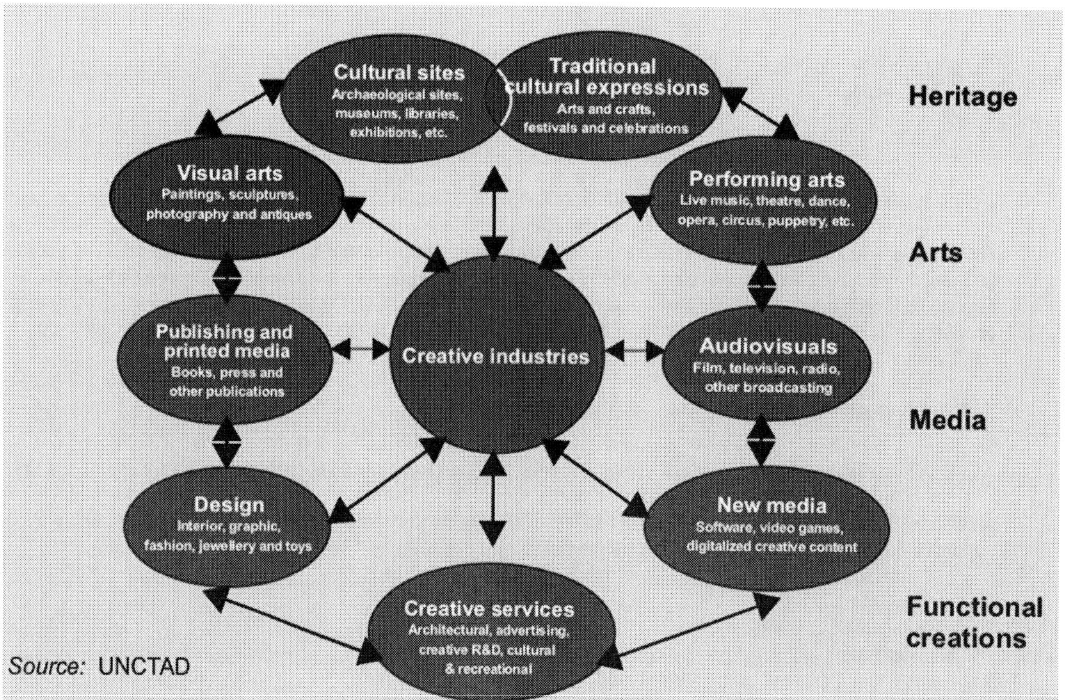

Fig. 1. Creative industry [8]

2 Creative industry in Slovakia

The creative industry in Slovakia is the responsibility of the Ministry of Culture of the Slovak Republic and the Ministry of Economy of the Slovak Republic. They also work closely with the finance, education, labor and regional development sectors. In 2011, the Ministry of Culture drew up a government paper: "The Concept of Support for the Cultural and Creative Industry in the Slovak Republic", which characterizes the creative industry and was approved by the government. The characteristic states that the creative industry is to be seen as an exploitation chain that is created by creating the conditions for creative activity, and subsequently includes creative activity, production, reproduction, presentation, representation including export, storage, archiving and recovery. Furthermore, the concept states that the structure of the creative industry can be divided according to the nature of goods and services into:

- an area of arts and culture;
- an area primarily related to the use of creative activities (advertising, design, entertainment),
- the area of selling or creating access to goods or services from previous areas to consumers (hardware production, electronic communications, archiving, information services). [9]

This definition is based on definitions used in official policies of other states or international organizations, and is adapted to Slovak conditions. The starting material from 2011 was followed by the "Strategy of Creative Industry Development in the Slovak Republic" and "Action Plan 2015-2017 on the Strategy of Creative Industry Development in the Slovak Republic 2014-2020". Both materials were prepared under the auspices of the Ministry of Culture and the Ministry of Economy.

According to the creative industry strategy, they include the creation, production and distribution of products and services that use creativity and intellectual assets as primary inputs. Also, a set of knowledge-based activities, focused but not limited to the arts, potentially generating revenue from trade and intellectual property, and tangible and intangible intellectual or artistic services with creative content, economic value and market objectives. [10]

3 The measurement of creativity-the index of creativity

Because creative economics is constantly on the rise, there is a need of its quantification. When measuring creativity, a question comes into existence: what are we measuring, when we are measuring creativity? From psychological, sociological and economical point of view, creativity is seen as an ability. It isn't possible to measure it, however what can be measured are indicators of creativity. Looking for adequate indicators is very difficult, because uniform definitions of notions related to creative economics don't exist: what is the creative industry, cultural industry, who belongs to creative class and so on. Another problem is the valuation of creativity. The problems of creating a benchmark for measuring creative sectors consist of many factors from the practical ones to the conceptual ones. On the conceptual level, there is a problem of determining the boundaries between art, culture and industrial production. „Generally, the knowledge, that economic indicators aren't suitable for capturing every output of creative economics is accepted, on the other hand creative economics has a deep impact, not only in other sectors of economics, but generally in the society". [11]

Most often we meet the approach of R. Florida. His approach is based on the valuation of the creative environment and the creative class. The author created the Index of creativity, based on the 3T, which consist of three domains, which are: technological skill, talent a tolerance. Together with his collaborators he created the Index of creativity made of

- The Index of talent with two indicators (the index of human assets and the index of creative class)
- The Index of technologies with two indicators (the index of innovation and the hi-tech index)
- The Index of tolerance with three indicators (the gay index, the index of arts and the index of migration) [12]

The approach of D. Hui and his colleagues differentiates significantly from the approach of Florida. The understanding of creative society Hui *et al* differentiates the most from the understanding of Florida with its classification of social assets. Hui et al created a so-called hongkong model 5C. According to their model (Outcomes of Creativity Index) the creative process continuously affects social, cultural, structural and human assets, therefore he also classified the social assets into the structure of the creative index. [13] Two of these indicators are concerned with charity contributions, the others include the indicator of generalized and institutionalized trust, reciprocity, the interest of public matters, holding modern or traditional values, contacts in the community and so on. It is possible to state, that the understanding of creativity by hongkong authors comes probably the closest to the content of the concept of krea-togenic society. [14] Graphical representation of model 5C is seen in the picture number 2.

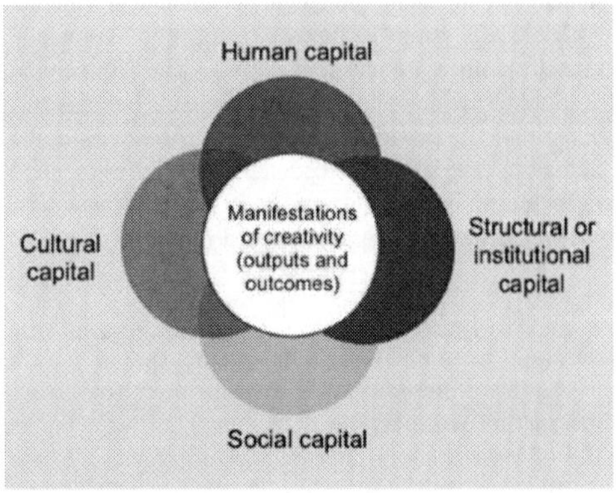

Fig. 2. Mode 5C [8]

In Slovakia, Antalová and Barta attempted to quantify the creative society. Based on the methodology of Florida and his collaborators they created their own creativity index, the so-called. Slovak-Creativity index, which is a synthetic indicator consisting of three partial indexes:

- the index of human assets,
- the index of scientific talent and
- the index of innovations. [15]

Author M. Balog and col. presented two ways of measuring the creative industry, one based on a sectoral perspective and the other on a career perspective. [16] In the Czech Republic, author J. Kloudová compiled a Czech creativity index and a situational matrix of regions of the Czech Republic, where individual regions were grouped into leaders, growth regions, delays and regions losing ground beneath their feet. [17]

4 The creative class

The term creative class is associated with the name of American economist and Professor Richard Florida. Florida distinguishes professions whose main function is to come up with new ideas, technologies or creative content and their bearers are representatives of the creative class, scientists, engineers, developers, architects, designers, educators, artists, musicians and many other professionals in the cultural industry - the central creative class. There is a space around the central creative class, bringing together a wider group of creative professionals working in business, finance, law, healthcare and other areas. Florida emphasizes that the prosperity of the entire locality depends on the creative class, as it boosts the growth of the entire economy of cities or regions, and therefore it is crucial to attract these people for the economic development of a particular locality. The creative class is freer, more flexible, with a higher degree of individuality than other classes. R. Florida sees the main difference between the creative class and the other working classes in what they are paid for. The working class and the class taking care of service are paid to follow a pre-prepared plan. However, the creative class is paid to be individual and independent. Richard Florida divided the creative class into two groups:

- Super-creative core is represented by the representatives of the sectors such as: scientists, engineers, education, programming, research, artists, designers, poets and the media workers. Their common feature is that they are fully involved in the creative process within their workload. The primary function of these workers is to be creative and innovative, while the production of products and services for commercial use is secondary;

- Creative professionals are classic knowledge-oriented workers, consisting of workers in the healthcare, business and finance, law and education sectors. The primary task of these employees is to solve specific problems with their knowledge and creativity. [18]

5 The potential of the creative industry

The Great Britain was the first European Union country to institutionalize the creative industry and include it in strategic documents at the highest political level in 1998. Howkins in his book. The creative economy. How people make Money from ideas'' was the first to name a newly created creativity-based society as a creative economy. This was done based on the finding that creative workers in the US account for 41% of the workforce and account for 48% of GDP. He was the first to come up with the idea of creativity as an engine for economic growth. [12] The European Parliament affirmed the potential of the creative and cultural industry with a resolution on the future progress and exploitation potential of cultural industries in 2003, the second resolution on the cultural industry was published in April 2008. 2009 was declared European Year of Creativity and Innovation, reflecting the importance of innovation and promoting creativity for the development of Europe. In 2010, the Green Book on unlocking the potential of the cultural and creative industry was published. The book was intended to start a debate on the requirements to create a stimulating environment for its development and to identify opportunities for its better use to boost economic growth in Europe. In the recent decades, many researchers and economists in the US and Europe have been dealing with the creative industry and economy. The creative industry is becoming part of more and more EU documents and policies. According to the European Parliament resolution from 12th of May 2011 on unlocking the potential of the cultural and creative industries, the EU's cultural and creative industries provide 5 million jobs and represent 2.6% of its GDP. In 2011, the European Creative Industry Alliance was established. The creative and cultural industries are among the fastest growing industries in the world. The Paris Conference in December 2015 reported on its significant contribution to the world economy, presenting the first ever survey mapping the impact of cultural and creative industry on the world economy. According to the conference, the cultural and creative industry makes up 3% of the world's GDP and are among the world's leading employers (1% of the world's active population). The development of a creative economy is linked to certain preconditions that must be met. According to Kloudová, "it is the achievement of a certain degree of maturity of the economical, technological, social and cultural environment that creates the conditions for the growth of the creative sector. The premise for the growth of the share of creative sector in the economy output is maturity and wealth of consumers, ability and education of the working force and environment, which is on a high economical and technological level, is opened to new solutions and where a good institutionalized base exists ".[17] The main feature of the global development of the creative economy is a fundamental change in the human factor in the economy, in the sense, that the production is moving increasingly to the area of innovations and ideas, to creative, original, non-traditional ways of thinking that produce new ideas and create the highest added value. The contribution of the creative and cultural industry lies in its inter-sectoral dimension with minimal territorial constraints. A complex study was brought by the company EY "Culture era - the first global map of the Creative Industry", where it analyses 11 sectors of creative and cultural industry across Europe, North America, Latin America, Asia and the Pacific, Africa and the Middle East. In every region in the world, the creative industry has an impact, as shown in the following table 1. [19]

Table 1. Creative industry in the world regions

Region	% worldwide sales to the creative industry	% jobs in the creative industry	Note
Europe	32	26	strong public support, high creative concentration
North America	28	16	a leading position in the cinematography, TV industry and performing arts
Latin America	6	7	TV, music and dance production
Asia and Pacific	31	43	clothing industry, cinematography growth and librarianship
Africa and Middle east	3	8	growth of film, television and music production

Conclusion

The conclusions of all the studies agree that the creative sector of the economy is growing at a faster pace than most of other sectors and it ranks in a leading position among other sectors in terms of value added to the country's GDP. We can now perceive this economy as a new globalization trend in a modern society that was established at the end of the 20th century. Creativity is the basis for developing the creative economy. Creativity has been seen as a new type of competitive advantage in recent years and we need to learn how to sell our competitive advantage. Slovakia has a creative potential for the development of the creative industry and with the view of the limited natural mineral resources it is appropriate to focus on a creative economy, independent of exhaustible resources. The term creative industry includes several areas and at the same time defines the social requirement for a change of the perception of the concept of industry, which is now understood only in terms of pure production. In this context, industry must be seen as an exploitation chain that is being created by creating the conditions for creative activity, which then includes creative activity, production, reproduction, presentation, representation, including export, storage, archiving and recovery.

References

1. J. Howkins, The Creative Economy: How people Make Money from Ideas, London: Penguin Books, 288 (2001)
2. R. Florida, Entrepreneurship, Creativity and Regional Development. In: Hart, D. (ed): The Emergence of Entrepreneurship Policy: Governance, Start-Ups, and Growth in the U.S. Knowledge Economy. Cambridge, MA: Cambridge University Press. Higgins, L.F. "Applying principles of creati- vity management to marketing research efforts in high-technology markets", In: Industrial Marketing Management, **28,** 305-317 (2002)
3. J. Gašparík, J. Krammer, A. Václavová, GAŠPARÍK, Príležitosti kreatívnej ekonomiky, online at https://www.ia.gov.sk/data/files/NP_CSD_II/Analyzy/Stat/Prilezitosti_kreativnej_ekonomiky.%20pdf (2017)
4. D. Sava, A. Badulescu, SAVA, The Creative Economy: Facts and Figures. In: The Annals of the University of Oradea, Economic Sciences, online at https://www.researchgate.net/publication/322631634_The_creative_economy_facts_and_figures (2017)
5. KEA European affairs, Promoting Investment in the Cultural and Creative (2010)
6. UNESCO, Creative Industries, online at http://www.unesco.org/new/en/santiago/culture/creative-industries/ (2017)
7. UNESCO, *The 2009 UNESCO Framework for Cultural Statistics (FCS).* Montreal: UNESCO Institute for Statistics. (2009)
8. Creative economy report, online at unctad.org/en/docs/ditc20082cer_en.pdf, (2008)
9. MK SR, Východiská koncepcie na podporu kultúrneho a kreatívneho priemyslu v Slovenskej republike. UV-40861/2011, online at http://www.rokovania.sk/Rokovanie.aspx/BodRokovaniaDetail?idMaterial=20575 (2011)
10. MK SR, Akčný plán na roky 2015 – 2017 k Stratégii rozvoja kreatívneho priemyslu v Slovenskej republike na roky 2014 – 2020. Uznesenie č. 11/2015. online at http://www.rokovania.sk/Rokovanie.aspx/BodRokovaniaDetail?idMaterial=24172 (2014)
11. UNCTAD, *Creative Economy: A Feasible Development Option.* Geneva. `online at http://www.un-ctad.org/en/docs/ditc20082cer_en.pdf. (2014)
12. F. Murgaš, M. Ševčíková, Geografický prístup k meraniu kreatívneho kapátálu. 37 – 56, (2012)
13. D. Hui, N.G. Chun, P. Mok, W.K. Chin, CH. Yuen, A Study on Creativity Index. Hong Kong: Home Affairs Bureau, online at ttp://http://www.hab.gov.hk/file_manager/en/documents/policy_responsibilities/arts_culture_recreation_and_spo rt/HKCI-InteriReport-printed.pdf (2005)
14. M. Kačírková, Kreativita v kontexte znalostnej spoločnosti. Bratislava: Ekonomický ústav SAV, 37. (2017)
15. M. Antalová, M. Barta, Kreatívna spoločnosť – prístupy k jej kvantifikácii v Slovenskej republike. Nová ekonomika, 157 – 165, (2013)
16. M. Balog, Možnosti rozvoja kreatívneho priemyslu na Slovensku. Bratislava: Slovenská inovačná a energetická agentúra, 66 (2017)
17. J. Kloudová, Kreativní ekonomika a její měření, online http://www.creative-economy.cz/dokumenty/9-ekonomicky-casopis32009pdf.pdf (2012)
18. R. FLORIDA, Tinagli, I., Europe in Creative Age , online at: https://www.demos.co.uk/files/EuropeintheCreativeAge2004.pdf (2004)
19. CISAC, Nová globálna štúdia o kultúrnom a kreatívnom priemysle, online at: ://www.soza.sk/aktuality/85/cisac-nova-globalna-studia-o-kulturnom-a- kreativnom-priemysle (2015)

Cooperation of Business Intelligence and Big Data in one Ecosystem

Matej Černý[1,*]

[1]University of Economics in Bratislava, 852 35 Dolnozemská cesta 1, Bratislava, Slovakia

Abstract. This paper is focused on the issue, how the business can analyze all data types (structured and unstructured) in one cooperative environment. With structured data handle Business Intelligence and with unstructured data on the other side Big Data. As a solution to this issue, we have suggested our Business Intelligence and Big Data ecosystem. This model - the ecosystem is based on already proven data processing processes running in Business Intelligence and in Big Data areas. Both processes are integrated into one unit. We have also described their common functioning.

1 Introduction

Business Intelligence (BI) is a modern term used to refer to a set of technologies and processes that use data to understand and analyze business results [1, 2] designates it as a process that involves gathering sufficient information at the right time and in a usable form and analyzing it so that it can be used for a positive impact on business strategy and tactics.

According to Rainer et al. [2], BI supports comprehensive and non-recurring decision-making with a focus on tactical management and knowledge workers, which as is defined above, can be extended to include strategic and operational management levels. Consulting company Gartner [3] and Mallach [4] also agree with the above definitions. Most authors associate this term mainly with the creation of a unified enterprise data warehouse, respectively data marketplace and also advises it in the area of knowledge management, which also concerns.

According to the authors Laudon & Laudon [5], is the term Business Intelligence used to describe: „the infrastructure for warehousing, integrating, reporting, and analyzing data that come from the business environment, including big data. The foundation infrastructure collects, stores, cleans, and makes relevant information available to managers."

Most of the data collected by businesses today are in the form of transactional data that can be easily written in the form of two-dimensional tables, in rows (records) and columns, and are stored in a relational database.

Such databases containing structured data can be used in the enterprise as a data source for further analysis. However, in order to use them, it is necessary to extract, cleanse and store this data first in a central data warehouse, where it will be possible to perform further complex analyses to meet the needs of the decision-making process. A data warehouse can be defined as a database or a set of databases that contain business data from various sources in the enterprise as well as from external systems that cover all aspects of business processes in all functional areas, all aspects of products and services and customers and partners [2]. The data warehouse contains both historical and current data [5]. This integrated data is used

* Corresponding author: matej.cerny@euba.sk

to support data analytics and decision-making [6]. The Creation and operation of a data warehouse is costly [7].

A data warehouse that contains only a subset of all data stored in a data warehouse is called a data mart [6]. Data mart is a cheaper and smaller version of the data warehouse and is intended to support only a specific group of users in the enterprise, e.g. sales department, or strategic business unit [7].

To load data from different sources and insert them in a data warehouse is used data extraction, transformation, and load (ETL) process. This process extracts information from internal and external databases, transforms it using a set of business definitions, and uploads it to the data warehouse.

Another term is Data Analytics (DA). This concept is more focused on tools and techniques for analyzing and understanding data. It includes techniques like OLAP, statistics, models and data mining (Data mining is a process for extracting hidden, unknown, but potentially useful information and knowledge from massive, incomplete, noisy, fuzzy, and random data [14]).

BI and DA are essentially about integrating all the information flows produced by a company into one coherent corporate data set (data warehouse) and by using modeling, statistical analysis tools (such as normal distribution, correlation, and regression analysis, Chi-square analysis, forecasts, and cluster analysis) and data mining tools give a sense to all this data so that managers can make better decisions and plans.

Summarizing the above knowledge, we will get a comprehensive overview of the functions and capabilities of BI. The entire BI data processing process could be described in these steps. Raw transaction data are obtained from internal systems such as CRM, ERP or other OLTP systems. This data can be enriched with data from external sources. Using the ELT process, data are transferred to the data warehouse, where they are stored. From data warehouse are distributed to the data mart according to the needs of functional areas of the company. In the data marts are performed the required analyzes with the help of decision-support systems and then are provided to the end-user in a visual form through the presentation layer.

Recent trends in organizations' data retention need point to the need to retain unstructured or semi-structured data, such as images, videos, sound recordings, or text files from various sources such as visitor web records, email messages, social media content, as well as machine-generated data from sensors or automated systems (e.g. stock systems, etc.) [13], and such data is not suitable for processing and storage using traditional relational databases. This data typically consumes a huge capacity of data storage at the level of petabytes or exabytes, so we are talking about billions or even trillions of records from various sources.

Data in such volumes and forms are called Big Data. Authors such as Laudon & Laudon [5], Stair et al. [2] and Baltzan [6] agree on the Big Data definitions. According to them, Big Data can be defined as the collection of large data sets, including both structured and non-structured data, which cannot be analyzed using traditional methods and database tools. In terms of processing such diverse and mainly unstructured data, it could not be longer use standard database systems and relational databases, therefore the use of NoSQL (and Not Only SQL) respectively non-relational database management systems [6]. These use a more flexible data model and are designed to manage large volumes of data on many distributed devices and allow easy scaling. They are useful for accelerating simple queries over large volumes of structured and unstructured data, including the Web, social media, graphics, and other forms of data that are difficult to analyze with traditional SQL-based tools. NoSQL also supports SQL queries.

A comprehensive software ecosystem for data processing and storage was brought by Hadoop. It is an open-source software platform that enables distributed parallel processing and storage of huge amounts of data on many commercial computers [5]. It also allows the interconnection of these unstructured data sets with relational databases [2]. It divides the big data problem into smaller sub-problems, distributes them among thousands of cheap computer processing nodes, and combines the result into a smaller data set that is easier to analyze. Hadoop can process large amounts of data, including structured transactional data, semi-structured data such as Facebook and Twitter feeds or complex data such as web server log files, or unstructured audio and video data.

For these types of data had to be developed new types of analyzes - Advanced Analytics. Advanced analytics can be characterized as an autonomous or semi-autonomous examination of data or content using sophisticated techniques and tools, usually beyond traditional business data analytics (BI) procedures to discover more detailed, deeper information, make predictions or generate recommendations. The main goal of advanced analytics is to extract useful information from large data sets and transform them into an understandable form. The major processes of Big Data include capture, curation, storage, search, sharing, transfer, analysis, and visualization [11, 12].

2 Methods

This article focuses on identifying data collection processes, which consist of storing, processing, analyzing, and compiling output for business decision making in the BI and Big Data areas. Based on the analysis of the identified processes, we try to find the intersection of these two processes. The goal is to create one common ecosystem of processing both structured and unstructured data in the enterprise. Easy to say provide the right information in the right way at the right time and for the right people at acceptable costs.

By achieving this goal, we focused on finding available scientific and professional publications that include the data processing processes in the areas of BI and Big Data and which defining these and other terms. Using the synthesis of acquired knowledge, we try to find their commonly used definitions. This is the content of the first part of this paper.

In the area of research results and discussion, we publish our proposed BI and Big Data ecosystem. This model – the ecosystem is based on already proven processes that are integrated into one unit and describe their common functioning. It requires validation in terms of its content and applicability in practice, which will require further work.

3 Results and Discussions

BI and Big Data can and should be used as one integral unit that will help an enterprise to economically accept, collect, and store any type of data in relatively any amount. Using this data, the organization can then support the decision-making process using various types of analyzes depending on the goal that it wants to achieve. It can then process, distribute and present the data for specific groups of managers, specialized employees or use it for automatized decision-making. The BI and Big Data ecosystem is shown in Fig. 1.

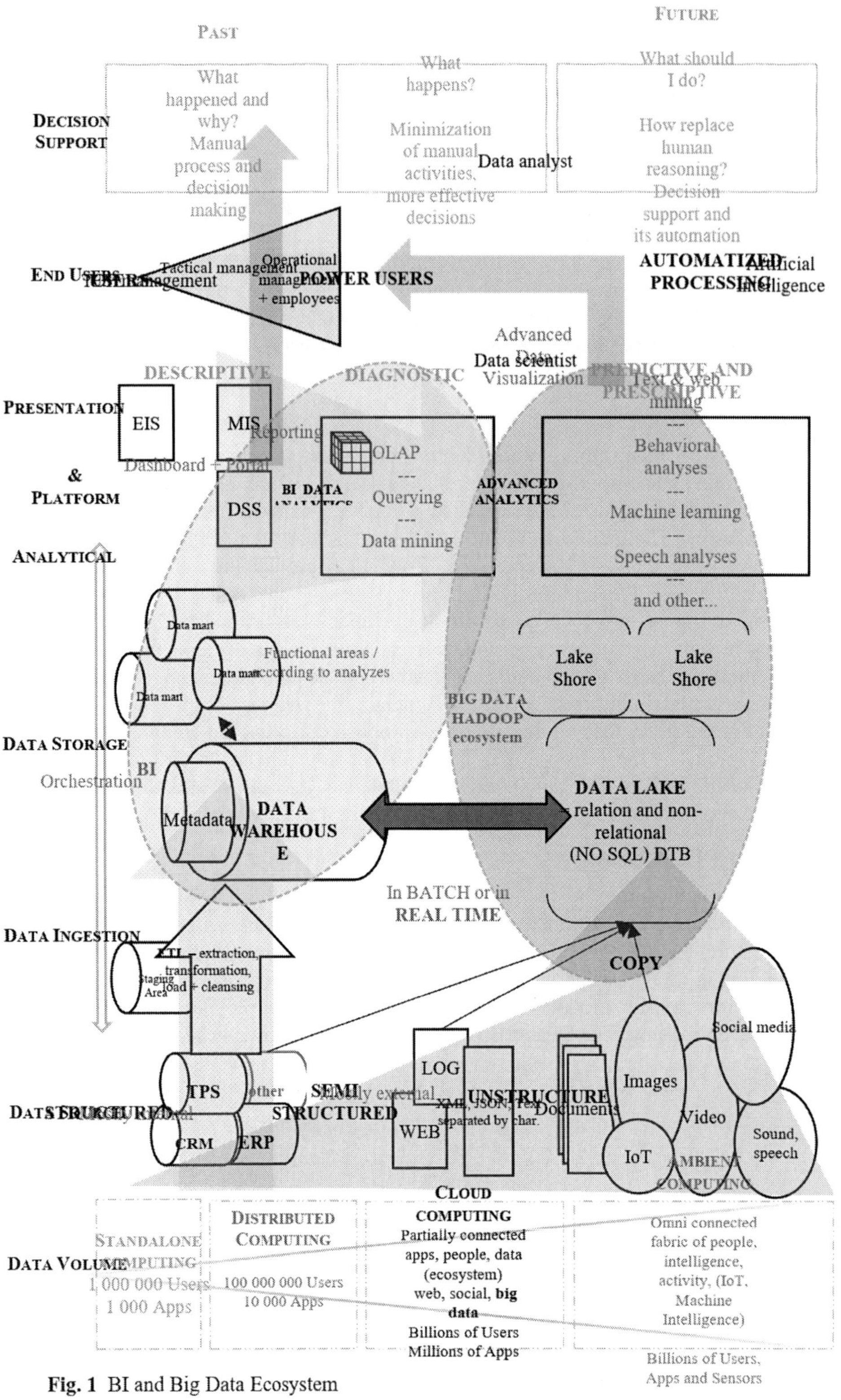

Fig. 1 BI and Big Data Ecosystem

The BI and Big Data ecosystem inherently provides all information system functions such as data collection, transmission, storage, processing, distribution and presentation. The ecosystem itself is based primarily on the BI ecosystem, which is enhanced by the specifics of Big Data.

The first column in Fig. 1 characterizes the evolution of data creation, from the beginning of computer use in enterprises to the latest ubiquitous technologies producing large volumes of diverse unstructured data.

In the second column, we are more specific, we define areas of origin of various types of data in organizations, where we go from structured more or less internal data to unstructured external data sources.

The next section deals with how the data is transferred to the central database of the company. Structured data uses a process called extraction, transformation, and load (ETL). In addition to these activities, the data in this process is also cleaned (schema-on-write). For unstructured data in large volumes, this process cannot be used and therefore the data is simply copied to the repository and before processing, they shall be edited only to the extent required for their analysis (scheme-on-read).

Structured data is stored within a central enterprise database called a data warehouse. From data warehouse it can be distributed to data marts, where it is analyzed for a specific purpose of a functional area of the enterprise. For unstructured data and not just for it, because it is also possible to store structured data at a low-cost, are used distributed storages based on relational and non-relational NO SQL (not only SQL) databases and so-called data lakes. The data lake is: „a methodology enabled by a massive data repository based on low-cost technologies that improve the capture, refinement, archival, and exploration of raw data within an enterprise“ [15] Data lake from the viewpoint of business domain load all data from source systems, no data is turned away and data are stored at the leaf level in an untransformed or nearly untransformed state. Both data warehouse and data lake are data repositories. However, they are different in many aspects from concepts, structures, and implementation [16]. For example, a traditional data warehouse stores data in files or folders, the data lake uses a flat architecture to store data. Each data element in the data lake is assigned a unique identifier and is marked with a set of extended metadata tags. When a business question arises, the data lake will be queried for all relevant data, thus creating and providing a smaller dataset that can then be analyzed to help answer the question. A data lake is something like a data warehouse for any type of data, while lakeshore marts are an image of data marts. They can be characterized, according to El Kaim [8], as data marts, each has a specific model for one limited context and processes and organizes data for analytical use.

Within the BI and Big Data ecosystem, we have identified a part of the process as an orchestration. This term could be defined according to Watts [8] as the automation (understood as running a single task on its own) of multiple activities at once. Orchestration includes automated organization, coordination, and management of computer systems, middleware and services. Orchestration uses more automated tasks to automatically perform a larger workflow or process. It could consist of multiple automated tasks and include multiple systems.

The analytical and presentation platform provides an overview of analytical methods by specific data volume levels. Here are mentioned IS-s, which use data from the data warehouse, respectively data lake, process them and shift to support the decision-making process at individual management levels. We also present analytics used in BI and Big Data.

Individual outputs from data analytics and individual types of analyses are designed for different end-users. The BI is used primarily by managers at all management levels and from power users they are data specialists. The Big Data is characterized by a high degree of sophistication in the analysis, and therefore their main consumers are data scientists who extract useful knowledge from the data for the enterprise, which can be then applied in different areas of business. The last area is automated systems that are used to automate decisions and thus relieve managers of certain types of decisions. The last column shows the time orientation of the analyses and also the way how these analyses help the company in the decision-making process. Big Data are at the last stage of a long-term transition from systematic reporting (IT approach) to the means of automated analysis (business approach) [17].

In the literature as well as in the professional community it is not always clear what the relationship between BI, Big Data and data analytics is. The relationships between these terms were well summarized by Dedic and Stanier [9] with the visualization shown in fig. 2. Based on an analysis of available literature and validation in practice, they consider knowledge discovery (KD) to be a top-level concept that uses data analysis techniques to discover and create new knowledge in organizations. Within KD, data analytics is its object and contains various disciplines and approaches, including BI and Big Data (advanced analytics). BI and Big Data are considered by the authors as disciplines at the same. We would like to add that Big Data is a logical extension of BI, especially in the field of unstructured data processing and the necessary data analysis, and within the ecosystem, these two approaches to data work complementary.

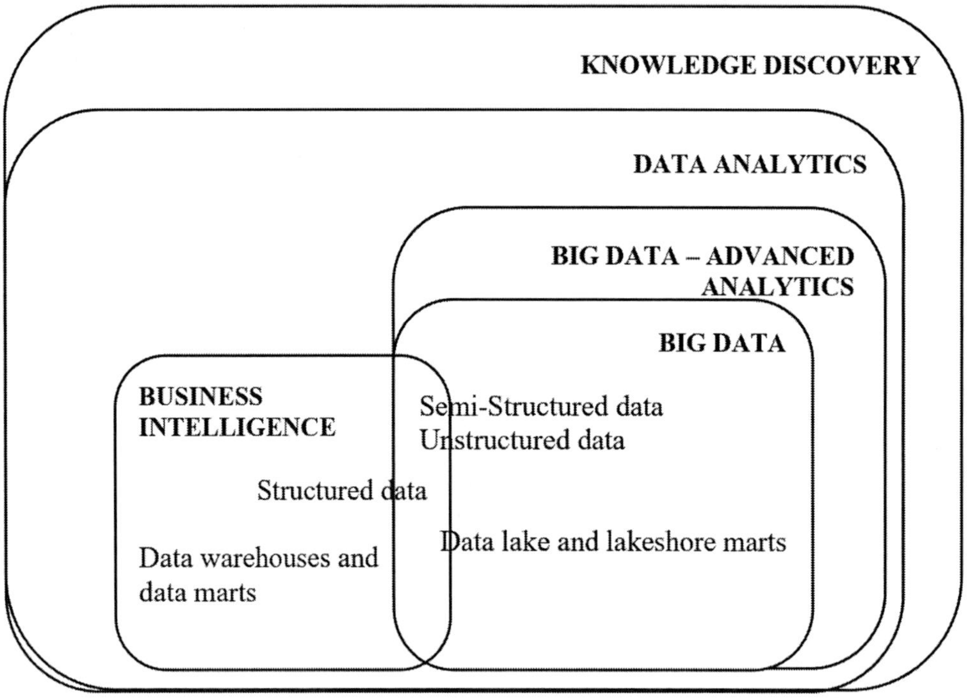

Fig. 2 Visual representation of the relationship between BI, Big Data, data analytics and knowledge discovery

4 Conclusion

Our proposed ecosystem (fig. 1) requires validation in terms of its content and applicability in practice, which will require further work.

Its compilation is based primarily on the conclusions of the authors Dedic and Stanier [9], as well as others. Our ecosystem is based on two independent frameworks, BI and Big Data, and their data processing processes, which we have integrated into a single unit. Between these two frameworks, we defined the possibilities for cooperation based on existing processes, technologies, solutions, research and practice.

Acknowledgement

The paper was elaborated within VEGA No. 1/0388/20 IT Management in Enterprises in Slovakia: International Standards and Norms Versus Individual Business Processes – proportion 100.

References

1. T. H. Davenport, J. G. Harris, *Competing on Analytics* (Harvard Business Review Press, 2007)
2. R. Stair, G. Reynolds, T. Chesney, *Principles of business information systems* (Cengage, United Kingdom, 2018)
3. Gartner. IT Glossary, Analytics and Business Intelligence (ABI). https://www.gartner.com/it-glossary/business-intelligence-bi/ [accessed 12.09.2019]
4. E. G. Mallach, *Information Systems*. (CRC Press, USA, 2016)
5. K. C. Laudon, J. P. Laudon, *Management Information Systems*. (Prentice Hall, New Jersey 2016)
6. P. Baltzan, *Information Systems*. (McGraw-Hill Education, New York, 2018)
7. R. K. Rainer, B. Prince, C. G. Cegielski, *Introduction to Information System*. (Wiley, USA, 2014)
8. W. El Kaim, Big data architecture: Hadoop and Data Lake (Part 1). https://www.slideshare.net/welkaim/big-data-architecture-hadoop-and-data-lake-part-1 [accessed 27.09.2019]
9. S. Watts, Difference between IT Orchestration and IT Automation Explained. https://www.bmc.com/blogs/it-orchestration-vs-automation-whats-the-difference/ [accessed 27.09.2019]
10. N. Dedić, C. Stanier, Towards Differentiating Business Intelligence, Big Data, Data Analytics and Knowledge Discovery. *ERP Future 2016: Innovations in Enterprise Information Systems Management and Engineering*. **285**, 114 (2016)
11. Neaga, Y. Hao, A Holistic Analysis of Cloud Based Big Data Mining. *International Journal of Knowledge, Innovation and Entrepreneurship*. **2, 56 (**2014)
12. D.P. Shukla, S.B, Patel, A.K. Sen, A Literature Review in Health Informatics Using Data Mining Techniques, *International Journal of Software and Hardware Research in Engineering*, **2**, 123 (2014)
13. B. M. Balachandran, S. Prasad, Challenges and Benefits of Deploying Big Data Analytics in the Cloud for Business Intelligence. *Knowledge-based and Intelligent Information & Engineering Systems*. **112**, 1112 (2017)
14. X. Wu, V. Kumar, JR. Quinlan, J. Ghosh, Q. Yang, H. Motoda, GJ. McLachlan, A. Ng, B. Liu, SY. Philip *Top 10 algorithms in data mining. Knowledge Information Systems*. **14**, (1):1–37 (2014)
15. M. Chen, SW. Mao, YH Liu, Big Data: A Survey. *Mobile networks & Applications*. **19**, 171 (2014)
16. F. Huang, Managing Data Lakes in Big Data Era: What's a data lake and why has it became popular in data management ecosystem, *2015 IEEE International Conference*

on Cyber Technology in Automation, Control, and Intelligent Systems (CYBER), 820 (2015)

17. PP. Khine, Wang ZS. Data lake: a new ideology in big data era. *4.TH Annual International Conference on Wireles Communication and Sensor Network*, **17** (2017)
18. S. Mitrovic. Specifics of the integration of Business Intelligence and Big Data technologies in the processes of economic analysis. *Biznes Informatika*. **42**, 40 (2017)

Control, controlling and its objectives in the organization

Nina Dědečková[1,*]

[1]University of Economics in Bratislava, Faculty of Business Management, 852 35 Bratislava, Slovak Republic

Abstract. The aim of the paper is to define and compare definitions and objectives of the managerial control and controlling based on research of scientific and professional literature through the opinions of domestic and foreign authors. By studying several domestic and foreign sources, through analogy, analysis, synthesis and comparison, we were able to create an overview of the research areas and defined the basic areas of managerial function of control and controlling. Our finding is that managerial control and controlling is still an actual topic. It is the subject of research by many theorists who focus on defining of these terms. Control is often perceived only as a comparison of two states, ie as the difference between the plan and the reality, while controlling in addition to comparisons also understands controlling as the implementation of corrective actions, thus supporting planning and decision-making. In this paper we identify the opinions of various authors and emphasize the importance of these terms.

1 Introduction

The business environment in which companies compete for competitive advantage is constantly changing and evolving. Once managers form plans and strategies, they must ensure that the plans are carried out. They must make sure that people are doing what needs to be done and are not doing inappropriate things. The word "control" leads to negative emotions. For many people, control means, above all, constraints, pressure, lack of independence. Managers continually look for ways to improve customer satisfaction, maintain relationships with suppliers, cut inventory costs, and develop the right products. As a result, every organization needs basic systems for allocating financial resources, developing human resources, analyzing financial performance, and evaluating overall profitability. Controlling is the management function in which managers establish and communicate performance standards for people, processes, and devices [2]. Control is defined as any process that directs activities toward organizational goals. It is how effective managers make sure that activities are going as planned. Some managers don't want to admit it, but control problems — the lack of controls or the wrong kinds of controls—frequently cause irreparable damage [1] . Controls should take place at all hierarchical levels of the organizational

* Corresponding author: nina.dedeckova@euba.sk

structure in order to identify all possible negative impacts on the business. The aim of our paper is to define and compare definitions and objectives of the managerial control and controlling based on the research of scientific literature through the opinions of domestic and foreign authors.

2 Methods

Literary review is an important part of the research work because it enriches the research with new knowledge and complements the current state of knowledge. Using literature review, we discuss and compare the views of experts in the perception of the definition of control and its objectives.

Basic scientific methods such as analysis and synthesis, comparison, induction and deduction have been used. Based on the findings, the theoretical basis of the knowledge of the issue was created. All the necessary information on the issue we obtained from the study literature, where a substantial part of the resources were foreign authors. Study literature was available mainly in the form of book publications, but we also drew on various scientific and research works, and we also used scientific articles in magazines.

3 Results and Discussions

3. 1 Definition of control and controlling

The roots of the term control date back to antiquity, when control was applied before the term itself was conceived. The concept of control consists of two Latin words „contra“ and „rotulus“. As reported by Majtan et al. [12], the simple translation can be understood as a counter-effect. At that time, it was written on parchment sheets that were curled into coils. In order to avoid misunderstandings about the original content of the report, parchment sheets were produced in two copies. It was based on these parchments that the correctness and truthfulness of the content of the report was checked. Since then, the term control has spread to all European languages and has become a very popular expression.

During its development it has acquired various meanings and still does not have a generally valid definition. Fayol [6] sees control as a process that begins with setting standards, continues with measuring performance, followed by comparing standards with performance, and taking corrective action in the final phase. Flamholtz [7] asserts, that control is a dynamic management function that has to be constantly adapted to changes in business. Merchant and Van der Stede [14] consider control as a critical management function in organizations, because its failure or neglect can damage the reputation of the company and can cause huge financial losses and, last but not least, the failure of the whole society and bankruptcy. Mockler [16] defines control as a systematic management effort to compare performance with predetermined standards, plans, or objectives whether performance is in line with these standards and presumably to take any remedial action required to see that human and other corporate resources are being used in the most effective and efficient way possible in achieving corporate objectives. Benowitz [2] asserts that organizational control is the process of assigning, evaluating, and regulating resources on an ongoing basis to accomplish an organization's goals.

Sedlák and Lišková [22] understand control as an accompanying phenomenon of planning that completes the overall management process. An essential variable are the plans that inspire action - control. They consider monitoring as the core of the control function and, in the case of the above-mentioned planning, monitoring the progress and compliance with the plan and its objectives. According to Králiček and Molín [10] it is necessary in any company and at any control to determine what we control, what meaning of control should be and what benefits or results we expect from the entire control process.

The German word "Kontrolle" was taken from the French "contrôle" or "contrerolle" (central France) in the 18th century. Unlike common definitions of the German term "Kontrolle", the English term "to control" includes not only a comparison of target and actual values, but also perfect control, management and regulation of processes [5]. While the German term "Kontrolle" fulfills its role only until the moment when the controlled facts are compared, the Anglo-American term "to control" is perceived more comprehensively. The term "Kontrolle" is therefore perceived as part of a complex process of control in the Anglo-American sense [4]. Mäder [11] understands the term Kontrolle as a systematic process of information processing in which two variables are present, which represent realized, required or predicted facts that consider each other. "Kontrolle" is limited to a pure comparison of the target and the actual state. Heuer [9] states that three contents of the term "Kontrolle" were found in German literature by the second half of the 1980s, being traditional, widespread and broadest. Traditional is derived from older German literature. It mainly included a comparison of predetermined and measured values. It was extended by analysis of causes of deviations. The broadest content of the term "Kontrolle", which is underrepresented in German literature, also includes remedial measures. The author draws attention to the fact that many German authors see the difference between "Kontrolle" and "control", because in their view this term means not only carrying out control but also planning at the same time. Therefore, "Kontrolle" is perceived as a subordinate planning process and is dependent on it.

There are some similarities in both languages (English and German). The term control / Kontrolle can acquire two important meanings for our purposes - from both languages we can translate control as comparing the differences between the actual and the desired state. Control can also be translated as to have the ability to control, to have power over something [18].

Later, the term "control" evolved into "controlling". While control refers to the one-time execution of this managerial function during the manager's work, controlling refers to the regular application of control in the enterprise. Control is considered to be a one-off act that captures the specific steps of the control process as well as the factors affecting that act. According to Mišún and Mišúnová [15] control is an ongoing process of drafting standards, measuring performance, benchmarking, and taking corrective actions to ensure effective and efficient activities within an organization. As Lester Bittel noted, controlling is the function that brings the management cycle full circle. It is the steering mechanism that links all the preceding functions of organizing, staffing, and leading to the goals of planning. The planning process determines the goals and objectives that eventually become the foundation of controls [19].

Robbins and Coulter [20] define controlling as a process of monitoring, comparing, and correcting work performance. All managers should control even if their units are performing as planned because they can't really know that unless they've evaluated what activities have been done and compared actual performance against the desired standard. Effective controls ensure that activities are completed in ways that lead to the attainment of goals. Whether controls are effective, then, is determined by how well they help employees and managers achieve their goals. Schermerhorn and Bach [21] explained, how does control fit in with the other management functions. Planning sets the direction of organization, organizing arranges people and resources for work, leading inspires people to put forth their best efforts and controlling sees to it that the right things happen, in the right way, and at the right time. If things go wrong, control helps to get things back on track.

Despite the diverse definitions of control, many have common characteristics. These basic common features include [23]:

Table 1. Characteristics of control.

Universality	The control process is one and the same time irrespective of the type of organization or function to be controlled. It is the foremost responsibility of a manager to regulate the operations on an on-going basis and keep the operation focused on goal accomplishments.
Continuity	Planning is a perpetual process. Koontz compares the control function to a navigator who takes readings to ascertain where it is related to a planned course. Likewise, a business manager has to constantly ensure that the enterprise is on course. It requires constant review and revision in response to behavior of environmental variables. Thus, it is not a one-time affair.
Dynamism	Control function needs to be changed in response to change in plans, objectives and to behavior of environmental variables. Therefore, it is a dynamic function.
Omnipresence	Managers across the levels, right from the CEO down to the first line supervisor, exercise control function. The nature, scope and limit of control functions vary. For example, top management is engaged in exercising strategic control, middle level managers are involved in tactical control and those at lower echelons are exercising operational control.
Future	Managers cannot control the past. But the benefits of bitter and better experiences of the past can be helpful in shaping a fruitful course of action in future. Preventive control guards against future loss, wastage and deficiency or deviation.
Final function	It follows all other functions. Planning is the basic function control. It is indispensable for ensuring accomplishment of organizational goals. It requires measurements of actual performance and comparing it against the pre-fixed standards. Deviation, if any, is addressed to reach level of desired performance. Thus, control is linked with other managerial functions.

Influencing	Control is supposed to structure the events and to condition the behavior. It is aimed at curbing undesirable behavior and shaping the pattern of future events. It includes people to conform to certain norms and standards.

3.2 Control Objectives

Mullins [17] considers control as one of the essential features of managerial work. Traditional views of control placed emphasis on the need for conformity of activities with the organization. Tannenbaum, for example, sees control as an inherent characteristic of the nature of organizations. The process of control is at the center of the exchange between the benefits that the individual derives from membership of an organization and the costs of such benefits. Organization implies control. It is the function of control to bring about conformance to organizational requirements and achievement of the ultimate purposes of the organization.

Boddy [3] asserts, that control has many positive meanings – standing for order, predictability or reliability. If things are under control, suppliers know what to deliver and when, people know what they are expected to do, customers know when to expect a delivery, staff know they will be paid. In this sense, control is an essential part of organizational life, because it helps to ensure that cooperative work by many people and units collectively adds value to resources. An absence of such control implies uncertainty, chaos, inefficiency and waste – an organization that is destroying, rather than enhancing, value. Hansen [8] considers feedback to be an important element of control. Feedback is information that can be used to evaluate or correct the steps taken to implement a plan. Based on the feedback, the manager or worker may decide to let implementation continue as it is, take corrective actions of a certain type to bring the activities back to compliance with the original plan, or to adjust the standards. Hansen considers feedback to be a critical part of the control function. Fayol [6] states that the aim of control is to point out errors that need to be corrected and prevented. As in the definition of Fayol himself, Mockler [16] considers that the objective of control is to take corrective action to ensure that business objectives are met as effectively as possible.

According to Robbins and Coulter [20] managers that didn't use control, they have no way of knowing whether their goals and plans were being achieved and what future actions to take. The second reason controlling is important is because of employee empowerment. Many managers are reluctant to empower their employees because they fear something will go wrong for which they would be held responsible. But an effective control system can provide information and feedback on employee performance and minimize the chance of potential problems. The final reason why managers control is to protect the organization and its assets. Today's environment brings heightened threats from natural disasters, financial scandals, workplace violence, global supply chain disruptions, security breaches. Managers must protect organizational assets if any of these things should happen. Comprehensive controls and back-up plans will help assure minimal work disruptions.

Tripathi and Reddy [24] defined the main objectives of control as increasing productivity, reducing mistakes and defects, facilitating communication, improving safety or reducing costs.

Benowitz [2] states that organizational control is the process of assigning, evaluating, and regulating resources on an ongoing basis to accomplish an organization's goals. The main purposes of control, inspired by Benowitz are shown in Table 2:

Table 2. Purposes of control.

Effectiveness of plans	Managers need to measure progress, offer feedback, and direct their teams if they want to succeed.
Consistency of organizational activities	Policies and procedures help ensure that efforts are integrated.
Effectiveness of organization	Organizations need controls in place if they want to achieve and accomplish their objectives. Effectiveness is a measure of how well the outcomes of an activity relate to the broader objectives of the unit – that is, how well it supports the achievement of broader goals.
Efficiency of organization	Efficiency probably depends more on controls than any other management function. It is widely used to show how productive a unit is, and how well people have managed it – the more output for the fewer inputs, the better, since that implies that value is being added to the resources.
Feedback	Not only do they measure progress, but controls also provide feedback to participants as well. Feedback influences behavior and is an essential ingredient in the control process.
Support for decision making	The ultimate purpose of controls is to help managers make better decisions. Controls make managers aware of problems and give them information that is necessary for decision making.

In some organizations, efficiency and cost savings have become primary goals. When this happens, the mission of an organization can become obliterated - noble goals such as providing safe vehicles, safe workplaces, and contributing to society take a backseat to maximizing human and organizational output and financial results. Second, when controls become the goal rather than an enabler of organizational success, organizational cultures can become toxic. In companies where people are constantly under pressure and stressed, it is difficult to sustain creativity and innovation, and people are often not able to do and be their best. Having said all this, the need of controls in organizations is obvious. Organizations welcome ways to work more efficiently, and all leaders want to produce quality products and services [13].

According to Williams [25] people may have objectives imposed on them, they may set them for themselves, or they may sit somewhere between those two extremes. Regardless of the situation, it will be needed to ensure that objectives and those you set for your team are sufficiently detailed and complete for you to know exactly what you and they are expected to achieve. To be truly useful objectives should be SMART, which means Specific, Measurable, Agreed, Realistic and Time-based. In other words, the outputs should be clearly defined such that performance can be quantified, agreed between employee and the organization, achievable within the constraints of time and resources and with a target date so that achievement can be measured. There are obvious motivational and practical reasons why objectives should be agreed and realistic. The SMART definition places so much emphasis on quantified and measurable objectives. That is because objectives are of little value on their own. They only become valuable as a basis for measuring performance.

4 Conclusion

At the end of the paper we state that managerial control at all levels of the management of organization is still an actual topic. It is the subject of research by many theorists who focus on defining the managerial control and its objectives. Control is a critical function within any organization because negative or even disastrous consequences can be associated with not meeting the established standards of performance. For example, poor inventory control can result in lost business because of a product shortage. Poor quality control may result in angry customers, lost business, and the necessity to provide customers with replacement products. Poor cost control can lead to negative profitability and perhaps even bankruptcy. The list of potential control problems is almost limitless. These problems all point to the fact that improving operational effectiveness and quality is virtually impossible without stringent control mechanisms. Except for very small organizations, the need for control is ubiquitous. Some people even claim that management shouldn't exercise any form of control whatsoever and should only support employee efforts to be fully productive members of organizations and communities. Along those same lines, some experts even use the word "coordinating" in place of "controlling" to avoid sounding coercive. All managers exercise control as they try to add value by transforming resources into outputs of greater value. However thoroughly they plan their objectives and how to meet them, internal and external events intervene. So, they need to supplement the activity of planning with that of controlling – checking that work is going to plan, and if necessary, taking corrective action. The sooner they note deviations, the easier it is to bring performance into line with the plan.

Acknowledgements

This article is a partial output of the research project of VEGA no. 1/0017/20 entitled „Changes in the application of managerial functions in the context of the fourth industrial revolution and adaptation processes of enterprises in Slovakia."

References

1. T. S. Bateman, S. A. Snell, R. Konopaske, *Management: Leading & Collaborating in a Competitive World.* (2019)
2. E. Benowitz, *Cliffsquickreview. Principles of management.* (2001)
3. D. Boddy, *Management: an introduction*. (2008)
4. R. Bramsemann, *Controlling*, 2. (1980)
5. C. Endenich, *Comparative Management Accounting: ein Vergleich der Controllingforschung und-praxis in Deutschland und Spanien*. p. 100. (2012)
6. H. Fayol, *General and industrial management, trans.* (1949)
7. E. Flamholtz, *Effective management control: theory and practice.* (1996)
8. D. R. Hansen, M. M. Mowen, *Managerial accounting 8th edition*. (2007)
9. M.F. Heuer, *Kontrolle und Steuerung der Materialwirtschaft*. (1988)
10. V. Králiček, J. Molín, *Vnější a vnitřní kontrola z pohledu managementu.* (2014)
11. O. B. Mäder, *Controlling klipp & klar*. (2018)
12. M. Majtán and col, *Manažment*. (2003)
13. A. McKee, *Management: A Focus on Leaders.* (2012)
14. K. A. Merchant, W. A. Van Der Stede, *Management Control Systems: Performance Measurement, Evaluation and Incentives.* (2007)
15. J. Mišún, I. Mišúnová-Hudáková, *Kontrolovanie v manažmente*. (2017)
16. J.R. Mockler, *Readings in Management Control*. (1970)
17. L. J. Mullins, *Management & Organisational Behaviour.* (2016)
18. D. Otley, K. Merchant, C. Emmanuel, *Accounting for Management Control.* (2013)
19. W. R. Plunkett, G. S. Allen, R. F. Attner, *Management*. (2008)
20. S. P. Robbins, M. Coulter, *Management*. Vol. **14**, 752 (2017)
21. J. R. Schermerhorn, D.G. Bachrach, *Exploring Management*. (2018)
22. M. Sedlák, C. Lišková, *Manažment.* (2016)
23. K. Sundar, *Principles of Management.* (2013)
24. P. C. Tripathi, P. N. Reddy, P*rinciples of Management*. (2012)
25. K. Williams, B. Johnson, *Introducing management: A development guide.* (2003)

Economic - cultural aspects of the situation arising from the worldwide spread of Covid - 19

Gabriela Dubcová [1*], *Lea* Pusztiová[2]
[1]BMF, UE in Bratislava Dolnozemská cesta 1, 852 35 Bratislava, Slovakia
[2]BMF, UE in Bratislava Dolnozemská cesta 1, 852 35 Bratislava, Slovakia

Abstract. The ongoing situation with the new Covid-19 virus has hit a huge number of countries whose populations and economies are under a tough test. No one expected such a catastrophic scenario in 2020. Coronavirus is a major topic in the whole world as no other up to now. Individual countries have chosen various approaches to fight the virus. These approaches have been influenced by a number of factors that either make the situation better or worse. Timely response, strict measures, discipline of the population, political position of countries, access or ability to produce or provide medical protective equipment - being the most desirable and precious goods of today, represent the salient role in coping with the situation.

Introduction

We are currently in a situation that greatly affects countries around the world. Occurrence and the speed of the spread of a new virus that has been transmitted throughout the world from China, called Covid-19, has caused major changes beyond our imagination. Countries affected by the virus have been fighting hard trying to mitigate its effects on the health and the economy [16]. Most countries have gradually tried to adopt measures, however, not always have the measures been successful. The virus is therefore multiplying at a rapid rate and penetrating into other countries that have perhaps learned from the mistakes made by other countries and shall be better prepared to cope with the situation. It is not yet certain how long the virus will persist and do harm, but we already know that it reveals and focuses on weaknesses and brings significant changes to which we will have to adapt [17].

Mankind remembers many momentous epidemics throughout its existence. Some of them had devastating consequences, and higher lethality of infected people. However, the population on the planet is growing causing the emergence of more densely populated city districts. They can be considered the source for the rapid occurrence and spread of viruses, Covid-19 in this particular case, which has, unfortunately, succeeded in a significantly negative impact on our lives today.

* Corresponding author: gabriela.dubcova@euba.sk

1 Status of infected people in individual countries and an approach to testing

The number of countries hit by the virus reached 171 as of 1.4.2020. The virus was firstly detected in China, namely in the city of Wu-chan at the end of 2019. Here is the beginning of the global pandemic. In just a few months, the virus managed to affect a huge part of the world's population. The first case in Slovakia appeared, or was confirmed on 6.3.2020. In the Czech Republic it was a little earlier - on 1.3.2020.

Table 1. Status in Slovakia since the first confirmed case (6.3. do 31.3. 2020) [13]

Date	Total Cases	Tests
6.3.2020	1	307
7.3.2020	3	392
8.3.2020	5	496
9.3.2020	7	548
10.3.2020	7	590
11.3.2020	10	634
12.3.2020	21	853
13.3.2020	32	985
14.3.2020	44	1191
15.3.2020	61	1436
16.3.2020	72	1595
17.3.2020	97	1913
18.3.2020	105	2138
19.3.2020	123	2439
20.3.2020	137	2807
21.3.2020	178	3247
22.3.2020	185	3489
23.3.2020	204	3736
24.3.2020	216	3940
25.3.2020	226	4416
26.3.2020	236	4751
27.3.2020	269	5664
28.3.2020	292	6411
29.3.2020	314	7131
30.3.2020	336	7532
31.3.2020	363	8220

Table 2. Status in the Czech Republic since the first confirmed case (1.3. - 31.3.2020) [14]

Date	Total Cases	Tests
1.3.2020	3	211
2.3.2020	3	262
3.3.2020	5	340
4.3.2020	5	407
5.3.2020	8	483
6.3.2020	19	594
7.3.2020	26	787
8.3.2020	32	928
9.3.2020	38	1193
10.3.2020	63	1358
11.3.2020	94	1816
12.3.2020	116	2353
13.3.2020	141	3094
14.3.2020	189	4065
15.3.2020	298	5068
16.3.2020	383	6302
17.3.2020	450	7664
18.3.2020	560	9402
19.3.2020	765	11619
20.3.2020	889	13704
21.3.2020	1047	15584
22.3.2020	1161	17377
23.3.2020	1287	19624
24.3.2020	1472	22600
25.3.2020	1763	26698
26.3.2020	2022	31127
27.3.2020	2395	36374
28.3.2020	2657	40700
29.3.2020	2817	43498
30.3.2020	3001	48811
31.3.2020	3308	55017

Tables 1 and 2 contain daily data since the confirmation of the first cases in Slovakia and the Czech Republic. To compare, it can be seen that we approach the testing as a state and country which is very close to us. Number of inhabitants in Slovakia is just half of that of the Czech Republic. Number of the tests made in the Czech Republic, however, exceeds the number of the tests made in Slovakia several times. Due to the low number of tests, the real numbers of the infected people can be unclear and deviate from reality. Wrong information can lead to wrong decisions about the further measures and steps. The only effective way how to prevent people from getting infected is their isolation, movement restriction, an order to wear protective equipment. All these measures, however, impact and weaken the economy. According to the Council for Budget Responsibility, the pandemic may have an impact on Slovakia's general government deficit in the range from 2 to 6% of GDP [12].

1.1 Development of positive cases by area

Some time has passed since the outbreak of Covid-19 in Asia, and the virus has spread to other continents, while Asia has successfully reduced the numbers of infected people.

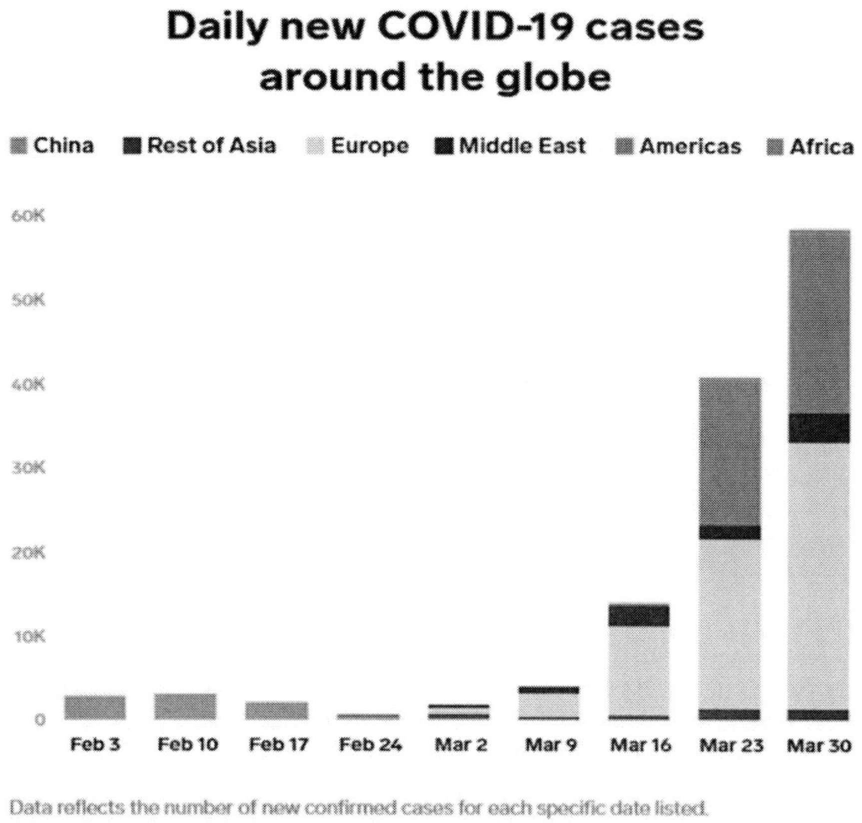

Fig. 1. Daily new COVID-19 cases around the globe [1]

Since February, the number of cases has been gradually decreasing in Asian countries which implemented strict measures to prevent inhabitants from being infected. The occurrences of Covid-19 have, however, increased in European states and in the USA. We can even talk about numbers higher than in Asian countries. There is a real threat that the corona virus will persist longer then in the Asian countries.

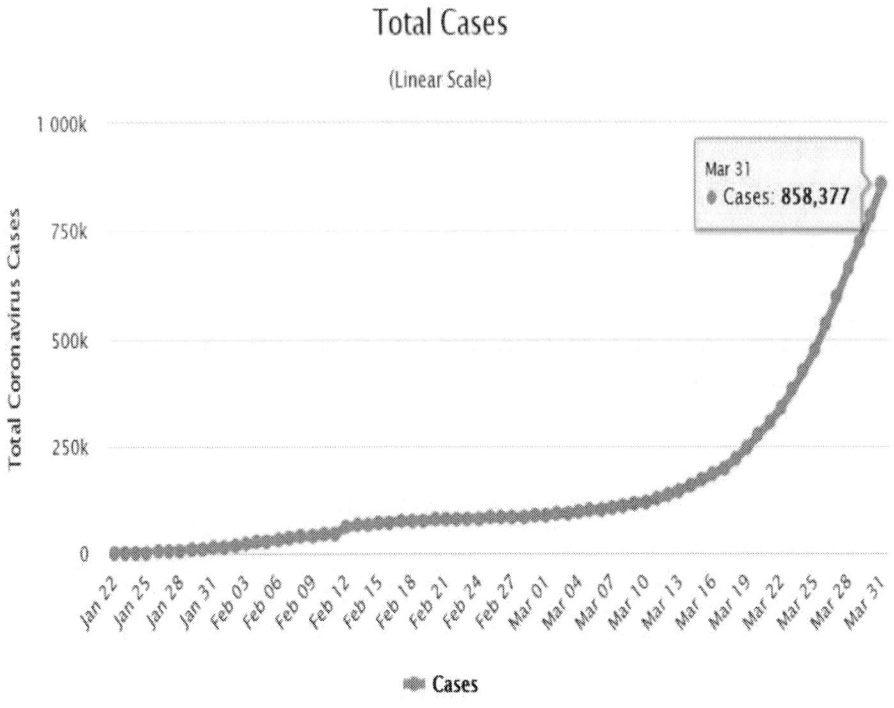

Fig. **1.** Total Covid-19 cases (22.1.2020 – 31.3.2020)

The Fifure 2 shows the total number of confirmed cases of Covid-19 virus for the selected period. Nearly a million cases in a few months. With this number we are approaching the most unpleasant epidemic so far in terms of the number of infected people in 2009 influenza A virus subtype H1N1 with a total of 1 million people infected.

1.2 Testing

The most common tests for COVID-19 are called "PCR tests". The first PCR tests for COVID-19 were developed very quickly - within two weeks after the disease detection and they are now a part of the World Health Organization (WHO) protocol to address this disease. The testing capacity of COVID-19 is, unfortunately, still too low in a lot of countries around the world. Furthermore, it should not be forgotten that not all tests are reliable. Also, for this reason, we still do not have sufficient knowledge of the spread of the pandemic. There is a positive correlation between performed tests and confirmed cases. This does not necessarily mean that countries which have run more tests do indeed have more cases, although in many cases it will be like that.

Several studies have shown that in countries such as South Korea, Singapore and Taiwan, where the epidemic has been brought under control, citizens have been massively tested. Most notably, the data show that South Korea has run far more tests than other countries. This means that we can expect that in South Korea the number of confirmed cases of positive tests is closer to the actual number of infected people than in other countries. It is therefore encouraging to see that the number of confirmed cases in South Korea had been decreasing daily. South Korea was able to quickly produce and use a huge number of tests, proving thus that it is possible. Testing is crucial; therefore, it is essential that other countries follow the South Korean leader.

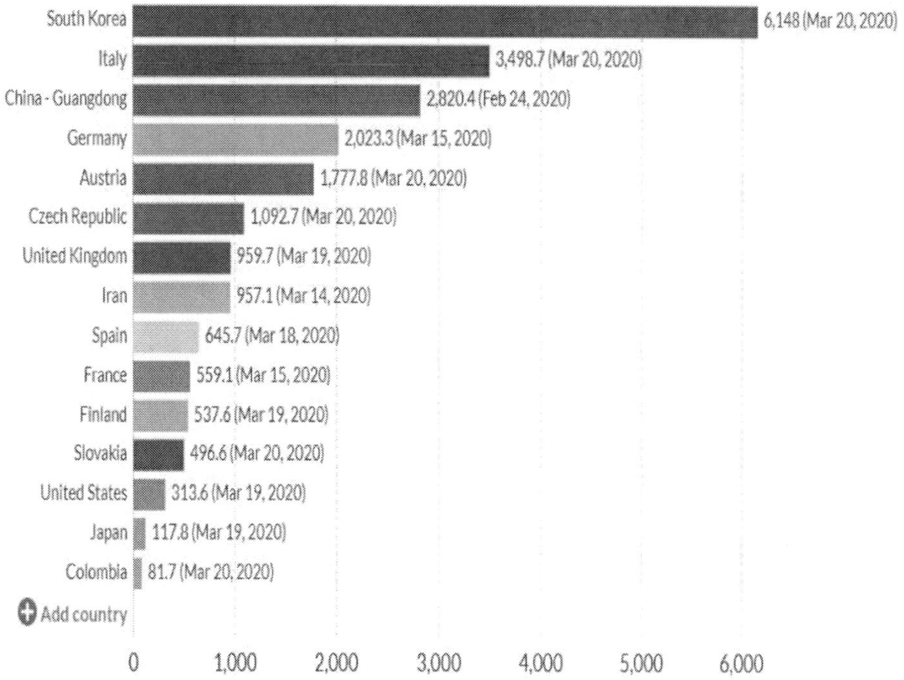

Fig. 3. Total tests performed per million people [3]

Figure 2 includes several countries where it indicates the number of tests performed per million inhabitants. Slovakia ran 496.6 tests per 1 million inhabitants as of March 20, 2020, which was more than in the case of the United States of America, but significantly less than in case of our Czech neighbors with a value of 1092.7 tests on the same date. Of the countries shown, North Korea has been really able to test intensively, up to 6,148 tests per 1 mil. residents.

2. Cultural-political aspect

The approach to the current situation and its solution by the national authorities and the attitude of the population to the pandemic is largely influenced by the culture and mentality of the population of the countries, or areas. These differences affect and transform scenarios in the development of the conditions of the countries concerned. Is it even possible to quantify how much influence the culture itself, or mentality of the people have on how the countries cope with this problem? [20,21]

The sharp fall in the number of infected people in Asia gives also the European countries a chance and hope that they can put the situation under control. The steps taken by governments and population behavior in the Eastern and Western world, however, differ significantly [22].

The differences can result in much worse course of pandemic in the Western world. Asian countries experienced a strong SARS epidemic in 2003, increasing thus their readiness for a new epidemic. Many of them, such as Singapore or Hong Kong, took strong precautionary measures a few days after the announcement of the Wu-chan virus outbreak. While prevention worked in many smaller countries, an epidemic, similar to the one in China, broke out in South Korea. Both states, however, immediately came up with the strict measures to limit the spread of the virus. Quarantine, digital monitoring of potential disease carriers and mass population testing. The Chinese government checked the mobile phones of citizens to see if they really comply with the quarantine regulations.

All who violated them had to pay a high fine and attend mandatory lectures on safety and personal hygiene. Democratic South Korea launched a similar citizen tracking by monitoring credit card transactions in addition to mobile phones. Several Asian countries set up systems that controlled the movement of each potentially infected person to warn the remaining population. By stringent and well-managed measures, Asian states have been able to prevent the further spread of epidemic. But it was at the beginning when the serious problem occurred, causing today's alarming situation around the world. Chinese doctor Li Wenliang had issued a warning about an unknown new virus before it actually posed a threat of today´s scope. He was summoned by health officials and police early in January and forced to sign a statement condemning his warning as unfounded and illegal gossip. The Chinese authorities ordered not to disclose any information concerning the disease. By suppressing the truth that a deadly coronavirus had appeared on its territory, the Chinese regime endangered not only its own people, but also the entire international community. [2]

Unfortunately, the situations just like that may arise as a result of the Chinese regime, which is significantly different from the one in Europe. [6]

The circumstances in Europe are fundamentally different. While in Asia, the family takes the priority over the state in the hierarchy of values, in Europe it is the individual[23]. It turns out that not only Europe's unreadiness for epidemic, but especially Western individualism is the greatest threat to the spread of the epidemic on the old continent. Europe has lagged far behind in prevention, planning, testing potentially infected, collecting and evaluating data, but also in the approach of individuals to the spread of the epidemic. Measuring citizens' temperature and monitoring their mobile devices would be considered a violation of GDPR. According to the protection of personal data, data about citizens belong to them and therefore their use must be subject to the consent of the citizen. [9]

In the United Kingdom, up to a quarter of the population absolutely refuses to change their habits because of the coronavirus. Therefore, also due to their attitude, the British government has not taken any stricter measures, such as the cancellation of social events, or the closure of schools. Government chief scientific adviser Patrick Vallance even said it was desirable that up to 60 percent of the British population would be infected with the virus, creating thus collective immunity against the disease.

Italian people are typical for their temperament and the strong need to socialize, which implies the current lack of discipline and disregard for the regulations which in any way had been introduced too late. Moreover, taking into consideration the seriousness of the situation and the fact that in Italy there is a large proportion of elderly people who are at-risk category and can develop severe symptoms, the measures were totally insufficient. According to statistics, Italy is the country with the second highest number of people over 65 years. These difficulties point to the fundamental problem of how to control a dangerous epidemic in a democratic country that values basic freedoms, such as freedom of movement. Italian measures are less restrictive than one might expect. [11]

Two mutually admiring allies: Donald Trump and Boris Johnson have also become allies in denial of potential problems. The reaction of the US President was initially expressed by the promise that the virus would disappear, by playing golf, blaming China, blaming the media, blaming the Democrats, deceiving and claiming that his respond to coping with the situation was perfect. B. Johnson was not as eccentric in his view as D. Trump, but just as irrational. His chief scientific adviser advised him on a strategy for the British population to gain collective immunity. Such a solution would, according to him, bring a drastic and rapid epidemic but would cause less economic damage.

A similar attitude was presented by the US government on the Spanish flu after the First World War. The President Woodrow Wilson did not mention the outbreak of the epidemic in the country, although nearly 700,000 Americans had died of it. The strategy to ignore the epidemic could have been implemented in times when 16 percent of the US population got their food by farming, fishing and forestry. Naturally, social distancing was relatively high. Nowadays, people live together in cities, which greatly increases the threat of rapid spread of disease to a large proportion of the population. [10]

European politicians start to take action only after a month and a half after the outbreak in China. Only days after the epidemic hit Europe.

The new situation of the world transforms various views, opinions, people and much more including economic or political powers of countries. Several countries benefit from positive international relations. There is a struggle to get medical supplies that everyone needs now. The positive relations with the countries that have sufficient stock of them, or can provide enough of them can be greatly beneficial for selected countries and their strengths. [19]

China is more or less coping with the disease and the government decided to work on a new task: to change the image of the country from which the disease spread into the world to the country which helps the world during the pandemic. [7]

The criticized positive relations of the Czech President Milos Zeman with China have now been turned and the country benefits in this difficult situation from this relationship in the form of an assistance with the import of necessary medical supplies. The Czech Republic gains precious sources for health care system and they helped also Slovakia with providing us with contacts in China, therefore also the necessary medical supplies for Slovakia could have been arranged. [14]

Serbian President Vucic switched to anti-European rhetoric when he received help from China and at the same time, he promoted Chinese President Si Jinping from a friend to a "brother". [7]

They are not the only ones. Italian government also asked officially for medical supplies, but the European allies had ignored the demand. China, however, has responded positively by supplying the required material. And it continues like that also in case of other countries. [24]

3. Rarity during crises

As already mentioned, political struggles prevail, the winners of which receive the most valuable goods at the moment - medical protective equipment. Under normal circumstances, these goods are available, not very expensive and not difficult to manufacture. The high societal demand caused by the crisis for inferior commodities as consumer medical supplies in particular facial masks, has made their value higher than under normal circumstances. There is a lack of protective equipment around the world, which makes it rare, although it is not minerals. In this crisis, human perception of rare goods has changed [18,25]. We have come to a situation when the rare goods (medical supplies and medical protective equipment) have higher value for the world population threatened by Covid-19 than diamonds or gold being more valuable under normal circumstances. Increased price of the masks on the market represents a profitable opportunity for entrepreneurs. Even those, who would not otherwise be willing to sell the masks are now making them at home and sell them over the Internet. For example, Zornice, a company that is a well-known domestic manufacturer of shirts from Bánovce nad Bebravou. The company adapted to market needs and sells the masks to the state at a price of € 2.25 per piece, which is definitely the price higher than the price in a crisis-free situation. Nevertheless, people are grateful and willing to pay for this opportunity. Well, ultimately, it is about saving human lives which are invaluable [5].

Conclusion

Who could have expected the current situation? Crises, whether epidemiological or economic, have at least one common feature: they can be stopped. Several states in Asia have shown already how to stop the spread of the dangerous virus and almost completely eliminated it. It is a tough fight. Either from managerial, or state point of view. But it is not impossible. If rapid and effective steps are taken and the spread of the virus is stopped in the first phase, as has been done, for example, by Singapore, the economic consequences are minimal. If the infection becomes an epidemic, however, the steps are much more complicated and more painful. South Korea also shows that such a situation can be managed. Well, in the end, also Wu-chan managed it. There is a simple global rule in relation to this crisis: the regime's ideology does not primarily matter, also not the level of democracy, or whether the regime is European or Asian. The governments that tell people the truth, act quickly and listen to experts stand a much better chance of coping with the coronavirus.

Acknowledgement

This paper is the outcome of projects: KEGA 002EU-4/2019 (50%) & VEGA No. 1/0309/18 (50%) - funded by the University of Economics in Bratislava; with the project membership of both authors.

References

1. *Businessinsider*, China used technology that aims to trace every single COVID-19 case. https://www.businessinsider.com/chinas-coronavirus-quarantines-other-countries-arent-ready-2020-3#the-country-implemented-large-scale-contact-tracing-in-the-early-2000s-8. (2020)
2. Čínska vojna proti pravde o koronavíruse, https://www.konzervativnyweb.sk/c/22355814/cinska-vojna-proti-pravde-o-koronaviruse.html. (2020)
3. E. Ortiz-Ospina, J. Hasell, How many tests for COVID-19 are being performed around the world ?, https://ourworldindata.org/covid-testing. (2020)
4. European Centre for Disease Prevention and Control. Situation update worldwide, https://www.ecdc.europa.eu/en/geographical-distribution-2019-ncov-cases. (2020)
5. *Hnonline*, Zornica: Rúška budú. Dostať ich majú všetci. https://finweb.hnonline.sk/ekonomika/2113450-zornica-ruska-budu-dostat-ich-maju-vsetci. (2020)
6. Ch. Buckley, *Nytimes*, Chinese Doctor, Silenced After Warning of Outbreak, Dies From Coronavirus https://www.nytimes.com/2020/02/06/world/asia/chinese-doctor-Li-Wenliang-coronavirus.html (2020)
7. M. Tóda, *Dennikn*, Orbán nie je sám. Autokratickí lídri využívajú koronavírus, aby si upevnili moc, https://dennikn.sk/1831960/orban-nie-je-sam-autokraticki-lidri-vyuzivaju-koronavirus-aby-si-upevnili-moc/?ref=tema. (2020)
8. P. Stach, *Euractiv*, Koronavírus ako argument v prospech európskej federácie. https://euractiv.sk/section/buducnost-eu/news/modra-vlna-koronavirus-ako-argument-v-prospech-europskej-federacie/. (2020)
9. R. Ižip, *Etrend*, Čierna labuť sa vrátila. Hrozí najväčšou ekonomickou krízou našej doby https://www.etrend.sk/ekonomika/cierna-labut-sa-vratila-hrozi-najvacsou-ekonomickou-krizou-nasej-doby.html?split=2 (2020)

10. R. Ižip, *Etrend*, Štúdia, ktorá otriasla politikmi, priniesla pre Slovensko dobrú aj zlú správu, https://www.etrend.sk/ekonomika/studia-ktora-otriasla-politikmi-priniesla-pre-slovensko-dobru-aj-zlu spravu.html?itm_site=etrend&itm_template=article&itm_area=article&itm_module=top-articles&itm_position=1. (2020)
11. *Dennikn*, Talianska karanténa testuje hranice demokracie, https://dennikn.sk/1794139/talianska-karantena-testuje-hranice-demokracie/?ref=list. (2020)
12. *Trend*, Slovensko a svet, 12-13, 6 (2020)
13. Úrad verejného zdravotníctva Slovenskej republiky, http://www.uvzsr.sk/index.php **1-7** (2020)
14. Úrad vlády Slovenskej republiky. Na Slovensko dorazilo lietadlo so zdravotníckymi *pomôckami* z Číny, https://www.vlada.gov.sk//na-slovensko-dorazilo-lietadlo-so-zdravotnickymi-pomockami-z-ciny/. (2020)
15. Základní přehledy dle hlášení KHS a dle pozitivních nálezů laboratoří, které jsou určeny pro další šetření. https://onemocneni-aktualne.mzcr.cz/covid-19. (2020)
16. N. Goda, M. Čiefová, Cultural Diplomacy of Selected Countries in a Comparative Perspective. Folia Geographica 61, 2, 172–189 (2019).
17. P. Puškárová, Assessing the magnitude of globalization-induced technology flows in expanded EU-sample: a multi-channel approach. Globalization and higher education in economics and business administration, GEBA 2013. Elsevier 20, 544-552 (2015).
18. K. Jančovičová Bognárová, R. Sivák, The Effect of financial and non-financial performance measures on managerial behavior. *SGEM* 443-448 (2015).
19. H. Majdúchová, B. Siváková, D. Rybárová, S. Šagátová, Quantification of the systematic risk in industries. *Ekonomický časopis* 65, 7, 602-617 (2017).
20. A. Remišová, A. Lašáková, Z. Kirchmayer, Infuence of Formal Ethics Program Components on Managerial Ethical Behavior. *Journal of Business Ethics* 160 , 151 - 166 (2019).
21. M. Bohatá, A. Putnová, M. Rašticová, M. Bédiová, A. Cebáková, Social Effectiveness of Public Institutions Through the Eyes of Citizens. *Proceedings of the 12th International Scientific Conference.* PUBLIC ADMINISTRATION, 18-27 (2018).
22. A. Lašáková, A. Remišová, On Organisational Factors that Elicit Managerial Unethical Decision-Making. *Ekonomický časopis* 65, 4, 334 – 354 (2017).
23. K. Beličková, Z. Staríčková, Current Problems of the Financial Position of Slovak Households. Economics and Finance 11, 4-12 (2018).
24. S. Ludwig, P. Markovič, Business acquisition in the People's republic of China: developments and conditions. *Liberec Economic Forum 2017.* International conference. Liberec Economic Forum 513-521 (2017).
25. J. Blštáková, Z. Joniaková, Z. Skorková, I. Némethová, R. Bednár, Causes and Implications of the Applications of the Individualisation Principle in Human Resources Management. *AD ALTA journal of interdisciplinary research*. MAGNANIMITAS 9, 2, 323-327 (2019).

Development of Tax Stimulation and Business Environment in Slovakia

Ivona Ďurinová[1,*], *Natália* Tačarová[2]

[1]University of Economics, Faculty of Business Management, Department of Corporate Finance, Dolnozemská cesta 1, 852 35 Bratislava, Slovakia

[2]Slovenská sporiteľňa, a. s., Tomášikova 48, 832 37 Bratislava, Slovakia

Abstract. Tax incentives are specific and targeted tax instruments in the economic and social policy of the state. With appropriate tools, the state can support the creation of new jobs and also the sustainability of human resources in existing companies. The article deals with the issue of stimulation of companies in Slovakia and the suitability of the applied incentives in connection with the business environment. The aim of the paper is to approach and analyze the impact of tax incentives applied in the Slovak Republic and evaluate them based on their impact on business entities, respectively business environment. The analysis of the overall development of provided tax incentives in the Slovak Republic is focused on the period from 2002 to 2018. The paper uses the method of analysis and synthesis together with the methods of tabular and graphical representation. Mathematical and statistical methods are used in individual calculations. On the one hand, it can be stated that tax incentives are important, even necessary, for countries, given the inflow of foreign capital, but on the other hand, domestic businesses are disadvantaged by which foreign investors disrupt the domestic business environment.

1 Introduction

The business environment also includes the country's tax system, which should be simple and transparent in order to create favorable conditions for doing business. It should be borne in mind that not only differences in individual countries but also any change in the field of income taxation in a given country affects the behavior of business entities. Many authors are researching this fact. Harumová [1] examines how the tax system affects work as well as business activity, the amount of savings and investor decision-making. Investors' decisions are often influenced by various constructed tax incentives. Tax incentives are specific and targeted tax instruments in the economic and social policy of the state, through which the state influences the behavior of economic entities in order to achieve its goals. With the help of taxes, the state can support investment and business, regional development, innovation and scientific and technical development, dampen production in certain sectors, etc. [2]. With appropriate tools, the state can support the creation of new

* Corresponding author: ivona.durinova@euba.sk

jobs and also the sustainability of human resources in existing companies. Incentives are often provided to large companies, but many authors point out that small and medium-sized enterprises (SMEs) are considered globally to be the driving force behind economic growth, enabling job creation and increased productivity [3]. Other authors [4] point to the fact that SMEs are facing internal markets with higher taxation and transfer price compliance costs as well as cross-border loss compensation problems. In particular, the works of Michael Porter [5], Matatková and Stejskal [6] focus on companies in terms of their possibilities for regional and local grouping. In the case of small and medium-sized enterprises, networking is usually one of the many ways to become competitive compared to strong and often highly sophisticated partners. Gajdová [7] sees the cluster as a joint union of independent, regionally unified, jointly supporting companies and institutions that have the potential to be competitive within EU regions. In this way, the merged companies can achieve a synergistic effect.

The article deals with the issue of stimulation of companies in Slovakia and the suitability of the applied incentives in connection with the business environment. The aim of the paper is to approach and analyze the impact of tax incentives applied in the Slovak Republic and evaluate them based on their impact on business entities, respectively the business environment. The partial goals are as follows: to perform an analysis of the applied tax incentives, to evaluate the state of the business environment in the Slovak Republic, to evaluate the impact of tax incentives in relation to the basic principles of taxation.

2 Methods

The paper uses the method of analysis and synthesis together with the methods of tabular and graphical representation. Mathematical and statistical methods are used in individual calculations. In the final phase, attention is focused on summarizing the data obtained. In this part, the method of synthesis (summary of the obtained results) and the method of comparison are used, through which the effectiveness of the applied incentives in the Slovak Republic is compared and analyzed.

3 Results

The analysis of the overall development of provided tax incentives in the Slovak Republic is focused on the period from 2002 to 2018. The total number of issued decisions for the monitored period was 211 and reached the amount of almost 2 billion €. However, the number of beneficiaries is lower, due to the fact that some companies have benefited more than once. Companies that have benefited twice are: Continental Automotive Systems Slovakia s.r.o., Dell s.r.o., Embraco Slovakia s.r.o., Holcim Business Services s.r.o., OSRAM, a.s., Mondi SCP, a.s., SAMSUNG Slovakia s.r.o., PCA Slovakia, s.r.o., Magneti Marelli, s.r.o. and Johnson Controls, s.r.o. Companies that have benefited from tax incentives even three times are: Kia Motors Slovakia s.r.o., T-Systems Slovakia s.r.o., ZKW Slovakia s.r.o. and SCA Hygiene Products Slovakia, s.r.o. The highest number of incentives was recorded in 2006 and this year the volume of these incentives was also the highest. In 2005, no incentives were provided by the state [8].

Figure 1 shows the development of the provided incentives for the total monitored period. In the last three years, there has been a growing trend in the incentives provided, to which the legislative changes adopted at the end of 2015 have also contributed. Legislative changes have simplified access to incentives for businesses in the least developed regions.

The largest volume of incentives provided was in 2004, 2006 and 2015. For these years, the number of planned jobs created was also the highest.

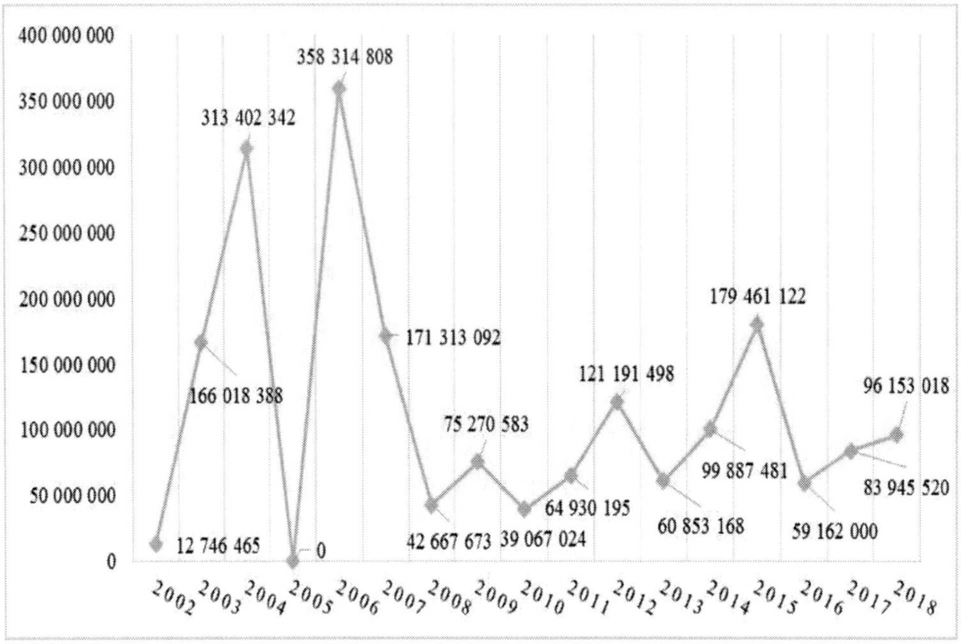

Fig. 1. Development of provided incentives for the period from 2002-2018 in € *[8]*

It was planned to create 64,539 jobs for the entire period under review. The average cost of such one job was € 28,463. We can note that most jobs were to be created in 2006, which was also due to the highest number of approved incentives for the whole period. On the contrary, the least jobs were to be created in 2002, when only one incentive was provided. In processing these results, it was necessary to take into account the fact that in 2014, out of the total volume of incentives in the amount of € 99,887,481, a tax incentive was provided to Duslo, a. s. amounting to € 58,560,000, with the company not planning to create any new jobs and the incentive being approved only in the form of a tax credit. The same situation occurred in 2018, when out of the total volume of provided incentives in the amount of € 96,153,018, a tax incentive was provided to Mondi SCP, a. s. This company was also not created by creating new jobs and the incentive was provided only as a tax relief. Figure 2 shows the cost per job in each year. The lowest cost per job was in 2018. We recorded a more significant decline in 2008, which can be attributed to the economic crisis. From 2009 to 2012, we see a growing trend in these costs.

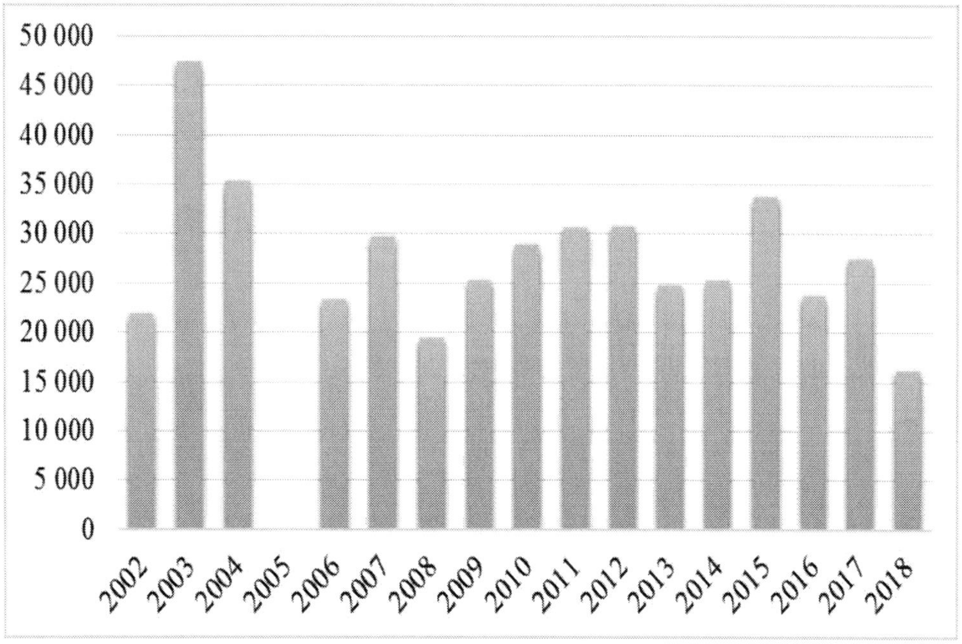

Fig. 2. Overview of costs per job in individual years *[8]*

Table 1 provides an overview of all beneficiaries by country of origin, together with the percentage on total incentives provided.

Table 1. Beneficiers of incentives by country of origin *[8]*

Country of origin	Sum of incentives in €	% share	Country of origin	Sum of incentives in €	% share
Belgium	6 797 500	0,35	Hungary	10 202 574	0,52
Brazil	1 991 635	0,10	Malaysia	38 268 000	1,97
Czech Republic	62 641 126	3,22	Norway	1 163 372	0,06
China	5 550 000	0,29	Poland	531 600	0,03
Denmark	9 110 333	0,47	Austria	85 606 121	4,40
France	193 277 550	9,94	Russia	3 160 061	0,16
GB	129 812 750	6,68	Slovakia	127 839 132	6,57
Netherlands	176 461 902	9,08	Germany	232 917 375	11,98
Hong Kong	38 800 372	2,00	Spain	37 700 047	1,94
Ireland	652 000	0,03	Switzerland	49 408 108	2,54
Iceland	3 264 883	0,17	Sweden	7 928 607	0,41
Japan	19 250 000	0,99	Italy	70 312 898	3,62
Cayman Islands	11 405 306	0,59	Ukraine	18 000 000	0,93
Republic of Korea	450 114 343	23,15	USA	109 814 425	5,65
Luxemburg	42 402 357	2,18	Total	**1 944 384 377**	**100,00**

The largest beneficier of the incentives provided is the Republic of Korea with a share of 23.15% for the entire period under review. Another country is Germany with an 11.98% share, followed by France, which has a 9.94% share. These three countries were thus granted incentives in the amount of € 876,309,268, which represents 45.07% of all incentives provided. Incentives in the amount of € 127,839,132 were allocated to Slovak business entities for the total monitored period, which represents 6.57%. Slovakia placed 6th. In addition to the already mentioned Republic of Korea, Germany and France, the Netherlands and the United Kingdom are in the top five. Businesses from Russia, Brazil, Norway, Ireland and Poland received the smallest volume of incentives provided.

The total amount of incentives consists of several parts. Table 2 shows the sums of these individual parts for the whole monitored period together with their percentage share.

Table 2. Breakdown of the total amount of incentives *[8]*

Incentives	Sum in €	% share
Subsidy for tangible and intangible fixed assets	821 613 753	42,26
Contribution to created places	126 115 264	6,49
Retraining allowance	52 512 165	2,70
Tax relief	897 447 568	46,16
Transfer of assets at a price below market value	46 695 630	2,40
Total	1 944 384 380	100,00

The most frequently provided incentive is the tax relief, which for the entire monitored period amounted to € 897,447,568 and its percentage share was 46.16%. The subsidy for tangible fixed assets and intangible fixed assets, which acquired the value of € 821,613,753 for the entire monitored period, has a 3.90% lower share.

Table 3 shows the impact of incentives that are applied in the Slovak Republic.

Table 3. Impact of tax incentives applied by the Slovak Republic

Year	The amount of incentives provided in mil. €	The total volume of collected corporate income tax in mil. €	Corporate tax rate in%	New corporate tax rate in%
2002	12,75	926,00	25,00	**24,66**
2003	166,02	1 118,00	25,00	**21,29**
2004	313,40	1 171,00	19,00	**13,91**
2005	0,00	1 344,00	19,00	**19,00**
2006	358,31	1 599,00	19,00	**14,74**
2007	171,31	1 848,00	19,00	**17,24**
2008	42,67	2 087,00	19,00	**18,61**
2009	75,27	1 582,00	19,00	**18,10**
2010	39,07	1 659,00	19,00	**18,55**
2011	64,93	1 666,00	19,00	**18,26**
2012	121,19	1 631,00	19,00	**17,59**
2013	60,85	1 975,00	23,00	**22,29**
2014	99,89	2 315,00	22,00	**21,05**
2015	179,46	2 860,00	22,00	**20,62**
2016	59,16	2 812,00	22,00	**21,54**
2017	83,95	2 995,00	21,00	**20,41**
2018	96,15	3 167,00	21,00	**20,36**

It is an overview of the amount of incentives provided in individual years, together with the total volume of corporate income tax collected and the relevant tax rate in individual years of the observed period.

If we compare the amount of provided incentives with the total amount of collected corporate income tax in individual years, it is possible to determine what "new" corporate tax rate could have arisen. The most significant difference can be seen especially in 2004. The difference between the tax rate applied at that time and the rate recalculated by us reached 5.09%. Other significant differences relate mainly to 2003 and 2006, when in 2003 the rate could have been lower by 3.71% and in 2006 by 4.26%. In recent years, such significant differences are no longer observable and the tax rate would fall by around 0.50%, which was due to the lower volume of tax incentives provided.

It can be stated that if the approved incentives were not provided in individual years, it would have been possible to reduce the corporate tax rate in 2004 by even more than 5%. Such a reduction would apply to all legal entities, and thus the advantage would benefit all entities, not just those to whom tax incentives have been approved.

Business entities have long been dissatisfied with the subsidy system in the Slovak Republic and consider it unfair. Peter Kremský from the Business Alliance of Slovakia also states that: "This should not be done through investment incentives for specific companies, but especially through improving the business environment in the region, such as improving infrastructure, developing local services, supporting education in the region, etc. The state would help investors the most by cutting red tape, corruption, simplifying tax rules and the labor market." [9]. A similar view is shared by Martin Vlachynský, who claims that he would welcome even more general measures, in the form of regional deductions, by adding in the construction of infrastructure, but also in the regionalization of the minimum wage [10].

4 Discussion

The development of the business sector is closely linked to the promotion of investments that would create more opportunities for economic growth and job creation [11].

According to Martin Vlachynský, we can state that: "In reality, there is a situation where some companies have consistently lower taxes compared to their competitors. In addition, there is a false impression that those who do not receive incentives from the government cannot do business in Slovakia, which does not benefit the business climate. "If companies receive such assistance repeatedly, the question may arise approve this support. However, the government should under no circumstances support companies that are unable to stay in the market. It is a waste of public resources and discrimination against entrepreneurs who "feed" other companies on taxes paid through incentives, with whom they are fighting for customers as well as potential workers [10].

On the one hand, it can be stated that tax incentives are important, even necessary, for countries, given the inflow of foreign capital, but on the other hand, domestic businesses are disadvantaged by which foreign investors disrupt the domestic business environment. The legal regulation of tax incentives is often incomprehensible and complicated for taxpayers, and therefore the state should start to pay more attention to their simplification. Markovič, Majdúchová and Rybárová [12] point to the fact that especially start-up small companies with high innovation potential encounter frequent and fundamental fluctuations of the business environment in which they try to plan and implement their activities. The state should focus mainly on improving the quality of the business environment, focusing on the elimination of bureaucracy, the stability of legislation and a level playing field for foreign and domestic entrepreneurs. In terms of economic efficiency, tax systems should ideally be "neutral", particularly as regards economic decisions. From this perspective, the location of investments in different countries in the international context should not face markedly different effect levels of taxation [13]. According to Clausing [14] and Mura [15] it is important to monitor the overall tax burden that represents the size of the enterprise's tax.

There are also options through which tax incentives could be replaced by solutions that could be used by all businesses and not just some. One of them would be, for example, to completely abolish tax incentives and instead reduce the tax burden across the board. Another alternative could be the auction of incentives, respectively. regional auction of incentives. Through this auction, the whole system would be more objective, clearer and companies that have not yet had such an opportunity would also be given incentives. The system would consist in the specified volume of funds being allocated to those applicants who would show e.g. the lowest possible cost per new job created. Another alternative could be to reduce the levies on low-income employees, which would compensate for the provision of investment incentives. Such a solution would affect a large part of the population and would also not significantly distort the competitive environment. The employer would have significantly lower costs per employee and at the same time the employee would have a higher net salary [16]. Dubcová [17] draws attention to the social and solidarity economy in Slovakia, where the stimulus system is lacking. She recommends the application of specific financial and economic instruments in this area (tax breaks, tax deductions, specialized system of the state supports, etc.).

5 Conclusion

Tax incentives are used by states in order to motivate businesses for a specific purpose, e.g. to support certain activities, to expand entrepreneurial activity to disadvantaged regions, to support research and development and others. However, the incentive function of taxes is contrary to the principle of neutrality. It can be stated that tax incentives distort the basic principles of market competition. If it is in the state's interest to attract foreign investors, it should focus primarily on improving the quality of the business environment and not just on offering incentives.

This paper is a partial output of the research project VEGA n. 1/0569/18: „ STABILIZATION OF THE HUMAN CAPITAL IN SLOVAK ENTERPRISES AS AN IMMANENT DETERMINANT OF THEIR SUCCESS, COMPETITIVENESS AND SUSTAINABLE DEVELOPMENT" in the range of 100%.

References

1. A. Harumová, Impact of Taxes on the Development of Business Sphere. *Journal of Economics* **50**(2), 277-292 (2002)
2. G. Lénártová, Tax Stimulation and Business Environment in Slovakia. In *Current Problems of the Corporate Sector 2014: Proceedings of International Scientific Conference.* (Bratislava: Vydavateľstvo EKONÓM, 2014)
3. D. Gajdová, G. Dubcová, M. Krechovská, New challenges of SMEs through clusters creation in Slovakia. In 3rd international conference on new challenges in management and business. International conference. *3rd international conference on new challenges in management and business : conference proceedings : Dubai, UAE, 2 may, 2016.* - [Tehran] : [Industrial Management Institute] (2016)
4. V. Solilová, D. Nerudová, M. Litzman, Application of the CCCTB and Safe Harbours to European SMEs: Can the Decrease in Compliance Costs Support better SME Performance? *Journal of Economics* **67**(6), 587 – 606 (2019)
5. M. E. Porter, Location, Competition, and Economic Development: Local Clusters in a Global Economy. *Economic Development Quarterly* **14**(1), 15-34 (2000)

6. K. Matatkova, J. Stejskal, Assessment of Shift-share Analysis Suitable for Identification of Industrial Cluster Establishing in Regions. *Journal of Economics* **60** (3), 935-948 (2012)
7. D. Gajdová, Clusters and Cluster Initiatives in the Slovak Republic. In *Current Problems of the Corporate Sector 2012: Proceedings of International Scientific Conference.* (Bratislava: Vydavateľstvo EKONÓM, 2012)
8. Ministry of Economy of the Slovak Republic, List of Entities to which Investment Aid has been approved. Avaliable at: <https://www.economy.gov.sk/podpora-investicii/investicna-pomoc/zoznamy/zoznam-podnikatelskych-subjektov-ktorym-bolo-schvalene-poskytnutie-regionalnej-pomoci> (2019)
9. P. Kremský, Investment Incentives for Large Companies Distort the Business Environment and Harm Entrepreneurs. Avaliable at: https://www.alianciapas.sk/2016/03/24/investicne-stimuly-pre-velke-podniky-deformuju-podnikatelske-prostredie-a-poskodzuju-podnikatelov/ (2016)
10. D. Krajanová, The Government Repeatedly Subsidizes the Same Companies, and then it looks like others have no Place here. Avaliable at: <https://iness.sk/sk/vlada-opakovane-dotuje-tie-iste-firmy-potom-vyzera-tak-ze-ini-tu-nemaju-miesto-dennik-n> (2017)
11. A. Harumová, K. Kubátová, *Business Taxes.* (Žilina: Poradca podnikateľa, 2006)
12. P. Markovič, H. Majdúchová, D. Rybárová, *Stress Techniques - Innovation of Risk Management Methods.* (Praha: Wolters Kluwer, 2018)
13. V. Konečná, A. Andrejovská, Investor Decision-making in the Context of the Effective Corporate Taxation. *E&M Economics and Management,* **23**(1), 111–120 (2020)
14. K. A. Clausing, Corporate tax revenues in OECD countries. *International tax and public finance,* **14**(2), 115–133 (2007)
15. L. Mura, Entrepreneurship internationalization – case of Slovak family businesses. *Ad Alta: Journal of Interdisciplinary Research,* **9**(1), 222–226 (2019)
16. N. Tačarová, *Application of Tax Incentives in Selected EU Countries.* University of Economics in Bratislava. Faculty of Business Management; Department of Corporate Finance. (Bratislava: FPM EU, 2019)
17. G. Dubcová, *The Social and Solidarity Economy - One of the Stabilization Ways of Human Capital in Slovak Business.* (Ljubljana: Dashöfer, 2018)

Sustainable human resource management in a time of global uncertainty

Zuzana Dvorakova[1,*]

[1]CTU in Prague, MIAS School of Business, Kolejni 2637/2a, 160 00 Prague 6, Czechia

Abstract. The paper characterizes the state-of-the-art knowledge of sustainable human resource management (HRM) based on on-the-desk analysis of articles in the Web of Science database published in the last two years in journals in Q1 and Q2. In the literature, sustainable HRM is considered an extension of strategic HRM with a focus on long-term human resource development, regeneration, and renewal. The current focus has been on economic, environmental aspects in HR processes, and it turns out that social sustainability remains on the edge of HR managers′ interests. The situation in the Czech environment illustrates two narrative cases from March and April 2020. It becomes a challenge for all actors in education to change their view of ways of education and to accept as a necessity to apply distance education on a mass scale.

1 Introduction

Recently, articles on sustainable human resource management (HRM) have published. This concept of managing people is promoted, especially by "green" organizations. For the community of HR professionals and managers, it is necessary to define what dimensions and characteristics this new approach will include so that they can use in HR practices. Corporations with commitments to sustainable development goals are open to HR practices that motivate sustainable development in work incentive systems.

Research on the benefits of HRM for business and the strategic dimension of HRM today lags what business expects of it. Kaufman [1] refers to the phenomenon is as the three-decade-long psychologization of HRM, which is further deepened by scholastic scientism and normative promotionalism. Psychologization reduces explanations of macro-level HRM outcomes to individual-level psychological-behavioral factors and individual differences. It is essential but often proclaimed in favor of an employer branding that the company creates a favorable environment for training and development. The key to deciding whether and how much to invest in people must be, what outputs will flow from the freedom given to employees to create and form the belief that their ideas will reflect in new products and services. Innovations mean a driving force to overcome the crisis, but it is necessary to ensure that the criteria for assessing their contributions to the business are a part of performance management and the basis for rewards.

* Corresponding author: zuzan.dvorak@gmail.com

The paper aims to characterize the understanding of sustainable HRM, to identify the place of education in social sustainability, to illustrate on narrative cases the transition to distance education in the CR, and to outline the prospects for online education.

2 Theoretical ground for sustainable HRM

2.1 Analysis of bibliographic data

Searching for relevant sources, it retrieved 13,110 bibliographic records that matched the query in database search in the period 1990-2020 (accessed on 2020/05/05): records from the Web of Science databases count for 5,129 and Scopus 7,981. The structure of bibliographic records shows Table 1. The enormous amount of literature of various sciences does not allow conducting a complete analysis of all of them. In this paper, the analysis is limited by the most significant works, considering two requirements:
- papers published in journals from Q1 and Q2 quartiles from last two years,
- the citation index related to strategic human resource management, green human resource management, SHRM, organizational sustainability, and social sustainability.

The study outlines the development of promising directions, mostly in management, business, applied psychology, industrial relations, economics, and social sciences.

Table 1. Bibliographic records by the query in Web of Science and Scopus database 1990-2020.

Query	Web of Science[1]	Scopus[1]
Strategic human resource management	641	917
Green human resource management	121	175
Sustainable human resource management	64	78
Organizational sustainability	256	492
Social sustainability	2 241	3 046
Massive open online courses	1 896	3 273
Total	5 129	7 981

[1] Source: Author. Accessed 2020/05/05.

Strategic HRM belongs to the most studied topics in the period 1990-2020 in the USA (208 publications in the Web of Science), Peoples Republic of China (118), Australia (61), England (51), Spain (32) and Germany (25). In the CR, it covers eight publications and Slovakia 3. Articles dealt with strategic HRM cover according to categories topics from management (475 publications), business (148), applied psychology (115), industrial relations and labor (85), economics (30), operations research management science, and social sciences interdisciplinary (both 29). In the case of green HRM, the order of the number of publications characterizes the following: Peoples Republic of China (25), Malaysia (20), England (16), and Brazil, France, India (each 14). The CR represents six publications and Slovakia one. Sustainable HRM is a relatively new topic and mostly written in publications from Peoples Republic of China (9), Spain (7), England (6), Lithuania and Slovakia (both 5) and only one in the CR. Publications about organizational sustainability mainly print in the USA (75), Brazil (22), England (21), Malaysia (15), India

(14), and Peoples Republic of China (13). In the same period in Slovakia, one and nothing in the CR. They focus predominantly on management (72 publications), green sustainable science technology (61), environmental sciences, and environmental studies (both 54), and business (36). Social sustainability mostly becomes a subject of publications in the USA (314), England (234), Peoples Republic of China (190), Italy (182), Australia (169), Spain, and Sweden (both 143), but in the CR 18 and Slovakia 5. They belong to environmental sciences (599), green sustainable science technology (548), environmental studies (492), engineering environmental (178), management (172), regional urban planning (149), and urban studies (139). Massive open online courses (MOOCs) are subjects of papers published in the USA (366), Peoples Republic of China (253), Spain (214), England (152), Germany (98), and Australia (87). The CR represents ten publications and Slovakia none. Topics of papers mainly cover Web of Science categories: education and educational research (917), computer science interdisciplinary applications (306), computer science theory methods (268), scientific education disciplines (228), computer science information systems (221), engineering electrical electronic (197), and computer science artificial intelligence (126).

The number of papers increases during the last 20 years, as shown in Table 2. It demonstrates prospective potentials for innovations and competitiveness.

Table 2. Bibliographic records by the query in Web of Science database 2000-2020

Query	Years								
	2000	2005	2010	2015	2016	2017	2018	2019	2020
Strategic HRM	12	21	24	41	47	47	41	55	8
Green HRM	-	-	-	5	14	14	21	34	25
Sustainable HRM	-	1	-	1	4	7	8	30	10
Organizational sustainability	1	1	3	29	28	36	49	37	15
Social sustainability	4	15	56	201	223	284	382	419	121
Massive open online courses	-	-	-	316	340	369	294	264	60

Source: Author. Accessed 2020/05/05.

2.2 Literature overview

2.2.1 Sustainable HRM – the concept and characteristics

Sustainable HRM is seen as an extension of strategic HRM and turns into a new approach to managing people with a focus on long-term human resource development, regeneration, and renewal. However, the attributes of sustainable HRM are not explicit [2]. Chams and Garcia-Blandon [3] summarize their findings of sustainable HRM, arguing that it plays a crucial role in developing a sustainable work environment and in facilitating the attainment of sustainable development goals. Therefore, the main themes in the field of sustainable HRM are environmental management, socially responsible HRM, and turnover [4].

Attentions for sustainability activities of corporations evoke green HRM. Amrutha and Geetha [5] characterize green HR practices, employee green behavior at the workplace, and organizational sustainability and argue that social sustainability is the least explored area than economic and environmental pillars of sustainability.

Macke and Genari [6] focus on the state-of-the-art of sustainable HRM. They believe that four categories emerge to describe its strategy, namely: sustainable leadership; the relationship among HRM, environmental sustainability, and organizational performance; the tensions and paradoxes between HRM practices and sustainability, and the link between HRM and the social dimension of sustainability. However, it is rare to find the managerial meaning of sustainable HRM. Finish top managers describe the theme in four dimensions - justice and equality, transparent HR practices, profitability, and employee well-being, and four broader responsibility areas: legal and ethical, managerial, social, and economic [7].

Roca-Puig [8] identifies on an extensive longitudinal data set that HR investment strategy negatively affects income inequality and that corporate decisions on social sustainability can influence social sustainability in society. So, the norm of a circular or two-way relationship between business and society builds social sustainability by increasing trust and cooperation in any group of people. One of the main drivers to generate social sustainability is knowledge sharing behavior [9]. In the case that employees are willing to cooperate and share in fair organizational contexts, when they feel satisfied and affectively committed, and when they get fair rewards for their contributions (Ibid.). However, HR professionals and managers must involve in developing sustainable HRM. If they perceive their role in the business as meaningful and going beyond the sole focus on economic performance, then they can achieve higher job satisfaction and have lower intention to leave [10].

Even though the number of papers about sustainable HRM and organizational sustainability is growing (as shown above in Table 2), literature review remains substantially unclear in which HR practices present a basis for sustainable leading and managing people. Aimed at innovation behavior, employee engagement, and retention, the significant practices for their long-term achievement consist of organizational fairness, the quality of relations between supervisors and subordinates, performance appraisal, pay-for-performance, and work-life environment.

Sustainable HRM underlines the involvement of partnerships and external relations in learning and personal development processes [11]. Based on the literature review, they believe that a crucial role must play innovations in education and development as their HR practices open mind for creativity and successful adaptation to changes in the labor world [Ibid.].

2.2.2 Sustainable HRM in the cultural contexts

Culture plays a significant role, and organizations demonstrate their influence in developing HR strategies and day-to-day activities. Challenges for implementing HRM activities across borders must mainly tackle multinational enterprises (MNEs) because they operate in culturally different contexts. They must decide whether to standardize or localize their processes and to improve the quality of working life and work environment [12, 13, 14, 15, 16]. In the expatriation strategies of MNEs, differences depend on the internationalization of the MNE and disparities in industrial relations between the parent and subsidiary's environment [17]. The most critical practices in managing people cover performance management, compensation, and industrial relations.

Findings of a large scale set of employer-employee data in 14 countries confirm a positive relationship between HR practices combined in high-performance work systems and organizational performance across societal cultures [18]. Based on the research mentioned above, it is evident two paradoxes, namely the co-existence of both process and result-focused orientations and both normative and pragmatic approaches to managing people [19, 20]. Findings achieved by Gu, Nolan, and Rowley [21] reveal about organizational justice in performance appraisal processes a necessity when analyzing three different types of banks in China, i.e., the state-owned bank, the city-commercial, and foreign-owned counterparts. They argue that it is time to re-evaluate the assumption that national culture variables are incredibly stable and have universal explanatory value in all organizational contexts.

HR practices labeled as high-performance work practices (HPWPs) are affected by national cultural dimensions [22]. However, these HR practices are not analyzed in literature from cross-cultural HR management aimed at understanding of the role of national culture on the effectiveness of HPWPs (Ibid.). Another example of a widely accepted perception can be about benefits stemming from performance appraisal, productive relationship between performance-based rewards, and innovative behavior. Research done by Curzi, Fabbri, Scapolan, and Boscolo [23] shows that formal performance appraisal probably reduces the promotion of individual innovation and creativity at work. Large scale research across ten countries about the relationship between performance-based rewards and innovative behavior confirmed no dependence of rewards on behavior that generates innovations [24]. Country specifics exist in the case of individualized pay-for-performance schemes for employees, even though this practice seems to be a standard tool for work motivation. Its implementation determinates either cultural and institutional factors at the country level, and senior managers' agency. So, in the USA, the adoption of these schemes is rather useful than in many other countries [25].

Summarizing: The least analyzed area of sustainable HRM belongs to social sustainability. The crucial role in HRM for mastering challenges must play innovations in education and development. Organizational preconditions for building the creative atmosphere include friendly and knowledge sharing environment, trust and cooperation between supervisors and subordinates, informal feedback about performance, massive open access to learning across borders, and opportunities for using new skills at work.

3 Methodology and findings

3.1 Methodology

The methodology insists on several methods that enable a triangular analysis. The research combines state-of-the-art knowledge identified by in-depth-analysis of secondary sources, i.e., papers indexed in Web of Science and Scopus in 2018-2020, with two semi-structured interviews held in March and April 2020. The semi-structured interviews use due to the possibilities to address one managerial respondent, one teacher, and the author´s observations. They provide information for writing two narrative cases.

The reasoning methods are first that ethical aspects of sustainable HRM as it deals with social phenomena, which is deeply rooted in history, culture, values, and behaviors, and the business reality loves to apply pragmatic HR practices. Secondly, based on bibliometric analysis, it is evident a need to formulate how and which tools can support a change in employees´ attitudes to MOOCs. Therefore, research collects and analyses data by an interpretative group of methods that reveal the meaning-making practices of human subjects to get a basis for achieving results, emerging questions for the practice, and ideas for future research.

3.2 Narrative cases

Case study 1. Helena, 57, professor, university.

From the second week of March, we learned how to teach online and work from the home office. The mistake was that we had no idea which platforms to use and which ones posed a cyber risk. In March, the Ministry of Education, Youth and Sport organized webinars about contactless teaching. The employer-provided two online pieces of training for beginners in MS Teams. However, a real benefit for acquiring new skills has been a shared experience among teachers. As of April, only Moodle and MS Teams can use it at the university. Most of the colleagues managed to teach remotely during the first three weeks after the declaration of the state of emergency. Teaching takes place according to the original schedule, and gradually both teachers and students have gained a routine in the use of Teams. During the lessons, we can monitor and save how many students and who connect to the ongoing lesson, participants can ask questions through a microphone, open a video camera, actively participate in the course of lessons, and use chat. In March and April, I followed students' attitudes to online teaching in the case of three subjects with a total of 150 students, totally, for five weeks using MS Teams, Moodle, and emails. The data monitor the number of students joined, the development of the connected students during the class, and their activities. The teaching took place according to the schedule from the time of contact teaching. For a given period, the number of connected ones remains constant. Roughly 1/3 of the students enrolled in the courses never joined the meetings, even if they have received an invitation, and a minimum of those joined meetings participated in the discussion. Nobody used the offer of individual online consultations.

Case study 2. Pavel, 56, managing partner, a multinational consultancy firm.

The consultancy firm operates in seven European countries and has got five branches outside Europe. The company specializes in corporate training and development, coaching, and designing compensation. The Czech subsidiary belongs to the micro-enterprise category and cooperates in the frame of an extensive network of freelance consultants. Clients stopped contracted pieces of training from 16 March. However, some of them, mainly clients outside the CR, asked for using online training, like appraisal interviews. Challenges were which applications can be useful, user friendly, and safe in such training. Clients preferred MS Teams, Google Hangouts, or Webex. The scope of clients willing to use online training is slightly rising as the Czech team consists of diverse consultants who speak fluent English, and the company has an excellent long-term reputation. On the side of the consultancy firm, the situation looks as opportunities for development in several areas; even it was tough to reduce personnel costs and find ways how to work from the home office effectively. All consultants struggle with an unpleasant form of training as they must speak and behave without social contacts, especially, it is difficult for less experienced trainers. The future will build on high staff retention, flexible working regime, home office, diverse team, and their language skills. The intention is to provide both traditional and online training plus to design some training as sophisticated MOOCs.

4 Discussion and conclusion

The Czech society places the primary responsibility for the failure of pupils and students on the shoulders of teachers: stakeholders have an opinion that teachers are responsible for if pupils and students do not know and have little interest in learning. During a state of emergency, both sides of the educational process entered the online environment unexpectedly. The families received an experience that children who go to school mean additional work in caring for them at home. In the process of transforming teaching, educators had to rely on themselves and their schools to handle challenges. They started to work from home because the employers often have trouble providing adequate technical support, such as PC, software, a reliable Internet connection. They often remained alone and have no one to turn to for advice or have no trainers on how to teach online. Those who were trainers at the beginning of March were very overwhelmed with questions and stopped gradually answered.

Teachers' experience can serve as a starting point for the creation and implementation of MOOCs in the Czech environment. They experience that the widening of social differences between people is accelerating. The future will depend on the technical equipment of households, the quality of the Internet connection, and the family care regarding education. The opening of the scissors will reflect on the level of education of the population. In addition to the technical side of prospective MOOCs, it is evident that the employability and job applicant success in the labor market will depend on the individual's motivation to learn. Students joined in meeting courses are mostly passive, reluctant to speak through the microphone and unwillingness to open the camera when speaking, i.e., they play a game the best thing is to remain anonymous.

There can identify several reasons for not accepting online teaching in the last two months. They range from the low attractiveness of the subject, imperfect forms of teaching, and the little experience of teachers to teach without impersonal contact when it is not possible to build social contacts and respond to non-verbal communication. Some of these negatives can address and eliminate. Another issue of contactless teaching exists that two groups of people polarize: those who want to learn and those who want to have fun enjoying the free regime at school. For university graduates, the level of motivation to learn may apply as personal responsibility for self-education when assessing job seekers. If anybody is interested in learning and see it as an inner motive, it is highly probable that such an orientation will remain as life-long characteristics. A definite group of students can distinguish for whom the study at a university is a burden; the only stimulus is the vision achieving an elite status thanks to the title, which becomes more accessible to get in these days than ever before by reducing knowledge requirements due to shortcomings of contactless teaching and testing.

References

1. B.E. Kaufman, The real problem: the deadly combination of psychologization, scientism, and normative promotionalism takes strategic human resource management down a 30-year dead end. *Human Resource Management Journal* **30**, 49-72 (2020)
2. Z. Stankeviciute, A. Savaneviciene, Designing sustainable HRM: the core characteristics of emerging field. *Sustainability* **10**, 4798 (2018)
3. N. Chams, J. Garcia-Blandon, On the importance of sustainable human resource management for the adoption of sustainable development goals. *Resources Conservation and Recycling* **141**, 109-122 (2019)
4. M. Santana, A. Lopez-Cabrales, Sustainable development and human resource management: a science mapping approach. *Corporate Social Responsibility and Environmental Management* **26**, 1171-1183 (2019)
5. V.N. Amrutha, S.N. Geetha, A systematic review on green human resource management: implications for social sustainability. *Journal of Cleaner Production* **247**, 119131 (2020)
6. J. Macke, D. Genari, Systematic literature review on sustainable human resource management. *Journal of Cleaner Production* **208**, 806-815 (2019)
7. M. Jaerlstrom, E. Saru, S. Vanhala, Sustainable human resource management with salience of stakeholders: a top management perspective. *Journal of Business Ethics* **152**, 703-724 (2018)
8. V. Roca-Puig, The circular path of social sustainability: an empirical analysis. *Journal of Cleaner Production* **212**, 916-924 (2019)
9. N. Cuguero-Escofet, P. Ficapal-Cusi, J. Torrent-Sellens, Sustainable human resource management: how to create a knowledge sharing behavior through organizational

justice, organizational support, satisfaction, and commitment. *Sustainability* **11**, 5419 (2019)
10. M. Guerci, A. Decramer, T. Van Waeyenberg, I. Aust, Moving beyond the link between HRM and economic performance: a study on the individual reactions of HR managers and professionals to sustainable HRM. *Journal of Business Ethics* **160**, 783-800 (2019)
11. K. Stachova, J. Papula, Z. Stacho, L. Kohnova, External partnerships in employee education and development as the key to facing Industry 4.0 challenges. *Sustainability* **11**, 345 (2019)
12. G.C. Banks, H.M. Woznyj, R.S. Wesslen, K.A. Frear, G. Berka, E.D. Heggestad, H.L. Gordon, Strategic recruitment across borders: an investigation of multinational enterprises. *Journal of Management* **45**, 476-509 (2019)
13. C.L. Jiang, D. Yahiaoui, French multinational companies' HRM in China: strategic orientation and integration approaches. *Asia Pacific Business Review* **25**, 3-18 (2019)
14. H.C. Chao, A.M. Morrison, B.H. Wu, All work and no play? Antecedents of the leisure involvement of Taiwan expatriate managers in Mainland China. *International Journal of Manpower* **40**, 1215-1238 (2019)
15. G. Hong, E. Kim, How to attract talented expatriates: the key role of sustainable HRM. *Sustainability* **11**, 5373 (2019)
16. P. Poisat, M.R. Mey, G. Sharp, Do talent management strategies influence the psychological contract within a diverse environment? *SA Journal of Human Resource Management* **16**, a1044 (2018)
17. J. Meuer, M. Kluike, U. Backes-Gellner, K. Pull, Using expatriates for adapting subsidiaries' employment modes to different market economies: a comparative analysis of US subsidiaries in Germany, the UK and Switzerland. *European Journal of International Management* **13**, 328-353 (2019)
18. Dastmalchian, N. Bacon, N. McNeil, C. Steinke, P. Blyton, M.S. Kumar, S. Bayraktar, W. Auer-Rizzi, A.A. Bodla, R. Cotton, et al., High-performance work systems and organizational performance across societal cultures. *Journal of International Business Studies* (Jan 2020)
19. A. Erthal, L. Marques, National culture and organizational culture in lean organizations: a systematic review. *Production Planning & Control* **29**, 668-687 (2018)
20. M. Furusawa, Ch. Brewster, Japanese self-initiated expatriates as boundary spanners in Chinese subsidiaries of Japanese MNEs: Antecedents, social capital, and HRM practices. *Thunderbird International Business Review* **60**, 911-919 (2018)
21. F. Gu, J. Nolan, Ch. Rowley, Organizational justice in Chinese banks: understanding the variable influence of guanxi on perceptions of fairness in performance appraisal. *Asia Pacific Business Review* (Dec 2019)
22. S. Nadeem, M. Raza, N. Kayani, A. Aziz, D. Nayab, Examining cross-cultural compatibility of high-performance work practices. *International Business Review* **27**, 563-583 (2018)
23. Y. Curzi, T. Fabbri, A.C. Scapolan, S. Boscolo, Performance appraisal and innovative behavior in the digital era. *Frontiers in Psychology* **10**, 1659 (2019)
24. K. Sanders, F. Jorgensen, H. Shipton, Y. Van Rossenberg, R. Cunha, X.B. Li, R. Rodrigues, S.I. Wong, A. Dysvik, Performance-based rewards, and innovative behaviors. *Human Resource Management* **57**, 1455-1468 (2018)
25. P. Gooderham, M. Fenton-O'Creevy, R. Croucher, M. Brookes, A multilevel analysis of the use of individual pay-for-performance systems. *Journal of Management* **44**, 1479-1504 (2018)

Consumer legal protection in the EU

Mária Dzúrová[1,*]

[1]University of Economics in Bratislava, Department of Marketing, Dolnozemská cesta 1, Bratislava, 852 35 Slovak Republic

Abstract. Consumer protection is a very wide-ranging issue and needs to be given due attention. It concerns the safety of consumers in the environment of individual countries of the world, it concerns the consumer safety of certain groups, such as the European Union. The basic frameworks of consumer protection are set by the guidelines of world organizations - the UN, WHO, but also the European Union and individual member states. In the area of consumer protection, attention is paid to major health problems caused by unsuitable food, such as food scandals, various types of diseases - mad cow disease, swine fever, covid 19.

1 European Union law and consumer

An effective consumer protection policy guarantees the proper and efficient functioning of the single market. It aims to protect consumers' rights vis-à-vis traders and to provide enhanced protection to vulnerable consumers . The financial crisis has shown that consumer protection rules can make markets fairer and the quality of competition can improve. Empowering consumers and effectively protecting their security and economic interests have become key objectives of European policy

Article 4.2 letter (f) and Articles 12, 114 and 169 of the Treaty on the Functioning of the European Union (TFEU) and Article 38 of the Charter of Fundamental Rights of the European Union.

In order to promote the interests of consumers and ensure a high level of consumer protection, the Union shall contribute to protecting the health, safety and economic interests of consumers, and to promoting their right to information, education and the formation of associations for the protection of their interests."

The EU's agenda for consumer policy consists mainly of the European Consumer Program and the Consumer Program 2014-2020.

The Consumer Agenda has various objectives:

- promote consumer safety through the 2013 Product Safety and Market Surveillance Package, and in particular through better product identification and traceability, measures to strengthen safety in the food chain and the new rules on cosmetic product safety introduced in mid-2013,

* Corresponding author: maria.dzurova@euba.sk

• raise awareness of consumer rights: Interactive tools have been developed, such as the Consumer Classroom, to inform and educate consumers and help them fully participate in the single market.

• strengthen the enforcement of consumer protection rules through coordinated action against breaches of EU consumer law through website inspections by a network of national consumer protection authorities; and simple, fast and inexpensive out-of-court redress procedures for consumers, available through the Alternative Dispute Resolution Directive (2013) and the Online Consumer Dispute Resolution Regulation (2013),

• integrate consumer interests into key sectoral policies through new legislation in sectors such as telecommunications, digital technologies, energy, transport and food, and new measures to increase transparency and access to retail financial services and facilitate the transfer of bank accounts.

Strengthening consumer protection, which is a key objective of the EU Consumer Policy Strategy set in the Commission Staff Working Document on Strengthening EU Consumer Rights, aims to empower EU consumers through consumer choice, information and awareness.

2 Examining consumer behaviour

The Consumer Agenda identifies challenges, such as moving towards more sustainable consumption and addressing the specific needs of vulnerable consumers.

The Consumer Program 2014-2020, with a budget of € 188.8 million, will support EU consumer policy for the coming years. It aims to help citizens enjoy their consumer rights and participate actively in the single market, thus fostering growth and innovation and helping the EU meet the Europe 2020 goals.

The Consumer conditions scoreboard monitors the consumer environment across Europe through three key dimensions: knowledge and trust; compliance and enforcement; complaints and dispute resolution. It also examines progress on the EU Digital Single Market from a consumer perspective

The consumer conditions scoreboard is based mainly on surveys of consumers' and retailers' attitudes towards cross-border trade and consumer protection.

It monitors national conditions for consumers in three areas

- knowledge and trust
- compliance and enforcement
- complaints and dispute resolution.

The conditions scoreboard also examines progress in the integration of the EU retail market and the uptake of e-commerce.

3 Product health and safety

European consumer protection measures are aimed at protecting the health, safety and economic and legal interests of European consumers, wherever they live, reside or shop in the EU. EU provisions govern both physical transactions and e-commerce and contain general rules, together with provisions for specific products, including medicines, genetically modified organisms, tobacco products, cosmetics, toys and explosives.

The basis is to ensure that all consumers in the Union, wherever they live, reside or shop in the EU, enjoy a high common level of protection against risks and threats to their security and economic interests, and increase consumers' opportunities in defending their own interests.

Consumer safety and health is essential for a healthy life. For this reason, high attention is paid to this issue in the EU environment. Directive 2001/95/EC introduced a system of general product safety. On the basis of this basis, all consumer products placed on the market must meet certain standards. These include information for consumers, measures to prevent threats to safety, monitoring product safety and traceability. Clearly even if they are not covered by specific sectoral legislation. Where a product poses a serious threat requiring urgent action, the Member State concerned shall immediately inform the Commission through THE RAPEX system, which is intended for the rapid exchange of information between the Member States and the Commission. When Slovakia joined the EU, we did not have dangerous products on the market. Over time, the situation has changed. Some have emerged in the Union environment, others have appeared on the Slovak market. The supervisory authorities shall comply with the information obligation for EU Member States. It is a paradox that products intended for children, i.e. a vulnerable group of consumers, and toys, often appear in the rapid information system. Therefore, separate directives are processed for areas where the rules are most frequently violated.

4 Consumer food safety

The objectives of European food safety policy are: protecting human health and consumer interests and promoting the smooth functioning of the Single European Market. The EU therefore ensures the establishment and compliance of control standards in the areas of feed and food hygiene, animal and plant health and the prevention of contamination of food by external substances. The Union also regulates the labelling of food and feed.

Food safety in the EU has undergone major changes at the beginning of the current millennium. This was due to shortcomings in both food and animal feed. This was an epidemic of BSE and dioxin threat and various food scandals. An approach called 'from farm to table', i.e. supply chain, has been established. It assumes and guarantees a high level of food safety at all levels of the chain. Thus, from the process of production and distribution of all food products marketed in the EU, regardless of whether they were produced in the EU or imported from third countries. It is a set of legislation which is a comprehensive and integrated system of rules applicable to the entire food chain. This means from feed and health to plant protection and food production to processing, storage, transport, import, export up to the retail sale.

In the case of various mediated food scandals, there is also an interest in consumers finding out why such a situation has occurred. Consumers begin to pay attention to the information on packaging, they begin to notice the country of origin of the product. They will start to consider how and where to obtain information on the quality of the products they buy, the country of origin is suddenly very important. Thus, in the environment of Slovakia, there is a reservation on food from Poland after the mediated scandals. Otherwise, consumers are not very interested in the issue of their own protection.

The EU pays sufficient attention to food law. It sets framework regulations declaring general principles and requirements relating to food and feed law. The precautionary principle applies. The Regulation specifies the risk assessment approach and the general provisions on the traceability of food and feed. In principle from field to table, this means that if in any way the unsuitability of a food for human use is detected the Rapid Alert System for Food and Feed, RASFF, is activated. This system also allows Member States, as well as the

Commission, to exchange information quickly and coordinate responses to health threats caused by food or feed.

The main idea was the establishment of the European Food Safety Authority (EFSA). This accident evaluates the risks associated with the food chain. And, of course, he informs the public about them. In some cases, current food law is assessed, depending on the different risks, confronts scientific studies and efforts are made to improve cooperation with Member States in providing experts and providing data. Regulations on novel foods, regulations on GMOs, pesticides, food contact materials and food ingredients are being examined in order to bring them into line with the review of general food law and to increase transparency. This is critical to the general security of the population.

Food hygiene is an important part of food safety entering the EU market. For this reason, legal measures and guidelines for food hygiene have been adopted in this area. Food hygiene rules deal with specific hygiene rules for foodstuffs of animal origin. The Community declares official controls on animal products intended for human consumption. This means that specific rules are laid down for fresh meat, bivalve mollons, milk and milk products. The set of rules imposes responsibility for food hygiene directly on the various actors in the food chain. This is using a self-regulatory system that uses the HACCP (hazard analysis and critical control points) method, which is monitored by official controls. Official controls must be carried out by the competent authorities.

The legal framework on food labelling is designed to guarantee consumers access to clear, comprehensible and reliable information on the content and composition of products in order to protect their health and best interests. For example, allergens such as soya, gluten or lactose must be clearly indicated on the packaging.

The provision of information is an important and important piece of information in the protection and safety of food for the consumer. As of 2016, the regulation has declared the information that the manufacturer must comply with and place on the packaging of products – i.e. food that enters the market. The basis is the presence of allergens in unpackaged foods, i.e. in restaurants and canteens, the origin of unprocessed meat. There was already an obligation to indicate the country of origin and the presence of imitation foodstuffs, such as plant products replacing cheese or meat. A separate chapter is the country of origin in food. Member States shall specify in the provisions on designation of origin the requirements (with certain exceptions) for the indication of the place of rearing and the place of slaughter of packaged fresh, chilled and frozen pig, mutton, goat's and poultry meat.

The rules prohibit misleading consumers, including all information. I mean, advertising and presenting food. The rules for permitted nutrition and health claims apply. For example, "low fat" or "high fiber" or statements about the relationship between food and health. All claims must be supported by scientific evidence contained in the EU public register.

It is important to respect groups of vulnerable consumers in the market that have their specific needs. Specific groups of vulnerable consumers are infants and young children, persons with special health problems and people on a reduced energy-reduced diet for weight control. Strict rules apply to food and vulnerable consumers

In recent years, there have been many activities and complaints about double quality food on the eastern European countries market. Various checks, surveys show that products also on the Slovak market are less valuable, their composition is not the same throughout Europe. This applies to food primarily, but also to drugstore goods. The declaration of differential quality of products with the same brand name and from the same manufacturer has led to changes in the rules. The Consumer Protection Directive also found in prohibited commercial practices that products must be correctly labelled. Double quality is prohibited and is considered a prohibited commercial practice. In order to be able to prove double quality, in particular for foodstuffs, it is necessary to follow the correct sampling procedures – i.e. one producer, one batch, one day. And so assess the difference in quality. However, since this is classified as a prohibited commercial practice, such situations of different quality on the European market must not occur.

Conclusion

Health policy is a very wide area. It deals with the normal commercial relationship of the consumer-trader. The legislation regulates the safety and quality of food as well as the safety and quality of ordinary non-food products. Many issues are solved through the European Consumer Centre. There is an agenda of vulnerable consumers who have specific needs and the possibilities of harming them are many. The protection of the economic interests of consumers is important. The alternative solution of consumer disputes is one way of shortening the resolution of consumer problems, which can only be dealt with in our courts. A separate chapter is financial literacy and consumers' ability to solve their economic problems.

The EU contributes to the achievement of the consumer protection objectives by measures taken to approximate the provisions of the laws, regulations and administrative provisions of the Member States in order to complete the creation of the internal market. One of the EU Member States' means of consumer protection is the European Consumer Programme – Promoting Confidence and Growth. The consumer is the most numerous market operator, but it is not organised. This means that it is not strong in its own defence against the unfair commercial practices it encounters in some areas. This implies that increased attention should be paid to targeted education and consumer information in each area. Consumers need to learn not only to know their own rights, but also to use them. Raise awareness for your own safety. To follow different campaigns, such as food or a healthy lifestyle, it is necessary to distinguish the plausibility of information that the consumer receives from different media. And without which, for example, he can't make the right decision.

References

1. D. Vokounová, Current ConsumerBuying Behavior. *Application of knowledge in process of business dynamization in CentralEurope : theproceedings of the 5th internationalscientificconference on trade, international business and tourism : 16. - 17. october 2014, Mojmírovce, Slovak Republic,* WOS, 514-520, (2014)
2. D. Vokounová, J. Kopaničová, Can I beopened? *Central and EasternEurope in thechanging business environment. International jointconference. Central and EasternEurope in thechanging business environment : proceedings of 16th internationaljointconference : may 27, 2016, Prague, CzechRepublic and Bratislava, Slovakia,* WOS, 347-355 (2016)
3. D. Vokounová, Generation Y – HealthRelatedAttitudes and Behaviour. *Central and EasternEurope in theChanging Business Environment. International JointConference. Central and EasternEurope in theChanging Business Environment : Proceedings of 18th International JointConference : May 25, 2018, Bratislava, Slovakia and Prague, CzechRepublic,* WOS,*467-476 (2018)*
4. G. Dubcová, D. Gajdová, K. Grančičová, Evaluation of the functioning system of thesocial and solidarity economy in Slovakia. *International conference on new challenges*

in management and organization: organization and leadership. ProcediaSocial and BehavioralSciences : the 3rd International conference on new challenges in management and organization: organization and leadership, 2 May 2016, Dubai, UAE. - [s.l.] : Elsevier, WOS, **230**, 254-263 (2016)

5. D. Hrušovská, Nové trendy v spotrebiteľskom správaní. *Aktuálne problémy podnikovej sféry : zborník vedeckých prác*, WOS, 143-148 (2012)
6. D. Hrušovská, Vnímanie prieskumu trhu z hľadiska podnikov v Slovenskej republike. *Podnikanie a konkurencieschopnosť firiem 2010 : zborník z medzinárodnej vedeckej konferencie organizovanej v rámci podujatí k 70. výročiu založenia Ekonomickej univerzity v Bratislave,* WOS, 108-111 (2010)
7. G. Dubcová, K. Grančičová, D. Hrušovská, Marketing performance assessment in businesses in Slovakia. *ISCOBEMM* 2017, WOS, 56-64 (2017)
8. D. Hrušovská, Stratégie prístupu k zákazníkovi - vnímanie hodnoty. *Aktuálne problémy podnikovej sféry 2013 : [recenzovaný] zborník vedeckých prác.* WOS, 187-192 (2013)
9. D. Hrušovská. Systémy vyhodnocovania dotazníkov. *Aktuálne pohľady na konkurencieschopnosť a podnikanie - nové výzvy : zborník [recenzovaných] vedeckých prác,* WOS, 173-177 (2011)
10. R. Madlenak, L. Madlenakova , MULTI-CRITERIA EVALUATION OF E-SHOP METHODS OF DELIVERY FROM THE CUSTOMER'S PERSPECTIVE *TRANSPORT PROBLEMS*, WOS, **15**, 5-14 (2020)
11. M. Pechota, M. Matuskova, L. Medlenakova, The trend of cost of universal services provided by national postal operator and correlation between price of letter mail and amount of sentlettermails in Slovakia . *12TH INTERNATIONAL SCIENTIFIC CONFERENCE OF YOUNG SCIENTISTS ON SUSTAINABLE, MODERN AND SAFE TRANSPORT,* WOS, **192**, 574-579 (2017)
12. P. Kolarovszki, Z. Kolarovszka, L. Medlenakova, MULTIMEDIA AS A SUPPORT FOR INTERACTIVE LEARNING PROCESSES . *EDULEARN16: 8TH INTERNATIONAL CONFERENCE ON EDUCATION AND NEW LEARNING TECHNOLOGIES,*WOS, 2126-2133 (2016)
13. R. Medlenak, L. Medlenakova, THE GLOBAL IMPACT OF E-MARKETING STRATEGIES ON SME IN THE SEGMENT OF TOURISTIC SERVICES . GLOBALIZATION AND ITS SOCIO-ECONOMIC CONSEQUENCES, 16TH INTERNATIONAL SCIENTIFIC CONFERENCE PROCEEDINGS, PTS I-V, 1235-1243 (2016)
14. R. Medlenak, L. Medlenakova, L. Svedlenka, Analysis of WebsiteTrafficDependence on Use of Selected Internet Marketing Tools. 2ND GLOBAL CONFERENCE ON BUSINESS, ECONOMICS, MANAGEMENT AND TOURISM,WOS, 123-128 ,**23** (2015)
15. M. Maciejewski, Ch. Ratcliff, K. Næss, Spotrebiteľská politika: zásady a nástroje. ONLINE, Available from: https://www.europarl.europa.eu/factsheets/sk/sheet/46/spotrebitelska-politika-zasady-a-nastroje.
16. European Commission, Consumer conditions scoreboard: consumers at home in the single market - 2019 edition. ONLINE. Available from: https://ec.europa.eu/info/publications/consumer-conditions-scoreboard-consumers-home-single-market-2019-edition_sk
17. European Commission, Opatrenia na ochranu spotrebiteľa, ONLINE. Available from: https://www.europarl.europa.eu/factsheets/sk/sheet/47/opatrenia-na-ochranu-spotrebitel
18. Z. Laky, Bezpečnosť potravín, ONLINE. Available from: https://www.europarl.europa.eu/factsheets/sk/sheet/51/bezpecnost-potravin

Models of Regulation as Tool for Risk Elimination in Network Industry Markets in Slovakia

Eleonora Fendekova[1*]

[1] University of Economics Bratislava, Department of Business Economics, Dolnozemská cesta 1/b, 852 35 Bratislava, Slovakia

Abstract. In Slovakia in the recent period of time we have seen an intensive discussion concerning the effectiveness of the price regulation methods in Network Industries and the objectivity and the labour social effectiveness of The Regulatory Office for Network Industries, as well. It is understandable as network industries in fact ensure the production and distribution of energy sources which play a key role in an effective operation of the developed economies. The discussions are usually focused on the question of a reasonable profit of the network industries companies and on the other hand on the question of prices which are determined by the reasonable and generally acceptable costs of their production. In relation with objectification and increases in nationwide operation effectiveness of natural monopolies in network industries[†] on the market the role of regulation mechanisms is increasing. Its aim is the closest approximation of proportions between price and level of network industry products supply to the situation which would occur in the conditions of market competition. In the paper we will analyze analytical scheme for monopoly price regulation – rate of return regulation.

1 Introduction

Modern market structures in economically developed countries are generally represented by imperfect competition. We need to realize, that monopoly, in consequence of its exclusive market position as a sole supplier of commodity or service on the relevant market, can set the market price and supply any level of production, which allows it to maximize profit above the level of capacity of competitive company.

---

[*] Corresponding author: eleonora.fendekova@euba.sk

[†] Under network industries in terms of Slovak Republic in accordance with Act No. 276/2001 Coll. on Regulation in Network Industries we understand the following industries

1. generation, transmission, distribution and supply of electricity and the related services
2. production, transport, distribution, storage and supply of gas and the services related to these activities
3. production and distribution of heat
4. water management and mass conduct of sewage by a public sewage system

Blum, Müller and Weiske [2] show, that monopoly is, of course, acting within the legitimate market conditions and its position is determined by the real market situation. On the other side it is natural for such means to exist, which allow redistribution of this profit in a way that part of it would support nationwide economy development objectives of the country it is operating in.

In principle, there are two schemes, which can be applied by the state in regulation of monopoly price policy.

a) Taxation of monopoly

There is a potent tool at states disposal to regulate the development of market equilibrium conditions on monopolistic market. It is a tax policy. In a simplified way, we can tell that the state is redistributing the profit of monopolistic firm through its taxation. This process of taxation in the imperfect competition environment has, in comparison with perfect competition environment, certain specificity.

Firm operating in perfect competition environment can respond to enforcement of particular scheme of tax burden only by modifying the amount of goods supplied on the market. Fendek and Fendeková [6] show, that monopoly has, however, broader room for decision making because in its competence there is not only setting the amount of goods supplied on the market but also, after taking into account consumer behavior, setting the price of good on relevant market.

b) Application of specific regulatory mechanisms affecting behavior of natural monopolies

In Rode [10] we can see, that or the regulation of subjects of the chosen industries, which are in light of their market share monopolies or natural monopolies, the state is creating institution, so-called regulator, which task, under the state authority, is to create such legislative environment and regulatory mechanisms, which will ensure market equilibrium for observed commodities while granting fair profit for regulated subject. Among these industries particularly belong network industries, which subjects often share characteristics of monopolies and state has therefore desire to regulate them.

2 Means of state to regulate price policy of monopoly

In Slovak conditions the role of the regulator is represented by Regulatory Office for Network Industries (RONI), which role is to issue licenses, regulate prices and quality norms for network industry products. Regulatory offices in relation with fulfilment of their main mission – technical and price regulation of enterprise in regulated activities of chosen industries – are facing the effective solution of two tasks:

(1) Preparation of working and, this needs to be stressed, competitive market environment for network industries goods while applying standard regulatory mechanisms, mainly in context of Slovak entry into the EU and gradual adaptation of Slovakia to the conditions on energy markets of united Europe;

(2) Preparation of such analytical apparatus for network industries price regulation, which would guarantee effective development of regulated subjects. In the first period it was equally important to eliminate deformations in prices of network industries products.

A need of the particular legislation of regulatory framework for natural monopolies in Slovakia has arisen in the beginning of 1990s. Legislation in this period has not been covering the whole area of regulation therefore in 1993 the Antimonopoly Office of the Slovak Republic prepared a bill for regulation of natural monopolies, which assumed delegation of powers to a single body. This proposal should have supported companies by stimulating competitive environment towards greater effectiveness and reaching better quality levels.

Regulation, according to this proposal, should have been related to generation, transfer and distribution of electricity, heat, gas, water, but also telecommunications. Subjects of regulations should have become prices, tariffs, expenses, trade conditions and entry to the market in form of issuing licenses. Bill on regulation of natural monopolies was worked out in two alternatives from the perspective of founding and status of the regulatory bodies.

First alternative considered founding of four independent regulatory offices, financed from the revenues of regulated subjects. The aim was to create basis for high expertise, objectivity and stability of regulatory system. Second alternative considered that agenda of regulatory office would be performed by central body of the state administration set by the government.

Regulatory office oversees following of obligation to provide a third party with access to the networks and adjudicates disputes. Obligation to provide a third party with network access is also related to full opening of the internal energy market and to consumer option to freely choose energy supplier. According to this, RONI regulations are limited to activities with natural monopoly character only, that is, prices or tariffs set by the office will not include any unregulated items.

In order for Regulatory office to meet this objectives it should be financially and personally independent from executive body for reasons of transparency. Additionally regulatory powers to pass mandatory regulatory statues must be at its disposal. It also needs virtue to enforce them.

RONI is carrying out regulation through the use of standard, in developed economies proven methods. Details about price regulation process in network industries and determination of scope of justified costs and reasonable profit are enacted by the office in office decrees, which are published in Collection of Acts, in Office bulletin and on the website of the office.

Efficient functioning of commodity and service market in network industries in conditions of Slovak republic can be insured only by quality and potent regulatory policy in the context with the Slovak Republic law and respecting the adopted legal acts of European Community and European Union. All consumers, with emphasis on "vulnerable" groups, mainly households, must have an opportunity to make use of high level consumer protection, referring mainly to households, while maintaining adequate guarantees and suitable incentives for regulated subjects.

3 Analytical Methods for network industries regulation

In Waldman, and Jensen [13] we can see, that traditional methodological tool for price regulation applied by price regulators to set maximum price of network industries products is the regulation on the basis of the rate of return – Rate of Return Regulation, by which are the prices of electricity, gas and other companies regulated in the most of the developed countries.

The aim is to ensure that the regulated entity will set the price of commodity or service for its consumers in a way the revenues will cover all its *reasonable and provident costs incurred* as well as regulated return on its *provident*[‡] investment.

Let us now analytically derive the allowable rate of cost return for investment, co called *RoR* parameter of the regulated entity. Let us suppose that the firm is producing a

[‡] It should be noted, that under term *provident* we understand situations, when certain decision about investment or expenditures has been made in conditions of verified and relevant information available in the time of decision making.

homogenous product in production volume q, which it realizes on a relevant market for the price p. Let us further suppose that the firm uses two production factors, namely labor force with consumption level L by labor price w and the capital with consumption level K by the capital price r.

The profit of the firm is generally defined as the difference between the yields and costs

$$\pi(q) = t(q) - n(q)$$

where

$t(q) = p \times q$ – function of revenues of the firm, t: $R \to R$

$n(q) = nv(q) + nf$ – function of the total costs of the firm, n: $R \to R$

$nv(q)$ – function of the variable costs of the firm, $nv : R \to R$

nf – fixed costs of the firm, $nf \in R$

If we substitute general cost function on the basis of consumption of production factors, we get a profit function in a following form

$$\pi(q) = p \times q - w \times L - r \times K$$

If we further express the production volume q on the basis of the production function in the form

$$q = f(K, L)$$

and the production price p on the basis of the price-demand function in the form

$$p = p(q)$$

then we can express the profit function in the form

$$\pi(q) = p(q) \times q - w \times L - r \times K$$

and after further modification in the form

$$\pi(q) = p(f(K, L)) \times f(K, L) - w \times L - r \times K$$

Fendek and Fendeková [4] show, that a non-regulated firm can set its endogenous decision parameters in any way. So it chooses an optimum output volume q^*, an acceptable optimum price p^* and corresponding consumption levels of the production factors labor L and capital K in a way to reach maximum profit. Optimum output and optimum price will be calculated by solving the following mathematical programming task

$$\pi(q) = p(f(K, L)) \times f(K, L) - w \times L - r \times K \to \max$$

$$K, L \in R_{\geq 0}$$

In this case the non-regulated firm has no formal boundaries for setting the parameters guaranteeing its maximum profit. On the other hand, the regulated firm must respect boundaries given by the regulator. Price regulation regime on the basis of the rate of return lies in a fact, that through the use of exogenously defined control variable RoR the allowable level of quotient of the revenues $p \times q$ reduced by its non-capital expenditures $L \times w$ and the volume of consumed capital K is regulated.

In other words, a firm can optimize or freely determine the consumption levels of labor L, capital K by the market prices of production factors w, r and on the other side the level of its production q but also the production price p. However firm has to respect the rate of return defined by the regulator i.e. the validity of the relation

$$RoR \geq \frac{p \times q - w \times L}{K}$$

Let us further in detail explore the relation between the rate of return of the capital expenditures and the profit of the regulated entity. Profit can be analytically expressed as the difference between the proceeds and the costs of the firm in the form

$$\pi(q) = p \times q - w \times L - r \times K \tag{1}$$

Let us deduct the price of the capital r from the both sides of the relation (1). We get the relation

$$RoR - r \geq \frac{p \times q - w \times L - r \times K}{K} \tag{2}$$

From the comparison of the relation (2) and (3) we get the relation

$$(RoR - r) \times K \geq z(q) \tag{3}$$

From the relation (3) we can see that the regulated entity can set its system parameters only in a way that profit it reaches will not exceed the value of the capital evaluated by the difference between the rate of return *RoR* defined by the regulator and the price of the capital *r*.

Regulated entity can set its controlled (endogenous) decision parameters only in a way respecting regulatory condition. It sets the regulated volume of output q_R, the acceptable regulated price p_R and the corresponding consumptions of production factors labor L and capital K in a way to reach maximum profit while respecting regulatory conditions (3) about non-exceeding the reasonable level of profit. The regulated output and regulated price are calculated by solving of the following mathematical programming task

$$\pi(q) = p(f(K,L)) \times f(K,L) - w \times L - r \times K \rightarrow \max$$

by the boundaries (4)

$$p(f(K,L)) \times f(K,L) - w \times L - r \times K - (RoR - r) \times K \leq 0$$

$$K, L \in R_{\geq 0}$$

Solution of this optimization task is optimum consumption level of production factors labor L^* and capital K^*. On their basis with the help of production function the regulated optimum level of output q_R^* is quantified according to the following form

$$q_R^* = f(K^*, L^*)$$

and regulated optimum price p_R^* with the help of price-demand and production function on the basis of the relation

$$p_R^* = p(q_R^*) = p(f(K^*, L^*)$$

While respecting the rate of return on capital defined by the *RoR* parameter, i.e. exogenous parameter set by the regulator.

In the situation when the firm would not be regulated and had an exclusive position on the relevant market it would choose such optimum volume of consumption of variable inputs labor L^* and capital K^*, to ensure maximum profit. On the basis of optimum consumption of variable inputs it would set its optimum supply volume and optimum price of production

$$p^* = p(q^*) = p(f(K^*, L^*)$$

In the case the firm is regulated, it can only choose the combination of input factors in such a way the corresponding production volume and price of production would generate *reasonable profit* i.e. to adhere to this form

$$(RoR - r) \times K \geq p(f(K,L)) \times f(K,L) - w \times L - r \times K$$

$$(RoR - r) \times K \geq \pi(q)$$

Thus the regulated firm can produce in a manner that its *reasonable profit* does not exceeds $RoR - r$ multiple of the variable capital input volume. This condition is called *reasonable profit margin of the regulated firm*. It is obvious that in case when the regulator sets rate of return on capital *RoR* in a way it will exceed its cost r, the firm reaches positive profit for any combination of inputs of production factor K with positive volume. This statement predicates the problems of exclusive position of production factor capital on decision making of the regulated firm.

However this form of price regulation is hiding one serious risk. It often motivates firm to use higher volume of variable input capital than non-regulated firm.

We can show, that firm can produce regulated production volume q^*_R with any combination of variable inputs labor and capital corresponding to isoquant $q^*_R = f(L, K)$. Barthwall [1] show, that if the regulation on the basis of the rate of return stimulates firm towards using large volumes of capital in order to reach maximum profit allowed, the firm is not motivated to use often more effective and from the point of view of production price and production volume of inputs equivalent combinations of inputs localized on isoquant (e.g. supporting employment) in comparison with purposeless investments into the equipment. However there are also other, more sophisticated examples of cost structure of the firm.

4 Conclusion

The aim of this paper was to analyze the effects of network industries price regulation from the point of achieved level of reasonable profit of the regulated entities.

On the basis of the firms behavior in the conditions of the rate of return on capital regulation we show that in this regulatory scheme the firm has the tendency to respond to the tightening of the regulatory conditions by increasing the volume of used capital. Increase of used capital volume is, however, not the goal that the regulatory system is following. Regulation aims to influence others, for the firm and for the economy more important indicators, such as the volume of production, level of sales or the volume of costs.

Therefore other price regulation forms were developed to influence regulated firm reasonable profit on the basis of its total costs. The aim is, with the help of the regulatory mechanisms, to encourage effective development of the regulated entity.

On the basis of the economic development analysis, we can state that the applied regulatory methods have not only the stabilizing effect in the network industries market environment but also the motivational aspect for the effective development of the regulated entities.

References

1. Barthwal, R.R., Industrial Economics. New Delhi: New Age International, 545 s. ISBN 81-224-1278-5 (2010)
2. Blum, U., Müller, S., Weiske, A., Angewandte Industrieökonomik. Theorien - Modelle – Anwendungen. Berlin: Verlag Gabler (2006)
3. Bodea, T., Ferguson, M., Segmentation, Revenue Management, and Pricing Analytics. New York and London: Routledge (2015). Carlton, D.W., Perloff, J. M., Modern Industrial Organization. Boston: Addison Wesley (2005)

4. Fendeková, E., Fendek, M., Mikroekonómia: Oligopoly a regulované monopoly. Bratislava: Wolters Kluwer, p. 230, ISBN 978-80-8168-765-5 (2018)
5. Fendek, M., Fendeková, E., Modely cenovej regulácie sieťových odvetví. *Ekonomický časopis: časopis pre ekonomickú teóriu a hospodársku politiku, spoločensko-ekonomické prognózovanie*. Bratislava: Ekonomický ústav. vol. **48**, no. **10**, pp. 1039 – 1055. (2010)
6. Fendek, M., Fendeková, E., Models of regulation of network industries in Slovakia. In: *International journal of economics and business research. - UK* : Inderscience Enterprises Ltd, vol. **1**, no. **4**, pp. 479-495 (2009)
7. Fendek, M., Fendeková, E., Cost-oriented models of network industries price regulation. In: *Operations research proceedings, selected papers of the Annual international conference of the German operations research society (GOR) University of Augsburg,* September 3-5, 2008. - Berlin: Springer Verlag, pp. 73-78 (2008)
8. Meehen, J. M. et al., Pricing and Profitability Management. Singapore: John Wiley & Sons (2011).
9. Pepall, L., Richards, D., Norman, D., Industrial Organization: Contemporary Theory and Empirical Applications. New York: Wiley-Blackwell (2008)
10. Rode, S., Modern Microeconomics. Ventus Publishing, 231 p. ISBN 978-87-403-0419-0. (2013)
11. Sherman, R., The regulation of monopoly. Cambridge university press. Cambridge (1990)
12. O'Sullivan, A., Sheffrin, S., Perez,P., Microeconomics: Principles, Applications, and Tools (New York: Prentice Hall) (2006)
13. Waldman, D. E., Jensen, E. J., Industrial Organization: Theory and Practice. New York: Addison Wesley (2006)

The process model of subjective quality assessment of videoconferencing in enterprise

Jana Filanová[1,*]

[1]Ekonomická univerzita v Bratislave, FPM, Dolnozemská cesta 1/b, 852 35 Bratislava, SR

Abstract. Videoconferencing is one of the most appropriate ways to transmit information online to participants not only during quarantine. This paper describes a novel process of evaluating the quality of videoconference. Time-consuming subjective measurements were supported by models and programs that simplified the preparation, testing, and processing of results. The process of quality assessment can help IT management to decide on the use of videoconferencing as a means of communication in business. This is especially important in times of pandemics and quarantine.

1 Introduction

Videoconferences represent a form of synchronous communication based on audio and video transmission with the possibility to integrate text and other forms of presentation of information at a distance. The quality of this communication is influenced by the used communication technologies and transmission characteristics of communication networks. Videoconferencing is one of the most appropriate ways of online transmission information to participants. The videoconferences could be recorded and it is possible to view the records even in off-line mode [1].

Videoconferencing allows to conduct meetings with several people at the same time. We can talk and chat with people from different parts of the world at the ease of your residence or office. Business owners can connect with clients and employees and hold discussions. Strategic planning and meetings which require several members to be present can be carried out with ease. We get the option of talking to multiple people at the same time and carry out your business strategies with ease. Thus projects get completed faster and deals get finalized within a short period of time.

As information overload becomes a more and more serious topic in modern society, many efficient tools have been designed to overcome information anxiety. Videos, the fastest-growing information carrier, will account for more than 80% of all Internet traffic by 2020 [2]. Video data has become easily accessible and dominant in the present time [3].

The visual perception of people is a highly complex matter that involves several mechanisms. It is influenced by their expectations and their previous experience. The view of the quality is linked to their mechanisms of imagination. In [4] the authors show that semantics has a significant impact on viewers' sensitivity to the quality of a video sequence

[*] Corresponding author: jana.filanova@euba.sk

for spatially separated parts of the sequence and, more importantly, that this difference in sensitivity can be changed by the presence of an audio signal. This result is important for any testing of subjects' responses to visual material. One example is the subjective assessment of the quality of video in an audio-visual communications system (such as television or videoconferencing) [4].

The results of the research [5] have shown that the presence or absence of audio has a significant impact on the overall subjective perception of the videoconferencing quality. It has also been found that the viewer is more sensitive to the quality of the image in the foreground of the speaking person than to the quality of the image in the background. If there are multiple people in the scene, even not speaking right now, the viewer is likewise more sensitive to the quality of the image of the captured people than to the quality of the image in the background. Video should supplement, not replace, the telephone, for which there is a considerable evidence base from research studies and some guidance [6].

Subjective quality cannot be represented by an exact figure. Due to its inherent subjectivity, it can only be described statistically. Even in psychophysical threshold experiments, where the task of the observer is just to give a yes/no answer, there is a significant variation in contrast sensitivity functions and other critical low-level visual parameters between 50 different video quality observers. When the artefacts become supra-threshold, the observers are bound to apply different weightings to each of them [7].

2 Research methods

The subjective assessment is based on the use of human observers (people) who watch the sequences and score the video quality. It is the most reliable way to determine the video quality and should not be replaced with an objective assessment. The disadvantage of this method is that it is time consuming and human resources are needed [8].

International recommendations for subjective methods of quality testing include specifications on how to implement different types of subjective tests. Some of these test methods are known as "double stimulus" methods where an observer evaluates quality or quality change between two (reference and test) video sequences. There are also "single stimulus" methods where the observer evaluates the quality of just one (test) video sequence [9, 10].

2.1 DSCQS method

The Double Stimulus Continuous Quality Scale (DSCQS) method is suitable for measuring the quality of the system that is related to the reference value as the observer is not familiar with the reference sequence order. DSCQS is quite sensitive to small differences in quality and is thus the preferred method when the quality of the test sequence and reference sequence are similar [9].

2.2 DSIS method

The Double Stimulus Impairment Scale (DSIS) method is suitable for assessing the extent of degradation of the test sequence as compared to the reference one, especially in case of visible/significant degradation. For example, it is used to evaluate the degradation of the sequence during transport. This method is faster than DSCQS since the sequences are displayed only once. Subjects rate the amount of impairment in the test sequence on a discrete five-level scale ranging from “very annoying” to “imperceptible”. The DSIS method is well suited for evaluating clearly visible impairments such as artefacts caused by transmission errors [9].

2.3 ACR method

The Absolute Category Rating (ACR) method is a single stimulus method; viewers only see the video under test, without the reference. They give one rating for its overall quality using a discrete five-level scale from "bad" to "excellent". The fact that the reference is not shown with every test clip makes ACR a very efficient method compared to DSIS or DSCQS, which take almost 2 or 4 times as long, respectively [10].

2.4 MPEG-4 H.264/AVC compression standard

The latest and today most used compression standard designed for a wide range of applications, ranging from mobile video to HDTV, is MPEG Part 10, called also MPEG-4 H.264/AVC. MPEG-4 H.264/AVC defines the Profiles and Levels, too, but its organization is much simpler than in MPEG-4 Part 2. There are only three Profiles currently defined (Baseline, Main, Extended) [8].

3 Process model of subjective quality assessment of video sequences

In this research, we simulated an environment of a real packet network and tested video sequences that would simulate the diverse content of video calls. We artificially degraded the quality of video sequences by packet loss and jitter. The objective of the test was to compare subjective methods, which was used. We also wanted to show that semantics has a significant impact on viewers' sensitivity to the quality of the video sequence [11]. We describe a novel process of evaluating the quality of videoconference. Time-consuming subjective measurements were supported by models and programs that simplified the preparation, testing, and processing of results.

The process of quality assessment can help IT management to decide on the use of videoconferencing as a means of communication in business. This is especially important in times of pandemics and quarantine [6].

Subjective video quality testing is difficult not only because of the time-consuming nature of testing itself but also due to the complexity of the steps that precede the actual testing. Figure 1 describes five steps of the process model we have designed for subjective evaluation of the quality of video sequences. It is based on the process model that we presented in the article [12].

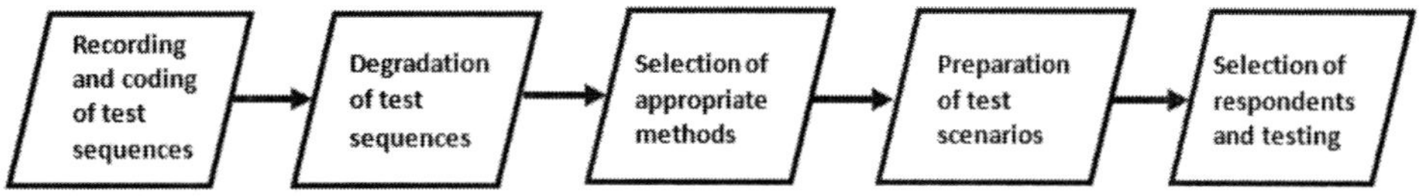

Fig. 1. The process model of video sequences quality evaluation

3.1 Recording and coding of test sequences

Reference video sequences were created based on real video calls. These sequences were recorded using the Logitech C270 web camera with HD resolution of 1270 x 720 pixels, utilizing the Logitech Webcam Software shipped with the web camera. Due to the purpose of the testing, it was important to create diverse demonstrations with a different emphasis on content, the importance of video or audio capture. Two types of reference video sequences are described below.

In the first test sequence (video sequence No. 1), the intention was to create a preview where the emphasis would be on the picture detail. The manager in this video preview informs their employees that if they have any questions, they can contact him at his e-mail address. The person in the preview does not pronounce this e-mail address but writes it on the board. So the only way this e-mail address information gets to the user of the videoconference is that the image quality will be sufficient to recognize it without difficulty.

In the second test video sequences, the aim was to create a demonstration where an emphasis would be placed on the quality of the audio during static image transfer. In the video sequence No. 2, a woman asks the recipient to contact someone by phone. She dictates her name and phone number.

Each video sequence was encoded, because the video and audio formats used, as well as bit rates, do not match those used in videoconferencing. Recording and coding technical parameters of reference video sequences are described in Table 1.

Table 1. Recording and coding technical parameters

parameter	recording	coding
pixel	1270 x 720	1270 x 720
frame rate	15	30
sequence length [sec]	10	10
video format	WMV2	MPEG-4 AVC
audio format	WMA	AAC
bit rate of video [kbit/sec]	3535	1024

3.2 Degradation of test sequences

To introduce degradations into the reference videoconferencing sequences, it was necessary to emulate the transfer environment through which the sequences were transmitted (Figure 2). Network emulation is a process by which we can control and repeatedly simulate network performance. The changes in network parameters such as latency and packet loss are provided by traffic shapers. They must be controlled according to predefined specifications to simulate the required features of the network.

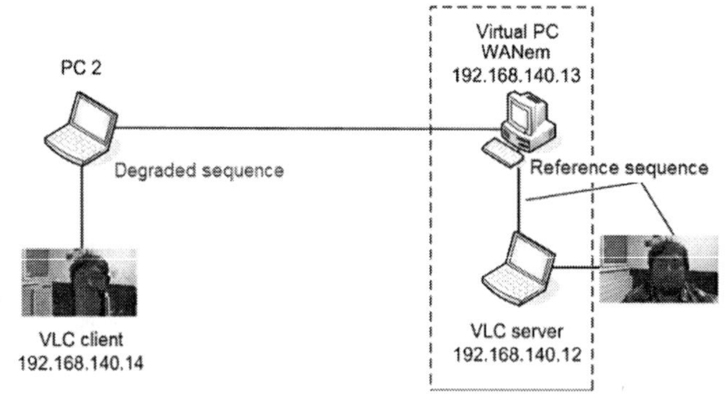

Fig. 2. Network model for creation of degraded sequences

Each of the four reference samples was degraded by packet loss (0.5 %, 1 %, 3 %, 5 %, and 10 %) and jitter (50 ms jitter at 100 ms latency).

3.3 Selection of appropriate methods

Absolute Category Rating (ACR) and Double Stimulus Impairment Scale (DSIS) methods were selected for the subjective evaluation of video samples. The ACR method has the advantage of being fast as the evaluator watches the sample only once and the length of the sample is relatively short (about 10 seconds). The DSIS method was also selected because of its time efficiency and the ability to capture more accurate differences between degraded samples, as we also have a reference sample for this method [9, 10]. The choice of suitable methods was also influenced by the fact that both the ACR's and DSIS's outputs are MOS scores with values ranging from 1 to 5, so the results can easily be compared [13].

3.4 Preparation of test scenarios and selection of respondents

Since the testing was performed within the VLC multimedia player environment, it was necessary to create playlists in which the individual video sequences were arranged appropriately. To prepare the scenarios and the course of the subjective measurements, a program was created in the C# programming language. To play a video sequence the program uses an open-source DmediaPlayer that is a modification of the VLC player [14]. The program consists of two parts: test manager part and tester part (Figure 3). Test manager part is an interface used to create structure of the test. You can choose the type of subjective method, test sequence, reference sequence (if necessary) and enable or disable sound step-by-step.

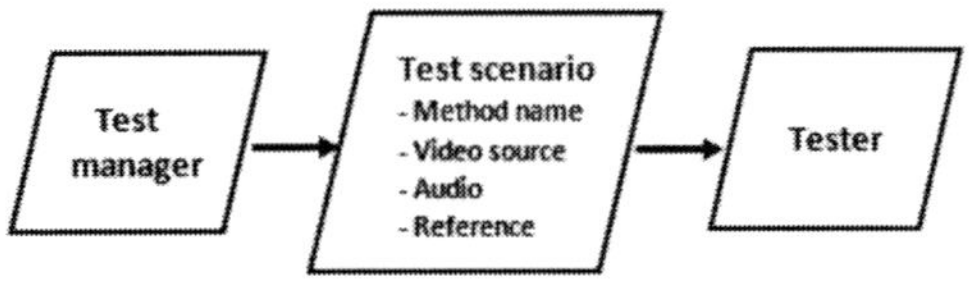

Fig. 3. The procces model of testing and test scenarios preparation program

The ITU-T Recommendations specify that the number of respondents for subjective quality assessment must be greater than 4 and less than 40 [9, 10]. Based on this, we chose 20 respondents (10 women and 10 men), aged 20-51.

The fifth step of the subjective quality evaluation includes testing. The course of testing, evaluation, and comparison of the results are described in the following section.

4 Comparison of subjective evaluation results

Our research aimed to compare subjective methods of real video calls testing and to determine the degree of impact of audio quality on overall video quality with respect to the semantics.

Due to the time-consuming manual processing of results, two programs were created. The first program was written in the C# programming language. The program has two outputs in the form of text files. In the first one, the results are processed according to the evaluation of the individual sequences. In the case of the DSIS method, the format is the reference sequence name and the test sequence name followed by five numbers. In the case of the ACR method, the format is the test sequence name and five numbers. The five numbers correspond to the evaluation scale of the given methods [9, 10]. If it has been chosen to take the statistical processing into account, the output is in the same file. In the second text file, the results are processed according to respondents who evaluated individual sequences. This output is needed for statistical processing of results for the DSIS method.

The second program is used for statistical processing of measured results. It was created in Matlab version R2008b. The average score $\bar{u}_{jkr}$ is calculated for each test sequence

$$\bar{u}_{jkr} = \frac{1}{N}\sum_{i=1}^{N} u_{ijkr} \tag{1}$$

where u_{ijkr} is the respondent score i for test condition j, sequence k and number of repetitions r. N is the total number of respondents. One of the output of the program is a text file with the names of the individual respondents who were excluded on the basis of the calculation and comparison of the standard deviation.

Table 2 lists the summary of video sequences quality measurement results. In the case of subjective evaluation, the video sequences were rated by MOS scores that range from 1 to 5 [9, 10]. A video sequence rated by the score of 4 or higher is considered to be of high quality [13, 15].

Table 1. Results of subjective quality evaluation (d.a. – degraded audio, r.a. – reference audio)

Video sequences	Degradation	ACR d.a.	ACR r.a.	DSIS d.a.	DSIS r.a.
No. 1	Packet loss 0.5%	4.70	4.65	4.80	4.75
	Packet loss 1%	3.55	3.90	4.15	3.90
	Packet loss 3%	2.25	3.15	3.15	3.30
	Packet loss 5%	2.15	2.65	1.95	2.85
	Packet loss 10%	1.00	1.95	1.05	2.35
	Latency 100 ms, Jitter 50 ms	3.65	x	3.75	x
No. 2	Packet loss 0.5%	2.45	3.25	3.25	3.70
	Packet loss 1%	2.10	3.40	2.30	3.80
	Packet loss 3%	1.55	2.65	2.05	3.20
	Packet loss 5%	1.70	2.20	1.35	2.25
	Packet loss 10%	1.10	1.70	1.00	1.85
	Latency 100 ms, Jitter 50 ms	4.15	x	3.85	x

The *ACR d.a.* and *DSIS d.a.* columns show the results for video sequences degraded by packet loss (0.5 %, 1 %, 3 %, 5 %, and 10 %) and jitter (50 ms jitter in 100 ms latency). The *ACR r.a.* and *DSIS r.a.* columns show the results for video sequences degraded by packet loss and in which the degraded audio track was replaced by the audio track from the reference sequence.

Comparing the evaluations for the ACR and DSIS methods, we found that video sequences were rated by a higher score when the DSIS method was used. This difference can be explained by the fact that in the case of the DSIS method the respondent was influenced by the reference sample. Even at 0.5 % and 1 % packet loss degradation, some video sequences with the reference audio track received higher ratings than those with the original disturbed audio track. The results also imply that, in general, the degradation caused by jitter (50 ms jitter in 100 ms latency) does not affect the quality ratings as much as the degradation due to packet loss.

The results of the subjective quality evaluation have shown that under the ideal conditions in the transmission network (without packet loss and latency) the quality of videoconferencing has been rated as "good" (MOS > 4). Therefore, from the perspective of the user, the video frame resolution, audio and video bitrate, and the used codecs provide the user with sufficient quality.

Based on the results of the subjective evaluation of the sequences with the original audio and the sequences in which the degraded audio was replaced by the reference, we see (table 2) that for the packet loss of 3 % and 5 % the sequences with the reference audio are rated much higher (often by more than 1 point on the MOS scale). The difference between individual sequence evaluations is much smaller in samples with the reference audio compared to sequences with the original audio track. In our research, we have confirmed that the quality of audio has a great impact on the overall quality of videoconferencing. In future work, we can investigate whether a similar trend is observed when changing the tasks, that is, if we gradually insert different deteriorated audio tracks into the reference video sequence.

5 Conclusion

Increasing bit rates in today's modern networks allow us to provide ever-new services to support enterprise communications such as video conferencing, on-demand streaming or online streaming [16]. Like voice services, moving image services need to be monitored to see if the service is of adequate quality to the customer. This quality monitoring must necessarily be automated, as it would of course be impractical, costly and error prone to employ people for these activities.

In this paper we describe a novel process of evaluating the quality of videoconference. Time-consuming subjective measurements were supported by models and programs that simplified the preparation, testing, and processing of results. The process of quality assessment can help IT management to decide on the use of videoconferencing as a means of communication in business. This is especially important in times of pandemics and quarantine.

Based on the results of the subjective evaluation of sequences with the original audio track and the sequences in which the degraded audio track was replaced by the audio track from the reference sequence, we have confirmed that the quality of the audio has a significant impact on the overall quality of videoconferencing and the understanding of its content. As a result, if any video information is supported by relevant audio information, we can compensate for the loss of video information by improving audio quality. We can also influence the quality of videoconferencing by ensuring correct pronunciation, intelligibility, and articulation.

The paper was elaborated within VEGA No. 1/0388/20 IT Management in Enterprises in Slovakia: International Standards and Norms Versus Individual Business Processes – proportion 100 %.

References

1. P. Bisták, et al.: *Utilisation of Videoconferencing for Education*. 1 st . ed., Elfa Kosice (2005)
2. J. Fajtl, H. S. Sokeh, V. Argyriou, D. Monekosso, P. Remagnino. *Summarizing videos with attention*. In ACCV Workshops (2018)
3. R. Kapoor, O. Mishra, O., M. Tripathi, Human action recognition using descriptor based on selective finite element analysis, *Journal of Electrical Engineering*, **70**, 443-453 (2019)

4. M. R. Frater, J. F. Arnold, A. Vahedian, Impact of audio on subjective assessment of video quality in videoconferencing applications, *IEEE Transactions on Circuits and Systems for Video Technology*, **11**, 9, 1059-1062 (2001)
5. O. O. Andriichenko, O. I. Denysenko, Subjective evaluation of the clarity of the noisy language in the lecture room, *Electronic and Acoustic Engineering*, **2**, 3, 55-60, (2019)
6. T. Greenhalgh, J. Wherton, S. Shaw, C. Morrison. *Video consultations for covid-19 BMJ 2020*, 368 :m998 (2020)
7. T. T. Hai Uyen, O. Kwon, S. Choi, I. Hussain, Subjective Assessment of 360° Image Projection Formats, in *IEEE Access*, **8**, 33588-33599 (2020)
8. M. Uhrina, J. Hlubik, M. Vaculík, Correlation between Objective and Subjective Methods Used for Video Quality Evaluation, *Advances in Electrical and Electronic Engineering*, **11**, (2013)
9. ITU-R Recommendation ITU-R BT.500-11, *Methodology for the subjective assessment of the quality of television pictures*, (2012)
10. ITU-T Recommendation ITU-T P.910, *Subjective video quality assessment methods for multimedia applications*, (2008)
11. S. Han, Z. Xi, Dynamic scene semantics SLAM based on semantic segmentation, *IEEE Access* (2020)
12. J. Filanová, M. Mardiak, Meranie kvality video signálu, *Elektrorevue*, **15**, pp. 32.1-6, (2010)
13. H. Rizek, K. Brunnström, K. Wang, B. Andrén, M. Johanson, Subjective evaluation of a 3D videoconferencing system. *Proc. SPIE 9011, Stereoscopic Displays and Applications XXV*, (2014)
14. W. Jackson, Digital Video: Streaming Video, MediaPlayer, and MediaController Classes. *Android Apps for Absolute Beginners*, 291-337 (2017)
15. S. Katsigiannis, J. Scovell, N. Ramzan, L. Janowski, P. Corriveau, M. Saad, G. Wallendael, Interpreting MOS scores, when can users see a diference? Understanding user experience diferences for photo quality. *Quality and User Experience*, **3**, (2018)
16. P. Kuppusamy, Modern Network Analytics Architecture Stack to Enterprise Networks, *International Journal for Research in Applied Science and Engineering Technology*, **7**, (2019)

Industry 4.0 brings changes in human resources

Hana Gažová Adamková[*]

University of Economics, Faculty of Business Management, Department of Management, Dolnozemská cesta 1, 852 35 Bratislava, Slovakia

Abstract. Industry era 4.0 is an era of robotization that assumes the unification and automation of processes in the personnel area: data collection, information analysis, evaluation, training, performance enhancement and decision making. The aim of the paper is to point out the upcoming trend in the era of industry 4.0, of which employees are an essential part. It focuses on the perception of implementation of changes from the perspective of employees and the need for systematic work of HR professionals. The paper focuses on identifying the types of changes that are most often implemented in organizations and also the satisfaction of employees with the impact of the introduced changes. We start from the fact that changes contribute to the future of an ever-developing organization. The more people are connected to change, the easier it is to encourage their positive engagement.

1 Introduction

Given the rapidly changing nature of business, it is necessary to accept the fact of the need to adopt and introduce changes in the organization. Change management is becoming an essential part of the human resources as organizations need to keep up with the new, emerging era of industrial digitization. Industry era 4.0 becomes an era of robotization. In the area of human resources, it assumes the unification and automation of the processes of data collection, information analysis, evaluation, education, performance improvement and decision making.

Industry 4.0 creates a number of new opportunities for organizations in the personnel field. It has an overall impact on human resource development, labor productivity growth, opportunities for creating new industries, shifting different tasks from man to machines and robots. These are tasks: routine, repetitive, dangerous or health-threatening. The combination of human potential and computer algorithms opens the way for the emergence of new industries of the future.

The general transformation of the global world creates new opportunities for many countries. For example, the role of some of the constraints on the labor market, such as geographical location, underdevelopment and other similar activities, is declining.[1]

In relation to the current need to apply changes in organizations, the article focuses on the process of implementing changes from the perspective of employees. It monitors how people perceive the change process in an organization, how communication and overall satisfaction with the outcome of changes take place.

2 Changes in industry 4.0 in human resources

Industry 4.0 is characterized by a changing environment. According to Lasi [2], organizational structure is characterized rather by decentralization, empowerment of individuals, fewer formal rules, horizontal communication and teamwork, which is a more appropriate approach for an environment full of innovation and change. Transformation and acceptance of industry 4.0 in personnel is essential. It is not possible to rely solely on the traditional approach. This means a new focus on the processes and employees themselves. It is a challenge for the entire company management. In terms of human resources needed targeted preparation. The following facts should be taken into account.[3]

1. Human resources specialists must be able to identify the skills needed for the sector in which their organization is active. At the same time, they must define the new skills that the 4.0 era can bring.
2. HR needs to be more digitally focused. HR specialists need to keep up with technology trends to stay ahead of the competition.
3. Use of interconnection and integration. Active reform of the way of communication with internal and external clients. Focus on new, modern and efficient communication tools.
4. Retraining. In addition to managing their own changes, the Human Resources Department will also have to cope with the challenge of retraining other areas within the organization.

[*] hana.gazova@euba.sk

Personnel professionals face many challenges in the 4.0 industry. These are in particular: large amounts of data handling, rapid onset of change, new business and personnel models and "intelligent services" that enable digital tools. The basic determinant of managing organizational change is the experience of managers who are beneficiaries of change. The amount of change these managers make is related to the size of the organizational changes they make. Experience from the success of previous changes in each organization has been positively influenced and supported by this relationship. [4]

According to experts, the development of industry 4.0 will bring many problems and risks. For example, adapting employees to new conditions and high levels of technological change, or risking the extinction of existing occupations, including the most qualified [5]. Sommer L. [6] points out that strong pressure on innovation could be particularly detrimental to small and medium-sized enterprises, and the smaller the companies, the greater the risk. Changes may result in an increased share of part-time work with flexible working hours in several companies at the same time. Labor-law relations will be more based on outsourcing the workforce to selected positions and creating new internal structures. [7]

In November 2017, McKinsey's report emphasizes that nearly 800 million jobs are at risk as a result of the introduction of new technologies. This huge change will push for automating work and eliminating low value-added jobs. However, the implementation of these technologies creates a lack of staff skills. The human resources department must decide whether it is beneficial for companies to acquire the necessary skills in the labor market or to train old employees from within.[8]

Some limitations and concerns about the increasing tendency to use robotics and artificial intelligence are causing fear in humans. Fear of more massive job cuts, job loss and security. On the other hand, it should be noted that in countries such as Japan or Germany, ie the countries that are robotics leaders, they have had the lowest unemployment rate in the last twenty-five years.[9] Slovakia also appears to be a country with an increasing tendency to use digitization and robotization in this area, while the level of unemployment has been decreasing in recent years.[10]

Manpower Group has focused on answering questions about the impact of technology on business in the coming years.[11] It focused on 18,000 employees from 43 countries. The results in the report entitled "The Skills Generation" showed a new direction that pointed out that the workforce must come with the right skills and adaptability. 60% of respondents pointed out the importance of investing in internal training to maintain skills 'fresh', in order not to create a large gap between existing and required skills. Another interesting fact has shown that 45% of current positions can be threatened in the next two years by the automation of some processes and operations.

For organizations, Industry 4.0 is a challenge in many areas. One of the first processes that need to be prepared in an organization is to inform employees about the need for new changes. The changes that will be necessary for the organization to be competitive in the market. Changes contribute to the future of an ever-growing organization. The more people are connected to change, the easier it is to encourage their positive engagement. Therefore, it is important that organizations involve their employees as soon as possible in the process of planned change. People's fear of something unknown creates a problem visualizing the outcome of the change. It is therefore useful for the management of the organization to be able to illustrate what the change will mean.

2.1 Model of implementing changes in the organization

According to surveys, 40% of Slovak companies are already taking steps towards digital transformation. On the one hand, the results of the survey speak of increasing the innovation potential of companies and a positive shift in the availability of information. On the other hand, they also talk about insufficient conditions for the education and development of workers, or the need to change the role of management in the transformation process.[12] The success and failure of the planned change is largely the responsibility of the leaders and managers whose task is to create a suitable environment for achieving the goal.[13] Through the use of various methods of developing capabilities and skills of employees are able to adapt well to new challenges in the period after the introduction of changes.[14]

An important role in promoting the successful implementation of the 4.0 sector is the active cooperation of managers with HR experts. The link with the different sectors of the organization is essential in ensuring good implementation of the new organizational models. HR is considered essential in promoting the improvement of operational processes in companies. This is also confirmed by a study[15], that highlights the importance of human resource flexibility and its impact on organizational innovation and culture.

Although Slovak companies are interested in information and exchange of experience, most of them lack the confidence and determination to gradually set up their activities to new procedures. It is therefore important to talk about what approach to take and what the first steps should be.

All changes that are made to an organization affect employees as carriers and implementers of change. The task of management in period 4.0 is to gradually and purposefully prepare employees for the necessary changes. The introduction of new procedures into the working stereotype cannot be accidental or rapid. It has to respect many factors, such as the effect of the original procedure, staff attitudes, the timing of change implementation or the expected difficulty of implementation.

The need for change in organization is broadly triggered by two basic groups of incentives that are based on internal and external stimuli. Internal incentives changes are within the organization. These include, for example, a decrease in the quality and quantity of performance, changes in key employees, changes in opinions and attitudes in the group, dissatisfaction with evaluation and evaluation, changes in the division of powers, changes in organizational culture. The basic internal stimulus of organizational change is the gradual development of organizational culture. External incentives contain external environmental factors that affect the organization. These include the competitive environment, business partners, customers, the economic, political, cultural, social and technological macro environment. Changes made on the basis of external incentives are essential in order to maintain the organization's position in a competitive market. Both internal and external incentives are closely interconnected in view of organizational changes related to Industry 4.0.

Creating the need for change in an organization is based on the requirement of innovation and maintaining its position in a competitive environment. This implies an active approach and optimal timing of change, taking into account the internal situation as well as external environment pressures. The success of introducing new procedures is primarily based on a change of attitude, which is a complex process. The goal of change cannot be just the current short-term change. For a long-term change of attitudes, it is necessary to use a targeted model of introducing changes. A model that takes into account the position of the organization as well as the needs of employees. Lewin's three-phase sequential change management model is based on the fact that the forces that lead to change continue to affect the organization as they respond to dynamics and change from a societal perspective.

This model examines two areas of human attitudes that occur during the implementation of any change in the organization. The first area is the drivers that support change. The second area is resistance forces that reduce the likelihood of acceptance of change. The forces of resistance consist mainly of emotional and psychological barriers. The essence of the Lewin model is based on the assumption that in managing change processes, an organization must "relax" its current state to a neutral position. Old behavior can be subdued and new behavior can be successfully accepted. The change should be made by using the driving forces and minimizing the resistance forces.[16]

The principles of the Lewin model are applicable at various levels of change in the enterprise, including the introduction of changes related to industry 4.0. The implementation has three important phases. The first phase is perceived as the most difficult. This is referred to as defrosting. It is based on the fact that employees have developed addictive practices that are increasingly fixed over time. Decision making and performance becomes routine and simple. This creates a high degree of resistance to everything new, regardless of whether they are positive or negative changes. Accompanying phenomena are tension, anxiety, fear, fear, passivity or open aggression. The aim is to gradually unblock old habits and place less emphasis on remuneration. Use of open communication, based on clarification and consultation. The aim of communication is to focus on: the reasons why change is necessary, what happens if there is no change, what is expected if there is a change. Employees have the opportunity to realize the need for change.

The second phase is already the actual implementation of change. It is the core of this model. Employees are gradually using new procedures to verify the correctness and effectiveness of the change. Managers are in the role of facilitators. They seek to facilitate the implementation of new approaches. They use positive elements of motivation and encouragement. The aim is to: motivate at a weaker performance, clarify ambiguities, encourage, adapt to change. It is a transition to a new status, which is realized through new forms of behavior, values and attitudes, the adoption of new organizational structures and working practices.

At a time when an employee achieves some performance through new procedures, the third stage begins. Freezing phase. The aim is to stabilize and strengthen the required workflows through motivation tools. Again, it is essential to create a sense of security and satisfaction, backed by favorable results. Freezing means stabilizing the organization in a new equilibrium state. This phase is very demanding in terms of time and implementation. This model can be applied not only from the internal perspective of change management for employees, it is also effective from the general perspective of the organization.

3 Perception of changes in the organization by employees

In terms of personality structure, more than half of employees prefer a stable working environment. They are well versed in it, they have enough information, enough time and space to communicate about the possibilities of solving work tasks. They prefer situations in which they are not exposed to risk or where they may influence the risk.[17] For this reason, changes in the organization and the working group in the first phase often bring about a defensive attitude of employees. Refusal may have a hidden or manifested course. A targeted approach to analyzing the status quo and implementing change is important.

Changes are generally perceived as disrupting established organizational structures. They change or take away the powers and responsibilities of employees or delegate them to others. Changes bring new responsibilities to employees that they did not have before. Organizational relations of superiority and subordination are changing. Such situations give rise to concern and disorientation. Gradually, interest in work is lost. This rejection of the new state is associated with anger and causes confusion. The most common reasons for resistance to change can be divided into three basic groups:

1. Lack of communication: lack of information on the need for change, inadequate imposition of change, predominance of informal communication channels, misunderstanding of the purpose of change.
2. Employee individuality: low self-confidence to handle new practices, fear of incapacity and failure, demolition of established practices, insecurity, preference for work and relationship stability, better management of routine and established practices, low tolerance of change.

3. Social status in the group: weakening power, fear of losing status, job insecurity, fear of the unknown, self-interest in maintaining the existing situation, lack of identification with change.

The general preparedness of employees for change is related to their past experience. Whether the results of previous changes were positive or negative. Changes in people's consciousness go through four phases. The first is the phase of satisfaction, when there is no need for change and satisfaction with the current situation prevails. At this stage it is necessary to present an interesting vision of the future state of the organization. This is followed by a phase of anxiety and fear that is related to incoming changes and employees are beginning to feel a negative attitude. They deny that they should change. At this stage it is necessary to provide enough information to support change, to listen and to obtain the necessary information. If employees accept the need for change, the next phase begins - confusion. The problem is how to make the change. The plan and methodology of the procedure need to be clearly presented. Identifying with the process and making changes is the next phase - acceptance. Employees can be presented with other possibilities of positive changes. If employees are already able to accept change, they are able to implement change and adapt to it. The degree of organization's readiness for change reflects employees' willingness to accept change and work towards its implementation.[18] Employees in this respect can be divided into three groups.

1. open to change, with a high degree of flexibility. They come with proposals and positively accept new approaches. They perceive change as a challenge and a means against stereotype.
2. reasonably accepting the changes. The resistance phase does not last long, they are open to open communication. If appropriate arguments are chosen, they adapt quickly to change.
3. favor stability, the ability to accept change is very low. The stereotype at work brings them certainty and therefore they are not willing to give up routine. The resistance phase lasts quite a long time.

Personal perception of employees' changes and their willingness to adapt will determine whether the new system will succeed or fail [19]. In order for employees to be ready to accept change, it is important to point out its importance in closing the gap between the current and the desired situation. This requires considerable effort to manage employees' attitudes to the expected change and to foster the sense that the organization supports innovation. [20]

4 How employees perceive change in organizations

The aim of the survey was to point out the perception of changes in the organization by employees. Employees are in the front line affected by change and transformation, whether in a positive or negative direction. Their insight into the change implementation process in an organization can help to better set up the entire change management process in the context of industry 4.0. Respondents (72 respondents) were employees of medium-sized companies, with no specific orientation. At least 2 years in the company. Data were collected through a structured questionnaire in both written and electronic form. The questions in the questionnaire were aimed at identifying the types of changes that are most common in organizations. What communication channel do organizations choose when communicating change to employees and employee satisfaction with the impact of implemented changes in the organization.

One of the monitored areas was their perception of the current situation in the company from the point of view of the changes made in recent years. The results show that employees perceive changes in the organization differently. In each group there are employees who are oriented towards change but also against change.

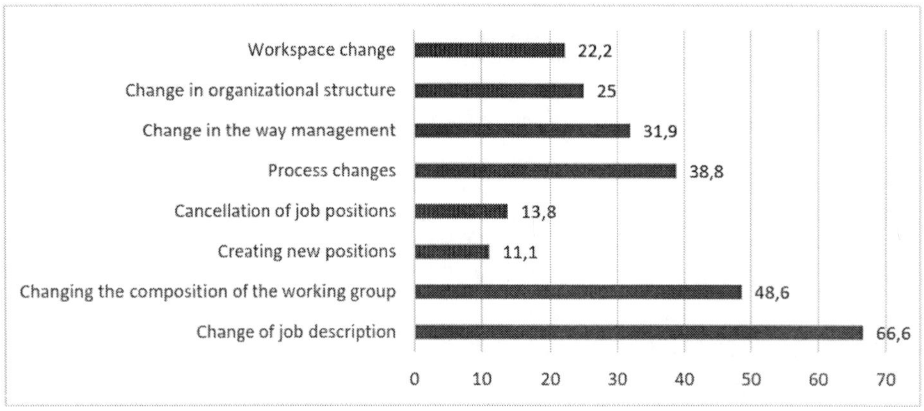

Fig. 1. Types of changes in the organization in %

Figure 1 shows that the changes most frequently registered by employees concerned changes in workload (66.6%) and changes in the composition of the working group (48.6%). It is possible that these changes influence each other. According to respondents, in their organizations, new jobs do not arise as a result of the implemented changes. Nor do they see a direct link to the abolition of certain jobs. Rather, they attribute this to changes in the organizational structure and changes

in the job description. According to respondents, the results of the changes also relate to changes in working processes (38.8%) and consequently changes in the way of management approach (31.9%), both from the direct management and the overall mood of the company. In connection with the advent of industry 4.0, as shown by the trends, it is possible to assume a higher incidence of changes related to the emergence of new and at the same time possible termination of some jobs.
Experience with implementing changes shows the importance of choosing how to communicate with employees about change planning, goals and vision of outcome. The survey shows (Figure 2) that organizations prefer their own, direct communication with employees over external collaboration to inform about changes. Although external collaboration is common in establishing change implementation processes, the survey shows that employees are generally communicated with employees through internal channels - the management of the organization (56.9%) and the direct supervisor, the manager (51.3%). Although the implementation of changes is directly related to the personnel area and therefore to the HR department, the communication line on the preparation of changes is used only in 26.3%. Information through HR departments is provided mainly in connection with adjustments to work and subsequent interpersonal changes in groups. Also interesting is the combination used to communicate about changes. The management of the organization focuses mainly on the communication of the vision and the expected outcome of the changes together with the time data. Subsequently, direct supervisors or manager specialists communicate about the changes. Their role is to communicate more closely and to explain the procedure and concrete expectations. Scheduling changes is set to shorter intervals. The manager cooperates more closely with the HR department, precisely because of dissatisfaction, fear or even resistance to change. HR specialists communicate change from the perspective of the positive impact of changes on employees.

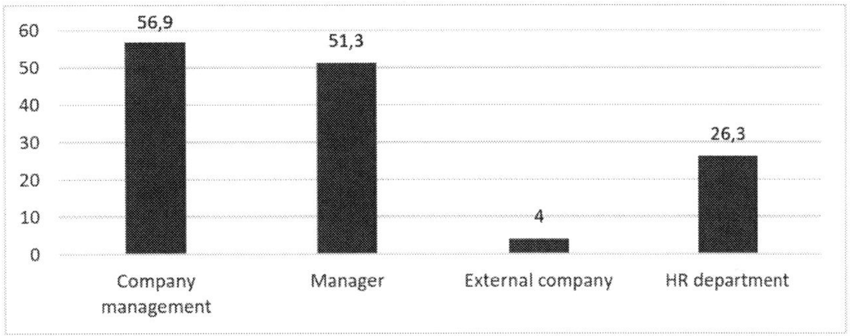

Fig. 2. Communicating change in organization to employees in %

Regarding the perception of the introduction of changes, it was mentioned that employees are initially cautious about changing settings. Feelings of fear of the unknown arise, especially from the impact of changes on their position in the organization. A feeling of work, social but also psychological insecurity arises. Respondents to the questions, which raised the most concerns during the planning and implementation of the changes, responded differently. Most were the fear of losing a job, which was associated with many other aspects. Based on the responses, the order of concern most perceived by employees when planning and implementing changes can be listed (Figure 3). The notion of change and its impact has a highly subjective charge. Prolonged exposure to a negative notion of change can create resistance and a negative attitude to the new situation, without adequate causes.
Concerns related to implementing changes in the organization can be divided into three categories. Working: loss of job, lack of new procedures, loss of responsibilities and competences, difficulty finding a suitable job, new superior. Social: loss of teammates, non-inclusion, lack of information. Personality: sense of uselessness, time pressure. It is clear from the results that employees are most worried that they will not be able to manage the professional side of organizational changes.

Order	Fears of change
1.	Loss of job
2.	Failure to master new procedures
3.	Loss of responsibilities and competences
4.	Problem finding a suitable job
5.	Loss of colleagues from the team
6.	New superior
7.	Non-inclusion in the team
8.	Feeling useless

9.	Time pressure
10.	Lack of information

Fig. 3. Fears of change

When introducing changes related to 4.0, an increased level of concern can be expected. The reason is that in this context, employees see a greater risk of the entire reorganization of work processes and the risk of job loss. Appropriate communication for the initial elimination of resistance and fear plays an important role here.

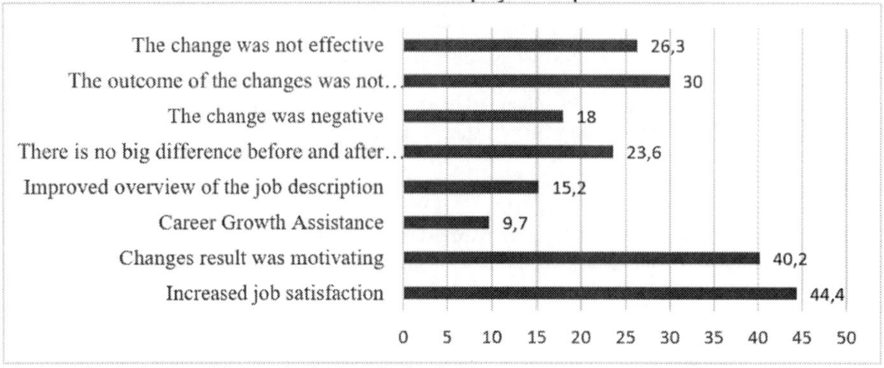

Fig. 4. Satisfaction with the impact of changes in the organization in %

Despite a cautious approach to change and overall concerns about the change process, respondents perceived the outcome quite positively. Figure 4 shows that, overall, 27.4% of employees evaluate the impact of changes in the organization effective. They perceive the changes as motivating (40.2%) and increasing their job satisfaction (44.4%). According to employees, the results of the changes had a positive impact on better transparency in their job description and overall work structure (15.2%). The positive impact of the changes also resulted in an improvement in career growth (9.7%). On the contrary, 24.8% of employees evaluated the negative impact of changes on work. They perceive changes in the organization as ineffective (26.3%) and consider the outcome of changes to be inadequate compared to expectations (30%). The change process had a negative effect on employees (18%). Neutral attitude with no significant differences before and after the changes was evaluated by 23.6%. The changes were assessed as unnecessary, insufficiently communicated and the overall effect was not significant.

Conclusion

Industry 4.0 has already started. More and more industries and work processes are being digitized and automated over time. It is important to adapt to this trend in order to maintain productivity growth and to enable organizations to respond quickly to environmental demands. Many organizations are already implementing new implementation strategies and systematic preparation for implementing the 4.0 related changes. Strengthening the capacity to accept change is essential throughout the organization. In terms of personnel, it affects all processes, from recruitment and daily operations to communication and remuneration systems [21]. In this context, many questions arise as to how these changes will affect our work and life. The answers can be different, but the position and tasks of the personnel area in organizations will certainly be different from the present.

The aim of the paper was to point out the upcoming trend in the era of industry 4.0, of which employees are an essential part. Targeted focus on employees is essential for the smooth implementation of emerging changes. It is necessary to know how employees perceive this situation, what they expect and what they are worried about. Targeted communication with employees should be a planned part of strategy 4.0.

The article is a partial output of the research project of VEGA MŠ SR no. 1/0412/19 "Human resource management systems in the era of the Fourth Industrial Revolution." In the range of 100%.

References

1. E. Brynjolfsson, A. Mcafee, A., *The second machine age.* New York: W. W. Norton & Company, Inc.(2014)
2. H. Lasi, P. Fettke, H. G. Kemper, T. Feld, M. Hoffmann, Industry 4.0. *Business & Information Systems Engineering* **6**,239-242 (2014)
3. A. Talerngsri, The impact of Industry 4.0 on HR professionals. https://www.seasiacenter.com/insights/the-impact-of-industry-4-0-on-hr-professionals/ (2018)
4. K. Ozawa, Managers' experiences as recipients: impact on organizational change. *Journal of Organizational Change Management* **33**,143-156 (2019)

5. Z.K. Chulanova, Professional standards as a factor of adaptation of human resources to the industry 4.0: approaches to development and implementation. *Journal of Human Resource management* 22, 12-20 (2019)
6. L. Sommer, Industrial revolution - industry 4.0: Are German manufacturing SMEs the first victims of this revolution? *Journal of Industrial Engineering and Management* **8**, 5 (2015)
7. G. Wisskirchen, The World of Tomorrow -Technological Changes: Impact on Labor Market, On Jobs and On the Organization of Work. https://www.americanbar.org/content/dam/aba/events/labor_law/2017/03/err/papers/tomorrow_kuhn.pdf (2017)
8. D. Trotta, P. Garengo, Industry 4.0 key research topics: A bibliometric review, *2018 7th Industrial Technology and Management (ICITM)*, 113-117 (2018)
9. W. Dauth, S. Findeisen, J. Südekum, N. Woessner, The rise of robots in the German labour market. VOX CEPR Policy Portal, https://voxeu.org/article/rise-robots-german-labour-market (2017)
10. K. Boroš, Nemci vyvracajú známy blud. Roboty prácu neberú. Finweb, https://finweb.hnonline.sk/komentare-a-analyzy/1675335-nemci-vyvracaju-znamy-blud-roboty-pracu-neberu (2018)
11. R. M.Varghese, P. Shaw, Industry 4.0 and future of HR, *Journal of Management* **5**, 96-113 (2018)
12. PC Revue, Podiel firiem aplikujúcich Industry 4.0, https://www.pcrevue.sk/a/Podiel-firiem-aplikujucich-Industry-4-0-dosiahol-40---Zmeny-zacinaju-prepisovat-nastavenia--firemnej-kultury-podnikov (2019)
13. B. Burnes, Introduction: Why Does Change Fail, and What Can We Do About It?, *Journal of Change Management* **11**, 445-450 (2011)
14. H. Bakari, A. I. Hunjra, G: S. K. Niazi, How Does Authentic Leadership Influence Planned Organizational Change? The Role of Employees' Perceptions: Integration of Theory of Planned Behavior and Lewin's Three Step Model, ?, *Journal of Change Management* **17**, 155-187 (2017)
15. B. R. Do, P. W. Yeh, J. Madsen, Exploring the relationship among human resource flexibility, organizational innovation and adaptability culture, *Chinese Management Studies* **10**, 657–674 (2016)
16. H. Chowdhury, Applicability of Lewin's Change Management Theory in Australian Local Government. *International Journal of Business and Management* 10, 53 (2015)
17. J. Šícha, Změny pricházejí. Pripravte se. *Moderni řízení* **9**, 60 (2008)
18. R. Gigliotti, J. Vardaman, D. R. Marshall, K. Gonzalez, The role of perceived organizational support in individual change readiness. *Journal of Change Management* **19**, 86-100 (2019)
19. A. R. Abdulaali, A Alnoor, B. Eneizan, A multi-level study of influence knowledge management small and medium enterprises. *Polish Journal of Management Studies* 19, 21– 31 (2019)
20. K. Y. Kwahk, J. N. Lee, The role of readiness for change in ERP implementation: Theoretical bases and empirical validation. *Information & Management* **45**, 474-481 (2008)
21. B. J. Weiner, A theory of organizational readiness for change. *Implementation Science* **4**, 97 (2009)

Application of optimal control theory on optimal advertising expenditure in monopoly

Nora Grisáková[1,*], *Peter* Štetka[2]

[1]Department of Business Economics, Faculty of Business Management, University of Economics in Bratislava, Dolnozemská st. 1/b, 825 35 Bratislava, Slovakia (EU)
[2] Department of Business Economics, Faculty of Business Management, University of Economics in Bratislava, Dolnozemská st. 1/b, 825 35 Bratislava, Slovakia (EU)

Abstract. Presented paper is being focused on Optimal control theory, Variation Calculus and its economic application. Aim of this research paper is to shortly describe Optimal control and Variation Calculus and to present how can we deal with these type of issues. The last part of this paper is presenting possible economic application of Optimal control, based on the maximization of profit in monopoly while introducing new product on the market. Our control variable is the advertising rate, which affects the profit of monopoly through advertising expenditures and as a state variable was the market share defined.

1 INTRODUCTION

Originally, the Optimal Control theory was a discipline focused on technical problems. But because of similarities between machine control and economic system management, optimal control theory was adopted by economists to solve economic problems. This adoption was very fast, because basic mathematic framework of optimal control theory was made without specific form of controlled system. Laščiak [1] says that optimal control is one that meets optimality criterion. Kamien and Schwartz [2] differ between management and decision making in optimal control. According to these authors, decision is a single action and management is an activity that directs system development over time. Optimal control is very close to variation calculus (we present the basics of this issue in next part of this paper). Both optimal problem theory and variation calculus are analytic methods of problems solving and planning in continuous time. Solution to the continuous time dynamic problem is a continuous function (or set of functions) indicating the optimal path to be followed by the variables through time (or space).

The indicators we want to put into the prescribed state, or we try to control, so these behave according selected criteria, are called system output and are denoted by $x(t)$. In the literature, so-called state variable is often mentioned instead of output. Control (or control variable) $u(t)$ is a variable that is being used to regulate the flow of control according to our criteria. Uncontrollable factors acting on the system are called failures and denoted as $p(t)$.

* Corresponding author: nora.grisakova@euba.sk

We can define control system through mathematical means without specific interpretations. Lincoln [3] defines control system as following:

Let P be a set of failure functions p and let U be a set of control functions u. System S is an operator defined on the Cartesian space $P \times U$*, assigning each pair* $(p,u)\epsilon P \times U$ *as function x,* $x = S(p,u)$*. In this case, the time, t, is represented in the operator S only by p and u, so it is an autonomous system. If the time in the operator S acts as a separate variable, we are talking about a non-autonomous system.*

If we use notation as a system, control variables or state variables we may imagine a machine. A lot of authors ([4], [5], [6], [7]) are using economic applications of optimal control theory. E.g. we may say that the national economy is a controlled system in the meaning of optimal control. State of national economy is indicated with a lot of economic indicators x. we try to set these indicators x to desired values using economic management tools u. Similarly, a company can be controlled system that seeks to achieve its objectives (x) with various tools (u).

2 OPTIMAL CONTROL PROBLEM AND VARIATION CALCULUS

In optimal control problems, variables are divided into two kinds, *state variables* $x(t)$ and *control variables* $u(t)$ in defined time period $t_0 \leq t \leq t_1$. State and control variables maximize objective function. The movement of state variable is governed by first order differential equations[†].

$$max \int_{t_0}^{t_1} f\big(t, x(t), u(t)\big)dt \tag{1}$$

$$x'(t) = g\big(t, x(t), u(t)\big) \tag{2}$$

$$x(t_0) = x_0 \; fixed, x(t_1) \; free \tag{3}$$

Where f and g are assumed to be continuously differentiable functions of three independent arguments, none of which is a derivative. The control variable $u(t)$ is a piecewise continuous function of time. The state variable $x(t)$ changes over time according to the differential equation (2) governing its movement. The control u influences the objective function (1), both directly through its own value and indirectly through its impact on the evolution of the state variable x (that also enter into the objective (1)). The highest derivative appearing in the problem formulation is the first derivative, and it appears only as the left side of the *state equation* (2). [2]

Optimal control problem could have several state and control variables. Each from state variables moves according differential equation (in continuous time period) or difference equation (in discrete time period). According to Dorfmann [8], differential equation condition presents the change rate in state variable in every time as a function of its present state and decision made $u(t)$. Number of control variables could be higher or lower than number of state variables.

If we define optimal control problem as in (1) – (3), then we need to find necessary conditions that lead to an optimal solution[‡] $u^*(t)$, $x^*(t)$, $(t_0 \leq t \leq t_1)$. These necessary conditions finding is similar to solving nonlinear programming problem with Lagrange multiplier. Condition (2) of problem (1) – (3) must hold at every point in time t. We have function of Lagrangian multiplier $\lambda(t)$ from optimality conditions. Function $\lambda(t)$ is continuous by a differentiable function of time t, $t_0 \leq t \leq t_1$.

[†] We denote derivative and partial derivatives by primes and subscripts. A prime label the derivative of the function with respect to its argument. Thus $f'(x) = df/dx, x'(t) = dx/dt, f''(x) = d^2f/dx^2,\; x''(g) = d^2x/dt^2$

[‡] Star shows optimal solution of the problem in this paper.

Sethi and Thompson say that the best way to find necessary conditions for function extreme is by applying Hamiltonian function. Work with this function is similar as with Lagrangian function for nonlinear programming problems (see example [9]). Before we start solving optimal control problem, we present fundamentals from variation calculus first. The reason is that optimal control theory comes from variation calculus and we need to apply it to some solutions [2].

2.1 Fundamentals of Variation Calculus

We can track the origin of variation calculus to Bernoulli [10] who was positing the brachistochrone problem and composed its solution. Later, other specific problems were solved and general theory was developed by Euler and Lagrange [11]. We can find the application of variation calculus in theoretical physics, in Hamilton´s principle (for example described in [12], [13], [14], [15] and others) or the Principle of Least Action (described for example in [16], [17], [18] and others). The first economic applications were made in 1920s by Roos [19], Evans [20], Hotteling [21], Ramsey [22] and others.

This part of paper is intended only to present clarification of some parts of variation calculus which use the optimal control theory. For this reason we don't change the designation used in literature. Some designation in variation calculus could seem as in optimal control, but they could have different meaning.

Laščiak [1] says that variation calculus is a mathematical discipline which examines methods of finding functional extremes'. He formulated an easy variation calculus problem, where unknown function is time dependent.

Let M be a set whose elements are functions. If we assign a single real number I(x) to each functionx ∈ M, we say that the set M is defined as a functional. The number I(x) is called the value of the functional I on the function x and the set M is a set of permissible functions. [1]

Let be a function $F(t,x,x')$ that has continuous derivatives of at least the second order for $x\epsilon\langle a,b\rangle$. From all curves K with the description $x = x(t)$, which have a continuous first derivative and pass through points $[a,x_1]$ and $[b,x_2]$, there must be found a curve for which the functional $\int_a^b F(t,x,x')dt$ has a minimum (maximum) value. We can formulate a necessary condition that this function must meet [2], [23]:

If the functional $\int_a^b F(t,x,x')dt$ takes extreme on the curve K with the description $x = x(t)$, then $x = x(t)$ satisfies the **Euler equation** $\frac{\partial F}{\partial x} - \frac{d}{dt}\frac{\partial F}{\partial x'} = 0$. According to Lagrange condition we find maximum or minimum of functional: If function $x = x(t)$ holds minimum (maximum) of functional $\int_a^b F(t,x,x')dt$, then $\frac{\partial^2 F}{\partial x'\partial x'} \geq 0 \left(\frac{\partial^2 F}{\partial x'\partial x'} \leq 0\right)$.

Let's have a task of variation calculus

$$max_x \int_{t_0}^{t_1} F(t,x,x')dx \qquad x(t_0) = x_0;\ x(t_1) = x_1 \tag{4}$$

For this task the Euler equation is

$$F_x\big(t,x^*(t),x'^*(t)\big) = \frac{dF_x\big(t,x^*(t),x'^*(t)\big)}{dt}, t_0 \leq t \leq t_1 \tag{5}$$

or

$$F_x = F_{x't} + F_{x'x}x' + F_{x'x'}x''\ t_0 \leq t \leq t_1 \tag{6}$$

In equation (6) we've used a substitution $x' = x'(t)$, $x'' = x''(t)$. Euler equation is then second order differential equation and we can rewrite it through two first order differential equations

$$p(t) = F_{x'}(t,x,x') \tag{7}$$

Now we can define new function – Hamiltonian function for the task (2):

$$H(t,x,p) = -F(t,x,x') + px \tag{8}$$

Where p is described from physics as general moment. In economic application p could be interpreted as shadow price. We can write

$$-\frac{\partial H}{\partial x} = p' \quad \frac{\partial H}{\partial p} = x' \tag{9}$$

which represents canonical form of Euler equation [2].

2.2 Necessary and sufficient conditions for the existence of extreme in optimal control

In this part of paper we derive the necessary and sufficient condition for optimal control problem. The necessary condition for optimal solution $u^*(t)$, $x^*(t)$, $t_0 \leq t \leq t_1$of problem (4) comes from Hamiltonian

$$H(t,x,u,\lambda) = f(t,x,u) + \lambda g(t,x,u) \tag{10}$$

and its derivative

$$\frac{\partial H}{\partial u} = 0 \rightarrow \frac{\partial H}{\partial u} = f_u + \lambda g_u = 0 \tag{11}$$

$$-\frac{\partial H}{\partial x} = \lambda \rightarrow -\frac{\partial H}{\partial x} = -(f_x + \lambda g_x) = \lambda' \tag{12}$$

$$\frac{\partial H}{\partial \lambda} = x' \rightarrow \frac{\partial H}{\partial \lambda} = g = x' \tag{13}$$

From initial conditions we know that $x(t_0) = x_0$ and we assume $\lambda(t_1) = 0^{\S}$. In every t is u a stationary point of Hamiltonian for given value of x *and* λ.Conditions (11) – (13) are necessary first order conditions for existing extrema for minimizing and maximizing problem.

Hamiltonian defined for optimal control problem (10) is similar to Hamiltonian defined for variation calculus (8). Also necessary conditions $-\partial H/\partial x = \lambda'$, $\partial H/\partial \lambda = x'$are similar to canonic form of Euler equation (9). Second order necessary condition for maximizing problem says, that $H_{uu}(t,x^*,u^*,\lambda) \leq 0$ and for minimizing problem $H_{uu}(t,x^*,u^*,\lambda) \geq 0$.

If necessary conditions of optimality are also sufficient conditions, in nonlinear programing these conditions hold together when maximizing (minimizing) objective function is concave (convex) on closed convex set. We can have same results for variation calculus and dynamic optimization [2]. λ could be interpreted as marginal value of continuous state variable $x(t)$ [24]. Extension of this process on more than one state variable x and more control variable u (number of state and control variable do not need be the same) could be find e.g. in [2] or [3].

3 Optimal advertising expenditure for monopoly – Optimal control application

Assume a company, which came to market with a new consumer durable product. This company is a monopoly till other companies do not come with similar product; therefore the company would like to maximize its profit from new product. We have to set a marketing plan during given time horizon in a way that total discounted profit from selling new product less advertising expenditures is maximized on diffusion constraint. We can formulate this model as:

§ Multiplier $\lambda(t_1)$ represents so-called a transversal condition if there is no rescue value introduced in the model below. Detail description of these conditions shows for example Kamien and Schwartz [2].

$$\max J = \int_{t_0}^{t_1} e^{-rt}\left[\left(p - c_0\left(\frac{x_0}{x}\right)^z\right)x' - (a + bu + cu^2)\right]dt$$
$$x' = g(x,u) = [\alpha + \beta u + (\gamma + \delta u)x](1 - x) \qquad (22)$$
$$x(t_0) = x_0$$

Where

J total discounted monopoly profit from selling new product less advertising expenditures

u advertising rate (control variable),

x reached market share (state variable),

p (p = 100) constant price,

$(a + bu + cu^2) = A$ quadratic rate of market expenditures, *a, b, c* nonnegative constants, $a = 0$, $b = 0.07$, $c = 0.04$,

r (r = 0.005) discount rate,

z (z = 0.1) rate of decline in production costs,

c_0 (c_0 = 70) constant production cost,

x_0 (x_0 = 0.4) Initial market share of monopoly,

α (α = 0.005) innovation coefficient,

β (β = 0.006) coefficient related to the effectiveness of advertising towards innovators,

γ (γ = 0.003) imitation coefficient,

δ (δ = 0.007) coefficient related to the effectiveness of advertising towards imitators.

(time interval of new product $0 \le t \le 31$ – 31 week of new product advertising campaign)

We are using brackets to show each coefficient initial values for finding a solution. In this case we didn't use Hamiltonian in continuous time, but we've used algorithm presented by Teng and Thompson [25]. This algorithm changes continuous time in discrete one and for monopoly case we've made an assumption that there is only one company in the market. All the other assumptions and steps of algorithm remained same.

Presented model includes 13 parameters, one state (market share) and one control variable (advertising rate). From advertising rate we've calculate according function $(a + bu + cu^2)$ rate of advertising expenditures. As convergence criterion we've chosen to compare the absolute difference in objective function value in algorithm iterations. This absolute difference has to converge to zero.

In analyse we at first find the solutio n for initial values of coefficients and then we *ceteris paribus* change one of them.

4 Discussion and Conclusion

Optimal control theory and variation calculus were originally applied in purely technical disciplines. But these dynamics were very early used also in other disciplines, such as medicine, chemistry, economy etc. After summarizing basic notations and calculus from this field of mathematics, we've presented an economic application of optimal control theory. We've demonstrated it using a case of advertising expenditures optimization in monopoly. We've made 26 simulations, using selected initial values, with *ceteris paribus* rule being applied when changing our model parameters. Based on these simulations we've come to following findings:

- In all the cases with changed parameters, the same direction of a change in advertising rate (u) and advertising expenditure rate (A) was found.
- Increase in a "rate of decline in production cost" (z) generates increase in all observed model variables.
- Increase in coefficient values in advertising rate function (A) decreases company profit and monopoly market share. It also increases advertising rate (u).

- Change in discount rate (r) causes opposite change in discount factor (e^{-r}) and all other parameters of the model except one: the monopoly market share (x), which changes in the same way as discount rate.
- Change in product price (p) generates change of all the observed parameters in the same way.
- Change in a constant production cost (c_0) generates opposite change in all other observed parameters.
- Initial market share (x_0) change generates opposite change in all other parameters except one: total market share of monopoly at the end of observed period of time. This has the same direction as the initial market share.
- Innovation (α) and imitation (γ) coefficient change generates the same direction change in monopoly market share (x) and monopoly profit (J). It also generates opposite changes in advertising rate and advertising expenditure rate (A).

Changes in advertising effectiveness coefficient towards innovators (γ) and imitators (δ) generate the same change in all model variables.

This paper is an outcome of research project VEGA 1/0646/20 „*Diffusion and consequences of green innovations in imperfect competition markets*“. Share of the project is 100%.

References

[1] Laščiak, A., Optimálne programovanie, Bratislava: Alfa, 1990, p. 597.

[2] Kamien, I. M., Schwartz, N. L., Dynamic Optimization: The Calculus of Variations and Optimal Control in Economics and Management, 2 ed., Dover: Dover Publication, 2012, p. 400.

[3] Lincoln, B., Dynamic Programming and Time-Varying Delay Systems, Lund: Lund Institute of Technology, 2003, p. 150.

[4] Goga, M., Ekonomická dynamika, Bratislava: Iura Edition, 2011.

[5] Yu, H., Bai, S., Chen, D., Dong, Ch., Feng, X., "Application of optimal control to the dynamic advertising decisions for supply chain with multiple delays," *Systems Science & Control Engineering,* vol. 8, no. 1, pp. 141-152, 2020.

[6] Berberich, J., Köhler, J., Allgöwer, F., Müller, M., "Dissipativity properties in constrained optimal control: A computational approach," *Automatica,* no. 114, 2020.

[7] Zhang, J., Garcia Fracaro, S., Chmielewski, D.J., "Integrated Process Design and Control for Smart Grid Coordinated IGCC Power Plants Using Economic Linear Optimal Control," *Processes,* pp. 288-, 2020.

[8] Dorfmann, R., "An Economic Interpretation of Optimal Control Theory," *The American Economic Review,* no. 59, pp. 817-831, 1969.

[9] Fendek, M., Nelineárne optimalizačné modely a metódy, Bratislava: Ekonóm, 1998.

[10] Bernouli, J., "Problema novum ad cujus solutionem Mathematici invitantur," *Acta Eruditorum,* no. 18, p. from 269, June 1696.

[11] Thiele, R., "Euler and the Calculus of Variations," in *Leonhard Euler: Life, Work and Legacy*, Amsterdam, Elsevier, 2007, p. 249.

[12] Salmon, R., "Practical use of Hamilton's principle," *Journal of Fluid Mechanics,* no. 132, pp. 431-444, 1983.

[13] Pingyuan, W., Ying, Ch., Jinqiao, D., "Hamiltonian systems with Lévy noise: Symplecticity, Hamilton's principle and averaging principle," *Physica D: Nonlinear Phenomena,* no. 398, pp. 69-83, November 2019.

[14] Steinboeck, A., Saxinger, M., Kugi, A., "Hamilton's Principle for Material and Nonmaterial Control Volumes Using Lagrangian and Eulerian Description of Motion," *Applied Mechanics Reviews,* vol. 71, no. 1, pp. 1-14, January 2019.

[15] Jinkyu, J., Hyeonseok, L., Jinwon, S., "Extended framework of Hamilton's principle applied to Duffing oscillation," *Applied Mathematics and Computation,* no. 367, 15 February 2020.

[16] Feynman, R., Feynman's Thesis — A New Approach to Quantum Theory, London: World Science Publishing, 2005, p. 144.

[17] Siburg, K. F., The Principle of Least Action in Geometry and Dynamics, New York: Springer Verlag Berlin Heidelberg, 2004, p. 130.

[18] Litvinov, V., "Topological Magnetoelectric Effect," in *Magnetism in Topological Insulators*, Cham, Springer, 2020, pp. 79-89.

[19] Roos, C. F., "A Dynamic Theory of Economics," *Journal of Political Economy,* no. 35, pp. 632-656, 1927.

[20] Evans, G. C., "The Dynamics of Monopoly," *American Mathematics Monthly,* no. 31, pp. 77-83, February 1924.

[21] Hotelling, H., "The Economics of Exhaustible Resources," *Journal of Political Economy,* pp. 137-175, April 1931.

[22] Ramsey, F., "Mathematical Theory of Savings," *Economic Journal,* vol. 38, no. 152, pp. 543-559, 1928.

[23] Tintner, G., "Monopoly Over Time," *Econometrica,* vol. 5, no. 2, pp. 160-170, 1937.

[24] Sengupta, J. K., Fox, K. A., Optimization Techniques in Quantitative Economic Models, Amsterdam: North - Holland Publishing, 1975.

[25] Teng, J. T., Thompson, G. L., "Oligopoly Models for Optimal Advertising when Production Cost Obey a Learning Curve," *Management Science,* no. 29, pp. 1087 - 1101, 1983.

Green innovations in logistics supporting the development of innovative green technologies, processes, products and services in companies in Slovakia

Klaudia Gubová[1,*]

[1] Department of Production management and Logistics, Faculty of Business Management, University of Economics in Bratislava, Dolnozemská 1/b, 852 35 Bratislava, Slovak republic

Abstract. Significant changes on global markets influence behaviour of businesses as well as position of logistics. As logistics represents one of the parts in business process, it is important to monitor its direction and development. Modern management in a business, application of new information, technologies and systems, stricter legislation requirements, they all move business logistics forward. Logistics systems are being constantly modernised. Modernisation greatly considers automation, robotics, development of artificial intelligence and digitalisation. Rapid development of technologies and systems causes that future form of logistics, together with technologies, will represent important source of competitive advantage on the market. The main objective of the scientific paper is to identify, analyze, evaluate and generalize the use of green logistics in Slovakia in order to achieve the competitiveness of the company on the market. Green logistics will be one of the most dominant trends in business logistic in forthcoming period.

1 Introduction

In today's dynamically developing world there are many economic, social, technical and political changes. Their nature and extent is determined by factors influenced by economic development and economic growth. As a result of globalization, there is a sharp increase in competition. Globalization of markets and business brings the same prerequisites for access to information, resources and technology. We see that the "world" is getting smaller and distances are minimized by information and communication technologies. [1].

Recently, there have been significant changes and advances in logistics. Its competences within the company have gradually expanded and have taken on more and more activities, thus beginning to fulfill an essential role in the management of the company. Globalization, economic changes, turbulent development of information and communication technologies,

* Corresponding author: Klaudia Gubová, Assoc.Prof. Ing. PhD., University of Economics in Bratislava, Department of Production management and Logistics, Dolnozemská cesta 1/b, 852 35 Bratislava, klaudia.porubanova@euba.sk

automation of production, development of industrial revolutions, especially the Fourth Industrial Revolution contributed to the new form of logistics [2].

The onset of digitalization in all business areas started to show major changes. New global markets, business models, advances in information and communication technologies and innovation are arising in each business area. Currently, Industry 4.0 dominates the world. It brings not only revolutionary changes in production and logistics, but fundamentally changes the society itself and the economy of the country. Industry 4.0 is based on the idea of linking digitization and automation using the latest technology [3]. Fourth industrial revolution and changes in production systems will bring changes in transportation logistics. New European legislation changes, industrial market is becoming more dynamic, innovations in logistics are its natural part. Part of ensuring effective transportation within the process of decision making, it is not only transportation distance that has an essential role, but it is also available transportation means. There are still new and new logistics centres being built by traffic junctions. This trend supports not only positive development in automobile segment, but also in dynamic sector of e-commerce. Emphasis on application of logistics innovations is not concentrated exclusively on production businesses, but it regards the entire chain – from suppliers through producer up to customer. Many businesses concentrate not only on optimisation of production processes, but also on implementation and development of innovations in logistics. They have understood that optimisation is possible also in logistics. Innovation is a necessity for all logistics processes not only via atomisation of transport; however atomisation needs to be transferred also to information processes and warehouses. The reason for automation is simple. The application of new technologies is pushing manufacturing to a new revolution. The essence of Industry 4.0 is connecting devices to communicate with each other. Industry 4.0 is mainly used in industry, but has a great impact on other areas of the economy as well as on society as a whole [4]. The outcomes of the so-called Industry 4.0, or smart manufacturing, are the object of high expectations. Today, many logistics enterprises take the attention to the importance of big data and claim that Data Analytics will be one of the key skills of the twenty-first century.

1.1 Logistics innovation

Currently, innovations represent factor influencing not only competitive ability of business, but also population quality of life, environmental behaviour of citizens and businesses, social sphere, informational literacy of business. Innovations are considered to be result of utilisation of different stimulus, tools, techniques, which in practical application represent, apart from satisfaction of basic needs, something more for consumer „over added value".

When we think about future, we believe in future of progress. Technology offers businesses new and better way to perform better and more effectively. After investigating different spectres for implementing innovation trends in logistics we might establish that in narrower sense technology development is one of the directions for expected progress in innovations sphere and globalisation is the second most considered factor. Business is able to utilise both ways of progress with an aim to achieve advancement in business at once, separately and all at once.

Nowadays, we can see huge differences between developed and developing countries in quality of life as well in technological progress. We can establish, that developed countries have discovered, developed and achieved everything the world needs, while developing countries are far behind. Here, globalisation did not assist to balanced development of different parts of the world. Globalisation profiles areas narrowly and has not moved its boarders to developing countries. Innovation in one sphere of business initiates chain of activities and changes in other areas with either positive or negative impact. Currently, logistics businesses note innovation development and gradually implement innovations into their logistics processes. Development of technologies and technics, implementation of new logistics innovations move and broaden boarders of globalisation, which are development of Internet and other communication, information and transformation technologies. There is polarisation between those able and willing to adapt to constant changes and those who are not. Also, there is growing difference between those with an access to new technologies, information and those who do not have this access. We can confirm that future development, progress will be in technologies. Globalisation is aimed to achieve sustainability; however, it would not be sustainable without technology development.

Entrance of internet and dynamic technics expansion supported growth of space for service innovations. Nowadays, the biggest competitive strength is hidden in information. Information moves the world of innovations. Innovation is an enormous power, an important factor for survival, growth and business prosperity. Innovation is now recognized as the key to bridging the gap between technology and business strategy.

The innovation concept can create confusion with other similar definitions, such as improvement, invention and discovery, because, according to Schumpeter (op. cit.) innovation is the launching of a new product or improvement of what already exists, or the introduction of a new method of production, i.e. innovation in the process, the opening up of a new market, a new source of raw materials or semi-manufactured goods, and a new form of industrial organization.[5] Fast changes in technologies and technics cause that companies do not have time to devote time to research and development of new products and services. Tangible assets are very quickly morally depreciated, which is a result of fast technological development. There is a rapid obsolesce of products and patterns. Innovation viability is conditioned by factors such as: dynamic technical progress, ever changing competitive environment and trends in behaviour of consumers. Majority of businesses lead their internal politics following „never try anything new as first, only things which are proved and certified.“ [6] The most suitable imitation sphere is technical sector; however there are many opportunities for imitating in all sectors.

The logistics process for the management of all physical flows both in and between companies. It is a transversal process that concerns all the services of a company and which includes in the reflection the partners. Innovation has been argued to be a strong driver of customer satisfaction in the logistics industry. [7] Industrial production and logistics fulfil customer demands. Since those change over time, production paradigms also change. With the current turn towards more individual products, new ways of production and logistics are needed in order to avoid an increase of costs and competitive disadvantages on global market. [8] At the moment, innovative management technologies are emerging, business models are being revised, latent management reserves are not used, the best world experience is being studied and the possibility of using it in Russian conditions is defined taking into account adjustments to existing realities. [9] Although manufacturing and logistics have gained maturity over the last decades, saturated markets and new customer demands put pressure on logistics systems, turning them from complicated to complex ones.[10] Important drivers are shortening product lifecycles, globalisation of markets, demographic change and customer demands to offer individualised products, also accompanied by sustainability aspects. [11] These drivers form a dynamic and challenging environment for enterprises, which, in manufacturing, can be addressed with Industry 4.0. For companies, indicators are indispensable to manage and control the plethora of innovative ideas and concepts that are submitted to them. The defined selection criteria are equally important for an efficient resource allocation and performance evaluation in each phase of the innovation process. [12]

To optimize the provision of logistics services, it is necessary to assess the quality of transport services using a system of indicators ranked according to their importance to passengers and minimize negative differences between expected customer and actual values of road transport quality indicators. This requires constant monitoring of environmental factors, competitors and own resources, quality control of logistics services, and passenger satisfaction with completed orders. [13] The management of the transport company should set a specific set of quality control goals for the staff, including the precise implementation of all aspects of customer orders, in compliance with transportation time, continuous monitoring of meeting passengers' requirements for the logistics service system, optimization of logistics costs to maintain the required level of transport services, integration of available information technology resources and the practical implementation of advanced technology. [14]

1.2 Green logistics innovations

Currently, development of innovations in logistics and ecologisation are connected with the income of new forms and attitudes of businesses towards decreasing environmental impact. Reverse logistics is an important area in achieving environmental goals in society and especially in the company. In general, the definition of reverse logistics is determined by its division into reverse logistics at macro or microeconomic level, their mutual coordination and integration [15]. Reverse logistics, from point of development and implementation of innovations into logistics businesses, creates one of the basic elements for circular economy. First impulse is „eco thinking" Life Cycle Thinking (LCT) which means considering environmental, social, economic aspects and their impact on product during their entire life cycle (from obtaining raw materials, through processing materials, production, distribution, utilisation, repairing and maintenance and disposal and recycling) and value chain from cradle to the grave. Main objectives of LCT thinking of life cycle are decreasing resource consumption and product emissions on environment and improving their social and socio-economic performance during their life cycle. Stated goals are achievable for logistics businesses in implementation of environmental politics and implementation of circular economy into managing relationship management with business subjects.

Green Logistics represents trend in logistics which deals with production planning, material management and goods distribution in sustainable ecological rate with regards of factors influencing environment and society. Green logistics concentrates on ecological, economic and social aspects.[16] Increasing environmental concern from customers, buyers, communities, and government regulations have forced companies to implement Green Supply Chain Management (GSCM) and green innovation. GSCM and green innovation have strategic interconnection in developing new green product.[17] Also "the construction sector constantly offers new products, more effective technologies and novel solutions aimed at improvement of the quality of human habitats and wider distribution of technologies. Currently, effective technologies that require less time and costs for production, installation and use are gaining greater significance. Among them are construction materials and technologies with increasingly popular sustainability features." [18] Green supply chain management (GSCM) is a strategy for sustainable development in today's competitive markets, which aims to simultaneously achieve financial benefits and reduce environmental risks.[19]

2 Methods

The main objective of the scientific paper is to identify, analyze, evaluate and generalize the use of green logistics in Slovakia in order to achieve the competitiveness of the company on the market.

Green logistics will be one of the most dominant trends in business logistic in forthcoming period. The reason behind it is mostly ecological thinking of businesses and formation of business programs for environmental protection as well as maintaining competitive ability for the future. It was confirmed also by tested hypothesis which states that on level of importance $\alpha = 0{,}05$ there is statistically important dependence between enforcement of environmental politics and improvement of Green logistics.

For a long time sustainability has been one of the most important questions of these days. The biggest contribution towards sustainability in the world is Green logistics. Green logistics contributes towards environmental protection with an aim to support ecological means of production and transportation. We believe that businesses promoting environmental politics try to improve green logistics in their business.

In the research, there were 135 businesses addressed, feedback was received from 95 businesses that filled out the questionnaire. We can appreciate high return 70,4% and willingness of businesses to fill out the questionnaire.

3 Results and Discussions

Innovations developing in logistics on the side of demand and offer, such as decreasing labour and growing importance of logistics in electronic stores, leading to wide acceptance of robotics and automation in warehouses. Even though logistics affectivity is rising, there are potential social consequences as a result of threat of losing millions of working places in logistics. Another innovation supporting development of logistic innovative trends is Blockchain. It is permanent digital record of transactions saved in decentralised computer network. It has advantages in many parts of this industry, such as cost savings (paperless transactions); data verification; „Intelligent contracts"; responsibility and consonance.

Many new technologies, daily entering markets in different areas of goods transportation, forwarding and storage increase development potential of other referring innovations in logistics as well as other market sectors. Devices, autonomously detecting when they need replacement parts, production systems, which perform their own quality control during operations and hence decrease control costs, robots, which autonomously recognize and move components. These scenarios are gradually becoming reality in industrial production. They are based on future technologies such as artificial intelligence (AI) and computing technics. These offer many opportunities for industry, because they open new business models and potential for productivity increasing. Hence, they are essential in order to secure competitiveness for industrial businesses in future world.

Implementation and application of selected logistics innovations leads towards reduction of costs, time, waste, decreases risks, simplifies decision making processes and leads towards development of new products with ecological aspect.

Green logistics requires particular attention. Changes in people's behaviour, technological advancement and industrial revolution significantly impacted environment. Current condition in form of climate changes, global warming, pollution of water, soil, air and future forecasts indicate we need to think ecologically – "green". Hence it is necessary for people to manage environment wisely and on business level prevent negative influences impacting environment to a great degree. From the research we learnt that businesses accept Green logistics very little, its assessment in Likert scale was usually in smaller numbers. Therefore, we would like to warn businesses about this situation. Businesses must realise their relationship to environment and act more considerable and develop more activities in sphere of Green logistics on business level.

Green logistics was dealt with also within questionnaire. Green logistics is very much applied only by 5,33% of Slovak enterprises. Based on research results 19,33 % of enterprises have neutral opinion towards application of Green logistics. Enterprises improve Green logistics with share of 9 % and neutral opinion was expressed by 18% of participants.

After summarizing these findings, enterprises should focus on activities in logistics leading towards more considerate behaviour to environment on business level. It is crucial for businesses to consider environment when starting enterprises, in individual parts of business logistics, certain internal logistics processes as packaging and packages. Last but not least, it is important to increase awareness of employees about Green logistics in form of lectures, employee activities and visual management in the business. Enterprises could organise employee Eco activities such as tree planting on premises, improved waste labelling in enterprises and others. It is not possible to create a singular model of how Green logistics should function in the enterprises, because it depends on its size and focus of activities. There is difference between Green logistics in enterprise offering services in transportation and Green logistics of a production business.

Out of mentioned business examples it is appropriate to focus on utilisation and operation of ecological transportation means. Within a enterprise it could be transportation means in form of electro mobiles. Outside of the enterprises it would be vehicles which release less $CO2$ into atmosphere. It would be suitable to discuss topic of electro mobiles in Slovakia more and to enforce them on business level, because in 2015 there were only 49 electro mobiles registered in Slovakia, however, number of all registered personal vehicles registered in 2015 reached almost 80 thousand. Also we would suggest to utilise also railway transport not only automobiles for transportation of goods and products. Enterprises could realise packaging and storage via plastics packaging which can be used until they are not physically deteriorating for internal logistics processes. Enforcement of Green logistics is possible also by disburdening businesses of paper administration. Its removal was mentioned back in 80th years of last century within business relationship B2B.

4 Conclusion

Changes in people's behaviour, technological advancement and industrial revolution significantly impacted environment. Current condition in form of climate changes, global warming, pollution of water, soil, air and future forecasts indicate we need to think ecologically – "green". If you want businesses to survive in a competitive environment, they must improve their business processes, to prosper, to make changes and be open to new opportunities [20]. Hence it is necessary for people to manage environment wisely and on business level prevent negative influences impacting environment to a great degree. From the research we learnt that enterprises accept Green Logistics very little, its assessment in Likert scale was usually in smaller numbers. Therefore, we would like to warn enterprises about this situation. Businesses must realise their relationship to environment and act more considerable and develop more activities in sphere of Green Logistics on business level.

Green logistics will be one of the most dominant trends in business logistic in forthcoming period. The reason behind it is mostly ecological thinking of businesses and formation of business programs for environmental protection as well as maintaining competitive ability for the future. Extending product life is an important part of a circular economy because it brings economic value and material savings. The product design approach to the system should take into account both forward and reverse logistics, as well as a new value that is based on maximizing customer service through multiple lives.

Industry 4.0 provides a technological drive for innovation, while a circular economy is seen as a driving force for the introduction of an industrial framework in 2030. It can not be a circular economy without the 4th Industrial Revolution - it can not be socially useful and sustainable 4th industrial revolution without improving the circular economy.

Acknowledgments

The article is a partial output of research project VEGA No. 1/0375/20 „ New dimension in the development of production management and logistics under the influence of Industry 4.0 in enterprises in Slovakia".

References

1. P. Richnák, Globalization and its impact on the present concepts in company management. *Globalization And Its Socio-Economic Consequences. Part 2: Proceedings: 15Th International Scientific Conference: 7Th-8Th October 2015, Rajecke Teplice, Slovak Republic*, 608-616.
2. P. Richnák, Nové smery rozvoja v logistike. *Journal Of Innovations And Applied Statistics: Vedecký Internetový Časopis*, **9**, 2, 95-103.
3. P. Richnák, The Current Trends in the Industry - Industry 4.0 in Slovak Enterprises. *Dokbat 2019: International Bata Conference*, 916-925.
4. P, Richnák, J. Sármány, Vývoj a postavenie Industry 4.0 vo svete a na Slovensku. *Manažment Podnikania A Vecí Verejných: Vedecko-Odborný Časopis*, **14**, 2, 3-9.
5. P. Hamilton, K. A. Getulio, T. Takeshy, Innovation and technology processes in micro and small business. *Cogent Business & Management*, **6** (2019)
6. O. Shenkar, *Imitátori*. Bratislava : EASTONE BOOKS, 2011. 1p. ISBN 978-80-8109-164-3.
7. S. Bellingkrodt, C. M Wallenburg, The role of external relationships for LSP innovativeness: A contingency approach. *Journal of Business Logistics*, **34**, 3 (2013)
8. E. Hofmann, M. Rüsch, Industry 4.0 and the Current Status as Well as Future Prospects on Logistics. *Computers in Industry,* **89** (2017)
9. U. Stephan, M. Patterson, C. Kelly, J. Mair, Organizations driving positive social change processes. *Journal of Management*, **42**, 5 (2016)
10. T. Bauernhansl, Die Vierte Industrielle Revolution – Der Weg in ein Wertschaffendes Produktionsparadigma. Industrie 4.0 in Produktion, Automatisierung und Logistik. Springer Vieweg, Wiesbaden. 2014, pp. 5–35. ISBN 978-3-658-04682-8.
11. M. Jubiz-Diaz, A. Santander-Mercado, J. E. Candelo-Becerra, A Multi-item Multi-packaging Model to Minimise Cost of Lost Units, Unpacking Cost and CO2 Emissions. *International Journal of Production Research*, (2019)
12. V. Dewangan, M. Godse, Towards a holistic enterprise innovation performance measurement system. *Technovation* **34**, 9 (2014)
13. C. E. Helfat, J. A. Martin, Dynamic managerial technologies for review, *Journal of Management* **41**, 5 (2015)
14. A. K. Vadera, P. Mishra, M. G. Pratt, Constructive deviance in organizations: integrating and moving forward. *Journal of Management* **39**, 5 (2013)
15. A. Dupaľ, P. Richnák, Vybrané trendy a koncepcie v procesnej orientácii podnikovej logistiky. *Ekonomika A Manažment: Vedecký Časopis Fakulty Podnikového Manažmentu Ekonomickej Univerzity V Bratislave*, **14**, 3, 8-19.
16. J. P. Rodrigue, C. Comtois, B. Slack, *The Geography of Transport Systems.* 4th Edition. New York: Routledge, 2017. p. 306. ISBN 978-0415483247.
17. N. A. Seman, N. Zakuan, A. Jusoh, M. S. Arif, M. Z. SAMAN, The relationship of green supply chain management and green innovation concept. *Procedia-Social and Behavioral Sciences.* 57 (2012)

18. S. Kildiene, E. K. Zavadskas, J. Tamošaitiene, Complex assessment model for advanced technology deployment. *Journal of Civil Engineering and Management* **20,** 2 (2014)
19. M. Hajikhani et al. Considering on green supply chain management drivers, as a strategic organizational development approach, Malaysian perspective. *Australian Journal of Basic and Applied Sciences.* **6,** 8 (2012)
20. P. Richnák, Usage of Logistics Technologies in Slovak Enterprises. *Logi – Scientific Journal On Transport And Logistics*, **10**, 2, 94-104.

Approaches to ITSM level measurement and evaluation

Anna Hamranová[1], *Mojmír* Kokles[1], *Tatiana* Hrivíková[2]

[1]University of Economics in Bratislava, Faculty of Business Management, Department of Information Management, Dolnozemska 1, 852 35 Bratislava, Slovakia
[2]University of Economics in Bratislava, Faculty of Applied Languages, Department of Intercultural Communication, Dolnozemska 1, 852 35 Bratislava, Slovakia

Abstract. The paper examines ITSM (IT Service Management) which is based on the internationally accepted ITIL standard (IT Infrastructure Library). Within the domain of ITSM special attention is given to constant improvements in IS/IT services. The aim of the paper is to identify and characterise ITSM metrics and metrics models among which ITSM Maturity Models play an important role. Standard methods of scientific work such as analysis, synthesis, comparison and selection are used for the paper. The main findings are based on the analysis of literature proving the importance of effective IS/IT service management in supporting the business goals of enterprises and organisations. Several authors criticised the extent and detailed character of ITIL.They suggested solving the problems of SME sector by simplifying the processes of ITSM implementation. This is closely related to the proposed system of ITSM metrics and metrics models. The paper provides a detailed elaboration of complex metrics models together with a derived, reduced model with 8 main metrics and other often used tools for IS/IT services measurement, namely ITSM Maturity Models. They are applicable for both a diagnosis of the level of IS/IT services in the companies and eventual correction of plans and for benchmarking the enterprises within their specific branch.

1 Introduction

Effective use of information technology systems and information technology (IS/IT) requires high-quality and systematic management. The area covering those activities is known as IT Service Management (ITSM). According to Winkler and Wulf: "Information technology service management (ITSM) has become the prevalent management approach to the provision of IT services worldwide" [1]. The management of IT services requires, besides the standard maintenance of technology and applications, a definition and description of the services the organisation offers to its consumers of the services, and discovery of their hidden value and reasons why they use the provided services. Subsequently, these services need constant administration and improvement at all levels of management [2-6].The concept of IT Service is based on the ITIL framework (IT Infrastructure Library) which describes the best practices and methods to achieve effective IS/IT services management [7-9].

ITSM represents a management method for information and communication technologies, their operation and development using the principles of management based on services from the perspective of both customers and providers of IS/IT services [10]. Development of IS/IT services, their implementation and support through the whole IS/IT services lifecycle are components of ITSM [11].

ITSM should be considered from 4 perspectives: organizations & people, information & technology, partners & suppliers, and value streams and processes [12]. Value streams and processes are the most important dimensions for the measurement and evaluation of ITSM. They contain all the organisational and process components of IT services, notably, definitions of terms, specifying activities, roles and responsibilities, the definition of input and output activities and processes, the definition of channels of communication, roles, metrics, reporting and documentation of the whole system.

According to Hunnebeck et al. [13], the main areas of ITIL for ITSM are:

- service strategy – representing the core of the services' lifecycle, it includes definitions, recipients, ways to measure the efficiency of the services, service portfolio management
- service design – contains operational processes such as management of services catalogue, capacity, availability, continuity, information security, service design package, …
- service transition – its function is to supply the proposed services to the premises, including management of change, service asset & configuration, release and deployment, transition planning and support, service validation and testing, change evaluation, and knowledge management
- service operation – all elements concerning the operation of IS/IT services: management of incidents, event, access, problem, request fulfilment, and service desk
- continual service improvement – includes service improvement, measurement and reporting.

To achieve the aim of ITSM, namely, supply and effective use of high-quality IS/IT services supporting entrepreneurial goals, it is important to measure and evaluate the level of IS/IT services. Several metrics models [14-15] and methods of measurement and evaluation [16 - 20] have been developed for that specific reason. ITSM Maturity Model [2], [5], [10], [21-22] belongs among the most often used ones.

2 Research Framework and Methods

The research framework consists of the formulation of the main aim of the paper, partial aims and the applied methods. The main aim of the paper is to examine the approach to measurement and evaluation of ITSM based on published scientific works at home and abroad, accentuating the synchronisation of the IS/IT aims with the business goals of the enterprise or organisation.

To support the main aim, the following partial aims have been set:

- analysis of relevant scientific literature,
- identification and characterisation of ITSM metrics and ITSM metrics models,
- identification and characterisation ITSM Maturity models,
- formulation of conclusions.

Standard methods of scientific work such as analysis, synthesis, comparison and selection have been used to elaborate the paper.

3 Results and Discussion

We provide the results in the following order: analysis of scientific literature, ITSM metrics and ITSM maturity models.

3.1 Analysis of literature

The importance of ITMS is accentuated by a high number of papers in renown journals. As the ITSM topic is quite extensive, the authors tend to focus either on solving a specific problem within the ITSM processes (Table 1) or on a chosen area, processes and specific research methods (Table 2).

Table 1. Survey of analysed literary resources focused on ITSM in general.

Author (s)	Area of research	Focus, method
Küller et al., 2012	ITSM for SME	All processes, SME, Maturity model
Machado et al., 2012	Towards a maturity model for ITSM for SME	All processes, SME, Maturity model
Iden & Eikebrokk, 2013	Implementing ITSM	All processes, literature review
Steinberg, 2013	ITSM metrics model	All processes, metrics model
Cots et al., 2016	Benefits of ISO 20000 ITSM	All processes, ISO 20000
El Yamami et al., 2017	ITIL Processes Performance Evaluation in SME	All processes, multicriteria decision making, focus on SME
Chunpir, & Ismailzadeh, 2019	Comparison of ITSM Practices in Public Administration and the Private Sector	All processes, analysis, comparison
El Yamami et al., 2019	Introducing ITIL Framework in SME	All processes, ITSM design for SME
Rocha et al., 2019	Theory of agency and outsourcing of ITSM	All processes, agency theory, outsourcing
Schmidt et al., 2019	ITSM Frameworks	All processes, simplification of ITSM for SME
Varga et al., 2019	Increasing Information Systems Availability Through ITSM	All processes, analysis, SME
Winkler & Wulf, 2019	Effectiveness of ITSM Capability	All processes, effectivity

Table 2. Survey of analysed literary resources focused on chosen ITSM areas and processes.

Author (s)	Area of research	Focus, method
Wan et al, 2011	Evaluation of ITSM Processes	7 processes, method analytical hierarchy process
Tang & Todo, 2013	Service Desk Setup in ITSM	Service desk, Maturity Model
Berrahal & Marghoubi, 2016	Lean Continuous Improvement of ITSM	Continuous Improvement, Maturity Model

Sukmandhani et al., 2017	Measurement Effectiveness and Efficiency to Improve ITSM	6 processes, Maturity Model
Ruiz et al., 2018	Optimization in the context of ITSM change process	Change Management, simulation models
Costa, et al., 2019	ITSM Automation	Incident Management, Machine Learning
Cusick, 2019	Survey of Maturity Models	Continuous Improvement, Maturity Model
Obwegeser, et al., 2019	Continual Process Improvement for ITIL Service Operations	Continuous Improvement,

All authors studying the topic of ITSM and ITIL agree that correctly implemented ITSM is an important element of the support of entrepreneurial goals in enterprises and organisations and they try to approach the topic from different perspectives such as implementation, metrics model, benefits, frameworks and standards, evaluation, effectiveness.
Several authors criticised the extent and excessively detailed character of ITIL and ITSM especially from the perspective of small and medium enterprises. In their published works, the authors proposed simplifications for SME [4], [7-8], [16], [21-22]. Varga also adds that the complexity of ITSM prevents a more holistic view on the enterprise/organisation and thus it makes the support of business goals more complicated [4].

3.2 ITSM Metrics Model according to Steinberg

To ensure the support of entrepreneurial goals of a specific enterprise, the implementation of ITSM requires both management of IT services and quality measurement of IT service efficiency. We can use a variety of approaches for the measurement of ITSM level. Application of ITSM metrics models published by R.A. Steinberg belongs among the most interesting approaches [14]. The purpose of using ITSM metrics models is the measurement of benefits ensuing from ITSM implementation for entrepreneurial activities.
The ITSM metrics model consists of several metrics categories: operational metrics, key performance indicators (KPI), tolerance thresholds, critical success factors (CSF), dashboards, and outcomes. The individual categories mutually influence each other so that the entry operational metrics are gradually transformed into output indicators based on which decisions can be made about the IS/IT management as well as about the whole enterprise (Fig.1).

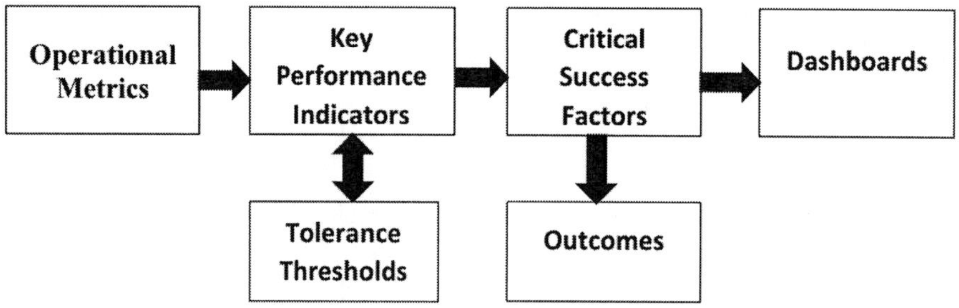

Fig. 1. ITSM Metrics model.

The operational metrics are the basic observations of the operational events in all areas of ITSM. They are the starting point of the model and are subsequently used for KPI calculations. Examples of operational metrics are overall number of changes, number of open incidents, number of problems with the execution of tasks, number of phone calls, evaluation of client satisfaction and total cost of IT.

The key performance indicators (KPI) are metrics used to mark the level of operation or process efficiency. KPI is derived from one or more operational metrics. The results are compared with the tolerance thresholds to establish whether the results are within the acceptable tolerance limits. Examples of KPI are the degree of efficiency of changes, changes in the use of labour force, degree of reoccurrence of incidents, level of capacity management or a total of reimbursed penalisation for services.
Tolerance thresholds are the top and bottom lines for acceptable and unacceptable values of KPI. They should be set by the IT manager based on an agreement with the managers of other departments of the enterprise. The tolerance thresholds are important because their role is to alert the company management to the threshold levels of KPI requiring measures to be taken.
The critical success factors (CSF) are those metrics which show the key requirements of operational efficiency. The CSF value shows whether the processes or operations are successful from the perspective of both the client and the enterprise. CSF is calculated or derived from one or more KPI by comparison of KPI values with the stated tolerance thresholds. CSF is usually indicated in levels required for success, such as high, medium or low.
Dashboards are used to display integrated metrics with the aim to unite information from several metrics into a united picture. Usually, those are important metrics suggesting the success, risks or failure of an operation. They are used for quick evaluation of the condition of IS/IT operation and for early measures to correct operational inadequacies. Dashboard metrics are usually derived from the results of CSF.
The results show the key indicators concerning general entrepreneurial risks. They are related to the efficiency indicators which define success, risk or failure of KPI or CSF. The results are used for quick evaluation of risk levels caused by process or operational inadequacies. In other words, the results category contains those facts which the area of IT tries to avoid.
In addition, Steinberg proposes examples of specific indicators for each process that an enterprise /organisation can choose from to create their portfolio.

3.3 ITSM Metric Model according to Rumburg

The approach of J. Rumburg, consultant of the MetricNet company is based on his experience collected from his consultancy. According to Rumburg [15], enterprises and organisations with effective IS/IT management declare that the rule of 80/20 is applicable in IT metrics and IT support. An average organisation shows around 20 IS/IT metrics in six areas (Fig. 2).

Cost	Quality	Productivity
Cost per Ticket	Customer Satisfaction	Tickets per technician per
Cost per Minutes of Handle	First Contact Resolution Rate	Technician Utilization
First Level Resolution Rate	Call Quality, Ticket Quality	Technicians as % of Total FTE´s
Technician	**Service Level**	**Ticket Handling**
Annual Technician Turnover	Average speed of answer (ASA)	Average Ticket Handle Time
Daily Technician Absenteeism	Call abandonment rate	User Self-Service Completion
New Technician Training Hours	% Answered within 30 Second	
Annual Technician Training	Mean Time to Resolve	
Schedule Adherence		
Technician Tenture		
Technician Job Satisfaction		

Fig. 2. Metrics for IS/IT services.

Based on empirical research of 4,000 completed benchmarks in the area of IS/IT services, it was found that a company needs only 8 criteria to collect enough information to manage IS/IT services related to support of the enterprise [15].
These important metrics are:

- Cost per ticket - is the best indicator of efficiency in service and support. It is calculated by dividing the total monthly operating expense of a service desk or desktop support group by the monthly ticket volume.
- Customer satisfaction - is the best indicator of effectiveness in service and support. One goal of every business is to achieve the highest possible quality at the lowest possible cost. It stands to reason, therefore, that cost per ticket and customer satisfaction should be measured on an ongoing basis. In fact, many would argue that cost and quality are the only two things that really matter.
- First-contact resolution - It turns out that customer satisfaction is affected by a range of other performance metrics, including call quality and mean time to resolve, to name just a couple. But the single biggest driver of customer satisfaction—by far—is first-contact resolution rate.
- Technician utilization - is the best measure of labour efficiency. Because labour costs represent the overwhelming majority of IT support expenses, if technician utilization is high, the cost per ticket will inevitably be lower.
- First-level resolution - the next KPI on the short list of metrics that really matter. This metric is a proxy for total cost of ownership (TCO) and is a critical measure of overall IT support efficiency.
- Mean time to resolve - Mean time to resolve (MTTR) is a service-level metric that measures the average elapsed time from when a ticket is opened to when it is closed. It is typically measured in business hours, not clock hours.
- Technician job satisfaction - is a bellwether metric that is strongly correlated with many other metrics in service and support. High levels of technician job satisfaction lead to lower turnover, lower absenteeism, lower handle times, and higher first-contact resolution rates.
- Balanced score - The balanced score aggregates the seven metrics we have just discussed and combines them into a single, overall measure of performance. The value of this metric, when tracked over time, is that it enables IT support to determine whether overall performance is improving or getting worse.

3.4 ITSM Maturity models

Maturity models show the level of IT service or process. Usually, it is the optimal level for IT service or process administration. The maturity models have usually four or five levels, but before reaching the next maturity level, the activities and processes of the lower level must be completed creating conditions for shifting to the higher level. In general, maturity models are very useful for both diagnoses of the level of IS/IT services in enterprises and potential corrections of plans and for benchmarking enterprises within a branch of industry.
Maturity models follow a method derived from a maturity model for software design CMM (Software Capability Maturity Model) [5]. It has spread into various areas. One of them is the ITSM Maturity Model serving for identification of the level of IS/IT services utilisation. The Model consists of 5 levels and Tang and Todo labelled them Chaotic, Reactive, Proactive, Service and Value [23]. Their model does not include all the ITSM services, only the Service Desk.
The complete version of ITSM Maturity Model, according to ITIL4, published by Cusick [5] (Fig.3) modified the names of the levels (Chaotic, Reactive, Stable, Proactive and Value

Driven), the characteristic features of each level, as well as the "important elements in the process of continual improvement", into which the levels are placed (Process Standardization, Business Alignment, Strategic Integration).
KPI based on a Likert scale is usually used to measure the achieved degree of individual indicators and then evaluated by implementing statistical methods [17].

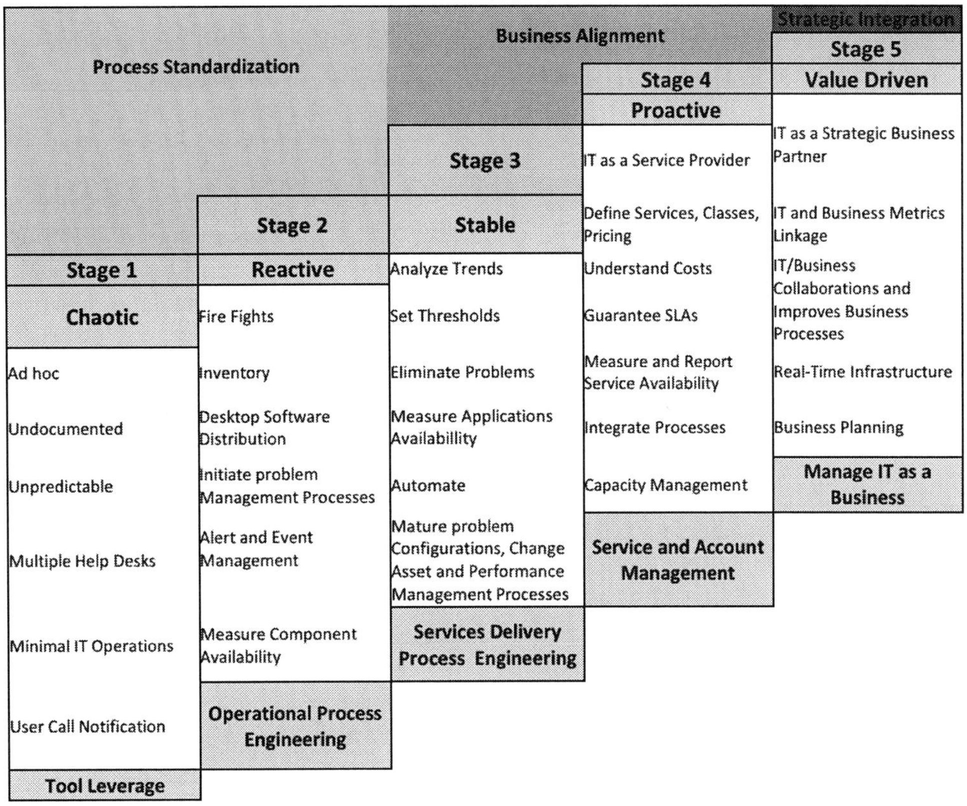

Fig. 3. ITSM Maturity Model.

4 Conclusion

The aim of the paper was to examine the approach to measurement and evaluation of ITSM with special emphasis on coordinating the IS/IT goals with the entrepreneurial goals of enterprises or organisations. The analysis of the literature has proven the importance of ITSM implementation in enterprises and organisations due to the necessity of effective management including measurement of IS/IT of the enterprise. The published metrics do not exclude each other, they rather differ in the extent of operation. Steinberg's complex model can serve as an example and inspiration for specific companies which can define their portfolio based on its metrics. On the other hand, Rumburg's metrics model is based on real entrepreneurial experience and consists of 8 main metrics (cost per ticket, customer satisfaction, first-contact resolution, technician utilization, first-level resolution, mean time to resolve, technician job satisfaction, balanced score) sufficient for effective ITSM and are applicable especially in the SME sector. The individual metrics specified in the metrics models can be also used in the ITSM Maturity Models that are appropriate for both diagnostics of the IS/IT services in the enterprises and potential plan corrections, and benchmarking of enterprises within their branch.

Acknowledgement

The paper was elaborated within VEGA No. 1/0388/20 IT Management in Enterprises in Slovakia: International Standards and Norms Versus Individual Business Processes – proportion 100 %.

References

1. T. J. Winkler, J. Wulf, Effectiveness of IT Service Management Capability: Value Co-Creation and Value Facilitation Mechanisms. *Journal of Management Information Systems*, **36**, 2, 639-675 (2019)
2. W. Berrahal, R. Marghoubi, Lean continuous improvement to information technology service management implementation: Projection of ITIL framwork. *In 2016 International Conference on Information Technology for Organizations Development (IT4OD)*, 1-6 (2016)
3. N. Obwegeser, D. T. Nielsen, N. M. Spandet, Continual Process Improvement for ITIL Service Operations: A Lean Perspective. *Information Systems Management*, **36**, 2, 141-167 (2019)
4. S. Varga, G. Barreto, P. D. Battaglin, Increasing Information Systems Availabiliy Through Accuracy, Awareness, Completeness and Manageability of ITSM. *In 2019 14th Iberian Conference on Information Systems and Technologies (CISTI),* 1- 4 (2019)
5. J. J. Cusick, A Survey of Maturity Models from Nolon to DevOps and Their Applications in Process Improvement. *arXiv preprint arXiv* 1907.01878 (2019)
6. J. Iden, T. R. Eikebrokk. Implementing IT Service Management: A systematic literature review. *International Journal of Information Management*, **33,** 3, 512-523 (2013)
7. A. El Yamami, K. Mansouri, M. Qbadou, E.H. Illoussamen, Introducing ITIL framework in small enterprises: tailoring ITSM practices to the size of company. *International Journal of Information Technologies and Systems Approach (IJITSA)*, **12,** 1, 1-19 (2019)
8. M. Schmidt, M. Brenner, T. Schlaaf, IT service management frameworks compared– simplifying service portfolio management. *In 2019 IFIP/IEEE Symposium on Integrated Network and Service Management (IM)*, 421-427 (2019)
9. M. Ruiz, J. Moreno, B. Dorronsoro, D. Rodriguez, Using simulation-based optimization in the context of IT service management change process. *Decision Support Systems* **112**, 35-47 (2018)
10. A. A. Sukmandhani, B.D. Wijanarko, E. Gunawan, D. Pratama, F. L. Gaol, I. Sutedja, Measurement effectiveness and efficiency to improve the IT services using ITSM. *In 2017 International Conference on Information Management and Technology (ICIMTech)*, 334-339 (2017)
11. S. Cots, M. Casadesús, F. Marimon, Benefits of ISO 20000 IT service management certification. *Information Systems and e-Business Management,* **14,** 1, 1-18 (2016)
12. Bestpractice. sk. Čo je to Service management a ITSM. https://www.bestpractice.sk/sk/Best-practice/service-management-itil-/co-je-to-service-management-a-itsm.alej, [accessed 15.2.2020]

13. L. Hunnebeck, S. Rance, R. Steinberg, V. Lloyd, C. Ruud, *Service design.* (London, The Stationary Office (TSO), 2011).
14. R. A. Steinberg, *Measuring ITSM* (Trafford Publishing, 2013)
15. J. Rumburg, *The power of KPIs. The 8 IT service management metrics that matter most.* https://techbeacon.com/enterprise-it/8-it-service-management-metrics-matter-most (2018)
16. A. El Yamami, K. Mansouri, M. Qbadou, Multi-criteria decision making approach for ITIL processes performance evaluation: Application to a Moroccan SME. *In 2017 Intelligent Systems and Computer Vision (ISCV)*, 1-6 (2017)
17. J., Wan, H., Z. Hui, D. Wan,. Evaluation on information technology service management process with AHP. *Technology and investment* (2011)
18. J. Costa, R. Pereira, R. Ribeiro, ITSM Automation - Using Machine Learning to Predict Incident Resolution Category. *In Education Excellence and Innovation Management through Vision 2020* (2019)
19. H. I. Chunpir, M. Ismailzadeh, Comparison of Information Technology Service Management (ITSM) Practices in e-Infrastructures, Libraries, Public Administration and the Private Sector. *In International Conference on Applied Human Factors and Ergonomics*, 32-41 (2019)
20. R.O. Rocha, M.C. Melo Silva Luft, M.E. Leon Olave, F.C. Heber Pedreira De Freita, Theory of agency and outsourcing: a proposal of a theoretical model for hiring and managing information technology services. *Revista Gestao & Tecnologia-Journal of Management and Technology*, **19**, 1, 201-224 (2019)
21. P. Küller, M. Vogt, D. Hertweck, M. Grabowski, IT service management for small and medium-sized enterprises: a domain specific approach. *Journal of Innovation Management in Small & Medium Enterprises*, **1** (2012)
22. R. F. Machado, S. Reinehr, A. Malucelli, Towards a maturity model for IT Service Management applied to small and medium enterprises. *In European Conference on Software Process Improvement*, 157-168 (2012)
23. X. Tang, Y. Todo, A study of service desk setup in implementing IT service management in enterprises. (2013)

Entrepreneur's Experience as Business angels' and Venture Capitalists' Decision criteria

Robert Hanák[1,*]

[1]Department of Information Management, Faculty of Business Management, University of Economics in Bratislava, 852 35, Dolnozemská cesta 1, Slovakia

Abstract. There are many business angels' and venture capitalists' decision criteria they apply when they are assessing business proposals of entrepreneurs asking for investment. Entrepreneur's experience and experience of his/her team are considered as the most important of all criteria related to the entrepreneur. Using theoretical literature analysis we investigated experience as decision criteria used by business angels and venture capitalists and some limitation of their usage. We found that there are many types of experience available and identified several limitations for applying them in the decision process. In the empirical literature, there are discussed various types of experience such as professional experience, industry experience, managerial experience, leadership experience, domain-specific experience, entrepreneurial experience, start-up experience, international experience, work experience, track record and functional experience (production, marketing, financial etc.).

1 Introduction

Business angel investors and venture capitalists evaluate business proposals from entrepreneurs who ask for their finance by many different evaluation criteria [1, 2, 3, 4, 5]. In some scientific papers are criteria listed explicitly [1], or all criteria are grouped into several key categories such as product&model, market, financial, team and other [6, 4]. Authors [7, p. 126, Fig. 2] in their systematic review of literature created a conceptual framework, where they have only three major investors' investment criteria groups: Management team, Business, Financial traction. These are very broad categories including tens of specific criteria. Based on a systematic literature search these findings are summing investment criteria from 54 scientific empirical studies. Entrepreneur and all related characteristics with him such as psychological profile, education, track record, experience, moral integrity, ability to cooperate and many others are generally considered very important. We are not aware of any scientific study describing the decision criteria of the

* Corresponding author: robert.hanak@euba.sk

business angel or venture capitalist investors, which is not considering entrepreneurs and team characteristics as important.

From many criteria related to the entrepreneur and his team, the experience is valued by investors highly. In general, it is assumed that any type of experience has positive effects on an entrepreneur's performance. Any experience is better than none but ten years in business are usually considered as extensive experience [8, 6, 9]. Importance of experience was proven by review study of authors [7], who list them at first place in decision criteria related to entrepreneur used by business angels and also by venture capitalists. For venture capitalists they [7, p. 126, Fig. 2] report them as "industry, domain-specific, leadership and professional experience" and for business angels, they report them as "experience of funding team". Very similar results of importance of experience were found by previous authors [5], who reports that from 13 studies describing angels investors' decision criteria, 11 had industry experience as criterion or 9 had a track record. More than half of them (7 from 13) had them both and none of 13 did report industry experience or track record as a criterion which is not taken into account. Specific terms used by [5] for describing experience were: "industry experience and track record".

From these review studies and individual empirical papers, we can draw two conclusions. Firstly we can say that that entrepreneur's experience is considered as very important by almost every angel and every venture capital group, which was investigated. Secondly, there is high variability in what is considered as experience and how are they measured.

Therefore our aim in this paper is at it follows: To describe the experience as decision criteria used by business angels and venture capitalists in the decision process about investment and to discuss several potential limitations for their usage.

2 Methods

We applied descriptive theoretical analysis, to outline the current situation in experience application as decision criteria in business angels and venture capitalists decision process. In the literature search, we focused on review studies and individual empirical studies. In scientific database search, we used terms entrepreneur, experience and combination of words entrepreneur and types of experience such as professional experience, industry experience, managerial experience, leadership experience, domain-specific experience, entrepreneurial experience, start-up experience, international experience, work experience, track record and functional experience (production, marketing, financial etc.)

3 Results

Our results section has two parts. In the first part, we will show what types of experience are used in literature. We could divide three blocks of literature which were used in this paper. First are theoretical studies, second are review studies describing business angels' and venture capitalists' decision process and third are empirical management studies.

In the second subchapter, we discuss the hypothetical limitation of experience use as decision criteria used by investors. There is also intensive theoretical – empirical limitation of experience effect. Business angels and venture capitalists should be aware of these effects when using an entrepreneur's experience as decision criteria. In sub-chapter 3.2 we will show some of the arguments.

3.1 Various type of experience

Firstly we will discuss the various type of experience. There are several proposals about how the experience should be classified and in many authors, we find overlapping in the contend. In some cases, the experience is defined vaguely and ambiguously. For example, some authors [10] when describing experience they are using a term such as "high credibility of entrepreneur/management team". From this statement, it is difficult to identify what specific experience are there and it indicates that there are many types of experience behind this term. In this section, we describe the most frequent types of experience used by various authors. We identified more than 10 different types of experience which we describe in this or next subchapters. For example [11] in her review study concluded, that experience can be classified into three categories: prior start-up experience, prior management experience and industry-specific experience. Prior start-up experience is usually considered the number of companies created in the past by entrepreneur, which is in line with other authors [12, 13, 14]. Because a startup is mostly defined as a fast-growing company, better will be to define these experience with condition that contains fast growth of the venture. There are serial [15], or by some authors [16], labelled as repeat entrepreneurs, are those entrepreneurs, who created and run several ventures. Launching and running number of ventures generate entrepreneurial learning and unique knowledge which others without it do not have [14]. In scientific literature, we can find also the term "track record" [1, 3] which is probably equivalent to start-up experience. So the first type of experience identified in the empirical and theoretical literature, which entrepreneurs have, are start-up experience.

The second type of experience is management experience, which is defined as an experience of managing business or team of people in a specified period. Some authors [8] combine managerial experience with start-up experience, labelled them as specific human capital, which is defined as years working with young firms. For better understanding effects of experience, we do not recommend merging different types of experience into summary blocs. We could also speculate about the term Leadership experience used by some authors [7]. These could be probably managerial experience. Industrial experience [11] define as working in a specific industry for a specified amount of time. Industry experience as a term is uniformly used by all authors in the field and [7] review they listed this decision criteria on the first place in the category management team. Also as reported above, authors [5] found industry experience in 11 studies from 13 investigating decision criteria. Being on the first places in the lists in the review studies and high frequency of use indicate, that industrial experience is considered by investors as more important than others. And for industrial experience, there is an agreement on the use of the term by various authors.

In the empirical literature, we can find also the term entrepreneurial experience. This labelling is overlapping with startup experience. Entrepreneurial experience is by some authors defined by years in self-employment, or by a number of ventures created [17], which is the same definition for startup experience. We also identified other types of experience which are reported in business and managerial literature and all of them have positive effects on venture performance [18]. They are: Functional experience such as marketing, financial, production, research and development experience is another type of experience [19]. International experience defines [20, p. 681] as: "whether executives had worked abroad for at least a year". In business literature, we can find also term working experience which is defined as wage work experience [21], or as work for others [22].

3.2 Problems with various types of experience

So many types of experience generate some specific problems. We will discuss them in the following paragraphs. The first issue is the parallel existence of various types of experience together. As we can see, there are many types of experiences an entrepreneur could have. We would like to stress, that these experience are not existing separately but the entrepreneur in his professional career gains different types of experience in a parallel way. To better understand the overlapping experience we could illustrate the problem with the following example. If the entrepreneur starts his venture and he is managing it and then after a few years starts another venture in the same industry he already has several different types of experience. First are industrial experience, then he has managerial experience, but also entrepreneurial experience. In the case of working with finance, or in production, the entrepreneur has plus functional experience in finance/production. If the company was active in foreign markets than he has also international experience. After all, this when entrepreneur asked what kind of experience he has, he had six different options to answer what type of experience he has (industrial, managerial, entrepreneurial, etc..) and all are true.

Second problem is that some authors merge different types of experience into broader categories. For example, term professional experience is used [7] as a broader term and could be interpreted as industrial experience probably with the combination of entrepreneurial experience, maybe with management experience. The same could be said about term specific human capital or leadership experience.

3.3 Limitations of applying experience as decision criteria

As was described above experience are considered as very important decision criteria with high weight in the decision process. But as with everything experience could have also negative effects. There are many different arguments about the limitation of applying experience as decision criteria. In the next paragraphs, we explain some of them.

First, we start with management experience, often operationalized as time spent managing people or resources is used as a proxy for general expertise in the business environment, had been often criticised for lack of validity and not only in the business domain [9]. Reasons for criticism are explained by the complex environment in which every enterprise exists [7]. In this complex environment, many very different factors have influence and some of them occur only once or very rarely, which does not provide necessary feedback. Infrequent events are more difficult to learn from due lack of repetitiveness and especially if they happened only once or a few times in their career. Running one company, of course, create specific managerial plus industry experience, but the knowledge of how to create another, different venture could be limited. Therefore the transfer of knowledge from one enterprise to another in a different industry is not a simple mechanical straightforward process. A large part of knowledge could not be applied to new condition. Applying uncritically experience from one enterprise to another, which operates in a different environment, could lead to serious setbacks and failures [15]. Launching the second venture in the different industry often seriously challenges the entrepreneur's knowledge and procedures from previous enterprise forcing him/her to revaluate many things which he/she took for granted [15]. Therefore we could speculate that managerial experience has a lower effect than industrial experience. But the situation is not only black and white for the managerial experience. Failure, or performing deep under expectations, can lead to a positive scenario to deeper thinking and deliberation. This could produce more sophisticated, structured and organized knowledge about how solid business opportunity and consequently business model looks like and which attributes it should have [14]. As was discussed above experience creates various types of knowledge. Specific knowledge about venture functions such as production, marketing, finance, research and development is very valuable, but are on the other side very limited to that specific domain. If an entrepreneur does have large experience in finance but lacks experience in production or all other domains, his/her knowledge is restricted. The same could be said about all other experience. This could be called not compensatory substitution of one type of experience by another. And it is our second limitation factor. The financial experience could not substitute industrial experience and vice versa. Therefore from entrepreneur, it is expected to have minimum experience from all types of experience because one type of experience could not compensate for missing another type of experience.

Third limitation factor for not relying on experience without careful consideration is not linearity of influence. The relationship between experience and performance is not straightforward and linear as mixed results show. Some authors [24] found out that previous start-up experiences were negatively correlated ($r = -.16^{**}$) with proceeding to due diligence. Or the ability to forecast future performance was weakly predicted by start-up experience $\beta = .068^{**}$ [25], even that previous start-up experience did not predict start-up survival [13]. On the other side Hsu et. al. (2014) found out that specific human capital (total years in start-up) is a significant predictor for positive angel decision, but others [13] did not found that it will predict start-up survival. Also, industry experience brings mixed results. Some authors [26] did not find that it, as a significant predictor, for passing first pre-screen evaluation process and did not correlate with status or level of any business characteristics and in other research [13] it was found out that industry experience has no significant effect on survival of a new firm.

These mixed results can be explained by the atypical relationship between experience and performance, which can be graphically illustrated with the U shape curve. In the large sample of 65 390 firms found out that financial performance of the second enterprise is falling, compared to first and then in the third, fourth starts to raise, thus forming specific U shape curve [15]. The very similar curve was found by other authors [15] in the relationship between the number of ventures founded and quality of business proposal applying for business angel investment.

4 Discussions

Entrepreneurs' experience is valued by business angels and by venture capitalist highly. Problem is that there are many types of experience which overlap in two ways. First is that entrepreneurs could have multiple types of experience together such as entrepreneurial, managerial, industrial and functional at once. And therefore the answer to the type of experience depends directly on how is the question about experience asked.

The second overlap is that many authors do not distinguish between some types of experience. Startup experience and entrepreneurial experience are theoretically different types of experience but several authors are using these terms interchangeably.

These problems with terminology and objective overlapping make it problematic in the identification of what experience are demanded. Not much is known about how business angels and venture capitalists interchange different types of entrepreneur's experience. Industrial experience seems to be considered as the most important, especially in the early stages of development. In the later phases of development, the managerial experience is valued also. But in summary, we could say that all types of experience are considered to be

positive. Experience should not be taken in the decision process without consideration and there are known several limitations of their influence. They are not interchangeable therefore compensatory decision process could not be applied. Also, there is empirical evidence that the influence of experience is not linear but has a U shape curve.
Recommendation for future research. For the better understanding role of experience as decision criteria, it is worthy to investigate all scientific papers were experience were listed and carefully and into detail check what kind of experience were described. Only after that, we could be sure what type of experience was demanded by investors and to conclude if their procedures are valid.

Acknowledgment

This paper is an outcome of one research project:"IT Management in Enterprises in Slovakia: International Standards and Norms Versus Individual Business Processes.", VEGA no. 1/0388/20- project share is 100%; led by doc. Ing. Anita Romanová, PhD

References

1. H. Landström, Informal investors as entrepreneurs: Decision-making criteria used by informal investors in their assessment of new investment proposals. *Technovation*, **18**(5), 321-333 (1998)
2. L. Feeney, G. H. Haines Jr, A. L. Riding, Private investors' investment criteria: insights from qualitative data. *Venture Capital: An international journal of entrepreneurial finance*, **1**(2), 121-145 (1999)
3. C. Mason, M. Stark, What do investors look for in a business plan? A comparison of the investment criteria of bankers, venture capitalists and business angels. *International Small Business Journal*, **22**(3), 227-248 (2004)
4. J. S. Petty, M. Gruber, "In pursuit of the real deal": A longitudinal study of VC decision making. Journal of Business Venturing, **26**(2), 172-188 (2011).
5. A. L. Maxwell, S. A. Jeffrey, M. Lévesque, Business angel early-stage decision making. *Journal of Business Venturing*, **26**(2), 212–225 (2011)
6. C. Carpentier, J. M. Suret, Angel group members' decision process and rejection criteria: A longitudinal analysis. *Journal of Business Venturing*, **30**(6), 808-821 (2015)
7. C. Granz, M. Henn, E. Lutz, Research on Venture Capitalists' and Business Angels' Investment Criteria: A Systematic Literature Review. In Contemporary Developments in Entrepreneurial Finance (pp. 105-136). Springer, Cham (2020)
8. D. K. Hsu, J. M. Haynie, S. A. Simmons, A. McKelvie, What matters, matters differently: a conjoint analysis of the decision policies of angel and venture capital investors. *Venture Capital*, **16**(1), 1-25 (2014).
9. K. A. Ericsson, Development of professional expertise: Toward measurement of expert performance and design of optimal learning environments. Cambridge University Press (2009)
10. C. M. Mason, R. T. Harrison, Barriers to investment in the informal venture capital sector. Entrepreneurship & Regional Development, **14**(3), 271-287 (2002)
11. D. Politis, The process of entrepreneurial learning: A conceptual framework. *Entrepreneurship theory and practice*, **29**(4), 399-424 (2005)
12. C. E. Eesley, E. B. Roberts, Are You Experienced or Are You Talented?: When Does Innate Talent Versus Experience Explain Entrepreneurial Performance?. *Strategic Entrepreneurship Journal*, **6**(3), 207-219 (2012)

13. F. Delmar, S. Shane, Does experience matter? The effect of founding team experience on the survival and sales of newly founded ventures. *Strategic Organization*, **4**(3), 215-247 (2006)
14. R. A. Baron, M. D. Ensley, Opportunity recognition as the detection of meaningful patterns: Evidence from comparisons of novice and experienced entrepreneurs. *Management science*, 52(9), 1331-1344 (2006)
15. R. Toft-Kehler, K. Wennberg, P. H. Kim, Practice makes perfect: Entrepreneurial-experience curves and venture performance. *Journal of Business Venturing*, **29**(4), 453-470 (2014)
16. J. O. Fiet, V. G. H. Clouse, W. I. Norton Jr, Systematic search by repeat entrepreneurs. *Opportunity identification and entrepreneurial behaviour*, 1-27. (2004)
17. S. Haber, A. Reichel,. The cumulative nature of the entrepreneurial process: The contribution of human capital, planning and environment resources to small venture performance. *Journal of Business Venturing*, **22**(1), 119-145. (2007)
18. M. Grežo, R. Hanák, , Entrepreneur's Experience And Venture Performance: A Meta-Analysis (under review) (2020)
19. H. Li, Y. Zhang, The role of managers' political networking and functional experience in new venture performance: Evidence from China's transition economy. *Strategic Management Journal*, **28**(8), 791-804 (2007)
20. A. N. Kiss, S. Fernhaber, P. P. McDougall–Covin, Slack, Innovation, and Export Intensity: Implications for Small–and Medium–Sized Enterprises. *Entrepreneurship Theory and Practice,* **42**(5), 671-697 (2018)
21. J. Iversen, N. Malchow-Møller, A. Sørensen, Success in entrepreneurship: a complementarity between schooling and wage-work experience. *Small Business Economics*, **47**(2), 437-460 (2016)
22. A. Laskovaia, G. Shirokova, M. H. Morris, National culture, effectuation, and new venture performance: global evidence from student entrepreneurs. *Small Business Economics,* **49**(3), 687-709 (2017)
23. J. Shanteau, Competence in experts: The role of task characteristics. *Organizational behavior and human decision processes*, **53**(2), 252-266. (1992)
24. Mitteness, C. R., Baucus, M. S. Sudek, R. Horse vs. jockey? How stage of funding process and industry experience affect the evaluations of angel investors. *Venture Capital*, **14**(4), 241-267 (2012)
25. G. Cassar, Industry and start-up experience on entrepreneur forecast performance in new firms. *Journal of Business Venturing*, **29**(1), 137-151 (2014)
26. C. G. Brush, L. F. Edelman, T. S. Manolova, Ready for funding? Entrepreneurial ventures and the pursuit of angel financing. *Venture Capital*, **14** (2-3), 111-129. (2012)
27. R. Hanák, M. Grežo, The effect of prior entrepreneurial experience on the quality of a business plan proposal in applying for angel investment. *International Journal of Entrepreneurial Venturing*. ISSN print 1742-5360. (accepted for publication). (2020)

Model H application to determine synergic effects on merger and acquisition

Anna Harumová[1,*]

[1] University of Economics in Bratislava, Faculty of Business Management, Department of Business Finance, Dolnozemská cesta 1, 852 35 Bratislava

Abstract. Synergistic effects arise as effects of joint action based on mergers or acquisitions. Bundling is a common part of the growth strategy. A merger may acquire forms of merger or merger. Acquisition represents the acquisition of the ownership and management value of one company over another. Theory distinguishes in this case from property acquisitions in which the acquisition of the company's assets and capital acquisitions occurs, in which the acquisition of a decisive share in the voting rights of the company is obtained. Reasons for merger and acquisition are to gain more market share, restructure entities, improve balance of payments, and so on. The success of the merger and the acquisition confirms the emergence of a synergy effect. In determining the value of synergies in this article was used newly created Model H, which is based on a valuation of the business enterprise.

1 Introduction

Over the last decades, natural forms of business integration in the form of mergers, acquisitions, joint ventures, and so on, have grown in the world. The main reasons can be found in the increasing liberalization and globalization of the world economy. Mergers and acquisitions are a well-known and often used term in the global economy. Especially for managers, these business combinations represent a vision of expanding their businesses and thus achieving higher profits, lowering costs, or for them can be the last resort to the crisis. Some experts focusing on the transactions of mergers and acquisitions in the financial sector regularly doubt on the creation of value or even revealing the destruction of values [6]. Several authors have dealt with this issue [17,15,13]. The motives for the concentrations are different and the successful execution of this process is much harder than it looks at first glance. Every year, they are appearing in the media about surveys in this field, whose statistics agree that the overwhelming majority of mergers and acquisitions will not reach the predetermined and expected targets [1,14]. For example, KPMG's study shows that only 31% of all mergers and acquisitions actually generate value [11]. In the English literature the M & M most frequently used matching and A, which is internationally recognized.

* Corresponding author: anna.harumova@euba.sk

A special form of M & A (Mergers and Acquisitions) is the Leveraged Buy-Out (LBO), which differs from the classic M & A in that a large portion of the purchase price is funded by debt. In practice, the LBO usually takes place by establishing a new company only for this purpose, and it merges with the target company. The owner of the company becomes the investor who bought the company and its managers. As a result of debt financing, a larger burden on capital resources will occur in the new company. In order to evaluate the effects of M & A, we must first be able to evaluate the business as a whole [3].

Business appraisal is an important part of the M & A effects assessment. It is therefore good to keep in mind a few principles. In particular, it is important to begin with the analysis from the target company, estimate its present value and determine the change in cash flows that this M & A will bring. Cash flows have a better disclosure value than accounting profit. Economic theory states that although profit is a very valuable indicator, it also hides the same risks of inaccuracies [19]. Before we determine the synergistic effects, we estimate the value of the overall economic effect of the M & A performed. To obtain its value, a model consisting of the following three steps [4,5]:

1. Separately, the two M & A enterprises are evaluated using the method discounting expected cash flows
2. The combined value of both companies is calculated, without any effect from M & A through a combination of the values obtained in the first step,
3. The combined value of both enterprises is calculated by including synergies.

The resulting value of all synergies indicates the difference between the combined value of an enterprise, including synergies and synergies. In the professional literature [8] and the practice we can meet the following methods used for the evaluation of the company:

- Methods based on state variables, t. j. methods for determining the cost of equity, including the book value method and the substance value method,
- Methods based on flow variables, t. j. the yield methods of which include the capitalized net return method, the discounted value of future financial flows method, the economic added value method,
- Combined methods that combine property and yield methods,
- Exchange benchmarking methods.

Determining the market value of an enterprise poses a complex problem in practice because each company is a unique unit that can not be compared to another company. Effecting a synergistic effect also causes the company's value to be higher than the sum of the individual components of its assets. When assessing a company as a whole, it is necessary to take into account not only the real value of the property, but also its yield potential and prospects for further development of the company, which largely determines the subjective factor.

2 The current state of the theoretical background of mergers and acquisitions

A merger can be defined as a process whereby two or more companies with legal personality merge into one company, although no definition of merger can be found in any regulation. There are several of them in the literature. The term merger is a close, proprietary form of merger of two firms, one in which two firms merge into one. A merger is a process in which two companies voluntarily merge [2,20]. It may take the form of a merger. The first concerns the merger of two equivalent undertakings, one of which retains its name, identity and acquires all the assets and liabilities of the purchased undertaking. The second ceases to exist as a separate business unit. In the second case, the merger of two enterprises into one new one, which takes place through the merger of share capital with the prior consent of the shareholders. The merger may also be effected by the fact that the new undertaking exchanges its shares under the favorable conditions for those of the companies which are being dissolved. The merger is usually conducted on a voluntary basis by agreement between the shareholders and the boards of the merging companies.

Acquisition is the process of engaging a smaller or economically weaker enterprise with a larger or stronger enterprise. Absorption is the purchase of 51% or more of the voting shares of the acquirer (acquirer) in the target undertaking (acquirer). This can be done by repurchasing the shares of the target company for cash, by offering bonds, respectively. a new issue of shares which are exchanged for shares of the target undertaking. Ingestion usually has an involuntary, forced character [21,22]. Confronted with the need to adapt to the common market, the emergence of the European Monetary Union and the increasing globalization of markets, European firms have focused strongly on acquisition strategies, especially since the mid-1980s [16]. Acquisition has been addressed by several authors [23,24] in their publications. In general, achieving a synergistic effect can be defined by a well-known equation 1 + 1 = 3. The equation implies that the joint value of the merged companies (in our case, equal to three) should be greater than the sum of individual enterprise values before the merger (1 + 1 = 2). It is this difference that increases the value of our business after the merger results from a synergy effect. The ways in which synergies can be achieved include, for example:

1. economies of scale - Cost savings are made by merging selected divisions of both companies, especially the production, distribution and marketing divisions or the entire management. Cost savings while maintaining revenue levels lead to higher profits and, therefore, higher corporate value. This effect is usually higher if they merge companies from the same industry (ie, in a horizontal merger). Merging companies with a different focus may entail even higher costs, especially administrative costs.
2. Financial savings - Business size also increases their financial stability and confidence in banks, making them cheaper and more profitable for lower transaction costs.
3. Differential performance - occurs when management of one company is more efficient than management of the other company, and can increase the value of the other company by more efficient management (eg in the area of costs and revenues).
4. Larger market share - With increasing market share, the firm's negotiating position is usually strengthened against suppliers and customers. Alternatively, there is the possibility to dictate market prices. In this case, however, the Competition Protection Office could have intervened and either the notified M & A would not allow or could impose conditions or sanctions on the company.
5. knowledge concentration - where the value of the target enterprise depends, in particular, on ownership of know-how and know-how. Small businesses often have some technology or products. However, they do not have such capital to fully exploit the potential of their uniqueness. This can be achieved by joining the large company that will provide it, which will be beneficial for both companies.

However, we should keep in mind that each M & A performed may not achieve the above synergies. It is usually the company's management, what goals it will try to achieve and whether they can do it. The most common M & A target is to achieve synergies from economies of scale. For comparison, we can say the division of synergies according to foreign, mostly English, specialized sources. We will most often encounter the following division on [25]:

- Operational synergies, including, for example, economies of scale, higher business growth or greater market share, and higher sales revenue, potentially higher profits.
- Financial synergies that we can combine, in particular, with the stronger financial position of the business and the less costly interest on borrowed foreign capital.

3 Determining the value of synergy effects on fusion and acquisition using Model H

Model H to determine the value of synergies has been created based on the use of an enterprise's business valuation method that can be included in the yield method of determining the asset's overall value. The model is based on the value of the cash flow in the initial period of the companies to be merged and the rate of growth of this cash flow in future periods (for the two-step model) and the capitalization rates. The evaluation procedure under this model is based on the market values of the A and B companies before the merger and the values of these enterprises after the merger. By comparing the values of merged enterprises without synergies and including synergies, we can calculate the resulting synergistic effect of the merger. The basic data needed to calculate the synergistic effect are shown in Table 1.

Table 1 Basic data needed to calculate the synergistic effect

Enterprises	**A**	**B**	**A+B without synergies**	**A+B including synergies**
Cash Flow at 0 in mill. Eur	300	150	450	450
Growth rate in the first 5 years in%	12,00	15,00	13,00	15,00
Growth rate in subsequent years in%	6,00	10,00	7,33	10,00
Capitalization rate (in%)	13,00	14,00	13,33	13,33

Source: custom processing

The columns labelled A and B represent data for individual evaluation of both businesses. Zero year cash flow and capitalization rate (i) would be determined from current financial statements and market information. Furthermore, we assume, based on estimates, that enterprises will grow at a constant rate of 12% (xi) for business A and 15% (xi) for business in the first five years, and 5% (xi) for business A and 10% (xi)) for business B. We apply this growth rate to the value of cash flow in the zero year to obtain cash flow values (pi) in the first five years and beyond. The calculation of the growth rate of the merged enterprises without synergies is as follows:

1. Calculation of growth rate (g) in the first 5 years in% for A + B without synergy:

2. Calculation of growth rate (g) in subsequent years in% for A + B without synergy:

$$g = \frac{\sum_{i=1}^{n} x_i \cdot p_i}{\sum_{i=1}^{n} p_i} = \frac{(0{,}06 \,.\, 300) + (0{,}10 \,.\, 150)}{300 + 150} = 7{,}33$$

3. Calculation of capitalization rate (i) for A + B without synergies:

$$i = \frac{\sum_{i=1}^{n} x_i \cdot p_i}{\sum_{i=1}^{n} p_i} = \frac{(0{,}13 \,.\, 300) + (0{,}14 \,.\, 150)}{300 + 150} = 13{,}33$$

The next step is to calculate the future cash flows of individual A and B companies, merged A + B companies without synergies, and merged A + B companies, including synergies. The calculation can be found in Table 2.

Table 2 Calculating Future Cash Flow (in millions of Euro)

Cash Flow 5 years	A	B	A+B without synergies	A+B including synergies
1	336	173	509	518
2	376	198	575	595
3	421	228	650	684
4	472	262	734	787
5	529	302	830	905
CF in subsequent years	560	332	892	996

Source: custom processing

Now we have all the necessary data to calculate the present values, using the formulas used from the business method of determining the general value of the business. The calculation is done in three points, namely the calculation of the present value of Cash Flow over a period of 5 years, in the second point is the calculation of the continuing value of the company based on Cash flow in the following years. continued value of the company. The calculation of the value of the synergy effect from the merger is in Table 3.

Table 3 Calculation of the synergy value of enterprises A and B (in millions of EUR)

	A Company			B Company			A+B without synergies			A+B including synergies		
Year	**KF**	**CF**	**SH CF**	**KF**	**CF**	**SH CF**	**KF**	**CF**	**SH CF**	**KF**	**CF**	**SH CF**
1	0,88496	336	297	0,87719	173	152	0,88238	509	509	0,88238	518	457
2	0,78315	376	294	0,76947	198	152	0,77859	574	575	0,77859	595	463
3	0,69305	421	292	0,67497	228	154	0,68701	649	650	0,68701	684	470
4	0,61332	472	289	0,59208	262	155	0,60621	734	734	0,60621	787	477
5	0,54276	529	287	0,51937	302	157	0,53490	831	830	0,53490	905	484
SH CF			**1460**			**770**			**3298**			**2351**
Value Continued (PH)			**4427**			**4194**			**8087**			**10823**
VŠH=SH CF + HT			**5888**			**4964**			**11385**			**13174**
The value of the synergy effect (13 174 - 11 385)												**1789**

Source: custom processing

This calculation is made for each company (A, B) separately, for linked enterprises (A + B without synergies) and for linked enterprises (A + B including synergies). According to the business method, the market value of company A calculated is EUR 5,888 million. EUR and market value of company B is 4 964 mil. Eur. The market value of the connected A + B companies, including synergies, is EUR 13 174 million. Eur. The synergistic effect of the merger represents an amount 1 789 Eur (13 174 – 11 385).

4 Achieved results and discussion

When calculating the value of the synergy effect, the other calculated values also serve as a measure for determining the contractual premium for the purchase of Company B, as can be seen in Table 4. We assume that the net present value of the investment must be positive or equal to zero . If the market value of target company B calculated by us is 4964 mil. EUR and the value of the total effect resulting from M & A in the amount of 1 798 mil. The purchase price should not exceed the sum of these values, ie 6 753 mil. Eur. The price is based on the market value of company B and the amount of the premium should not exceed the value of the synergy effect.

In practice, besides determining the value of the overall synergy effect of M & A, it is also necessary to focus on determining the value of specific synergies (synergies from sales, synergies from cost savings, synergies from financial savings, etc.). In assessing the value of these partial synergies, we will again base on the plans and estimates that management has to make before executing the M&A itself. The selected synergies are not the only ones that can be estimated. However, these are the most common economic effects of M&A and the calculation of their values is based on a similar basis. If we would like to estimate the value of other effects, these models can be easily modified for the given situation [18]. For example, if we would like to estimate the value of tax or financial savings resulting from M&A, we would use the cost savings value model [9,10]. Figure 1 compares the number of M&A transactions in CSE (Central and South-Eastern Europe), where we can also compare V4 countries.

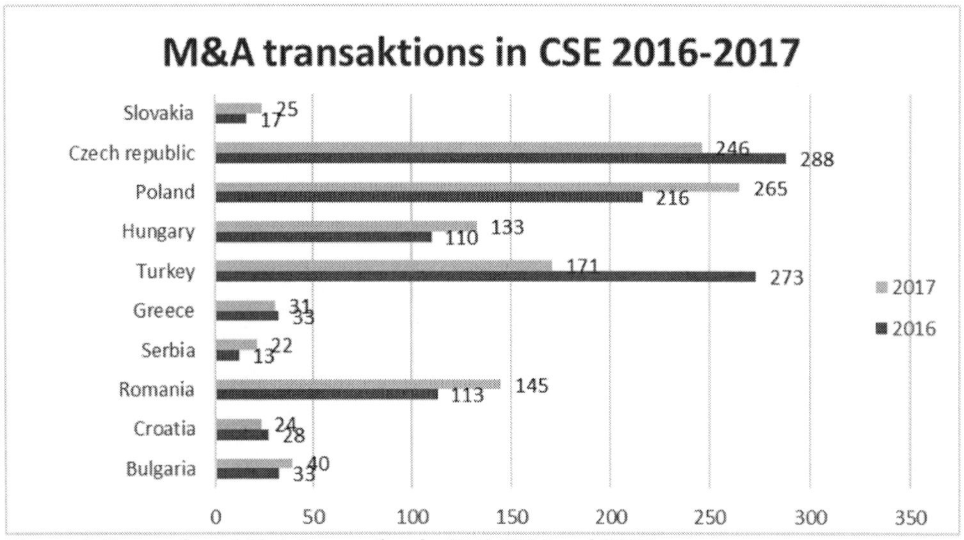

Figure 1 Comparison M&A transaction in CSE 2016 and 2017

M&A Barometer from Ernst & Young provides an overview of M&A developments and summarizes and analyses publicly available information collected from reputable databases. Methodologically, the M&A Barometer includes transactions in the private sector. It excludes data on acquisitions of minority interests below 15%, further acquisitions of minority interests by majority shareholders, real estate transactions (except where the target or buyer was a real estate company or real estate fund), capital market transactions (except for transactions that resulted changes in control), acquisitions of licenses, joint ventures, greenfield investments, listed companies for the first time, privatizations, multinationals (their value is ignored in each country but counted), and internal reorganization [7]. The graph compares the number of mergers and acquisitions transactions in CSE (Central and South Eastern Europe), where we can also compare V4 countries. In comparison with the V4 countries, Slovakia achieved the smallest number of mergers and acquisitions between 2016 and 2017.

Poland was the most active country in terms of deal volume in 2017, closing 265 transactions during the respective period. It was followed by the Czech Republic and Turkey, closing 246 and 171 deals respectively. Although in 2017 the total number of transactions in the CSE market was 1,132, which represents a 2.8% decline compared with 2016, six of the countries – Bulgaria, Hungary, Poland, Romania, Serbia and Slovenia – experienced increased M&A activity in terms of the number of transactions in 2017. The slight decrease was mainly caused by the significant drop in the volume of deals in Turkey, where the number of closed deals was 171, representing a 37.4% decrease compared with 2016.

5 Conclusion

World globalization of the economy forces the company to pool capital in order to achieve higher efficiency. With all the variety of reasons and intents to merge businesses, the overriding objective of most of them is to achieve a synergistic effect arising from the complementary activity of the assets of two or more enterprises. The volume of M & A transactions in Slovakia is decreasing. Practical experience shows that far from all companies achieve the expected effect in the M & A transaction process.

Assessing the potential synergy is one of the most complicated tasks in analyzing the effectiveness of the merger. The authors propose an express analysis that allows to evaluate the potential synergistic effect and to choose the optimal concept of the functioning of the future society at the decision-making stage. The success of fusion is influenced by many factors. The most important could include managing rationalization, gaining access to new banking techniques, optimizing economies of scale, portfolio diversification, synergy, and last but not least, the way of financing the merger.

Evaluating the success of the merger is very difficult after a few years, but the easiest way is to use different analytical methods and ratios. Indicators need to be tracked and judged in the timeline, and it is important to underline that even slight changes in the level of indicators can identify changes in clientele and changes in the financial market. Prior to a merger, the value of a synergy effect from a future merger can be an important element in decision making.

Currently, the activity on M & A markets is declining due to a discrepancy in the actual value of potential sales businesses and their price offered. China, the United States, India, Brazil, Germany and Indonesia remain the largest investment destination, with M & A being the most common in manufacturing, financial services, the oil and gas industry, and consumer goods. In terms of sectoral orientation, there are horizontal links between enterprises operating in the same industry, vertical links of enterprises operating in one production chain, and conglomerate links of enterprises engaged in business in different sectors in order to diversify business risk.

The volume of M & A transactions worldwide is growing steadily. Practical experience, however, shows that far from all companies achieve the expected effect in the M & A transaction process. Assessing the potential synergy is a challenging task in analyzing the effectiveness of the merger. The role of the proposed H Model is to contribute to determining the effectiveness of mergers based on the quantification of synergy effects.

Acknowledgement

The paper is the outcome of the grant project of the Ministry of Education of the Slovak Republic KEGA project No.025EU-4/2018 "Systematic knowledge transfer in the field of economic knowledge and forensic expertise.

References

1. Berger, P. G. – Ofek, E. Diversification's Effect on Firm Value. *Journal of Financial Economics*, 37, No. 1, pp. 39 – 65. (1995)
2. Brealey, R.A. – Myers, S. C. *Teorie a praxe firemních financí*. Praha : Victoria Publishing.(2003)
3. Brunner, R. F. *Applied mergers and acquisitions*. Hoboken: John Wiley & Sons Inc., 1029 p. (2004)
4. Damodaran, A. *This model estimates the value of synergy in a merger*. [cit. 2020-03-24]. Dostupné z www: http://pages.stern.nyu.edu/~adamodar/. (2006)
5. Damodaran, A. *The Value of Synergy* [online]. Stern School of Business. [cit. 2020-03-24]. Available from www: http://pages.stern.nyu.edu/~adamodar/pdfiles/papers/synergy.pdf. (2005).
6. Deyong, R. – Evanoff, D. D. – Molyneux, P. Mergers and Acquisitions of Financial Institutions: A Review of the Literature. *Journal of Financial Services Research*, 36, No. 2 – 3, pp. 87 – 110. (2009)
7. Ernst & Young. M&A *Barometer 2017* [online]. [cit. 2020-03- 24] Available from www: https://www.ey.com/Publication/vwLUAssets/ey-csema-barometer-2017/$FILE/ey-csema-barometer-2017.pdf. (2018)
8. Harumová, A et al. *Stanovenie hodnoty majetku : hlavné aspekty ekonomickej znaleckej a expertnej činnosti*. Bratislava : Iura Edition, Ekonómia. (2008)
9. Harumová, A. *Stanovenie hodnoty IV : stanovenie hodnoty synergických efektov* Lewes : Ecoletra.com. ISBN 978-1-947099-90-6. (2017)
10. Harumová, A. *Teória finančného riadenia holdingu*. Lewes : Ecoletra.com. (2019)
11. KPMG International. *A new dawn: good deals in challenging times*. In: [online]. [cit. 2020-03- 24]: http://www.kpmg.com/CZ/cs/IssuesAndInsights/ArticlesPublications/Pressreleases/D ocument s/KPMG_MA-New-Dawn-2011.pdf. (2011)
12. Kráľovič, J. – Vlachynský, K. *Finančný manažment*, Bratislava : IURA EDITION. (2002)
13. Lang, L. – Stulz, R. Tobin's q, Corporate Diversification, and Firm Performance. *Journal of Political Economy*, 102, No. 6, pp. 1248 – 1280. (1994)
14. Lelyveld, I. – Knot, K. Do Financial Conglomerates Create or Destroy Value? Evidence for the EU. *Journal of Banking & Finance*, 33, No. 12, pp. 2312 – 2321. (2008)
15. Links, K. V. – Servaes, H. Is Corporate Diversification Beneficial in Emerging Markets? *Journal of Financial Management*, 31, No. 2, pp. 5 – 31. (2002)
16. Lissauerová, D. Euro and the Process of Constitution of the European Monetary Union. In: *Integracja europejska a rynek*. PSB-SW Kraków 1999, s. 47 – 62. (1999)
17. Montgomery, C. A. – Wernerfeld, B. Diversification, Ricardian Rents, and Tobin's q. The RAND *Journal of Economics*, 19, No. 4, pp. 623 – 632. (1998)
18. RAY, Kamal G., *Mergers and acquisitions: Strategy, valuation and integration*. PHI Learning Private Limited, 870 s. ISBN-978-81-203-3975-0. (2010)
19. Soukupová, J. a kol. *Mikroekonomie*. Praha: Management Press. (2000)

20. Smolková, E. *Strategická partnerstvá ako fenomén globálnej ekonomiky*, Bratislava : Spoločnosť autorov vedeckej a odbornej literatúry – Savol. ISBN 80- 7079-558-1. (2005)
21. Schertzinger, A. *Value Creation by European Insurance M&A.* [Dissertation Thesis.] Oestrich-Winkel, Germany: European Business School. (2008)
22. Schoenberg, R. Measuring the Performance of Corporate Acquisitions: An Empirical Comparison of Alternative Metrics. *British Journal of Management*, 17, No. 4, pp. 361 – 370. (2006)
23. Staikouras, S. K. An Event Study Analysis of International Ventures between Banks and Insurance Firms. *Journal of International Financial Markets, Institutions and Money*, 19, No. 4, pp. 675 – 691. (2009)
24. Stoyanova, R. – Grundl, H. Solvency II: A Driver for Mergers and Acquisitions? *The Geneva Papers on Risk and Insurance – Issues and Practice*, 39, No. 3, pp. 417 – 439. (2014)
25. Tan, F. Mergers and acquisitions finance. Innovation Ventures. [cit. 2020-03- 24]. Available from www: http://www.innovationventures.sg/finance-for- entrepreneu rmanagers/mergers---acquisitions-finance. (2019)

Evaluation of Slovak Automotive Industry Competitiveness Based On Market Concentration Indicators

Vladimír Hojdik[1*]

[1]University of Economics in Bratislava, Faculty of Business Management, Department of Business Economy, 852 35 Bratislava, Slovakia

Abstract. This paper evaluates the concentration of Slovak automotive industry, and its potential impacts on competitiveness and innovation activities of businesses operating within the industry. It analyses the concentration of automotive industry by using relevant metrics as market share, concentration ratio CR_3, CR_5, CR_{10} and Herfindahl index. The aim of this paper is to analyze financial results of Slovak automotive businesses, then measure the degree of concentration of whole industry in order to explain industry competitiveness in context of Slovak economy. Finally, the paper should inspire a reader in order to make him consider, what is the impact of strongest automotive companies on whole business environment in Slovak Republic.

1 Introduction

Automotive industry represents a key sector of the Slovak economy. This industry has some specifics as it includes not only producers, but also suppliers and partially resellers (resellers officially belong to sector of commerce/retail in Slovak environment). For the production of the final product (automobile), wide range of components is necessary and that is why automotive industry is closely connected with the other sectors of the economy. Businesses operating in this sector are cooperating together and create complex supply chains, which fosters the importance of automotive even more.

Sector of automotive has gradually became the most important driver of the Slovak GDP. In the mid-nineties, the first newcoming automotive business was Volkswagen Slovakia in Bratislava. A decade later, Kia Motors Slovakia established their factory in Žilina and PSA Slovakia did the same nearby Trnava, both in 2006. Additional investments helped to build the supply chain what resulted to raising number of businesses producing parts for automakers [14].

By the time, automotive sector became the "backbone" of the Slovak economy. Thus, it is important to assess the competitiveness and concentration ratio within the industry. Both of them, competitiveness and concentration of the automotive sector are closely related to innovation potential of automotive companies what has an enormous effect on the

[1] Ing. Vladmír Hojdik, PhD. : vladimir.hojdik@euba. ORCID : 0000-0001-6955-1677.

performance of whole Slovak economy. All in all, this paper aims to build on the results of past studies in order to enrich this research area with new knowledge.

2 Methods

The aim of the paper is evaluation of Slovak automotive industry concentration, based on relevant indicators. In the process of the research, we used general and specific research methods. General methods – analysis, comparison, synthesis were predominantly used in introductory part of the paper. Introductory part had mainly theoretical character as it describes current situation and structure of Slovak automotive sector. However, in order to achieve defined aim of the paper, the use of specific methods was the main part of the research.

Specific research methods include:

- mathematical and statistic methods,
- data processing with use of computer editor MS Excel,
- application of special indicators: concentration ratio and Herfindahl index.

Mathematical and statistical methods enabled to quantify the key metrics that allow to assess the competitiveness of industry – in this case automotive industry. First, we had to review market share of businesses (expressed in %). Based on the market share, we could proceed to quantification of another important indicator – concentration ratio. Concentration ratio depends on market shares of businesses within the industry, and most often is identified for three (CR_3), five (CR_5) and ten (CR_{10}) businesses with highest market share (also expressed in percent). The final indicator was Herfindahl index (also known as Herfindahl-Hirschmann index) which measures absolute concentration of industry. Concentration ratio and Herfindahl index are two most frequently used indicators of industry competitiveness measurement [3]. Results obtained from CR_3, CR_5, CR_{10} and Herfindahl index quantification became the basis for evaluation of competitiveness within automotive industry.

3 Results and discussion

3.1 Current situation of Slovak automotive industry

Considering the automotive industry, it is crucial to realize that the sector is not only about automakers. Supply companies have an enormous impact on the sector operation, despite indirectly involved in the car production process. Automotive industry in Slovakia is thus formed by manufacturers (four at the time, mentioned above) together with suppliers and there are close relations between these subjects [14].

According to financial website finstat.sk, the number of businesses operating in Slovak automotive reached up to 350 for year 2018. Among the suppliers with highest revenue we can include Mobis Slovakia, SAS Automotive, Faurecia Automotive or ZF Slovakia which are located on western Slovakia. The strongest suppliers of eastern Slovakia are currently Lear Corporation Seating Slovakia and Marelli Kechnec Slovakia [7].

As listed in the report published by SARIO agency in February 2020, up to 250 000 jobs are created directly and indirectly by the automotive industry. Moreover, the share of automotive sector revenues on the total revenues of machinery industry climbed nearly to the level of 50% in year 2015 [19]. Slovak automotive sector also accounted for more than a third of total exports, reaching totally 35 percent [20]. Table 1 listed below represents a database needed for the research execution, as it completes industry revenues for period 2016 – 2018. All data were obtained from finstat.sk [7].

Table 1. Revenues of top 20 Slovak Automotive Businesses in years 2016-2018 (mil. EUR)

			2018	2017	2016
1.	Volkswagen Slovakia	producer	10 380	7 549	7 586
2.	Kia Motors Slovakia	producer	5 186	5 185	5 566
3.	PCA Slovakia	producer	2 792	2 694	2 499
4.	Mobis Slovakia	supplier	1 480	1 421	1 433
5.	SAS Automotive	supplier	910	644	670
6.	Faurecia Automotive	supplier	864,3	578	582
7.	ZF Slovakia	supplier	478,1	471	424
8.	Yura Corporation	supplier	381,1	363	368
9.	Adient Slovakia	supplier	346,6	454	183
10.	Lear Corporation Seating	supplier	333,4	270	184
11.	Continental Automotive Systems	supplier	326,8	314	255
12.	Marelli Kechnec Slovakia	supplier	297,9	354	398
13.	Faurecia Slovakia	supplier	285,0	X	X
14	HBPO Slovakia	supplier	244,4	196	191
15.	Plastic Omnium	supplier	232,8	228	182
16.	Sungwoo Hitech Slovakia	supplier	212,1	206	244
17	Hyundai Transys Slovakia	supplier	197,2	188	205
18.	Matador Automotive Vráble	supplier	193,8	165	149
19.	Golde Lozorno	supplier	190,4	200	248
20.	Jaguar Land Rover Slovakia	producer	186,8	70	28

3.2 Characteristics of key indicators

Market concentration and its changes inform about the quality of competition in the economy. Concentration ratio values give institutions an opportunity to monitor the competitiveness of businesses operating in specific sector of economy [16].

The literature is home to numerous methods that evaluate the degree of industry concentration. For purposes of this paper, indicators of absolute concentration were calculated (absolute concentration indicators quantify the degree of concentration in all subjects). Among absolute concentration indicators we include concentration ratio and Herfindahl index [9]. Both indicators are based on the calculation of market share. The fundamental data for indicators quantification are the revenues of individual businesses.

3.2.1 Market share

The basic indicator of concentration ratio is the market share. It is share of the individual enterprise revenues on the revenues of whole sector. Market share measures, how the manufacturer performs in comparison to the other market participants. Analytically, it is possible determine it by the following formula [9]:

$$r_k = \frac{q_k}{q} = \frac{q_k}{\sum_{j=1}^{n} q_j} \tag{1}$$

where n – number of companies operating in the sector,
q_k – revenues of k company, $k = 1, ..., n$,
q – total revenues for whole sector,
r_k – market share of k company.

Market share can be expressed as relative or as percentage result. It takes values from the interval <0; 1> for the first, and <0; 100> for the latter. Extreme values indicate either full concentration in the industry – monopoly for r_k=1, or the absence of market operators in the sector for r_k = 0 [9].

3.2.2 Concentration ratio (CR)

Concentration ratio is an indicator of the market share sum held by few of the strongest industry companies. Formula of the concentration ratio is as follows [9]:

$$CR_{\psi} = \frac{1}{q} * \sum_{k=1}^{\psi} q_k \tag{2}$$

in which ψ lays in the interval <1;n> and CR_{ψ} takes values from range <0;1> when calculated in relative numbers, or <0,100> if we calculate percentage results. In practice, the concentration ratio is commonly quantified for three, five or ten strongest companies in the industry. It is therefore quantification of indicators CR_3, CR_5 and CR_{10}. In some other studies, authors prefer indicators CR_4 and CR_8 quantified for four and eight strongest subjects [15, 17, 18].

3.2.3 Herfindahl index

Herfindahl index measures the absolute concentration in the industry. In the literature we may also be encountered with the concept of the Herfindahl-Hirschmann index [10]. It is used for sector analysis in many countries and it is well-known and commonly accepted indicator of market competition [5]. In this paper, it explains, how financial power (revenues) of individual subjects impacts the concentration of whole industry. Herfindahl index calculates the square of the market share of each subject in the industry and sums up the resulting numbers [6].

Analytical expression of Herfindahl index is as follows [9]:

$$H = \sum_{k=1}^{n} \left(\frac{q_k}{q}\right)^2 \tag{3}$$

and the meaning of the variables remains identical, as in the previously defined formulas.

Herfindahl index H takes values from <1/n, 1>. Extreme low value $H = 1/n$ means that every company has equal market share in the industry. For extreme upper level, for which $H = 1$ there is a pure monopoly on the market – one company produces 100% of whole industry production [4].

In practical use, the Herfindahl index may not be used as a number of the interval mentioned above. It is because in sectors with a higher number of enterprises we would operate with very low numbers. The value of H therefore tends to be adjusted by the appropriate multiplier, most commonly used multiplier is 10,000. Depending on what is the value H after conversion, we can classify the degree of concentration of the industry, according to the following criteria defined by the US Federal Trade Commission (FTC) :

- for $H < 1500$, the industry is concentrated,

- for *H* is in the range of <1500; 2500>, the industry is moderately concentrated,
- for *H*> = 2500, the industry is not concentrated [22]. Values of Herfindahl index also impact how government sets and enforces competition policy [13].

The current form Herfindahl index (as we know it today) was firstly applied in 1950 in dissertation thesis of O. C. Herfindahl. Later, in 1959 he used it in a study about international copper industry [11]. The similar concept (sum of squared market shares), however is known since 1940 when A. O. Hirshmann executed a trade study. Hirschmann in his paper suggests that original idea of index belongs to him [12]. Therefore, in the literature we can find concepts of Herfindahl index or Hirchmann-Herfindahl index, but both of them represent the same indicator.

Herfindahl index has been widely used in multiple studies to consider competition level in variety of industries, for example in banking, insurance or automotive sector [1, 5, 8, 18, 23]. Its construction, however, has some limitations – it can measure concentration only on geographically isolated markets, but cannot be applied for international companies, as the results may be misleading [2].

3.3 Quantification of indicators

In this chapter, we calculate and discuss calculated results. Results are presented in Tables separately for different indicators – market share, concentration ratio and Herfindahl index. Table 2, Table 3 and Table 4 complete all data needed for purposes of the research.

3.3.1 Market share quantification

Table 2. Market share of top 10 Slovak automotive businesses in years 2016-2018 (%)

			2018	2017	2016
1.	Volkswagen Slovakia	producer	34,14	28,38	29,40
2.	Kia Motors Slovakia	producer	17,06	19,49	21,57
3.	PCA Slovakia	producer	9,19	10,13	9,69
4.	Mobis Slovakia	supplier	4,87	5,34	5,59
5.	SAS Automotive	supplier	2,99	2,42	2,60
6.	Faurecia Automotive	supplier	2,84	2,17	2,26
7.	ZF Slovakia	supplier	1,57	1,77	1,64
8.	Yura Corporation	supplier	1,25	1,36	1,43
9.	Adient Slovakia	supplier	1,14	1,71	0,71
10.	Lear Corporation Seating	supplier	1,10	1,02	0,71

The balance of power between ten strongest businesses based on their market share is reflected in Table 2. Three producers keep the distance while Volkswagen Slovakia, as leading company, clearly holds top position. The revenues progression of producers remains steady during analyzed period, there was no any radical change. Volkswagen Slovakia holds approximately one third of whole industry and controls about half of the market among producers.

In the suppliers subsector, Mobis Slovakia holds its position as the strongest supply company. It also places fourth in the total revenues ranking of the whole industry and nothing has changed on that from 2010 to the present [14]. Even thought Mobis Slovakia has the strongest position among suppliers, its revenues are almost ten times lower in comparison to market leader, Volkswagen Slovakia.

Trinity of producers clearly dictate conditions of the Slovak automotive sector. Production among suppliers is significantly fragmented and none of them comes close to the level of producers revenues. The main reason is the character of the products sold – while suppliers sell only parts (B2B model), producers sell the finished product to the customer (B2C model). Producers thus sell products with incomparably higher added value which translate into higher revenues.

3.3.2 Concentration Ratio quantification

Concentration ratio of industry is dependant on market share values. As already mentioned before, concentration ratio in the industry is most often calculated for three, five and ten strongest companies in the industry. Table 3 contains already calculated concentration levels CR_3, CR_5 and CR_{10} for the automotive industry in Slovakia, based on revenues achieved by enterprises in the relevant period.

Table 3. Concentration Ratio CR_3, CR_5 and CR_{10} of Slovak Automotive Industry (%)

	2018	**2017**	**2016**
CR3	60,39	58,00	60,66
CR5	68,25	65,76	68,85
CR10	76,16	73,79	75,60

The results show stable development of concentration ratio during the period. Concentration ratio CR_3 represents the share of three strongest producers which control about three fifths of the local automotive industry. Their position is dominant and probably anything can threaten it in the near future. Moreover, their hegemony is not short-term at all – they stand on the top of the market since 2010 [14].

Concentration ratio CR_5 also does not indicate any rapid fluctuations, what applies on CR_{10}, too. What is more interesting, however, is comparison of concentration ratios CR_5 and CR_{10}. Analysing CR_{10}, we can notice significant asymmetry comparing first 5 strongest and second 5 strongest players. In 2018 the top 5 companies recorded 68,25% of Slovak automotive production, while the second strongest quintet achieved only 7,91% of total industry production. Such imbalance is highly inclined to elite five companies and this trend is evident throughout all the years under review. All in all, the data above confirm relatively large differences between financial power of top 10 industry participants.

3.3.3 Herfindahl index quantification

A first step before Herfindahl index quantification, a key metric evaluating a degree of industry concentration, was to determine the number of businesses in the sector to 51 (n = 51), as it would be difficult to include all industry participants (more than 350). The first 50 strongest companies are considered as autonomous subjects, other companies in the industry will be merged into the aggregate entity with serial number 51 (other). This simplification will cause slight distortion in Herfindahl index quantification. However, the aggregate market share of other producers reached relatively low values in 4 – 7% range. Therefore, the distortion of final result is almost negligible as it does not affect results of Herfindahl index negatively.

Table 4. Herfindahl index of Slovak Automotive Industry in years 2016-2018

	2018	**2017**	**2016**
Herfindahl index H	1651	1400	1506

Herfindahl index is a complex indicator of the degree of industry concentration. During the analyzed period, values of Herfindahl index stay in the range of <1506; 1651>. According to the criteria of the Federal Trade Commission (FTC), the industry can be classified as moderately concentrated (because Herfindahl index is in the interval <1500; 2500>. This suggestion is underlined by other criterion where Herfindahl index for moderately concentrated industry takes values from interval <1000; 1800> [4].

Results of Herfindahl index specify Slovak automotive industry as moderately concentrated industry, however, it is necessary to interpret these results from different perspective. On the one hand, a large number of entities operate in the industry which supports the idea that the sector is relatively open and thus not concentrated. On the other hand, there are three dominant businesses in the sector as they account for two thirds of the total industry revenue, and that favors the thesis of a highly concentrated industry. What is important and positive, that none of the entities possesses the power to use or abuse in their favor. The development of Herfindahl index remains steady between years 2016 – 2018 what points out stable conditions within the industry without significant market fluctuations.

4 Summary/Conclusion

Results obtained from the research prove that Slovak automotive industry is highly dependent on performance of three strongest producers. Based on thorough analysis of selected indicators it can be stated that automotive sector in Slovakia has signs of concentrated and also not concentrated industry. According to methodology for Herfindahl index calculation, Slovak automotove industry ranks in the category of moderately concentrated industries what means relatively good competition and suitable conditions for enterpreneurship within the sector.

Succesful development of Slovak economy heavily relies on optimal operating conditions of automotive businesses. Maintenance and improvement of environment within this sector are substancial preconditions for further economic growth. While in 2012 automotive industry contributed to the overall industrial production by 28%, in 2015 this rate reached up to 32% [20]. The share was continuously rising in following years, reaching the level of 35% in year 2018 [21]. Table 5 is a summary of facts mentioned above.

Table 5. Share of automotive industry production on whole Slovak industry production in years 2012, 2015 and 2018 (%)

	2018	**2015**	**2012**
Share	35	32	28

Even other financial data confirm the major role of automotive industry. Data obtained from financial website finstat.sk prove, that automotive producers Volkswagen Slovakia, Kia Motors Slovakia and PCA Slovakia placed very high according to net profit or generated value added, when compared to other Slovak companies [7].

Such operation of automotive sector has multiple interpretations. On the one hand, its further expansion should have positive short-term impact on growth of Slovak national economy. From long-term point of view, however, Slovakia should focus on diversification of its economy – national GDP is mostly generated by revenues of automotive sector. Moreover, automotive businesses in Slovakia (especially suppliers) are performing mainly production and operational activities, with lack of investment into innovation or development. The given claims points to fragile and possibly unstable Slovak economy to the future, because of its over-orientation on one sector. Serious financial or investment complications in automotive industry may trigger rising unemployment, decrease of corporate revenues and finally, weakening whole Slovak economy.

References

1. F. M. Aijde, J. O. Ajileye, Market concentration and Profitability in Nigerian Banking Industry : Evidence from Error Correctiion Modeling. *International Journal of Economics, Commerce and Management* **3**, 1-12 (2015)
2. P. Aghion, N. Bloom, R. Blundell, Competition and Innovation : An Inverted-U Relationship. *Quarterly Journal of Economics* **120** (2), 701-728 (2005) DOI : 10.1093/qje/120.2.701
3. J. A. Bikker, K. Haaf, Competition, concentration and their relationship : an empirical analysis of the banking industry. *Journal of Banking and Finance* **26**, 2191-2214 (2002) DOI :10.1016/S0378-4266(02)00205-4
4. I. Brezina, A. Oršulová, J. Pekár, Analýza absolútnej koncentrácie vybraného odvetvia pomocou Herfindahlovho-Hirschmannovhio indexu, *Ekonomický časopis* **57** (1) 77-94 (2009)
5. I. Brezina, J. Pekár, Z. Čičková, M. Reiff, Herfindahl-Hirschmann index level of concentration values modification and analysis of their change, *Central European Journal of Operations Research* **24**, 49-72 (2016) DOI : 10.1007/s10100-014-0350-y
6. M. C. Cavalleri, A. Eliet, P. McAdam, Concentration, market power and dynamism in the euro area. *ECB Working Paper Series, Discussion Paper No. 2253*, European Central Bank (2019) DOI :10.2866/379250
7. Database of Slovak Businesses and Organizations – automotive industry. (2016) Available at <https://www.finstat.sk/databaza-firiem-organizacii?activity=Automobilov%C3%BD%20priemysel&sort=sales-desc>
8. J. Fedderke, D. Naumann, An Analysis of Industry Concentration in South African Manufacturing 1972-2001, *Applied Economics* **43**, 2919-2939 (2011) DOI : 10.1080/00036840802631835
9. M. Fendek, E. Fendeková, *Mikroekonomická analýza* (Iura Edition, Bratislava, 2008)
10. G. Grullon, Y. Larkin, R. Michaely, Are US industries becoming more concentrated ? *Forthcoming, Review of Finance* (2017)
11. O. C. Herfindahl, *Copper Costs and Prices : 1870-1957* (Baltimore, The John Hopkins Press, 1964)
12. A. O. Hirschmann, The Paternity of an Index, *American Economic Review* **54**, 761-762 (1964)
13. P. MacKay, G. M. Phillips, How Does Industry Affect Firm Financial Structure ? *The Review of Financial Studies* **18** (4), (2005) DOI : 10.1093/rfs/hhi032
14. Š. Majtán, V. Hojdík, R. Šlosár, Financial performance of automotive companies and its impact on concentration of automotive industry in Slovak republic, *Finance and performance of firms in science, education and practice : proceedings of the 8th international scientific conference : april 26 - 27, 2017*, Zlín, Tomas Bata University in Zlín. (2017)
15. E. Nauenberg, K. Basu, H. Chand, Hirshmann-Herfindahl Index Determination under Incomplete Information, *Applied Economics Letters* **4** (10), 639-642 (1997) DOI : 10.1080/758533291

16. S. Rhoades, The Herfindahl-Hirshmann index, *Federal Reserve Bank of St. Louis* **52** (3), 188-189 (1993)
17. Z. E. L. Rolim, R. R. Oliveira, H. M. Oliveira, Industrial Concentration of the Brazilian Automobile Market and Positioning in the World Market, *Working Paper* (2019) DOI :10.13140/RG.2.2.11210.11207
18. M. N. Al Arif Rianto, T. B. Awwaliyah, Market share, concentration ratio and profitability : Evidence from Indonesian Islamic Banking Industry. *Journal of Central Banking Theory and Practice* **2**, 189-201 (2019) DOI : 10.2478/jcbtp-2019-0020
19. SARIO, *Automotive Sector in Slovakia* (Bratislava, SARIO, 2020)
20. Statistical Office of Slovak republic, Yearbook of Industry of the SR 2015 (Bratislava, 2015)
21. Statistical Office of Slovak republic, Yearbook of Industry of the SR 2018 (Bratislava, 2018)
22. U.S. Department of Justice and Federal Trade Commision, Horizontal Merger Guidelines, (2010)
23. D. M. L. Williams, P. Molyneux, J. Thornton, Market Structure and Performance in Spanish Banking. *Journal of Banking and Finance* **18**, 433-443 (1994) DOI :10.1016/0378-4266(94)90002-7

Acknowledgement

This paper is one of the partial outputs of the research grant no. I-20-103-00 entitled "Assessment of innovation potential in terms of new managerial trends on the base of proposed determinants within automotive industry", led by Ing. Monika Raková. PhD., University of Economics in Bratislava, Faculty of Business Management, Department of Business Economy.

Practical usability of CAT Model – a case study

Jaroslav Hradílek

Masarykova Univerzita, Brno, Faculty of Economics and Administration, Department of Corporate Economy, Lipová 507/41a, 602 00 Brno, Czech Republic; 10955@mail.muni.cz

Abstract. The average life span of an organization, irrespective of its size, is 12½ years. The majority of companies which get into a crisis find themselves in this situation because of internal factors. To survive in the evolving and more and more demanding environment, organisations have to focus on their internal processes, leadership and management. However, there is no reliable and easily usable tool or model evaluating such criteria of an organisation in place. Following the need to have a lean, yet reliable tool to evaluate a level of Leadership, Management Maturity and Culture a new Performance Model has been created. The CAT Model (acronym stands for Complex Assessment Tool) can be used to evaluate and assess its strong and weak sides using structured interviews. The usage of the tool proved useful, giving owners/management valuable feedback and areas with potential for improvement. This article describes its practical usability in an assessment carried out on a Czech company.

1 An introduction into CAT model

Despite the fact that enormous effort has been deployed in the field of strategy management, the "Search of excellence" does not seem to be any closer to its objective than in 1982, when the book from Peters & Waterman has been published. Thought significant development has taken place, the evolving environment, competition, society and technology still force organisations to constantly adapt or die. Ever since the eighties business writers have been claiming that the world was getting more dynamic, turbulent, unpredictable and competitive. As it turned out, in retrospect the 1990s were 'a piece of cake' compared to what is happening now in the world [1].

There are many reasons attributable for company success of failure and both business community and academic sphere are in constant search of them and their impacts. Some of the facts presented by de Waal (based on Booz Allen Hamilton survey among 1,200 large companies) indicate that majority of them are internal:

- from the 360 worst performers 87 percent of the value destroyed was caused by strategic missteps and operational ineffectiveness
- More than 50 percent of managers take decisions based on their gut feeling not on hard facts, and 36 percent has black boxes in the organisation of which they know hardly anything
- four out of every ten organizations do not actively work on increasing employee loyalty

- Return rates and warranty costs are dramatically rising while at the same time customer satisfaction levels are steadily decreasing, a strong indication of the deteriorating quality of products
- The majority of companies which get into a crisis find themselves in this situation because of internal factors, of which dysfunctional management (48 percent of the cases) and inadequate management information systems (42 percent) are the most common causes [1]

As a matter of fact, despite strategies, tools and processes, individual biases persist in every organizations [2]. To increase the chance that an organisation is successful (or at least survives) in a long run - the focus has to be on internal processes and management of a company. Yet, are we able to measure the quality and maturity of an organisation's managerial ability?

Many researchers and business community members focus on the field of company performance's measurement. There are also important criticisms of PMS. Hamel [3] criticises the suitability of performance measurement theories and practices for modern organisations. Suggesting that the measurement and management of performance in organisations increases fear, reduces trust, promotes "command and control" systems, diminishing employee engagement. Thought it may seem a bit too strict, a recognition of a democratic, open, participative and fear-free management is being essential for the twenty-first century organisation. However, the interplay among performance measurement, performance management, employee engagement and performance remains poorly understood. Understanding this interplay is now considered critical for further development of the performance measurement theory [4, p. 1207].

Collins [5] proposes to identify a sole key performance indicator (e.g. sales per customer visit) as one of key drivers to outperform competition. On the opposite, Kaplan & Norton [6] compare managing an organisation to the piloting an airplane, arguing that it is impossible to focus on one sole indicator (e. g. ground speed) ignoring the others (altitude, fuel level, engine temperature...). Neely et al. [7] pinpoints the enormous efforts spent on controlling (according to the Hackett Group an average corporation with turnover over 1 bln. USD spends 25 000 man-days by planning and budgeting, Ford alone spends for the same 1.2 bln. USD yearly). Still there is very rare to find a person responsible for a quality and improvement of planning, budgeting and performance measurement. We can observe that no consensus can be given to answer the questions like: Are we measuring the right things? Do we measure enough/not too excessive? Is the cause-effect description in place?

An effort to overcame the gap and improve organisation's PMS offers Van Aken et al., by developing a tool assessing PMS [8]. However, no widespread usage can be recorded. In non-financial indicators seems the situation even worse – thought their importance is generally acknowledged by [9-11] and many others, a set of criteria or a tool, which provides lean and easy evaluation is missing. Hálek [9, p. 9-10] states that "*In a global environment, where a major part of a business operates in an international environment, financial indicators seem insufficient, because for the most part they are historical indicators and do not reflect future developments*". Therefore, the focus on performance models and management techniques grows. An attempt to find a tool evaluating non-financial criteria in a measurable and comparable way was conducted by the author. The only performance models fulfilling the need for comparable results are EFQM and Baldrige award criteria. Yet there is a serious obstacle for broader usage of them – the usage of these tools is complicated and especially small and medium organizations do not have enough resources for it [11]. Its deployment requires to undertake a multiday training and hiring an external assessor. The assessment itself takes several days, which represents an effort that SMEs are usually not willing to invest. Model is also primarily intended for improving operational management [1], not as a benchmarking tool. Moreover, as identified by Rusev & Salonitis, [12, p. 277] "*...the existing*

assessment tools are generally either biased towards process improvement or towards results. Assessment tools developed by Institutes that award prizes such as the MBQA and EFQM are results driven with little focus on culture and process efficiency". This might change with an introduction of new EFQM Model 2020 version.
As no suitable tool exists, a new model serving mentioned purposes was developed. It serves the management/owners of a company to better understand their company and to identify its strong/weak sides. It enables them to compare the results over a certain time period or compare with other organizations. This paper deals with practical usability and results of the model.
The CAT model was created by author of this paper to give both researchers and practitioners an easy-to-use tool, which enables them to assess organisational maturity using "soft" factors of an organisation like leadership and culture. It was tested on 12 independent organisations, the extent should be considered adequate [13]. It consists of 17 criteria grouped in three categories - Maturity, Stakeholders & Others. They are: for **Management** – Leadership, Corporate culture, Quality, Performance, Controlling, Processes, for **Stakeholders** - Communication, People, Employee selection process, Customers, Suppliers, Owners, Public authorities and for **Resources** - Financials, Exclusivity, Technology, External factors. The scaling of criteria is in a range from 1 to 5 with 0.5 step. Each functional category has its subtotal to better identify strong/weak sides of an organization Total achievable score is 17 – 85, where upper quartile score - above 68 points indicates well managed, mature organization, lower quartile (under 34) on the other hand, poorly managed entity likely to fail in a near future. The results are visualized for every respondent (and an average result) on a web chart indicating strengths and weaknesses. The results are scalable, comparable over time and enable benchmarking.
The usage of the model is based on a set of structured interviews with three respondents from an organisation. It shall include a top manager or owner; in case he is managing actively. Second respondent shall be another high-level manager, who has best overview of processes and daily business. Preferably executive director, financial director or similar. Third respondent shall be an opinion maker, a person, who is close to the front-line employees, but has a good overview over the organisation. This can be HR manager, head of trade unions, compliance officer, ombudsman or alike. Each participant is interviewed separately.
Both graphical output and text results and recommendation is provided by the tool.

2 A practical usage of a model

2.1 A target company description

As the results are rather sensitive, it was decided not to disclose the name of a target company. It was founded in 1991 in Czech Republic, is fully privately owned and is a manufacturing company. It employs approximately 180 people and yearly turnover is around 16 mil. EUR. Let's refer to it as a COMPANY.

2.2 COMPANY analysis using CAT model

According to the CAT model methodology was the assessment conducted with three company representatives. Firstly, the representative of employees was questioned, it was a head of Human resources department, acronym VH will be used for him. As a second interviewed was an executive director (LK), lastly both company owners (S+S) The strong/weak sides of a COMPANY were identified and areas requiring immediate attention were discovered.

2.3 Graphical results of an evaluation

On Figure 1 there are the results of evaluation clearly visible on a web chart. All participants and their average are shown on the figure. It was expected, based on previous experience that high-level management (owners) ranking is higher than those of middle management. In this case the opposite was true, in most of the criteria the favourable answers were given by HR manager (VH), reflecting also opinions of the employees. The biggest difference can be seen at criterion Employee selection, where the HR manager was satisfied (with own work) at maximum level, whereas the owners received it as absolutely unsatisfactory. Together with criterion Performance were these two results most divergent, what suggests low amount of communication among owners and employees (when the owner is unsatisfied with respective departments performance, say for not fulfilling KPI's, it is necessary to implement some corrective actions and search for improvement). However, if the employees do not recognize the unsatisfaction, it is not reasonable to expect any improvement. Author's note: this resulted in parting of the COMPANY and both HR manager and executive manager.

Fig. 1: CAT model result for a COMPANY Source: own elaboration

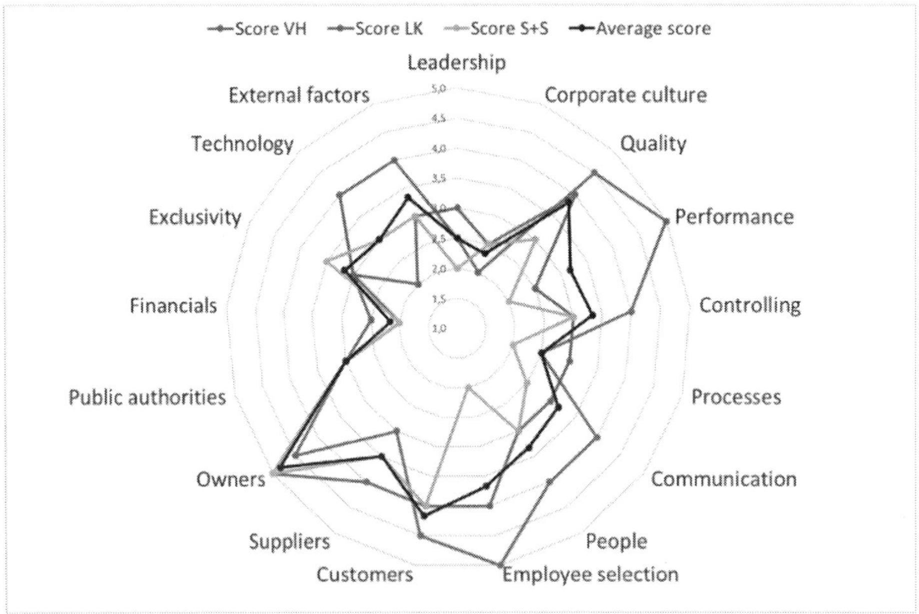

The main weaknesses can be seen in three criteria. First of them is Leadership, second is Corporate culture and third is Finance. For the last, it may be considered as a result than the root cause and you cannot trigger finance in a long run solely without improving in the criteria Leadership and Corporate Culture. This is also confirmed by Evaluation comments which the Model provides, see Chapter 2.4.

Results of Categories Maturity and Others were below average, an average result was achieved in Category Stakeholders, better result was not permitted by low score in criteria Communication, People and Public Authorities.

Table 1. Assessment results in respective categories

Category evaluation	Management	Limits	Stakeholders	Limits	Resources	Limits
Maximum		30		35		20
Upper quartile		24		28		16
Average		18		21		12
Lower quartile		7,5		9		5
Minimum		6		7		4
Score:	**17,7**		**25,7**		**11,7**	

If we compare overall results of each interviewed, the lowest evaluation was achieved surprisingly by owners (S+S). All other results were above-average, in some criteria we can expect deeper knowledge of owners than HR manager, resulting in informed pessimistic view in comparison to the view of HR or executive manager.

Table 2. Assessment results of each participant

COMPANY evaluation	Average score	Score VH	Score LK	Score S+S	Limits
Maximum:					85
Upper quartile					68
Average					51
Lower quartile					34
Minimum:					17
Score:	**55,0**	**62,0**	**54,0**	**49,0**	

2.4 Overall evaluation comments

- Total achievable score is 55, what is slightly above average (55 points). The average result indicates an organisation with fair level of management maturity.
- The highest ranking was provided by VH, LK was on average, S+S were below the average.
- The best results achieved category Stakeholders, from criteria both Customer and Owners. Categories Management and Others scored below-average.
- Significantly below average are the criteria Leadership, Corporate Culture, Processes and consequently Finance.

- A remarkable difference was identified in criteria Performance and Employee Selection.
- The company excels in focus on its customers and quality
- The downsize is in setting and living its vision, values and objectives into operations, relation of its strategy to targets and performance evaluation
- Exceptional relations with stakeholders, beside owners also customers
- Overall result is average, the recommendation is except running improvement projects to focus more on soft factors of an organisation, on sharing common values, to motivate on joint success (ownership culture). To focus more on internal communication among departments.
- It is recommended to formalize management processes (minutes of meetings, tasks setting and evaluation, managing of managers instead of front-line employees.

2.5 Evaluation results and next steps

The analysis discovered a weakness in internal communication and managing style as itself. Consequently (this is author deduction only) non-fulfilling of financial targets in last two years with worsening trend. The strategical meeting on mid-level management was organised by the executive manager to identify quick wins and to implement mid- and long-term measures.

Just half of the measures were not of financial or technological nature but dealt with issues like communication, motivation or organisation. This confirms the findings from CAT model that main problem within the organisation is communication, management style and company culture.

The organisation did not fully utilize the innovation potential of its employees (no improvement process was established, the employees did not come with own proposals), no coherent system of internal projects and changes management existed. It caused ineffective resources utilisation, duplicability, moderate progress and projects delays resulting in overall low motivation. The process did not meet expectations and did even not earn sufficient management.

Measures implemented by the company owners were among others much closer attention on company performance and processes, personal changes, profitability analysis on customer level and cost optimisation. This lead quickly to improvement in economic results see monthly Gross profit development as most important of them in figure 3. Even though seasonal effect in summer months the improvement can be identified, confirmed also by trend line.

However, little was done in regards of communication, company culture. Managers tried to focus on their leadership style, the future development and future assessment will confirm, if the changes were sufficient and the improvement did not exposit a small adjustment only.

Fig. 2 Development of Gross Profit with improvement since summer 2019. Source – own calculation on COMPANY data.

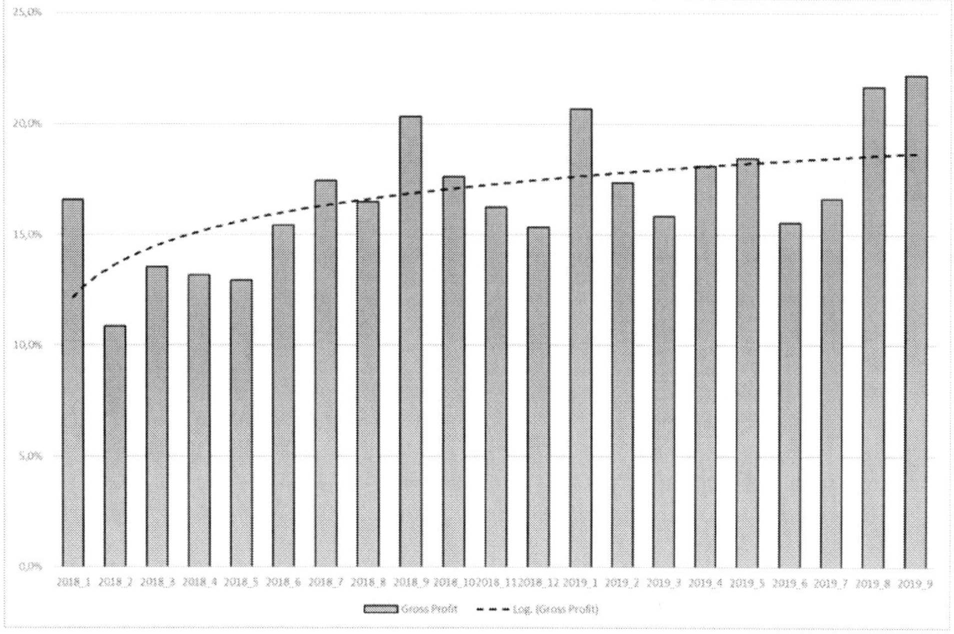

3 Conclusion, further application

The CAT model was used in order to help COMPANY when it faced deterioration of their market position and consequently worsening their financial results. It identified as most important findings:

- for several areas, there is a significant distortion in perception of company owners and their management, resulting in personal changes on positions of executive manager and HR manager
- the COMPANY is poor in maintaining its company culture, internal communication and leadership style. Presumably the financial problems can be attributed to these flaws
- excellent results were achieved in criteria Customers and Owners and also Quality

After the assessment the performance of COMPANY improved. There has been, beside personal changes, increased attention paid by the owners and many corrective actions were taken. As no significant change occurred in order to improve leadership style and internal communication, the long-lasting effect of these measures can be questioned. The new executive manager and his leadership style will be decisive in forming the COMPANY success in next years.

The usage of CAT model proved to be helpful in identifying weaknesses of an assessed organisation and proposing further actions. However, it still delivers identified weaknesses and recommendations to rectify only, the actions and measures have to be taken and implemented by executive management.

It would be reasonable to conduct the same assessment in 1-2 years period to track possible changes and evaluate effect of measures taken. Also, it would be helpful to conduct more assessments in similar organisations to allow benchmark for a COMPANY.

References

1. Waal, A.a. De. "The Characteristics of a High Performance Organisation." *SSRN Electronic Journal*, (2006). Doi:10.2139/ssrn.931873.
2. Powell, T. C., Lovallo, D. and Fox, C. R. Behavioral strategy. *Strategic Management Journal,* (2011). **32**. doi:10.1002/smj.968
3. Hamel, G. (2009), "Moon shots for management", Harvard Business Review, Vol. 87 No. 2, pp. 91-98.
4. Smith, M. and Bititci, U.S. (2017), "Interplay between performance measurement and management, employee engagement and performance", International Journal of Operations and Production Management, Vol. 37 No. 9, pp. 1207-1228
5. Collins, J. C. *Good to Great: Why Some Companies Make the Leap... and Others Don't*, USA: William Collins, 2001
6. Kaplan, R. S., & Norton, D. P. Linking the Balanced Scorecard to Strategy. *California Management Review.* **39**/1 (1996)
7. Neely, A. D., Adams, C., & Kennerley, M. *The performance prism: The scorecard for measuring and managing business success*. (Prentice Hall Financial Times 2002)
8. Van Aken, E., Letens, G., Farris, J.A., Coleman, G.D., & Van Goubergen, D. Application of an integrated tool for assessing performance measurement system effectiveness and maturity. *International Journal of Productivity and Performance Management.* **54**/5-6 (2005)
9. Hálek, V. *Non-financial indicators in the valuation process.* (Sciemcee Publishing, 2016)
10. Ittner C.D. & Larcker, D. F. Coming Up Short on Nonfinancial Performance Measurement. *Harvard Business Review*, (November 2003)
11. Aschenbrennerova, H. Měřením a řízením výkonnosti ke zvyšování konkurenceschopnosti malých a středních průmyslových podniků v období globální finanční krize. *Journal Úspěch.* **10**, 1 (2010)
12. Rusev, S. J., Salonitis, K. Operational Excellence Assessment Framework for Manufacturing Companies. P*rocedia CIRP. 55, 5th CIRP Global Web Conference - Research and Innovation for Future Production* (CIRPe 2016). ISSN: 2212-8271.
13. Long, T.B., Blok, V., Dorrestijn, S. & Macnaghten, P. The design and testing of a tool for developing responsible innovation in start-up enterprises, *Journal of Responsible Innovation,* **7**/1, 45-75 (2020)

Enclosures

https://doi.org/10.1051/shsconf/20208301023

Enclosure 1: The results of the assessment and full CAT model:

Category	No.	Criterion	Description	1	3	5	Average score	Score VH	Score LK	Score S+S
Management	1	Leadership	Has the organization defined a vision, values, purpose and they are clearly communicated? Is the strategy aligned with values and successfully incorporated in operational activities?	Vision, values, purpose are not existent or not communicated, strategy is not defined in alignment with values and purpose/measurable objectives do not exist.	Values and vision exist and the team can find them. Unit strategy exists, there is no direct connection among values/strategy/objectives, objectives are not met.	Values and vision/purpose are regularly communicated and most employees know them well. Vision and mission contain more than profit only. Strategy is aligned with them; its implementation and objectives fulfilment are regularly evaluated. In strategy is clearly defined, what will/will not organization do.	2,5	2,5	3,0	2,0
	2	Corporate culture	Is the culture of organization open, friendly and inspiring?	Nothing unites employees, internal rivality among departments/employees exists, there are signs of competition, fear or despotism. Nobody is interested in an opinion of front-line staff.	Cooperation is encouraged, there exists no favouritism/unfair behaviour. No shortcuts or suboptimization of individual interest on the costs of organization is accepted.	Culture is inspiring, it is usual to listen to and argue different opinions, high grade of employees' engagement. The culture of freedom and accountability leads to personal responsibility for results.	2,3	2,0	2,5	2,5
	3	Quality	Is the quality the cornerstone of a company?	No, quality is either considered unimportant or no effective measures are in place.	There are initiatives and measures regarding quality, effect not measured or not apparent.	The quality of work/processes/results is the main cornerstone, each individual is committed to it, a principle of personal responsibility for quality, results being evaluated on a regular basis	3,8	4,5	4,0	3,0
	4	Performance	Is individual performance measured/managed? Is the compensation based on performance?	No, the individual performance od units/individuals is not measured, no goals set, no effect on remuneration.	Goals are set, but no real influence on salary/not measured regularly, no direct connection to an organization's strategy.	Goals are set based on the strategy objectives of the company and its units. Organization strives to improve. There exists a direct link between performance and pay.	3,2	5,0	2,5	2,0
	5	Controlling	Company is managed based on hard data; key performance indicators are evaluated.	There is no structured controlling concept in place, sporadic interpretation of accounting data might exist.	Company has KPIs defined, there is either unclear cause-effect between KPI's and objectives or a way how to achieve them or organization does not pay respective attention to them or the significance is lost in huge amount of other data	Comprehensive controlling concept evaluates most important KPIs, describes the value flow incl. cause-effect relations and provides information for key decisions. It provides important information in advance. Bad news and unsatisfactory results are received and solved.	3,3	4,0	3,0	3,0
	6	Processes	Core processes are described, the effectivity of processes is being monitored	Processes are not described, no process effectivity monitored.	Core processes described, key processes have its owner, no effective evaluation/improvement in place	Processes sufficiently described, regularly tested, evaluated and improved based on KPI's.	2,5	2,5	3,0	2,0
		Subtotal Category Maturity					17,7	20,5	18,0	14,5
Stakeholders	7	Communication	The communication and cooperation with key stakeholders are an important topic.	Internal communication is poor the cooperation among departments is poor (specifically management <=> employees), cooperation among departments insufficient, communication with customers/public is not of importance.	Regular formal means of internal communication exist, the cooperation issues are being solved systematically. There are baselines for external communication (customers, suppliers, owners, public).	Very good internal communication both formal and informal, cooperation improvement initiatives supported. Communication concept exists based on organisation's objectives and desired relations to key stakeholders.	3,2	4,0	3,0	2,5
	8	People	Organisation is a partner for its employees.	Employees are just human resources, easily replaced, the main goal are personal expenses. Key employees are not unified with company, main reason for stay is the salary.	Employees are remunerated in such a way that they do not leave for salary purposes. They are motivated by non-financial factors like company culture, openness, a possibility of personal development, safety.	Employees are able and have opportunity to develop, they are top experts in their trade with great results. Thanks to their motivations and fair remuneration they do not leave. Fluctuation rate is under the industry average. People have freedom, authority and responsibility for their actions.	3,3	4,0	3,0	3,0
	9	Employee selection	Employee selection process is transparent and respects values and objectives of an organisation.	For a selection or stay of certain employees exist no reasonable reasons. Despite of low personal contribution they stay e.g. because of personal relations.	There are clear selection criteria, the employee initial training /adaptation process works well. Employees are encouraged in innovations and improvements.	Employees are hired (and fired) acc. to their identification with company values, their abilities, performance and efforts. Organisation develops necessary competencies and ensures successors at key posts. It stresses the employee retention.	3,7	5,0	4,0	2,0
	10	Customers	Is customer the key element in company's life?	No customer dedication, internal issues more important.	Frequent customers contact from sales people, demands being delivered by them.	All positions have a frequent contact with customer, R&D, sales, manufacturing, services are fully aligned with a customer requirements, which a company often knows sooner than they realize them.	4,2	4,5	4,0	4,0
	11	Suppliers	The total cost of ownership is preferred over short term benefits (low price), long term relationships are favoured	Main selection criterion is price/discount. A reasonable share of purchasing decisions is not transparent	All purchasing criteria are evaluated including lifecycle costs, OPEX, warranty. Some purchasing decisions are not transparent	TCO is key decisive factor, long term relations (frame contracts) established, cooperation with supplier (e.g. in R&D) in place. The sourcing process is transparent and documented	3,5	3,0	4,0	3,5
	12	Owners	The interest of the owner is the same as of the company	Owners follow short term financial/other interests only	Owners are ready to sacrifice (a part of) remuneration in order to further develop the company	The owners and company interests are closely aligned, owners support the company (investments, development, PR)	4,8	4,5	5,0	5,0
	13	Public authorities	The relationship is mutually beneficial.	There are disputes between company and authorities	The relationship with public authority is without any negative connotations.	The company is a good corporate citizen, working in mutual benefit with a local authority, environment and locality.	3,0	3,0	3,0	3,0
		Subtotal Category Stakeholders					25,7	28,0	26,0	23,0
Resources	14	Financials	Financial targets of an organisation (sales, costs, productivity, EBIT...) in three consecutive years are fulfilled or exceeded.	Organisation has no targets of a financial nature, do not follow them or targets were not met by more than 10 % in last three years (on average).	Realistic financial targets are set at the beginning of fiscal year, organisation fulfils on average +/- 5%.	Key financial targets of the company are exceeded on average by more than 10%, set goals are fulfilled/exceeded.	2,2	2,5	2,0	2,0
	15	Exclusivity	Has the organization any resource at disposal, which is exclusive and not easily copiable (e.g. tradition, brand, customer's passion; costs, logistics or other advantage)?	Company is average, surpasses the competition in no way. R&D is not oriented to gain/maintain an exclusivity.	Organisation has and important advantage over the competition and aware of it. R&D reflects this, but is not systematically directed to keep / improve exclusivity.	The exclusivity is actively utilized for a long-term development, also R&D systematically works on its maintaining/development.	3,2	3,0	3,0	3,5
	16	Technology	Organisation disposes with all technologies and infrastructure available for ensuring long term stability/development.	There exists significant underinvestment, the results of organisation could be far better. The lack of technologies threatens the future existence of an organization.	Organisation disposes with necessary technologies. Some inefficiencies exist. No long-time concept in place covering future needs and investments planning and their profitability. The intellectual property of the organization is adequately protected.	Technologies enable an organization to gain or maintain a lead over others, future needs are addressed well in advance, and there is a regular assessment of the effectiveness of the technologies used.	3,0	4,0	2,0	3,0
	17	External factors	Was there a significant external factor influencing the company performance within last three years (Change in regulations, significant market change, political/economical shifts)?	Yes, the performance was positively influenced significantly	No/not relevant	Yes, the performance was negatively influenced significantly	3,3	4,0	3,0	3,0
		Subtotal Category Others					11,7	13,5	10,0	11,5
Total							55,0	62,0	54,0	49,0

Usage of indicators of innovativeness in marketing performance assessment in companies in Slovakia

Dana Hrušovská[1,*] and *Katarína* Grančičová[1]

[1] University of Economics in Bratislava, Faculty of Business Management, Dolnozemská cesta 1, 852 35 Bratislava, Slovakia

Abstract. At present, measuring of marketing performance should be an integral part of a comprehensive evaluation of the performance of a company. In this context, the paper points out the trend of the increasing number of indicators used in the area of marketing performance assessment and summarizes the most commonly used indicators in this area, with an emphasis on the innovativeness indicators. Following the theoretical bases it subsequently presents the results of the research more broadly aimed at the issue of marketing performance assessment in companies in Slovakia. The presented results are focusing mainly on the usage of indicators of innovativeness, which have been in companies in Slovakia identified as the least preferred category of indicators in the context of marketing performance assessment. The overall results of the implemented research suggest that the general trends in the examined area to a large extent apply also to companies in Slovakia.

1 Introduction

At present, measuring and evaluating marketing performance should be an integral part of a company's marketing activities. The importance of this issue increases with the amount of resources invested in marketing and is gradually becoming a competitive factor of the company. Only with the correct assessment can a company acquire an idea if financial resources for marketing activities were used in an adequate manner and positively influenced the results and the overall performance of the company [5, 21]. The problem is that there is no uniform and universal method for this purpose. In this regard, the paper points out the trend of the increasing number of indicators used in the area of marketing performance assessment. Based on research from foreign scientific sources it also summarizes the most commonly used basic categories and indicators of marketing performance assessment. Following these theoretical bases, it presents the results of research focused on the assessment of marketing performance in companies in Slovakia, with an emphasis on the category of innovativeness indicators.

* Corresponding author: dana.hrusovska@euba.sk

2 The theoretical bases of the examined issue

The interest in the issue and conceptual development of marketing performance assessment can be dated back to the 1960s [9]. Numerous studies of this concept have been carried out since then, but it is still debated and is considered elusive. Thus, marketing performance has been and remains the center of marketing interest for the vast majority of organizations [10]. Therefore, both academics and practitioners are constantly looking for clear and reliable tools for assessing the merits of marketing [13].

2.1 Basic approaches in the field of marketing performance assessment

Overall, it can be noted that most sources summarize the developments and approaches in marketing performance assessment into two basic approaches. The traditional approach – that uses mainly financial or accounting measuring tools and the approach that uses non-financial measuring tools – approach based on marketing audit.

In the traditional approach, measuring marketing performance has been limited to determining marketing productivity, i.e. determining the amount of output achieved through a single unit of marketing input, in order to quantify how marketing contributes to the success of the company. The analysis on the input side included the individual expenditures on marketing activities and these were compared with the outputs – sales volume, profit, CF, i.e. mainly accounting measuring tools. Several studies were focused on finding the optimal allocation of marketing resources with the aim to maximize profitability. Pajtinková et al. [12] states that the limitations of this approach can be seen particularly in the false assumption that marketing outputs and inputs can be precisely evaluated and that, over time, they remain stable and also that productivity is primarily based on the quantity of inputs and outputs without evaluation of their quality. Marketing productivity thus underestimates the long-term value that marketing creates. Monitoring business strategies also requires measuring tools that can capture the potential from the point of view of future performance. On the other hand, it points out that this approach has also been beneficial in terms of important information that constitutes a conceptual model of marketing productivity providing a comprehensive view of the effectiveness of dimensions that enter into marketing performance and current understanding of the importance and the quantification of marketing expenses and revenues.

A significant shift in the evaluation of marketing performance occurred in the 1980s and especially in the 1990s when the need and the awareness of the importance of non-financial indicators started to get stressed. Clark [4,11] stated that in this period the strongest trend in marketing performance evaluation in various sectors was a step from measuring the success of products to indicators oriented on customer value. In this context, he indicated the gradual addition of non-financial measuring tools to traditional accounting indicators – market share, quality of provided services, customer satisfaction and loyalty, brand value, as well as broader considerations regarding marketing audit.

One of the aspects that caused this shift was also the gradual transition from transactional to relational marketing. While transactional marketing aims to maximize the number of one-off transactions (passive target orientation and marketing activities seen only as cost items) [16], the objective of relational marketing is to develop long-term relationships with profitable customers (focusing on quality and value of long-term customer relationships or other decisive interest groups). Customer expectations in both pre-sale and post-sale processes are a key. Gradually, there has been an increase in the interest in defining and improving the "correct" marketing inputs, with an emphasis on whether companies actually conduct proper marketing activities in a way that creates valuable assets contributing to the long-term financial performance of the company. The result of this gradual evolution in the field of marketing performance assessment is an approach based on marketing audit, which makes it possible to evaluate precisely the quality and efficiency of marketing inputs [12]. This approach is currently being developed in an effort to identify modern methods for optimizing the allocation of resources for marketing activities, while the issue of marketing assets is coming to the forefront [17,18]. At the same time, some authors suggest the usefulness of a holistic approach to marketing performance evaluation [19].

2.2 Basic categories of indicators of marketing performance assessment

The aforementioned development in the field of approaches is also reflected in the individual indicators used to measure marketing performance. At present, therefore, the number and diversity of tools for marketing performance assessment available to companies have increased considerably. For example, Clark [4] identified around 20 key indicators; Davidson [14] selected 10 key marketing efficiency metrics out of the hundreds of indicators identified by Mayer [15]. Farris [7] currently lists nearly 120 basic marketing indicators for the needs of marketing managers in 9 main categories.

Based on several surveys [1,2,3] carried out in business practice Ambler et al. determined 6 basic categories of marketing indicators. They are listed in Tab. 1.

Table 1 Overview of basic categories of indicators for measuring marketing performance

Category of marketing indicators	Basic indicators in the respective category
Consumer perception/attitude	Awareness; Perceived quality; Customer satisfaction; Importance of the brand for the customer; Perceived brand difference; Knowledge of the brand
Consumer behavior	Number of new customers; Loyalty; Conversion rate
Business customer	Customer satisfaction; Number of complaints
Competition-oriented indicators	Relative customer satisfaction; Perception of quality compared to competitors
Innovativeness	Number of new products; New product revenues; New product margin
Financial indicators	Revenues; Margin; Profit/profitability

Source: Processed according to [1,3]

The overview is primarily an expression of how practitioners approach marketing performance assessment and the grouping of marketing metrics, and also offers the opportunity to list the most commonly used indicators. Thanks to that it can be a useful indicator for managers in measuring and evaluating marketing performance.

The development in this area also poses some problems. One of them is the creation/formulation itself of suitable marketing indicators. On one hand, there are some demands on marketing indicators that arise from the general requirements of evaluating marketing performance, but on the other hand, business practice points out that many of them are often very specific for the field of marketing evaluation and require data that is difficult to access, is incomplete and complicated [6,7]. As the number of indicators for measuring marketing performance increases, there is often unnecessary tracking of too large a range of indicators. Therefore, in order for companies to avoid an information overload, it is necessary to select only those that are important to the company/marketing, i.e. they will be sufficient to effectively monitor marketing, but also its impact on the overall results and performance of the company [20]. The choice of indicators thus becomes as important as the performance measurement itself. Of course, in parallel to this, it is necessary to develop techniques and tools that will allow this monitoring. Finally, it should also be noted that the identification of suitable measuring indicators does not automatically improve performance, but merely enables managers to find and identify areas where there is a potential to improve performance [22].

3 Research design

The aforementioned theoretical foundations became the basis for formulating the objectives and research questions of the presented part of the research, which is overall aimed at identifying the current state and knowledge in the field of marketing performance assessment in companies in Slovakia. This research is to some extent following a similar research carried out in the United Kingdom, China, Spain [1, 2, 3], thus allowing a possible comparison with abroad. The presented part of the research aims to determine the state of the usage of the category of indicators of innovativeness in marketing performance assessment in companies in Slovakia. The individual research questions were formulated as follows:

RQ1: What is the extent of usage of the category of innovativeness indicators in comparison with other categories in marketing performance assessment in companies in Slovakia?

RQ2: What is the extent of usage of basic indicators in the category of innovativeness to measure marketing performance in companies in Slovakia?

RQ3: For the individual basic indicators in the category of "innovativeness" to find out: What is the importance, the frequency of tracking and the way of evaluation of a specific indicator in the examined category in companies in Slovakia?

At the same time on the basis of statistical survey, the aim was to verify which of the examined factors/characteristics of companies influence the tracking of innovativeness indicators (within the framework of marketing performance assessment) in companies in Slovakia.

In addition to general methods of scientific work, a questionnaire was used as a basic tool for data collection as part of the research task. Its first part was conceived in such a way as to enable obtaining of the basic characteristics of the respondent/company. The second part of the questionnaire dealt with areas such as need, importance, approaches, surveyed indicators, but also problems in the field of marketing performance assessment. Data collection took place between November 2016 and March 2017 in companies in Slovakia, using simple random sampling and the sampling units were companies. The sample consisted of 500 companies, of which 286 were subsequently included in the database (57.2% return). The sample of the companies involved in the research was examined and characterized in several respects - the size of the company, the structure of the company, the affiliation and time of activity in the sector, the prevailing market orientation of the company and the importance of marketing in the company. The selection of these identifying characteristics resulted from their possible impact on the field of marketing performance assessment. The obtained data were evaluated using mathematical-statistical methods (Chi-Square Test and Cramer's coefficient).

4 Results and discussion

This part is focused on presentation and interpretation of partial results obtained from the survey carried out in companies in Slovakia, with emphasis on research questions mentioned in the section – research design of this paper.

From the results of the solution of the overall research task, it can be noted that out of the total number of companies included in the survey (286), more than 2/3 (68.8%) consider it necessary to measure and evaluate the marketing performance in the company. At the same time, most of these companies also carry out this *assessment* – 62.7% which can be seen as positive. Companies that do not measure marketing performance gave as the most common reasons: size of the company; lesser importance of marketing in the company; company structure – i.e. evaluations are carried out at the headquarter level and also the difficulty of this issue both in terms of data and specialists. At the same time, it can be noted that, according to the examined characteristics, marketing performance is measured and evaluated to a great extent by – large companies (85%); companies with their own individual marketing department (87%); companies operating in consumer markets and in the retail sector (75%) and also companies that have been operating in the industry for a long time – i.e. according to the survey results, the longer the period of activity in the industry the higher the rate of assessment of marketing performance (more than 50 years – 83%, up to 1 year - 33%).

In the 1st research question (RQ1), the subject of the research were the basic categories of measuring instruments of marketing performance and their preference by companies in Slovakia in evaluating their marketing success. The basic categorization of marketing performance assessment tools was based on theoretical bases and it divided the indicators into 7 groups – financial indicators, competition-oriented indicators, indicators evaluating consumer perception and attitude, indicators evaluating consumer behavior, indicators focused on business customer, indicators of innovativeness. The results in the area of preference of basic categories of marketing performance assessment tools of companies in Slovakia are shown in Fig.1.

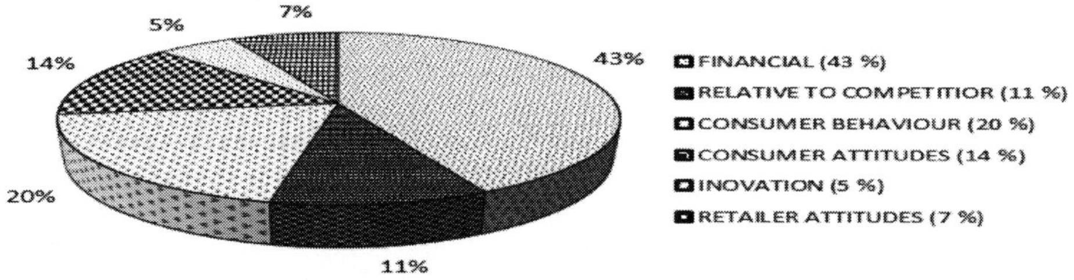

Fig.1. Preferences in terms of basic categories of marketing performance assessment indicators of companies in Slovakia Source: own research

The most preferred and also most frequently tracked category is financial indicators, which are preferred by 43% of companies that operate in Slovakia and assess marketing performance. On the other hand, the category of indicators of innovativeness, which are also the subject of further examination in this part of the research, are among the least preferred indicators (5%) to assess marketing performance and success compared to other categories of indicators. These findings also correlate with the theoretical bases [4] and the results of the studies that inspired this research [1,2,3]. As a rule, the basic reason for favoring financial indicators is the simple acquisition of data for their evaluation and comparison, while the indicators of innovativeness are perceived as "lifeblood of marketing" [8]. The presented results can also be supplemented by a % of tracking of the individual marketing indicators for marketing performance assessment in companies in Slovakia. The first 5 most tracked indicators are listed in Table 2.

Table 2 Extent of usage/tracking of individual marketing indicators in the field of marketing performance assessment in companies in Slovakia

Indicator	% of tracking
Customer satisfaction	95.74%
Sales/Revenues	93.12%
Profit	91.62%
Perception of quality	85.19%
Margin	76.60%

Source: own research

These results were followed by a survey within the individual categories of indicators, while in the category of indicators of innovativeness (RQ2) were identified as the most frequently measured indicators in companies in Slovakia – the number of new products, revenues from new products, the margin from new products and the customer satisfaction with new products. It can be noted that companies that track this category of indicators in general track the majority of the most basic primary metrics at approximately the same rate. Table 3 provides an overview of their usage in terms of marketing performance assessment in companies in Slovakia.

Table 3 The extent of usage of the basic indicators of innovativeness in marketing performance assessment of companies in Slovakia

Indicator	% of tracking
Number of new products	58.70%
Revenue from new products	69.60%
Margin from new products	68.25%
Customer satisfaction with new products	69.15%

Source: own research

At the same time, for the individual indicators of innovativeness, the following characteristics were examined: the importance that respondents/managers of the surveyed companies attribute to these indicators; the frequency of their tracking and the method of assessment (RQ3). The basic findings are given in Table 4. Overall, the extent of usage, the frequency of tracking and the way of evaluation of the tracked indicators meet the importance that the respondents give them as well as the nature of the indicators.

Table 4 Predominant assessment of tracked characteristics of individual indicators of innovativeness (% of companies in Slovakia)

Indicator	Importance	Frequency of tracking	Method of assessment
Revenue from new products	High 54.1%	Regularly/monthly 47.4%	Comparison plan/reality 60.4%
Customer satisfaction with new products	High 50.4%	Regularly/quarterly 36.1%	Comparison over time 35.3%
Margin from new products	High 59.1%	Regularly/monthly 47.0%	Comparison plan/reality 67.9%
Number of new products	Rather high 48.7%	Regularly/yearly 41.0%	Comparison plan/reality 43.6%

Source: own research

Finally, this survey was followed by a statistical survey of the impact of factors/examined characteristics of companies on tracking of the individual indicators of innovativeness in companies in Slovakia. With the help of the chi-square test, the assumptions about the existence of a statistically significant correlation between the marketing performance assessment of a company through the corresponding indicator and the individual examined attributes/characteristics of the company. In the case of confirmation of dependence, the Cramer's coefficient (Cramer's V) was used to measure its intensity. All tests were realized at a significance level of $p>0.05$. The results are presented for the margin from new products indicator. Only in the case of this indicator and only for one variable - the size of the company, the statistical dependence was confirmed, while its strength is moderate (Table 5). In case of the other examined indicators, the statistical dependence was not confirmed.

Table 5 Overview of the results of the chi-square test and Cramer's coefficient values for the examined variables and the margin from new products indicator

The margin of new products				
factor/variable	Value Chi-Square	Df	Asymp.sig. (2sided)/p	Cramer's V Value
company size	**12.96**	5	0.024	**0.26**
company structure	1.79	3	0.617	0.10
business focus	9.17	5	0.103	0.22
market orientation of the company	1.83	1	0.176	0.10
period of activity in the sector	3.48	4	0.480	0.14

Source: own research

Overall (not only on the basis of statistical findings, but also on the basis of qualitative assessment and assessment in other categories of marketing indicators as well) it can be stated that the size of the companies is a factor that has some influence on the tracking of the individual indicators for the marketing performance assessment of companies in Slovakia. That means as it would be expected that large companies rather than smaller companies evaluate majority of the indicators.

5 Conclusions

The results of the survey show that most companies in Slovakia are not only aware of the importance of measuring and evaluating marketing performance, but are also dealing with it. As assumed, this applies mainly to: large companies; companies that are oriented on the consumer market; companies that have been operating on the market for a long time, have a separate marketing department and place great emphasis on marketing. It was also confirmed that they mainly use the category of financial indicators for this purpose, while from among the individual basic marketing indicators they most often use mainly the indicators: customer satisfaction, revenues, profit, quality perception and margin.

On the other hand, the innovativeness indicators examined in this paper represent the least preferred and used category for measuring and evaluating marketing performance in the surveyed companies. However, the fact that, if they already track this indicator they do so on a regular basis and in a manner corresponding to their nature can be viewed positively. The statistical verification of dependence did not confirm, however, that the tracking of the innovativeness indicators (as opposed to the indicators in other categories) depends on the characteristics of the surveyed companies. The only exception was the margin from new products indicator, which showed a slight connection to the size of the company.

Overall, the results of the survey confirm that the general trends of marketing performance assessment are also to a significant extent relevant for companies in Slovakia. In many areas, as well as in the area of indicators of innovativeness, they also correspond with the results of the realized researches we use as bases.

In conclusion, the creation, improvement and selection of appropriate marketing indicators in an effort to create a functional model for effective monitoring of marketing performance are a constant challenge for both marketing theory and practice. The way to solve these issues can be found in research, which also involves conducting surveys in business practice.

This paper is the outcome of the VEGA project no. 1/0646/20: "Diffusion and consequences of green innovations on imperfect competition markets" with a 100% proportion.

References

1. T. Ambler, D. Riley., *Marketing Metrics: A Review of Performance Measures in Use in the UK and Spain.* London Business School, London, working paper, 1–30 (2000)
2. T. Ambler, W. Xiucun, Measures of Marketing Success: A Comparison between China and the United Kingdom. *Asian Pacific Journal of Management* **20,** 267 – 281 (2003)
3. T. Ambler, F. Kokkinaki, S. Puntoni, Assessing Marketing Performance: Reasons for Metrics Selection, *Journal of Marketing Management* **20**, 475-498 (2004)
4. B. H. Clark, Marketing Performance Measures: History and Interrelationships, *Journal of Marketing research* **15**, 64-73 (1999)
5. D. O´Sullivan, A. V. Abela, M. Hutchinson, Marketing performance measurement and firm performance. *European Journal of Marketing* **43**, 843-862 (2009)
6. P. W. Farris, et al., *50+ Metrics Every Executive Should Master.* (EDP Pearson Education, Inc., Upper Saddle River, New Jersey, 2006)
7. P. W. Farris, N.T. Bendle, P. Pfeifer, D. J. Reibstein, *Marketing metrics: The Managers Guide to measuring Marketing Performance.* (EDP Pearson Education, Inc., Upper Saddle River, New Jersey, 3rd edition., 2016)
8. K. Simmonds, Marketing as Inovation - The 8th Paradigm. *Journal of Management studies* **23**, 479-500 (1986)
9. R. A. Feder, How to Measure Marketing Performance. *Harvard Business Review* **43**, 132 -142 (1965)
10. N.A. Morgan, B.H. Clark, R. Gooner, Marketing Productivity, Marketing Audits and Systems for Marketing Performance Assessment. Integrating Multiple Perspectives. *Journal of Business Research* **55**, 363 – 375 (2002)
11. B. H. Clark, A Summary of Thinking on Measuring the Value of Marketing. *Journal of Targeting, Measurement and Analysis for Marketing* **99**, 357 -369 (2001)
12. G. Pajtinková-Bartáková, K. Gubíniová, Ľ. Cibák, Teoretické a praktické aspekty hodnotenia marketingovej výkonnosti. *Marketing Science and Inspirations* **8**, (2013)
13. M. Pont, R. Shaw. *Measuring marketing performance: A criticque of empirical literature.* ANZMAC, Conference Proceedings Adelaide 1-3 December 2003, 2064 – 2073 (2003)
14. J.H. Davidson, Transforming the Value of Company Reports through Marketing Measurement. *Journal of Marketing Management* **15**, 757-777 (1999)

15. M.W. Meyer, *Finding Performance: The New Discipline in Management. Performance Measurement – Theory and Practice.* (EDT Center for Business Performance, Cambrige, England, 1998)
16. L. Lamberti, G. Noci, Marketing Strategy and Marketing Performance Measurement System: exploring the relationship. *European Management Journal* **28,** 139-152 (2010)
17. R. K. Srivastava, T.A. Shervani, L. Fahey, Market-Based Assets and Shareholder Value: a framework for analysis. *Journal of Marketing* **62**, 2-18 (1998)
18. R. K. Srivastava, L. Fahey, H.K. Christensen, The resource-Based View and Marketing: The role of Market-Based Assets in Gaining Competitive Advantage. *Journal of Management* **27**, 777-802 (2001)
19. N. A. Morgan, B. H. Clark, R. Gooner, Marketing productivity, marketing audits, and systems for marketing performance assessment – Integrating multiple perspectives. *Journal of Business Research* **55** , 363-375 (2002)
20. B. H. Clark, A.V. Abel, T. Ambler, An Information Processing Model of Marketing Performance Measurement, *Journal of Marketing Theory and Practice* **14**, 191-208 (2006)
21. D. O´Sullivan, A.V. Abela, Marketing performance measurement ability and firm performance. *Journal of Marketing* **71**, 79-93 (2007)
22. M. Oreský, *Hodnotenie efektívnosti marketingu – východiská tvorby a uplatnenia merných ukazovatelov v marketingu*. Zborník vedeckých statí Obchodnej fakulty 2011, Vydavateľstvo EKONÓM, Bratislava, 1-13 (2011)

Performance analysis firm operating in the sector of IT services through the model HGN

Eduard Hyránek [1,*], *Branislav* Mišota [2]

[1] Faculty of Business Management, University of Economics in Bratislava, Dolnozemská cesta 1/b, 852 35 Bratislava, Slovakia
[2] Institute of Management, Slovak University of Technology in Bratislava, Vazovova 5, 812 43 Bratislava, Slovakia

Abstract. In this paper, we will briefly explain the application of the HGN model designed to measure performance, on the example of a selected non-financial corporation operating in the Slovak Republic. The aim of the paper is the application of the HGN model to IBM Slovakia Ltd. for the period from 2017 to 2019.At the same time, we will point out some difficulties in the application and the possibilities of improving the model.

1 Introduction

The HGN model [1] is the result of scientific research by the Faculty of Business Management of the University of Economics in Bratislava developed within the scientific project on the topic Verification and implementation of business performance modeling in financial decision making tools as a tool for financial decision-making [2, 3, 4]. In the field of research on the implementation of key determinants of performance models, there is a wide range of approaches at home [5, 6] and abroad [7 - 10].

From the application of the HGN model so far in our scientific research, we have gradually found that it is necessary to adjust some financial ratios operating in the model and adapt them to industry specifics. A skilled workforce has a major role to play in this sector [11, 12]. The information technology industry is a highly specific performance assessment industry [13,14 ,15]. For the purposes of the research, we selected a company from the information technology industry of IBM Slovakia Ltd. From the most recent data available as of April 2020, obtained from the register of financial statements for the period from 2017 to 2019, we analyzed the application of the HGN model.

2 The aims and methods

The aim of the paper is to assess the performance of IBM Slovensko s.r.o. based on an analysis based on the HGN model. We currently use two versions of the model. In this article, we focus on the HGN1 version. We will point out the possibility of its practical application. In addition to the practical application of the HGN model, we will evaluate the

* Corresponding author: eduard.hyranek@euba.sk

possibility of modification of some financial indicators contained in the model on the example of this company.

3 Results and Discussion

In current scientific research, we have been dealing with business performance modeling since 2015. The performance models and predictions used in the theory and practice used so far do not sufficiently take into account the picture of the financial performance of the company in the current conditions of the Slovak economy. Several prediction models are created for completely different conditions than in the Slovak Republic [7 - 10].

The current economic situation of companies in the Slovak Republic is characterized by considerable indebtedness and, as a result, a high level of funds in receivables. Another factor is the insufficient use of the company's assets. It is these factors that adversely affect the efficiency of the company are emphasized in the HGN model. The more the company is indebted to long-term debts and has a high commitment of funds in receivables, the lower its final value.

3.1 Model HGN1 version

Out of 47 ratios, we created two sets of indicators (Figure 1), the first contains three efficiency indicators (xi) and the other three demandingness indicators (yi). We interconnect these indicators; or rather synthesize them, into one comprehensive performance indicator.

The aim is to objectively reflect the financial situation of an enterprise, its performance and to enable comparative assessment of the financial results of enterprises. At the first stage, we verified the following six financial ratios for the HGN1 version [2, 3, 4].

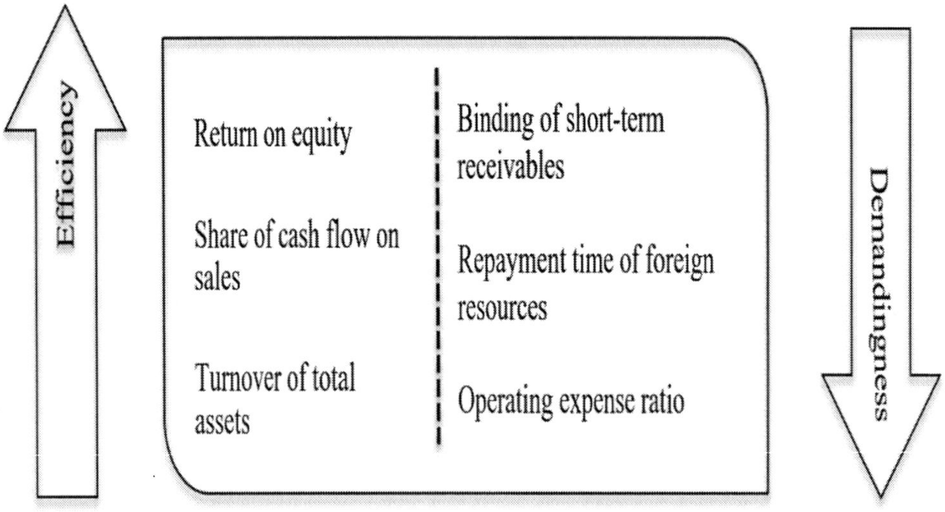

Fig. 1. The input indicators for creation of the HGN1 version

Source: Own research

The aggregate efficiency indicator xi :

$$\sum_{i=1}^{3} x_i \qquad (1)$$

will provide a partial picture of the company's performance in terms of efficiency.

The aggregate demandingness indicator yi:

$$\sum_{i=1}^{3} y_i \qquad (2)$$

will provide a partial picture of the company's performance in terms of demandingness.

The synthetic indicator (SI) is to be influenced by the values of six indicators. It is defined as follows:

$$SI = \sum_{i=1}^{3} x_i - \sum_{i=1}^{3} y_i \qquad (3)$$

3.2 Application of the HGN1 performance model to IBM Slovakia Ltd.

Table 1 lists selected IBM Slovakia Ltd. Assets entering into the calculations of the ratios of the determinants of the HGN1 performance model.

Table 1. Assets - Data from the company's financial statements for the years 2017-2019.

	Assets		
	Data from the fin. stat. (in thousands of €)		
Year	**2017**	**2018**	**2019**
Total Assets	39 816	38 537	39 477
Current receivables - total	30 308	25 352	30 902

Source: The Register of Financial Statements (2020)

Table 2 lists selected Liabilities and Equity of IBM Slovakia Ltd. entering into the calculations of ratios of determinants of the HGN1 performance model.

Table 2. Liabilities and Equity - Data from the company's financial statements for the years 2017-2019.

	Liabilities and Equity Data from the fin. stat. (in thousands of €)		
Year	**2017**	**2018**	**2019**
Equity	13 517	10 994	12 828
Net profit / loss for the accounting period after tax	11 003	8 705	10 539
Liabilities	32690	28822	25709
Current liabilities - total	20699	15992	15289

Source: The Register of Financial Statements (2020)

Table 3 presents selected indicators from the Income Statement of IBM Slovakia Ltd. entering into the calculations of ratios of determinants of the HGN1 performance model.

Table 3. Income Statement - Data from the company's financial statements for the years 2017-2019.

	Income Statement Data from the fin. stat. (in thousands of €)		
Year	**2017**	**2018**	**2019**
Revenue	105 683	98 565	98 414
Amortization	1 337	1 651	2 038
Added value	44066	47071	49444
Cost of merchandise sold	5545	5631	7092
Production costs	56070	45862	41877
Personnel expenses - total	38395	39483	40650
Taxes and fees	11297	9837	10376

Source: The Register of Financial Statements (2020)

Table 4 lists the IBM Slovakia Ltd. financial ratios included in the calculations of the HGN 1 ratios in group xi, (efficiency indicators).

Table 4. Aggregate indicators xi

	Aggregate indicators xi		
	Year		
Year	**2017**	**2018**	**2019**
Return on equity	0,7918	0,8216	0,8778
Share of cash flow on sales	0,0950	0,1237	0,1877
Turnover of total assets	2,6543	2,5577	2,4929

Source: Own research

Table 5 shows the financial ratios of IBM Slovakia Ltd. entering into the calculations of the ratio indicators of the HGN 1 model in group yi, (demandingness indicators).

Table 5. Aggregate indicators yi.

	Aggregate indicators yi		
	Year		
Year	**2017**	**2018**	**2019**
Binding of short-term receivables	0,2868	0,2572	0,3140
Repayment time of foreign resources	1,2776	0,8548	0,4747
Operating expense ratio	1,0659	1,0396	1,0368

Source: Own research

Table 6 shows the financial ratios of IBM Slovakia Ltd. entering into the calculations of ratios of the HGN 1 model in group xi i. efficiency indicators and yi, i. efficiency indicators and a synthetic indicator SI.

Table 6. Synthetic indicator for model HGN1.

	Synthetic indicator for model HGN1		
	Year		
Year	**2017**	**2018**	**2019**
$\sum_{i=1}^{3} x_i$	3,5411	3,5029	3,5584
$\sum_{i=1}^{3} y_i$	2,6303	2,1516	1,8255
$SI = \sum_{i=1}^{3} x_i - \sum_{i=1}^{3} y_i$	0,9108	1,3514	1,7330

Source: Own research

Figure 2 shows the development of aggregated indicators xi and yi and a synthetic indicator from IBM Slovakia Ltd. for the HGN 1 model.

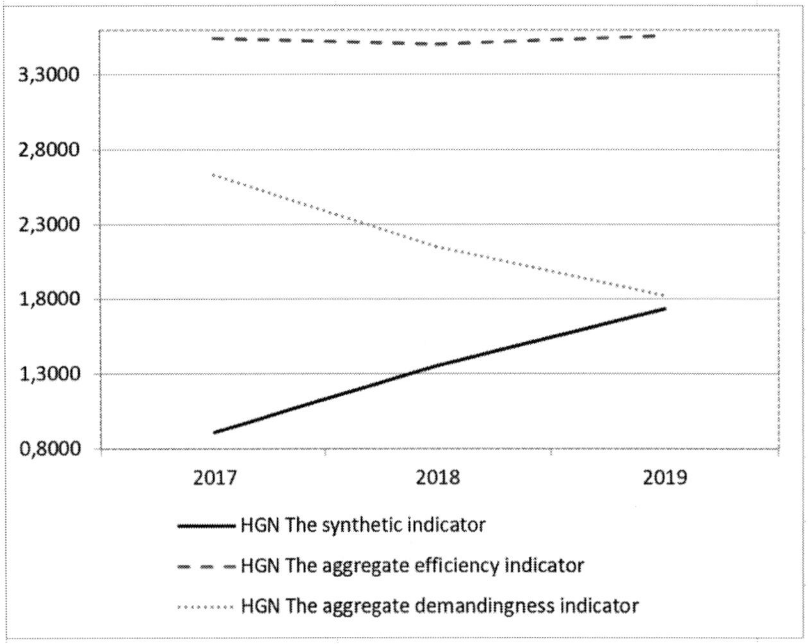

Fig. 2. Graph of aggregate indicators xi and yi and synthetic indicator for model HGN1.

Source: Own research

Based on the calculated data, the resulting values of aggregated indicators and coefficients, and finally from the synthetic indicator HGN, we can easily evaluate the performance of the surveyed company. The development trend of the company's performance measured by the synthetic indicator was favorable for the evaluated period. It can be stated that the indicators of efficiency xi had a balanced trend without significant fluctuations in the individual years evaluated. This can also be seen in individual indicators (Return on equity, Share of cash flow on sales, Turnover of total assets). Their development trend is basically balanced.

The synthetic indicator is significantly improved by the yi indicators, - indicators reducing efficiency. Their trend is decreasing in the evaluated years. This is a very favorable trend. This also confirms the correctness and objectivity of the principle of the HGN model. It contains indicators that generally have an adverse effect on efficiency. At IBM Slovakia Ltd, reducing them has a positive effect on overall performance and efficiency. The exception is indicator y1 (Binding of short-term receivables), which deteriorated slightly in 2019. Its deterioration did not have a significant effect on the positive trend.

An important indicator - net profit, there was no growing trend for the period under review. Due to the improving trend of efficiency-reducing indicators, the synthetic indicator did not deteriorate.

The HGN model emphasizes the long-term indebtedness of the company. The surveyed company does not have long-term debts, so the synthetic indicator is also positive.

4 Conclusion

Using the analysis presented in the research results in the previous section, we verify the accuracy and objectivity of the inclusion of individual indicators in the model. However, it should be noted that some indicators would require more precise calculation of their values. The absence of data drawn only from the annual financial statements does not allow this. We also verified the correctness of the inclusion of individual indicators in the model at this company. By obtaining more detailed data from the accounting records, it is possible to refine the results. The data contained in the model resulting from the flow variables are for the whole year and the stock variables as of the last day (31.12.) of the year.

We see a certain inaccuracy in this. For example, indicator x3 Turnover of total assets would express a more objective assessment of the average values for individual months of the year. Between the average of the year and the final state as of 31.12. there may be a difference. Similarly, in indicator y1 The binding of short-term receivables can be analogously different. During the year, the amount of short-term receivables may be much lower than at the end of the year or vice versa. The calculation methods used in the model can clearly affect the overall result. Every company can eliminate this problem, because it also has data from monthly financial statements.

The analysis of indicators contained in the HGN model in various companies provides us with the opportunity to verify the correctness or even incorrectness of the method of their calculation. This is also confirmed by the surveyed IBM company. In some indicators, it will be necessary to modify the calculation method.

The article is a partial result of research within the VEGA project. 1/0462/19 "Implementing key determinants into performance models as tools financial management under the current conditions".

References

1. E. Hyránek, M. Grell, L. Nagy, I. Ďurinová, The Economic-mathematical Nature of the HGN Model Concept as a Tool for Measuring Performance of Enterprises. *Ekonomický časopis* **66**, 3, 309-325. (2018).

2. E. Hyránek, M. Grell, L. Nagy, *Nové trendy merania výkonnosti podniku pre potreby finančných rozhodnutí.* Bratislava, Ekonóm (2014).
3. E. Hyránek, M. Grell, L. Nagy, I. *Implementácia modelu výkonnosti vo finančnom rozhodovaní podniku: model HGN ako nástroj finančného plánu*. Praha, Nakladatelství Baset (2017).
4. M. Grell, E. Hyránek, P Maticové modely na meranie výkonnosti produkčných systémov. *E a M: Ekonomie a Management* **15**, 1, 73-87 (2012).
5. B. Gavurova, et al, (2017). Predictive potential and risks of selected bankruptcy prediction models in the Slovak business environment. *Journal of Business Economics and Management* **18**, 6, 1156-1173. (2017).
6. R. Delina, M. Packová, Prediction Bankruptcy models validation in Slovak business environment. *E a M: Ekonomie a Management* **16**, 3, 101-112 (2013).
7. E. I. Altman, Financial Ratios, Discriminant Analysis and the Prediction of Corporate Bankruptcy. *Journal of Finance* **23**, 4, 589-609 (1968).
8. E. I. Altman, M. Iwanicz-Drozdowska, E. K. Laitinen, A. Suvas, Financial distress prediction in an international context: A review and empirical analysis of Altman's Z-score model. *Journal of International Financial Management & Accounting* **28**, 2, 131-171 (2017).
9. E. I. Altman, E. I., A. Saunders, Credit risk measurement: Developments over the last 20 years. *Journal of banking & finance* **21** (11-12 (1997).
10. S. Ashraf, E. GS Félix, Z. Serrasqueiro, Do Traditional Financial Distress Prediction Models Predict the Early Warning Signs of Financial Distress? *Journal of Risk and Financial Management* **12**, 2, 55 (2019).
11. J. Brown, D. A. Matsa, Boarding a sinking ship? An investigation of job applications to distressed firms. *The Journal of Finance* **71**, 2, 507-550 (2019).
12. P. Bolton, N. Wang, J. Yang, Optimal contracting, corporate finance, and valuation with inalienable human capital. *The Journal of Finance* **74**, 3, 1363-1429 (2019).
13. M. Jain, Y. Karmakar, Capital structure: An empirical study of choices and strategies for Indian service sector companies. *Finance India* **32**, 4, 1227-1252. (2018).
14. S. Mithas, A. Tafti, , I. Bardhan J. M. Goh, Information technology and firm profitability: mechanisms and empirical evidence. *MIS Quarterly* **36**, 1, 205-224 (2012).
15. K. Merkley, R. Michaely, J. Pacelli, Does the scope of the sell-side analyst industry matter? An examination of bias, accuracy, and information content of analyst reports. *The Journal of Finance* **72**, 3, 1285-1334 (2017).

Corporate relations and strategic subcontracting

Abdelkrim Inhaddou[1], and *Peter* Markovič[2,*]

[1]Faculty of Management, Comenius University in Bratislava, Odbojárov 10, 820 05 Bratislava, Slovakia
[2]Faculty of Business Management, University of Economics in Bratislava, Dolnozemská cesta 1/b, 852 35 Bratislava, Slovakia

Abstract. Technological progress, such as that associated with robotics and computer-aided manufacturing, is increasingly delivering machines that operate at lower variable costs, but often with a different optimum degree of utilization. If we restrict our analysis to companies that use two different types of machines to manufacture a product, a change in the optimal degree of utilization of machines of one type resulting from technological developments may make it necessary to adapt the entire manufacturing process in order to fully exploit the cost advantage. If the cost advantages cannot be fully achieved by reconciling internal company structures, this may be achieved by adjusting inter-company structures. Such an adjustment can take various forms, depending on whether the enterprise in question offers unneeded capacity to other enterprises or complements insufficient internal production resources with available capacity from another enterprise. The aim of the article is to analyse important trends in the context of subcontracting. In order to keep the subject of the study easily comprehensible, the important empirical methods are used, which can discover new insights in the scientific sources. The resulting generalization is intended to show when the question of possible outsourcing should be placed on the negotiating table.

1 Introduction

Looking at a company's decision, two parameters can be identified that lead to outsourcing, which are fundamental to the design of an outsourcing arrangement. These two parameters concern, on the one hand, the extent of the outsourcing and, on the other hand, the design of the inter-company relationship associated with it.

The first parameter, the extent of outsourcing, is the degree to which outsourcing should take place. It thus describes whether the external procurement of a work step is limited or whether a work step is completely removed from stock. In the first case, a work step is carried out in the company itself. In the second case, the company transfers its execution completely to another company.

* Corresponding author: peter.markovic@euba.sk

The second parameter to be defined by a company in the context of an outsourcing is the form of the relationship with the company that will provide the work step. If possible, exogenous restrictions are neglected, the number of different forms of contracts available to a company is unmanageable.

If it is assumed that a company that carries out outsourcing and determines the design parameters under consideration interacts with other companies, the question of the resulting economic implications arises. Especially the effect on other companies, which results from the outsourcing, is important for the answer to the question, because it can influence the optimal design of the above parameters.

The influence on other companies relevant for an optimal design can be divided into two directions of action or consideration. On the one hand the effect on competitors and on the other hand the effect on a supplier or customer. The resulting effects of outsourcing on competitors can further be divided into those resulting from the outsourcing itself, i.e. direct, and those resulting from the interaction with other companies. The latter are called indirect or strategic effects.

2 Outsourcing as an alternative in production design

In the scientific literature, there is agreement on the definition of outsourcing to the extent that the term means the use of external resources. The word itself represents a concept of art composed of outsourced resource using. It thus aims to restructure the internal production of goods or services. Furthermore, there is little agreement in the literature and in the use of the term. [1, 2]

In the following, it is assumed that outsourcing a work step that was formerly carried out in-house in return eliminates all the conditions necessary for this work step. Outsourcing includes the decision itself as well as the process of implementation. In addition, it is assumed that the outsourced work steps will continue to be a necessary part of the creation or disposal process of the goods concerned. A central assumption in outsourcing is that a company has previously carried out the outsourced work step internally, i.e. there is not only a change in the source of supply.

Since a rationally acting company carried out the considered work steps internally before outsourcing or subcontracting, it must be assumed that this form was optimal under the given environmental conditions. Thus, the various effects acting on the company must have been in equilibrium if the outsourcing could have been carried out previously but did not take place. Thus, for outsourcing to be beneficial, a shock of some kind must have occurred.

It is assumed that the triggering moment is to be found in the internal or inter-company technological framework. For this purpose, two possible forms of a shock that changes the framework conditions are considered. On the one hand, a shock that affects the available production technology, and on the other hand, a shock that changes the underlying information and communication technology.

2.1 Shock in production theory

Production technology refers to the transformation process by which a company or part of a company converts input factors into output. Assuming that the shock was caused by an innovation, a company will need or at least be able to replace its existing production technology for reasons of profit maximisation. However, since such a technological change is often associated with a change in the optimal degree of capacity utilisation, it may also be necessary to adjust capacities at those stages of production that are not directly affected by the shock.

For example, outsourcing can be advantageous for cost reasons if the capacities at the various production stages can be adapted to the new conditions at different speeds or not at all. Thus, outsourcing is a way of adapting the variable and fixed costs of production to the new conditions in order to reduce production costs. A different speed of adaptation to new technologies can also play a role.

At the same time, such a shock may also simply change existing fixed costs without affecting variable costs. However, even in this case there would be an impact on competition, as a change in fixed costs influences the market entry behaviour of companies. For example, a change in fixed costs may result in a change in the market power of the company under consideration or of one of its suppliers, which would also have an impact on the outsourcing behaviour of a company. This is also because it influences the possibility of opportunistic behaviour.

Similarly, a change in production technology can directly enable the use of more flexible and versatile inputs, thereby reducing factor specificity and thus reducing the risk of opportunistic behaviour. [3] This would also speak in favour of outsourcing in response to a corresponding shock in production technology.

2.2 Shock in information technology

An alternative triggering factor is a shock in the context of information and communication technology and thus to the internal or inter-company coordination costs associated with the execution of the task. It can be reflected in the costs that arise in the search for a suitable supplier or the conclusion of a contract with the supplier, i.e. the costs that arise for an inter-company division of labour and thus have a positive effect on outsourcing.

At the same time, however, such a shock can have a decisive influence on outsourcing, even with an already known supplier. For example, greater specialisation in the course of an inter-company division of labour also leads to the isolation of information. This is not a problem as long as the different work steps are completely independent of each other. If, however, there are positive interactions, so-called synergy effects, between them, these synergy effects can decrease or be completely lost through outsourcing. A more cost-effective transfer of information can reduce these negative effects by extending or improving communication possibilities and compensate for a loss of synergy effects in the case of an inter-company division of labour. Reducing the costs of information procurement can also influence coordination costs and thus allow for an extension of control possibilities. This can go so far as to increase the amount of control tools available. Thus, after such a shock, it may be possible to implement control instruments that previously could not be used for cost reasons. [4, 5]

Overall, better availability of specific information is the result of such a shock. Correct use of this additional information together with better asset utilisation can then lead to higher profits. [6] Conversely, however, the need for greater specialisation may arise for capacity reasons, without this having a positive effect on profits. [7, 8]

3 Subcontracting as a form of corporate relationship

If you look at more than one company and assume that they interact, there must be a direct or indirect relationship between them. The production of many goods or services requires a certain sequence of work steps. Such a successive sequence of work steps is also called a vertical chain. The vertical chain is also associated in the literature with the concept of value added. The focus here is on the inherent change in value that results from the execution of successive work steps for the goods or services to be produced. If one groups the resulting stages with regard to the change in value, a sequence of value creation stages is created, the so-called value chain.

By contrast, a horizontal relationship is a relationship between two companies or parts of companies that only perform work steps at the same stage of the creation process. In the context of horizontal relationships, this often refers to interactions between enterprises that produce an almost or completely identical good and usually compete for the sale of this good. If these goods are homogeneous from the point of view of the consumer, they are also called substitutes.

In contrast, a vertical relationship describes all relationships that exist along the direction of creation of a product or service between companies at different levels of the vertical chain. The products or services considered in the context of such a relationship are thus less homogeneous, but rather heterogeneous. Along the sequence of steps, a direction can be defined additionally for a more precise description.

A backward relationship or upstream direction is one that runs counter to the direction of production of the good or service, i.e. represents the receipt of an input for the enterprise under consideration. Conversely, all companies that process the output of a company under consideration, i.e. use it as an input, are in a downstream relationship with it from the company's point of view. It should therefore be noted that these terms are always relative. [9]

3.1 The essence of subcontracting

Subcontracting describes the outsourcing of a work step, where this work step is carried out within the company to a limited extent even after the outsourcing. In order to maintain the same production level, it is therefore necessary to procure additional units from external companies. [10] Subcontracting therefore requires a division of the required quantity between internal and external supply sources, so that capacity aspects play an important role.

This form of outsourcing thus represents a decision from the space of inter-company segmentation of a work step that is to be assumed as continuous. In subcontracting, this space is limited on the one hand by the fact that the company completely outsources a work step, and on the other hand by the fact that it executes it exclusively internally. If there is no marginal solution, subcontracting can also be seen as a special case of dual or second sourcing, in which one of the sources is internal. At the same time, however, subcontracting also includes the consideration of additional factors such as a change in the control and sanction mechanisms for controlling and coordinating these various sources of supply. Overall, both outsourcing and subcontracting can be interpreted as the substitution of the internal execution of a work step by external procurement. The scope of the execution of a work step is reduced or completely discontinued in the company under consideration and additionally required quantities are procured from other companies. While subcontracting focuses on the scope of outsourcing, outsourcing focuses on the structure of procurement.

3.2 Strategic Subcontracting

If one looks at a sector in which the production of a good or a service is carried out by a sequence of work steps that can be divided up between companies, the determination of the balance between the operating companies is usually based either on the transaction costs incurred or the underlying market structure. The basic work, which deals with the consideration of transaction costs as a starting point for the division of the production process, is by Williamson and Grossman & Hart. [11]

Based on this, many authors devoted themselves to an investigation of the role of transaction costs, asset specificity and the impact of incomplete contracts. All of this work is aimed at presenting statements for a decision on the optimal intra- and inter-company allocation of the creation process.

However, in many cases the focus was on a bilateral relationship between an individual outsourcing company and its potential supplier. Thus, the results obtained are not fully transferable to the analysis of industry trends, since the attractiveness of an available decision alternative for a company may also depend on how competitors react to such behaviour. [12, 13]

On the other hand, strategic aspects are taken into account in work that aims to examine the resulting market structure. A central work in this context is the paper by Spengler. There he shows that integration forms a balance when two successive monopolies would be the alternative. Building on this, Greenhut & Otha show that even in the case of a duopoly in a multi-level Cournot competition, vertical integration leads to equilibrium. This equilibrium results from the fact that the dissolution of double marginalisation through vertical integration reduces the costs for the integrating firm.

Based on the work of Greenhut & Otha [14], Abiru, Nahata, Raychaudhuri & Waterson [15, 16] investigated the case of a Cournot competition on two levels with a different number of companies. The elimination of double marginalization provides a cost advantage and thus an incentive for a company to integrate. However, the cost advantage also depends on the difference in the number of companies at both levels. Abiru, Nahata, Raychaudhuri & Waterson [17] conclude that the elimination of double marginalisation does not necessarily provide an incentive for integration, but rather concludes that if there is an unequal number of companies at both levels, there is generally no clearly dominant strategy.

The change of an industry equilibrium was also investigated by Salinger [18], who, however, in contrast to Abiru, Nahata, Raychaudhuri & Waterson, considers the reasons for a static equilibrium. In his investigation he uses a two-tier oligopoly, whereby the independent downstream companies take the price of the primary product as a given. Thus, the non-integrated upstream firms face a derived demand curve resulting from the equilibrium in the final product market. The price in an assumed quantity competition at this level is then derived from the resulting Cournot equilibrium. Through a merger and the resulting elimination of double marginalisation, a company can increase its final output.

This in turn leads to a backward shift in the residual demand curve, to the non-integrated firms, thus reducing their demand for the intermediate good. The stability of the result is crucially dependent on the assumption that changes in quantity remain endogenous. In his analysis Salinger, for example, assumes that the vertically integrated companies do not participate in the intermediate goods market, which means that the market equilibrium remains stable even with vertical integration.

It can now be shown that a vertical merger may lead to either an increase or a decrease in the price of the intermediate and final good or service. Salinger takes up this approach and shows in an adapted model that, in order for the price of the final product to rise, the upstream stage must be more concentrated than the downstream stage as a result of a vertical merger.

In contrast to the previous works, which deal with quantity competition, Bonanno & Vickers [19, 20] consider price competition. Like Lin [21], they assume that either demand is sufficiently inelastic, or franchising is possible. They can thus show that vertical separation results in a reduction of oligopolistic competition and thus an increase in profits, since the introduction of a wholesale market enables companies to credibly commit themselves to a higher price. Ordover, Saloner & Salep [22] generalizes this result to a vertical separation in the complete absence of vertical integration.

Cyrenne [23] and Economides [24, 25] also consider the effect described by Bonanno & Vickers in their study of strategic forces in vertical relationships. In a model with differentiated products they show that the decision on a vertical merger for an upstream company in price competition depends on the degree of product heterogeneity on the final product market. They consider the trade-off for an upstream company between double marginalisation and competitive weakening through vertical separation. The problem of double marginalisation results in a price increase at the wholesale level, which causes the volume to fall below the desired level. At the same time, however, price competition produces the positive strategic effect at the final stage described by Bonanno & Vickers.

They therefore conclude that, in the case of close substitutes, outsourcing makes sense for strategic reasons. In most of the work described above on the inter-company division of labor, however, it is striking that the phenomenon of subcontracting is not taken into account. The absence of subcontracting in these models results on the one hand from the fact that when using price as a strategic variable, only the costs of production are usually relevant.

4 Discussion

This means, however, that only the source of supply that allows cheaper delivery is used. If, on the other hand, the basic structure of the models allows subcontracting to be taken into account, they are often omitted anyway, as they often do not provide any gain in knowledge within the approaches used, but make the analysis more difficult due to the additional decision alternatives in the selection of the corporate structure, thus reducing clarity.

Notwithstanding the modeling problems discussed above, there are a few papers in the literature that deal with the investigation of a simultaneous internal and external reference of a working step or its result. One area where this occurs is production. The advantage of subcontracting as a means of production and capacity planning is particularly emphasized. It is argued that subcontracting allows short-term capacity adjustments under temporary fluctuations in demand.

In order to analyze the resulting behavior, Kamien, Li and Samet and Kamien & Li consider a multi-periodic, game theoretic aggregated planning model with given capacity constraints. The capacity constraints result implicitly from the underlying convex cost structure. In this way, you can show that the ability to subcontract provides an opportunity for production smoothing.

All the works mentioned once had in common that the justification for subcontracting was based on a capacity decision made under uncertainty. However, this means that the analysis of a strategic effect is distorted by the risk transfer associated with uncertainty. What would be important, on the other hand, would be a fundamental investigation that does not assume uncertainty with regard to relevant model parameters.

Furthermore, the majority of this work focuses on the vertical relationship associated with subcontracting. A consideration of horizontal effects is missing so far.

The article is a partial result of research within the VEGA project. 1/0240/20 „Financial Aspects of Sustainable Business - Enterprise Succession Solution for Small and Medium-sized Enterprises“.

References

1. W. Matiaske, T. Mellewigt, Motive, Erfolge und Risiken des Outsourcings – Befunde und Defizite. *Zeitschrift für Betriebswirtschaft* **72**/6, 641-659 (2002).

2. M. Pankowska, Information Technology Outsourcing Chain: Literature Review and Implications for Development of Distributed Coordination. *Sustainability* **11**, 1460 (2020).
3. W. Gick, Vertikale Integration und informations- und kommunikationsintensive Dienstleistungen. *Diskussionspapier* Serie B **08** (1999). M. Slade, Strategic Motives for Vertical Separation: Evidence From Retail Gasoline Markets. *Journal of Law, Economics, & Organization* **14**, 1, 84-113 (1998).
4. D. Elmuti, Y. Kathawala, M. Monippallil. Outsourcing to Gain a Competitive Advantage. *Industrial Management*, **5/6**, 20-24 (1988). B. Leavy, Outsourcing strategy and a learning dilemma. *Production and Inventory Management Journal* **4**, 50-54 (1996). C. Lynch, *Logistics Outsoucing – A Management Guide*. Council of Logistics Management, Oak Brook (2000).
5. E. Arrigo, Global Sourcing in Fast Fashion Retailers: Sourcing Locations and Sustainability Considerations. *Sustainability* **12**, 508 (2020).
6. J. Quinn, F. Hilmer. Strategic Outsourcing. *Sloan Management Review* **3**, 43-55 (1994).
7. M. Aoki, Horizontal vs. Vertical Information Structure of the Firm. *American Economic Review* **76**, 5, 971-983 (1986).
8. Y. Wei, Ch. Junwu, Q. Zhifeng, T. Sang-Bing, Ch. Hong, X. Yu, Operational Decisions on Remanufacturing Outsourcing Involved with Corporate Environmental and Social Responsibility—A Sustainable Perspective, *Sustainability* **10**, 1132 (2018).
9. D. Besanko, D. Dranove, M. Shanley, *The Economics of Strategy*. New York, John Wiley & Sons (1996).
10. J. Day, *Subcontracting Policy in the Airframe Industry*, Working Paper, Boston, Harvard University, Grad. School of Business Administration (1956).
11. O. Williamson, The Vertical Integration of Production: Market Failure Considerations. *American Economic Review*, **61**, 2, 112-123 (1971). O. Williamson, *Markets and Hierarchies: Analysis and Antitrust Implications*. New York, The Free Press (1975). S. Grossman, O. Hart, The Costs and Benefits of Ownership: A Theory of Vertical and Lateral Integration. *Journal of Political Economy* **94**, 4, 691-719 (1986).
12. S. Shaheen, L. H. Young, M. S. Muhammad, A Sustainable Outsourcing Strategy Regarding Cost, Capacity Flexibility, and Risk in a Textile Supply Chain. *Sustainability* **8**, 234 (2016)
13. G. Grossman, E. Helpman, *Integration vs. Outsourcing in Industry Equilibrium*, Working Paper, Princeton University (2001).
14. M. Greenhut, H. Otha, Vertical Integration of Successive Oligopolists. *American Economic Review* **69**, 1, 137-141 (1979).
15. J. Spengler, Vertical Integration and Antitrust Policy. *Journal of Political Economy* **53**, 347-352 (1950).
16. F. M. Battagello, L. Cricelli, M. Grimaldi, Prioritization of Strategic Intangible Assets in Make/Buy Decisions. *Sustainability* **11**, 1267 (2019).
17. M. Abiru, B. Nahata, S. Raychaudhuri, M. Waterson, Equilibrium structures in vertical oligopoly. *Journal of Economic Behavior & Organization* **37**, 4, 463-480 (1998).
18. M. Salinger, Vertical Mergers and Market Foreclosure. *Quarterly Journal of Economics* **103**, 2, 345-356 (1988).
19. G. Bonanno, J. Vickers, Vertical Separation. *Journal of Industrial Economics* **36**, 3, 257-265 (1988).

20. H. Wei, L. Lan, W. Kai, Economic and Environmental Implications of Quality Choice under Remanufacturing Outsourcing. *Sustainability* **12**, 874 (2020).
21. Y. Lin, Oligopoly and Vertical Integration. *American Economic Review* **78**, 1, 251-254 (1988).
22. J. Ordover, G. Saloner, S. Salop, Equilibrium Vertical Foreclosure. *American Economic Review* **80**, 1, 127-697 (1990).
23. P. Cyrenne, Vertical Integration versus Vertical Separation: An Equilibrium Model. *Review of Industrial Organization* **9**, 3, 311-322 (1994).
24. Economides, N. (1994): The Incentive for Vertical Integration, Working Paper EC-94-05, New York University, Stern School of Business, New York
25. A. M. Abubakr, A. T. Abbas, T. Italo, M. S. Soliman, M. Luqman, H. Hegab, Sustainable and Smart Manufacturing: An Integrated Approach. *Sustainability* **12**, 2280 (2020).

Normative and descriptive perception of strategic decision making

Nadežda Jankelová[1,*], and *Diana* Puhovichová [2]

[1] University of Economics in Bratislava, Faculty of Business Management, 852 35 Bratislava, Slovakia
[2] University of Economics in Bratislava, Faculty of Business Management, 852 35 Bratislava, Slovakia

Abstract. During historical development a greater number of decision theories were gradually conceived. Their differences arise from the normative and descriptive characters of individual theories. They differ in basic assumptions such as rationality, available information or choice of a decision. In practical life, they are manifested by accentuation of rational, or even intuitive decision making. This article is devoted to the definition of both types of theories and subsequent examination of the level of rationality in strategic decision making of managers. The findings have shown, that managers use more rational approach than intuitive approach. The most important factors that influence the level of rationality in strategic decision making were external factors. The most important of them has been assigned to a group of competition factors. Other factors influencing the level of rationality in strategic decision-making have been identified, namely the factors of the decision itself and factors of the internal environment. Managers combine rationality with strategic decision-making activities to take advantage of opportunity of an environment. The emphasis in the degree of impact on rationality in strategic decision-making is significant in favor of using the factors of the decision itself. The dependence of the perception of factors of the decision itself was significant in the variable economy.

1 Introduction

During historical development, a greater number of decision-making theories have been conceived, and they all differ in their understanding of the decision-making process, in different procedures for measuring the usefulness of variants, and in different approaches to rational decision-making. These differences arise from their normative and descriptive character. They differ in such basic assumptions as rationality (perfect or limited) available information (full access or limited access), choice of decision (principle of optimization or principle of satisfaction).

Within the framework of constantly developing opinions on individual theories, our goal is to examine, critically evaluate, and at the same time provide our own views on

* Corresponding author : nadezda.jankelova@euba.sk

rational-logical (normative) and descriptive areas of decision-making in the current conception, and subsequently verify the acquired knowledge in strategic decision-making in Slovakia.

The normative direction in decision theory focuses on providing guidance on how to solve decision-making problems, what procedures and models to use for each type of decision-making situation in order to achieve the required quality of decision-making processes. According to Bačová [2], normative decision-making models have long been recognized as a measure, criterion, standard against which the degree of "rationality" of individual and group decisions is compared and evaluated. This plays an important role in the eyes of many researchers even nowadays. They are based on the assumptions of rationality, which according to Robins and Coulter [22] are the exactness of the problem, the clarity of objectives and preferences, the knowledge of variants and the stability of preferences.

A rational approach to decision-making is understood by the authors:

- as a sequence of steps to achieve the goal [6, 9, 15], where the difference of authors is in the number of steps leading to rationality.
- as Evidence Based Decision Making, with use of this the authors are arguing, that it is currently important to direct the decision-making process in order to get the most out of the information that is available, thereby ensuring a competitive advantage [4, 7, 19, 21, 23, 24].

The descriptive direction in decision theory tries to clarify how are decisions actually made in real life. Psychology enters economic thinking, and decision-making uses knowledge of behavioral economics [3]. Besides psychology, the role of emotions and affects is one of the areas of research for making decisions. De Martino, Kumaran, Seymour, Dolan [10] emphasize the importance of emotional processes in decision-making models, emotions are considered as states of brain functioning, and represent how the brain can change and influence the effect of impact of deviations on achieving rationality. Many authors are involved in descriptive theories of intuitive decision-making [5,18,20,25].

Rational and descriptive approaches also appear in strategic decision-making. Strategic decision-making deals with poorly structured decision-making problems, where there is no clear procedure for resolving them, which leads to accepting the decision. Therefore, the approaches for strategic decision-making are also highly heterogeneous, and on one hand, they differ in terms of applying the level of rationality and exactness, on the other hand they use the intuition and experience of managers. Elbanna and Child [13] have developed an integrated model of rationality of the strategic decision-making process, which consists of three perspectives affecting rationality - characteristics of the environment, enterprise, and the decision itself. The various factors mentioned by the authors influence to a different extent, whether the manager uses more rational or more intuitive approaches. There is a strong empirical support for the impact of these factors on decision-making. For example, the importance of decision as one of the factors within its specific characteristics significantly influences the way of decision in terms of rationality. Similarly, activities undertaken in strategic decision-making due to environmental threats are more rational, than activities undertaken to exploit its opportunities. Internal factors, e.g. formal structures or centralization of power have a greater impact on the rationality of the process. Of course, there are other opinions, that rationality is limited in the process of strategic decision-making and is limited by the amount of obtained information [8,14,16].

Similarly, according to the above-mentioned authors, this process is not only characterized by analysis, but also by continuous evaluation, intuition and political behavior.

2 Methods

The importance of strategic decision making, which belongs to the competence of top-notch managers, is characterized mainly by a fact, that mangers' decisions affect the whole company and all subjects. However, strategic decisions are not taken uniformly. On the one hand, there is a need to apply rationality, which is emphasized in terms of using of rational analysis, exact methods, and procedures. On the other hand, the widespread application of rationality in strategic decision-making is questioned in view of its specificities. Theoretical considerations, as well as practical experience, in issues of rationality level application, are full of controversy when accepting strategic decisions. We were interested in the situation in Slovak companies. The research sample consisted of top 90 managers in the strategic level. Data were collected by a questionnaire. Besides the identification questions, the questionnaire also included questions concerning approaches for strategic decision-making, and factors that influence the degree of rationality in strategic decision-making as well as the context in which strategic decision-making takes place (opportunities - threats). The collected data were analyzed in the Microsoft Excel extended by a package of analytical tools (arithmetic mean, mode, median, minimum value, maximum value, standard deviation, etc.) and in the statistical software IBM SPSS v.23. For data analysis we used methods of descriptive statistics, regression analysis and paired t-test to compare the data. From a set of items, we have created aggregate variables from average values of items in each set. The established hypotheses were tested at the significance scale $\alpha = 0.05$. To verify the hypotheses, we used the following methods - paired t-test of the means of agreement - the significance level was set to 5% bilaterally, the linear regression of the dependent variable in the set of independent explanatory variables - also at the significance level of 5% bilaterally. By paired t-test of average values match, we also determined effect size in order to determine the magnitude of the effect based on the formula:

$$r = \sqrt{\frac{t^2}{t^2 + df}} \tag{17}$$

The first area was focused on the prevailing influence of selected groups of factors on the level of rationality of strategic decision-making. We set out the following hypotheses:

- null hypothesis:

H0: There is no significant difference in the degree of impact on rationality in strategic decision making among the factors of the characteristics of the decision itself and the factors of the internal environment of the company.

- alternative hypothesis:

H1: There is a significant difference in the degree of impact on the rationality of strategic decision making among factors of the characteristics of the decision itself and factors of the internal environment of the company.

The second area was focused on the fact, which factors influence the level of rationality in strategic decision making more significantly – the factors of characteristics of the decision itself or internal business factors and whether there is a dependency of perception of factors of decision itself in the context of rationality impact with respect at age, education, economic results, and other respondent identification data, what is the connection with other aspects of decision-making as well as other managerial functions. We set out the following hypotheses:

- null hypothesis:

H0: There is no significant difference in degree of impact on rationality in strategic decision making among the factors of the characteristics of the decision itself and the factors of the internal environment of the company.

- alternative hypothesis:

H1: There is a significant difference in the degree of impact on the rationality of strategic decision making among factors of the characteristics of the decision itself and factors of the internal environment of the company.

3 Results and Discussion

Managers said, that in strategic decision making they use a more rational approach than intuitive. To our question, in order to identify their approach to strategic decision-making on a scale of 1 to 5, where 1 meant purely rational and 5 meant purely intuitive, they answered an average score of 2,1. In strategic decision making, managers therefore prefer to acquire and analyze relevant information, use analytical techniques and methods in strategic decision making.

The factors that influence the level of rationality in strategic decision-making were one other field, that we have investigated in our research. Opinions that mainly appear in foreign scholarly literature are contradictory. There are alternating statements about the significant impact of the external environment on the degree of rationality in strategic decision-making, or on the impact of decision-making characteristics. Other authors point to the fact, that the external environment can be controlled only to a limited extent, so they focus their efforts primarily on the control of the internal environment and adapt the decision-making process accordingly.

In our research, we divided the factors into three groups, based on the integration model of the rationality of the strategic decision-making process, that we described in the introduction. The first group consisted of the factors related to the decision itself, such as. importance of decision, time pressure, severity of postponement of the solution, severity of consequences in case of wrong decision, setting parameters for subsequent decisions, uncertainty about activities to be carried out, clarity in the necessary information for the decision, clarity of objectives for participants, adequate freedom of access decision making, initial perception of the problem, motivation to make a decision, and confidence to make the right choice. The second group included factors related to the company - its size, financial and business results, organizational structure, planning and control system, corporate culture, informal power status of individuals and groups, social responsibility, employee satisfaction. The third group consisted of environmental issues - economy: inflation, interest rate, economic growth trends, unemployment rate, tax burden, policy: stability of government, laws, social policy, regulatory measures, competition: changes in prices, markets, strategies, entry of new competitors, product: customer preferences, demand, changes in components, product quality in production process.

Respondents evaluated the importance of each factor within each group of factors on a scale of 1 to 5, where 1 meant a very important factor and 5 meant that the factor was not important at all in the context of its impact on the rationality of decision-making. We calculated the average score of each factor. Graph 1 presents the results for a group of environmental factors. Graph 2 presents the other two groups of factors.

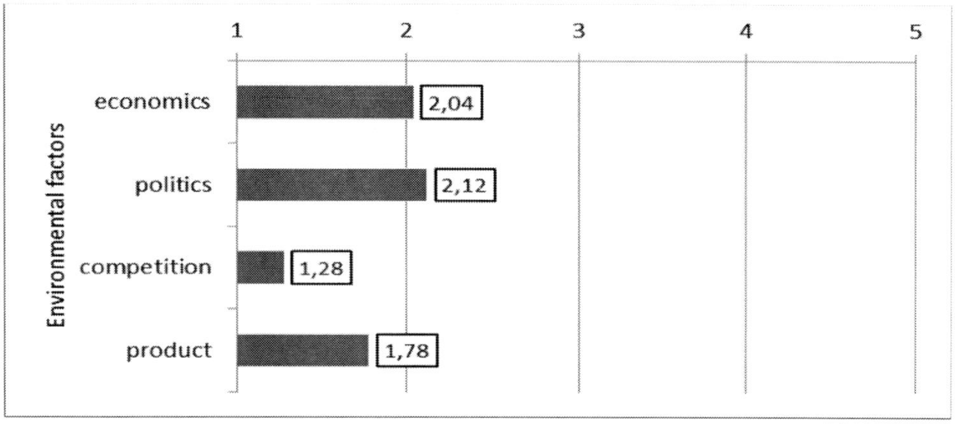

Graph 1 Scores of environmental factors in the surveyed enterprises (1- very important factor and 5 - not important at all)

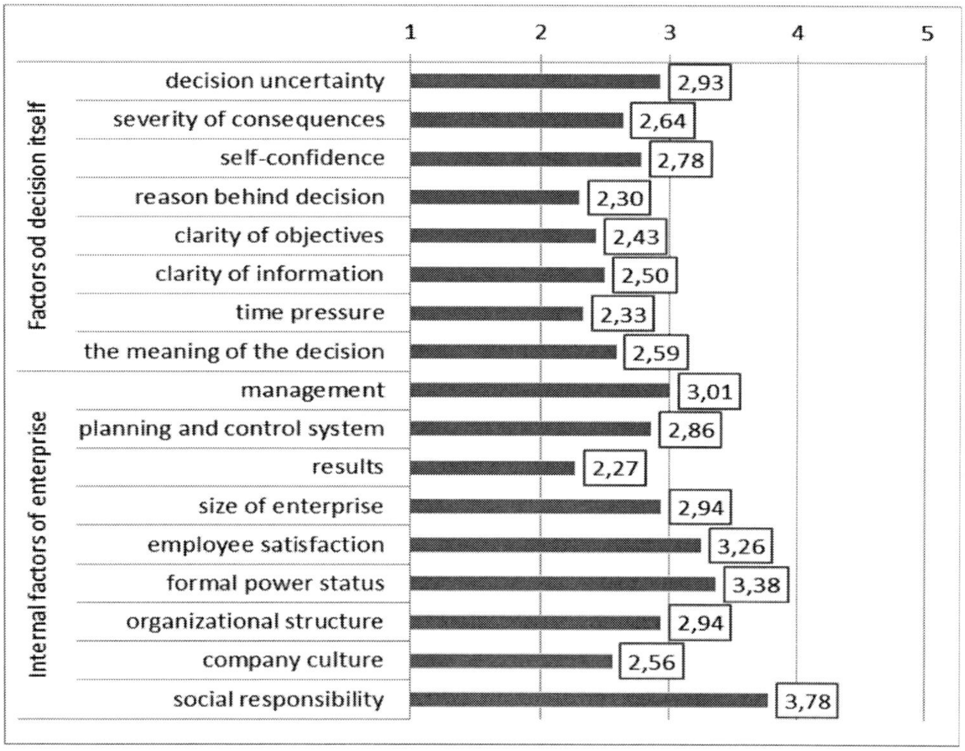

Graph 2 Score of the factors of the characteristics of the decision itself and the factors of the internal environment of the enterprises in the surveyed enterprises (1- very important factor and 5 - the factor is not at all important)

We assumed that the managers would perceive the environmental factors as a significant factor influencing the level of rationality in strategic decision-making. The most important of them was assigned to a group of competition factors, while others received the same average rating of importance in identifying the degree of impact on rationality. Our results coincide with foreign studies, where the environmental factor was identified as the most significant influence on the level of rationality [1,11,12,13]. However, from our results we have identified other factors influencing the level of rationality in strategic decision-making, namely the factors of the decision itself and factors of the internal environment of enterprises, in which the results of their impact on rationality were not negligible. We explain this by the fact, that the limited possibilities of controlling political or environmental variables can in some cases cause greater influence of internal factors on the rationality of strategic decision-making.

A fact that we find interesting was that up to 82% of managers associate rationality with strategic decision-making activities to take advantage of opportunities of environment. The remaining 18% decide rationally, because of environmental threats. From a theoretical point of view, hostility and the generosity of the environment significantly influence strategic behavior [2]. Although, little empirical research has been done in this area, the results suggest the importance of these factors. Rational decision-making processes are associated with the performance of organizations in a generous environment. In a hostile environment, organizations are under intense pressure, which is the reason why political tactics are important to secure the current position, increase power, and gain benefits. A team of authors [2] also found that uncertainty/ambiguity of the environment are determinants of rationality according to pivot theory. They state, that synoptic/aggregate processes (e.g. rationality) should be used in a stable environment while incremental / incremental processes (e.g. intuition) should be used in an unstable environment. This is because reliable data is available in a stable environment, there is less pressure to obtain new data and the cost of data collection is reasonable. Fact-based decisions can lead to better performance, than decisions based on hunch or judgment.

Considering the fact, that the two other groups of factors (the decision itself and the internal environment) are not examined in deeper in the context of the impact on the rationality of strategic decision-making, we have focused on this fact in our research. We found out which factors influence the level of rationality in strategic decision making more importantly - factors of the characteristics of the decision itself or internal factors of enterprise, whether there is a dependence of perception of the factors of decision itself in the context of the impact on rationality with respect to age, education, economic outcome and other identification data of the respondents, what is the connection with other aspects of decision-making as well as other managerial functions. We set out the following hypotheses:

- null hypothesis:

H0: There is no significant difference in degree of impact on rationality in strategic decision making among the factors of the characteristics of the decision itself and the factors of the internal environment of the company.

- alternative hypothesis:

H1: There is a significant difference in the degree of impact on the rationality of strategic decision making among the factors of the characteristics of the decision itself and the factors of the internal environment of the company.

Respondents evaluated the importance of each factor within the above two sets of factors on a scale of 1 to 5, where 1 meant a very important factor and 5 meant that the factor was not important at all. We calculated the average score of each factor. The results are presented in graph 2 above.

We statistically verified the hypothesis using the following steps:

We calculated the average score of a set of factors for the characteristics of the decision itself for each respondent. We repeated the same for a set of factors of the internal environment of an enterprise. To compare the overall score of both types of factors, we used a paired average t-test. The significance level was 5% on both sides. From table 1 is clear, that the score of the factors of the decision itself is lower, which means, that according to respondents, these factors have greater importance. Table 2 points to the fact, that the test is significant (t = -4.607 p = 0.000) with a median magnitude of effect r = 0.439 (formula given in the methodology).

Table 1. Paired Samples Statistics

Impact factors on the level of rationality		Mean	N	Std. Deviation	Std. Error Mean
Pair 1	Decision itself_f	2,56	90	0,739	0,078
	Internal_f	3,00	90	0,690	0,073

Table 2. Paired Samples Test

		Paired Differences					t	df	Sig. (2-tailed)
					95% Conf.Interval of the Difference				
		Mean	Std. Deviation	Std. Error Mean	Lower	Upper			
Pair 1	Decision itself_f Internal_f	-0,435	0,895	0,094	- 0,622	-0,247	-4,607	89	0,000

The completed test denies the null hypothesis, which means that we accept the alternative hypothesis. The emphasis in the degree of impact on rationality in strategic decision-making is significant in favor of the use of the factors of the decision itself.

Subsequently, we examined the dependence of perception of the factors of the decision itself in the context of the impact on rationality with respect to age, education, economic outcome and other respondent identification data. We have worked with the overall score for using these factors, listed above. The overall score of use factor was considered as a dependent variable. Independent variables were years in position, education, legal form, focus of activity, foreign participation, number of employees and economy for the period under review. The significance level of all tests was 5%.

The enterprises did not share their economic outcome with us, but only whether they had made a profit or loss over the last five years. We adjusted the variable economic outcome for the last years before processing into a variable with the name of economy, and we also created 6 categories based on the 5-year result, which are mentioned in table 3.

Table 3. Categories of enterprises by profit or loss

Category	Category content
0	5x loss and 0x profit
1	4x loss and 1x profit
2	3x loss and 2x profit
3	2x loss and 3x profit
4	1x loss and 4x profit
5	0x loss and 5x profit

With the use of mentioned way a new variable came out with a scale of 0-5 from the worst to the best economic result for the whole period. The number of enterprises, in each category, is shown in table 4.

Table 4. Number of enterprises in profit / loss categories

		Frequency	Percent
Valid	0	8	8,9
	1	16	17,8
	2	21	23,3
	3	16	17,8
	4	10	11,1
	5	19	21,1
	Total	90	100,0

The observed dependence of the perception of the factors of the decision itself in the context of the impact on rationality with respect to age, education, economic result and other identification data of respondents is shown in Table 5.

Table 5. Regression ANOVA table - Tests of Between-Subjects Effects

Source	Type III Sum of Squares	df	Mean Square	F	Sig.
Corrected Model	23.450[a]	10	2,345	9,799	0,000
Intercept	58,960	1	58,960	246,362	0,000
Legal form	0,672	2	0,336	1,405	0,251
Activity focus	0,191	3	0,064	0,2660	0,850
Number of employees	0,121	1	0,121	0,506	0,479
Economy	**20,699**	**1**	**20,699**	**86,489**	**0,000**
Years in position	0,115	1	0,115	0,481	0,490
Foreign participation	0,005	1	0,005	0,019	0,891
Education	0,007	1	0,007	0,030	0,862
Error	18,906	79	0,239		
Total	851,679	90			
Corrected Total	42,357	89			

a. R Squared = .554 (Adjusted R Squared = .497)

From Table 5 we can see that the effect on the economy criterion is significant ($F = 86.489$ $p = 0.000$). Other effects are not significant ($p > 0.05$). Table 6 provides estimates for found significant parameter, which is economy (other parameters are insignificant, so we consider them to be zero).

Table 6. Estimates for the significant parameter – economy (Dependent Variable: factors of the decision itself)

Parameter	B	Std. Error	t	Sig.	95% Confidence Interval	
					Lower Bound	Upper Bound
Intercept	4,013	0,381	10,523	0,000	3,254	4,772
Economy	**-0,323**	**0,035**	**-9,300**	**0,000**	**-0,392**	**-0,254**

From Table 6 we can see, that the effect acts in a negative direction (B = -0.323 t = -9.3 p = 0.000). From the selected explanatory variables, a significant impact of variable economy was found, which is equal to economic outcome for the period under review (t = -9.3 p = 0.000). While this dependency has been shown to be negative (B = -0.323) – that means that a coding of factor variable of decision itself, that hire efficiency of enterprises is connected with a greater emphasis on factors of decision itself by its impact on rationality of strategic decision making (that means with lower score). The following graph 3 illustrates the result:

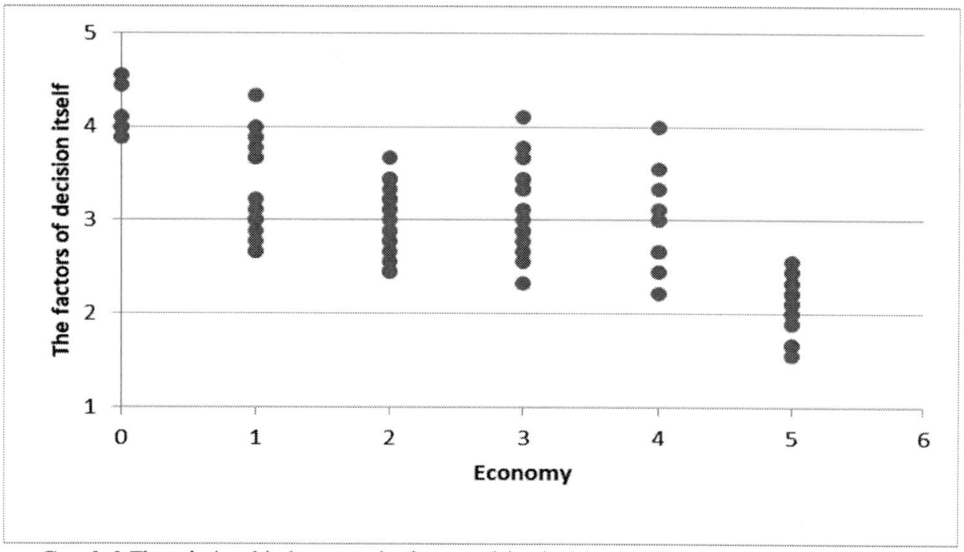

Graph 3 The relationship between the factors of the decision itself and the economic outcome

4 Conclusion

Decision making is a key characteristic of a successful organization and one of the most important roles of a manager. The ability to make the right choices is a complex phenomenon and represents the interplay of rational and intuitive elements, taking place in social groups in the context of social relationships in enterprises. The quality of the decision-making process is conditioned by the normative and descriptive frameworks, which are necessary to interact with a complex and variable external environment. In the article, we presented the results of research on the level of rationality in strategic decision-making. We have found out, that managers use more rational approach than intuitive approach while making strategic decisions. The most important factors, that influence the level of rationality in strategic decision-making were external factors of environment. The most important of these factors was assigned to a group of competition factors. Other factors influencing the level of rationality in strategic decision-making were identified, namely the factors of the decision itself and factors of the internal environment of enterprises. Managers are increasingly connecting rationality with strategic decision-making activities to take advantage of opportunities of environment. Emphasis on the degree of impact on rationality in strategic decision-making is significant in favor of the use of the factors of the decision itself. The dependence of the perception of the factors of the decision itself in the context of the impact on rationality was statistically significant in the economy variable. Other factors, such as the legal form of the company, focus of activities, number of employees and others did not have a significant impact on the level of rationality in decision making of managers.

Acknowledgement

This article is a partial output of research project VEGA no. 1/0017/20 entitled Changes in the application of managerial functions in the context of the fourth industrial revolution and the adaptation processes of enterprises in Slovakia.

References

1. CH. D. Allen, B. R. Coates, *Strategic decision making paradigms: a primer for senior leaders* http://www.carlisle.army.mil/usawc/dclm/Strategic%20Decision%20making%20%28Allen%20Coates%20July%202009%29%20final.pdf . (2009)
2. V. Bačová and colective, *Rozhodovanie a usudzovanie II* (2011)
3. V. Balaz, *Riziko a neistota* (2009)
4. E.Barends, D. M. Rousseau, R. B. Briner 2014. *Evidence-based management: The basic principles* (2014)
5. L.R. Beach, T.H. Mitchel, *Decision Making in the Workplace: A Unified Perspective* (2014)
6. J. Bridge, J.C. Dodds, *Managerial Decision Making* (2018)
7. R. B. Briner, D. Denyer D. M. Rousseau, *Evidence-based Management: Concept Cleanup Time? In The Academy of Management Perspectives*, vol. **23**, 19-32 (2009)
8. G. Calabretta, G. Gemser, N.M. Wijnberg, *The Interplay between Intuition and Rationality in Strategic Decision Making: A Paradox Perspective, In Organization Studies*, Vol. **38** (2017)
9. J. Dědina, J. Odcházel, *Management a moderní organizování firmy.* (2007)
10. B. De martino, D. Kumaran, B. Seymor, R.J. Dolan, *Frames, biases, and rational decision-making in the human brain*. 313(5787):684-7 (2016)
11. K. M. Eisenhardt, M. J. Zbaracki, *Strategic Decision Making, Strategic Management Journal*, Vol. **13**, 17-37, Special Issue: Fundamental Themes in Strategy Process research (1992)
12. S. Elbanna, *Strategic decision making: Process perspectives, International Journal of Management Reviews*, Vol.**8**, 1-20 (2006)
13. S. Elbanna, J. Child, *The Influence of Decision, Environmental and Firm Characteristics on the Rationality of Strategic Decision-Making, In Journal of Management Studies*, Vol. **44**, 561-591 (2007)
14. S. Elbanna, Y. Fadol, *The Role of Context in Intuitive Decision-Making, In Journal of Management & Organization*, Vol. 22, 642-661 (2016)
15. J. Fotr and colective, *Manažérske rozhodování. Postupy, metody a nástroje.* (2006)
16. J. W. Fredrickson, J. W. Iaquinto, Ine*rtia And Creeping Rationality In Strategic Decision Processes, In Academy of Management Journal,* Vol. **32**, 516-542 (2017)
17. R. Hanák, *Štatistika v PSPP*, https://statistikapspp.sk/author/robohanak/ (2015)

18. G. A. Klein, *Intuition at Work: Why Developing Your Gut Instincts Will Make You Better at What You Do* (2003)
19. B. Marr, *The Intelligent Company: Five Steps to Success with Evidence-Based Management* (2010)
20. J. B. Miner, *Organizational Behavior 2. Essential Theories of Process and Structure* (2015)
21. K. Morrell, M. Learmonth, *Against evidence-based management, for management learning. Academy of Management Learning & Education*, Vol. **14**, 520-533 (2015)
22. S. P. Robbins, M. Coulter, *Management.* Vol. **14**, 752 (2017)
23. D. M. Rousseau, B. Gunia, *Evidence-based practice: The psychology of EBP implementation. Annual Review of Psychology*, Vol. **67**, 667-692 (2016)
24. S. L. Rynes, J. M. Bartunek, *Evidence-based management: Foundations, development, controversies and future. Annual Review of Organizational Psychology and Organizational Behavior*, Vol. **4**, 235-261 (2017)
25. B. Schwarzt and colective, *Maximizing versus satisficing: Happiness is a matter of choice, Journal of Personality and Social Psychology,* Vol. **5** ,1178-1197 (2002)

Reward system design supportive to human capital stabilization in a company

Zuzana Joniaková[1], *Katarína* Grančičová[1], *and Jana* Blštáková[1,*]

[1]University of Economics in Bratislava, Faculty of Bussiness Management, Dolnozemská cesta 1, 852 35 Bratislava, Slovakia

Abstract. The stabilization of human capital at the enterprise level presupposes the systematic and effective use of human resources management tools. One of these tools is the employees' reward system. The paper in its theoretical contribution defines the conditions under which rewarding serves as an instrument for stabilizing human capital in an enterprise. Consequently, on the basis of its own survey data, the paper contains analysis of reward systems in companies in Slovakia in order to describe their development, current status and trends in the given context. Summary conclusions contains findings, that despite the significant and positive shift in several areas of remuneration systems in companies in Slovakia, the situation is still not optimal and offers room for further development.

1 Introduction

The starting point for addressing the issue of stabilizing human capital in an enterprise is to understand what keeps people at work. According to several authors, one of the decisive factors is the economic factors, namely the rewarad system. Remuneration can thus be seen as a critical function of human resource management, which has an impact on the individual's motivation to enter, learn and stay in society [12]. It is also an important tool for maintaining a high level of individual performance and promoting employee engagement [17, 21]. The paper therefore points out how the reward system should be designed to support the stabilization of human capital in the company and consequently examines what incentive approaches are used by companies in Slovakia. Based on the results of the research, the paper contains conclusions with summary evaluations that characterize the development and current trends in reward systems in Slovakia in the context of the stabilization of human capital in the company.

2 Literature review

* Corresponding author: jana.blstakova@euba.sk

Rewarding, as well as other human resource management functions, is subject to continuous development. The most recent changes in this area are summarized in Table 1 [4].

Table 1 Development in rewarding

Historical approach	Contemporary approach
Reactive wages	Strategic wages
Outlined reward systems	Integrated reward systems (into HRM systems)
Supporting business of the company	Interactive to the company strategy, employees' needs and conditions of the environment
Low flexibility, mechanistic, complicated reward system	Organic, variable, relatively symple in process of rewarding
Isolated initiatives	Integrated reward management
Unrealistic expectations	New realismus
Orientation to financial rewarding	Orientation to total compensation, engagement and lojalty
Administrative managing of wages	Wages reflecting entrepreneurship
Byrocracy in reward system	Flexible reward system
Hierarchical structures	Orientation to continuous developemnt
Performance based rewarding	Contribution based rewarading
Priority in inner fairness	Market orientation
Rewarding related to the content of the job	Rewarding related to people
The best „practice" principle	Adjustment to characteristics and culture of the company
Managing by HR department	Involment of line managers into rewarding
Multiple grade systems	Wide range systems
Planning	Delivering, realization
Top down information flow	Communication and employee involvment
„Big bang" style changes	Evolutionary changes

Source: own proceedings [4]

Cotemporary rewarding can therefore be characterized by features such as complexity, flexibility and a systemic approach. Companies benefit from a wide range of monetary, non-monetary benefits and employee benefits to create an integrated and mutually supportive set of tools that, when combined, will have a much stronger and longer-term impact on individual, team and organizational efficiency [3]. As part of their reward, employees themselves are perceived as partners with which they communicate openly and actively participate in reward processes [18]. There is a clear shift from situational responses to market and internal pressures and a shift towards reward not only in the human resources management strategy but also in the business strategy. Compatibility with culture and market orientation are also a prerequisite for successful reward system. Contemporary reward systems are evolving along with business, through evolutionary changes [6].

Schuster and Zingheim [23] introduce into contemporary theory the the concept of "New Pay". Its basic principles were defined as follows:

- employees are a source of competitiveness and efficiency of the company,
- communication with workers is crucial to the success of rewarding;
- the reward system is designed to reward results and work behavior in line with the company's objectives;
- variable (risky) part of rewarding is introduced into the reward system;

- rewarding is about employee relations, employees have the right to determine whether they are satisfied with the values, culture and reward system in the company,
- rewarding is understood as an element of the whole management process, it can be a positive force of organizational change,
- direct and indirect financial rewards are not separate and independent;
- emphasizing the importance of both team and individual rewards in new reward systems;
- partnership and participation as important elements in rewarding.

Rewarding of employees can now be seen as a strategic tool of human resources management, linked to the goals, strategy and culture of the company [10]. Trevor [19] sees the strategic approach to remuneration as "*a means of improving business performance and ensuring a competitive advantage by aligning reward strategies, systems, policies and practices with the business strategy.*" Financial reward no longer represents only the cost of hiring the necessary workforce, but is a means of reconciling the unique and inimitable asset of the enterprise - its employees - with the strategic direction of the enterprise '. Also, according to Brown [5], strategic reward is a way of thinking that links reward practices to business goals, as well as employee values and needs, which are critical to achieving business goals.

Taking reward as a strategic tool of human resources management in a company therefore means to see it as an integrating and innovative process that can support meeting the needs of the company and its employees, making it an important tool for stabilizing human capital in the company. However, emphasis should be placed on rewarding of the right things, which are valued and perceived by the company as important [15]. In this way, through a reward system, it can spread the message of what it values and considers to be value. Armstrong [3] stresses that the impact of the reward strategy on staff performance can only be maximized by linking it to the human resources development strategy and in particular to the recruitment strategy. Rewarding is no longer an isolated activity, but integrates with other human resource management processes and functions. As a human resource management function, it has become an increasingly organic, flexible process that reflects the business, rewards individual employees' contribution to the business and is increasingly geared to their lateral development [19].

The contemporary reward system is therefore determined by the business strategy and is intended to ensure that remuneration is managed in favor of the business, in line with its direction and objectives. In this context, Armstrong [4] sets out the main objectives in reward management:

- supporting the achievement of business goals through talented and committed people;
- rewarding people according to the value they create, that is, differentiation in rewarding of individuals according to the extent to which they meet the company's expectations;
- defining the desired behavior and outcomes to define performance expectations.
- promoting high performance, where the remuneration system should ensure that high performance is recognized, valued and supported;
- promoting and developing a corporate culture, ie interconnecting rewards with behaviors that are consistent with the company's value orientation.

Such an approach places great demands on the management of enterprises and, in particular, on their personnel managers, who have to cope professionally with the creation and functioning of reward systems, as otherwise the potential offered by this area of human resources management remains fully unused [14]. Rewarding is a complex and demanding task. As Fempel and Zander [8] point out, "*the perfection of rewards systems is in practice limited by their progressively increasing complexity. This limits operationality and transparency, which means that reward systems achieve their objectives only to a limited extent.*"

Contemporary rewarding often implicates and considers progressive and holistic approach dealing with the concept of so-called total compensation. Such rewarding, according to Dvořák [7], involves linking wages for work done with various forms of non-financial recognition and other human resource management tools. Some authors talk about the three main categories of rewards that co-create the total compensation of external, internal and social [22,13, 20]. Essentially, external rewards consist of tangible benefits, including pay, career development opportunities and a reasonable workload, while internal rewards are intangible forms of appreciation, such as the ability to participate in job definition, meaningfulness and collaborative support. Social rewards include those derived from positive social interaction in the workplace, such as the use of supportive relationships with colleagues and superiors [1]. Also, according to Peluso [17], cash rewards are a necessary but insufficient condition for the development of an engaged workforce, therefore the concept of total compensation should also include non-monetary rewards, such as career opportunities, a supportive working environment that allows employees to balance personal and professional spheres. Lyons and Ben-Ora [11] explain total compensation system using four components:

1. basic wage and variable components,
2. various types of contributions in cash - health care contributions, retirement benefits, savings, including additional days off;
3. education and development opportunities, such as: career management, talent management, succession planning, development programs, etc.,
4. a working environment that includes intangible elements such as corporate culture, leadership styles, performance promotion, or work-life balance.

Several research studies dealt with the impact of total pay on employee behavior and performance. These confirmed the positive link between the total compensation, employee performance and improved business performance [16]. The concept of total compensation can therefore serve as a source of competitive advantage for businesses, allowing them to respond more flexibly to internal changes and align employee performance with business objectives [17], since engaging and combining different forms of pay is a source of overlapping and mutually reinforcing effects on employee performance [9].

3 Research design and methodology

These knowledge in theoretical sources became the basis of the research focused on the reward systems used in companies in Slovakia. The main objective was to assess their development and contemporary situation in the context of the instrument of stabilization of human capital in the company. Research itself addressed a number of research questions: what is the degree to which a strategic approach to rewarding is used; what is the impact of individual subject on the design of reward systems and their implementation; what is the impact of trade unions on the rewarding employees; what is the use of the 'total compensation' principle, at what level the employee's wage is determined; which of the managers and to what extent they are involved in deciding the individual's wage; what is the importance and time-consuming nature of reward system within HRM; which factors affect the level of wages, which methods of assessing the rewarad system are used; what are the key issues in contemporary reward systems and the reasons for their adjustment; in companies in Slovakia.

In addition to general methods of scientific research, a questionnaire survey was used as a basic tool for data collection. Data collection was carried out in the period of 2019, using simple random sampling and the units of sampling were companies operating in Slovakia. A total of 148 companies were included in the research. The sample of companies was characterized in terms of size, legal form, affiliation to the sector. However, the specific structure of the sample of enterprises is not given due to the limited scope of the contribution.

The obtained data were then evaluated using mathematical-statistical methods. At the same time, the current survey followed a similarly focused regional survey conducted in 2006 and an international survey conducted within the CRANET research network in 2013, focusing more on the HRM functions. This made it possible not only to evaluate the development and to identify the basic trends of rewad systems in companies in Slovakia, but also to monitor their comparison in the international context.

4 Research results and discussion

This part of the paper presents and interprets the results obtained from a survey in companies in Slovakia focused on their reward systems. It focuses only on selected aspects of the reward system that can be considered essential in the context of the stabilization of human capital. These include assessing the use of the strategic approach to reward system design- formulating strategy, developing and implementing the reward system, the degree of influence of individual entities on the creation of the reward system, applying a holistic approach to reward - the concept of 'total compensation'. In conclusion, the summary evaluation formulates also other findings characterizing the development and current trends in the reward systems of companies in Slovakia.

The first of the areas examined was the assessment of the use of the strategic approach to rewarding. The results of the survey suggest that at present 70% of companies in Slovakia in the creation of reward systems are based on their own reward strategy, the definition of which is predominantly within the competence of top management in cooperation with personnel departments. Compared to the past (2006), there has been a clear increase in the use of formalized reward strategies (14%) and a decrease in the number of companies that do not create their own reward strategies (21%). (Table 2).

Table 2 The form of reward strategy in Slovak companies

Rewarding strategy	2006	2019
Formal design/documentation	56%	70%
Unformal	23%	23%
No strategy	21%	7%

Source: own proceedings

Most often, the companies have defined reward priorities and targets in their reward strategies; reward system components; attitude to competition; responsibility for individual parts of the reward system. It is also interesting that almost three quarters of the companies surveyed (71%) report that they do not directly link their reward system / strategy to corporate culture, which is one of the basic conditions for its effectiveness. Only 23% declare such interconnection within the category of employees as manager; for the other categories of employees, only 16% of companies support corporate values through rewarding. These results can be considered insufficient and are evidence that the management of companies in Slovakia is not sufficiently aware of the link between these two areas and does not work with them in a targeted manner. The reward system is considered to be one of the strongest instruments for promoting the corporate culture and therefore it is beneficial to use it in this regard. This area can thus be considered as a space for eficinecy improvement of reward systems in practice.

Research of the influence of individual entities in enterprises on the creation of the remuneration system shows minor changes during the researched period. During the creation of the system, the influence of HR managers as a professional component was strengthened, the involvement of employees' representatives increased significantly (up to double) and the use of external know-how in the design of reward systems was even greater. (Table 3)

Table 3 Involvement of particular subjects into reward system design in Slovak companies (2006-2019).

Subjects	2006	2013	2019
Top management	49%	74%	50%
HR management	47%	42%	68%
Line management	20%	20%	7%
Employee ambasodors	13%	10%	23%
External subjects	3%	4%	16%

Source: own proceedings

These changes indicate the pursuit of professionalism and a professional approach to rewarding. The increased involvement of workers 'representatives, in particular trade unions, is a manifestation of the pursuit of a more transparent, partnership-based approach and consideration of employees' requirements, which may support the perception of the rewarad system as fair and fair. On the other hand, a significant decrease in the involvement of line managers in the creation of the reward system may seem surprising. Taking a closer look at the involvement of entities in each stage of the reward process - from strategy development to system implementation (Table 4), the influence of line managers is most noticeable at the stage of implementation of the reward system (39% of companies). On this basis, it can be stated that even in our conditions we can see a gradual transfer of competences of line management in rewrding, which is in the context of current trends, although not yet fully implemented.

Table 4 Influce of particular subjects upon forming the reward strategy and its implemnation (2019).

Subjects	Reward strategy	Reward system design	Reward system implementation
Top management	80%	50%	25%
HR management	30%	68%	48%
Line management	11%	7%	39%
Employee ambasodors	11%	23%	14%
External subjects	5%	16%	9%

Source: own proceedings

After 1990, trade unions and employee representatives had a low impact on the rewarding in comparison with EU countries. At present, their impact has been strengthened mainly at the stage of system creation (to 23%), but at the stage of strategy formulation and implementation of the reward system remains weaker (11% and 14%). The low involvement of external subjects, ie. j. consultants into development of reward systems (16%) has been over come. However in the formulation of strategies and the implementation of systems these options are still used to a small extent.

The contemporary concept of total compensation is holistic, ie. j. it includes not only monetary and tangible components but also various intangible components. Armstrong [2] refers in this respect to relational rewards, which he classifies, for example. relationships in the workplace, corporate culture, management styles, talent promotion, etc. The aim of the research was therefore to find out whether companies operating in Slovakia approach rewarding in this way, and what importance they attach to instruments that, although long-term part of human resources management, have traditionally not been directly related to rewarding. The companies surveyed declare that they use these intangible instruments to a large extent as part of rewarding and consider them meaningful. The use of intangible rewards is mainly the responsibility of direct superiors, who must be qualified and trust their effects. Managers make the most of personal awards (92%) and give them the greatest importance (Table 5)

Table 5 The aplication of intengible forms of rewarding in Slovak companies (2019).

Intengible rewards	Number of companies	Significance (scale 1 – 5)
Personal appreciation	92%	4,0
Public appreciation	86%	3,4
Common time	78%	3,6
Talent support	70%	3,7
empowerment	71%	2,7

Source: own proceedings

5 Conclussion

Based on the overall results of the researching remuneration systems in companies in Slovakia in the context of the instrument of stabilization of human capital in the company, the most important findings can be formulated as follows:

- the reward system is gradually being professionalized;
- the trend towards a comprehensive and strategic approach to rewarding is stronger;
- the principle of "total compensation" in rewarding employees is widely applied;
- the focus of rewarding is gradually shifting from employee to company level;
- businesses make rewarading adjustments primarily based on performance improvements;
- the importance of knowledge and experience is increasing as the basic determinants of rewarding;
- there are weaknesses in particular in methods of examining the effectiveness of reward systems;
- the importance of active communication of rewarding has increased significantly.

Survey results also showed that the companies surveyed continuously update their reward systems, which is particularly important in an environment of constant and rapid change. Businesses are aware of this and count on changes in systems and actively work with them. However, it should be noted that interventions in reward systems need to be made sensitively and all changes actively communicated. Especially in communication about reward systems there was a significant positive change in companies in Slovakia. While communication has been assessed as a weakness in rewarding in the past, businesses have recently focused their attention on it. Not only the intensity of communication has changed, but also its ways and tools used.

This paper is the result of the research project VEGA project no. 1/0569/18: “Stabilization of human capital in Slovak enterprises as an immanent determinant of their success, competitiveness and sustainable development” with a 50% and the project VEGA no. 1/0412/10: „ Human resources management systems in 4.0 Industry Era“.

References

1. A. Alhmoud, H. Rjoub, Total Rewards and Employee Retention in a Middle Eastern Context. *SAGE open* **9** (2019)

2. M. Armstrong, *Řízení lidských zdrojů – Nejnovější trendy a postupy.* (EDP Grada Publishing, Praha, 2007)
3. M. Armstrong, *Odměňování pracovníků.* (EDP Grada publishing, Praha, 2009)
4. M. Armstrong, *Řízení lidských zdrojů.* (EDP Grada Publishing, Praha, 2015)
5. D. Brown, *Reward Strategies: From intent to impact.* (EDP Gardner Books, Chartered Institute of Personnel and Development, London, 2001)
6. S. Corby, C. Stanworth, G. White, No news is good news? Evaluating new pay systems. *Human Resource Management Journal* **15**, 4-24 (2006)
7. Z. Dvořáková, *Řízení lidských zdrojů.* (EDP C. H. Beck., Praha, 2012)
8. K. Fempel, E. Zander, *Integrierte Entgeltpolitik, Handbuch Entgeltmanagement.* (EDP Verlag Franz Vahlen, Munchen, 2005)
9. L. Innocenti, M. Pilati, A.M. Peluso, Trust as moderator in the relationship between HRM practices and employee attitudes. *Human Resource Management Journal* **21**, 303-317 (2011)
10. E. Lawler, *Rewarding Excellence: Pay Strategies for the New Economy.* (EDP Sage Publications, Inc., 2002)
11. F. H. Lyons, D. Ben-ora, Total rewards strategy: the best foundation of pay for performance, *Compensation and Benefits Review* **34**, 34-40 (2002)
12. J. W. Medcof, S. Rumpel, High technology workers and total rewards. *The Journal of High Technology Management Research* **18**, 59-72 (2007)
13. J. C. Morgan, J. Dill, A. L. Kalleberg, The quality of healthcare jobs: Can intrinsic rewards compensate for low extrinsic rewards? *Work, Employment and Society* **27**, 802 – 822 (2013)
14. F. P. Morgenson, M. A. Campion, C. P. Maertz, Understanding Pay Satisfaction: The Limits of a Compensation System Implementation. *Journal of Business and Psychology* **16**, 133–149 (2001)
15. J. Olsen, Societal values and individual values in reward allocation preferences. *Cross Cultural Management: An International Journal* **22**, 187-200 (2015)
16. J. Purcell, N. Kine, J. Swart, B. Rayton, S. Hutchinson, *People Management and Performance* (EDP Routledge, London, 2009)
17. A. M. Peluso, L. Innocenti, M. Pilati, Pay is not everything: Differential effects of monetary and non-monetary rewards on employees' attitudes and behaviours. *Global Forum for Empirical Scholarship* **5**, 311-327 (2017)
18. K. Tae-Yeol, et al., Equal or equitable pay? Individual differences in pay fairness Perceptions. *Human Resource Management* **58**, 169-186 (2018)
19. J. Trevor, *Can Pay be Strategic?* (EDP Basingstocke: Palgrave Macmillan, 2011)
20. J. M. Twenge, A review of the empirical evidence on generational differences in work attitudes. Journal of Business and Psychology 25, 201-210 (2010)
21. M. Vest , K.D. Scott, J. M. Vest, et al. Factors Influencing Employee Beliefs That Pay Is Tied to Performance. *Journal of Business and Psychology* **14**, 553–562 (2000)

22. I. O. Williamson, M. F. Burnett, K. M. Bartol, The interactive effect of collectivism and organizational rewards on affective organizational commitment. *Cross Cultural Management: An International Journal* **16** ,28-43 (2009)
23. P. K. Zingheim, J. R. Schuster, *Pay People Right. Breakthrough strategies to create great companies*. (EDP Jossey-Bass, San Francisco, 2000)

Analysis of the use of venture capital in Central and Eastern Europe countries

Lenka Kalusová[1,*]

[1]University of Economics in Bratislava, Department of Business Finance, 85235 Bratislava, Slovakia

Abstract. We classify venture capital as a non-standard, alternative source of financing. Venture capital has an important position especially in financing innovative project activities. For this reason, the main goal of the paper is to analyze the use of venture capital in the countries of Central and Eastern Europe with the focus on the Visegrad Group countries. We used scientific methods, both from the general - logical methods, as well as from exact methods (in the form of statistical methods and comparison). We analysed the years 2010 – 2018. The results of our study show that the use of venture capital is very low in the Central and Eastern Europe countries.

1 Introduction

Venture capital can be in terms of classification of funding sources included in the category of alternative sources. The authors Srpová and Řehoř [14] consider venture capital as capital that "participates in the financing of private growth companies by increasing their share capital. It is a partnership of the entrepreneur and the investor." Kleinschmidt [16] considers venture capital to be a type of capital used to finance the growth of private companies over a defined period, and according to his definition of venture capital, the provision of venture capital also includes the provision of non-financial support to the company. The author's definition includes four characteristic attributes of venture capital, which are equity financing, medium-term investments, a focus on growth companies and active investing. Based on the above definitions of venture capital, we can therefore conclude that risk capital is considered risky because it finances projects for which it is difficult to obtain funding because they are more risky than other projects.

The use of venture capital and the factors affecting venture capital are the subject of research by several academic. The authors Groh and Liechtenstien [2] analyzed determinants of venture capital allocation in CEE countries. The authors examined the impact of 6 main determinations. The research results point to the fact, that investors are satisfied with the economic potential and entrepreneurial opportunities, the problems in venture capital investments are caused mainly by the investor protection as well as the poor level of the depth of capital markets. The impact of the depth of the capital market in Poland on the use of venture capital was examined by Soloma [5]. The author found that although Poland is roughly equal in terms of the ratio of venture capital investments to

* Corresponding author: lenka.kalusova@euba.sk

GDP to other Central and Eastern European countries, the author found that it is more attractive to investors than other countries, mainly due to the well-developed Warsaw Stock Exchange (about 40% of all primary emissions in Europe was in 2012 in Warsaw The role of institutions supporting the development of venture capital in 18 European countries has been the subject of research by Bedu and Montalban [17]. The results of the study point to a strong correlation between the growth of the size of the capital market and the growth of venture capital investments, which is also confirmed by the previous findings of the author Soloma. The foundations emphasize the importance of country regards to development of venture capital investments. The use of venture capital as one of the possible alternatives to standard sources of funding in Romania is discussed in an article by Prelipean et al. [9]. Studies by several authors [11, 6, 7] show that the financial crisis has had an impact on the use of venture capital. According to Block and Sanders [12], the financial crisis has caused a reduction in venture capital to finance businesses. Ning et al. [19] report that macroeconomic factors significantly affect the use of venture capital. Likewise, Gomes Santa Felix et al. [8] confirm that the level of venture capital utilization affects the environment in which businesses operate as well as the degree of asymmetric information. This is also confirmed by a study by Jeng and Wells [15], Groh and Liechtenstein [1] who noted that there are significant differences between the uses of venture capital across countries. Another study Groh and Wallmeroth [4] adds that there are different factors influencing the use of venture capital in developed and developing economies.

Based on the domestic and foreign literature search, the main goal of our contribution is to analyze, evaluate and appreciate venture capital funding in Central and Eastern Europe Countries.

2 Methodology

We obtained information for our research from both domestic and foreign literature. The main source of statistics on the use of risk capital in the Central and Eastern Europe Countries was the Central and Eastern Europe Statistics. We analyzed the years 2010 - 2018.

The processing of research requires the use of several scientific methods, both from the group of general - logical methods and from the group of exact methods (in the form of statistical methods and comparison). The method of comparison was used both in the theoretical part of the paper and in its application part. In the theoretical part, we compared the views of different authors on the use of venture capital. We also analyzed the views of various authors on the use of venture capital depending on the economic cycle. In the practical part, we compared the use of venture capital in the Central and Eastern Europe countries and Europe as a whole. At the same time, we also compared the use of venture capital in the Visegrad Group countries. The results of our research are presented in tables and figures.

3 Results and discussion

The use of venture capital varies from one European country to another. Table 1 shows the total amount of risk capital used in Europe and especially in the countries of Central and Eastern Europe. The table shows these data according to the specific forms of risk capital used. The year 2018 was subjected to analysis.

Table 1. The value of venture capital investments according to particular types of venture capital in CEE countries and Europe (2018).

Type of investments	**Value of investment in CEE countries**		**Value of investment in Europe**		**CEE countries´ s investments ratio to Europe as a whole**
	EUR Thousand	**%**	**EUR Thousand**	**%**	**%**
Seed capital	38 097	1.42%	721 445	0.90%	5.28%
Start-up capital	97 786	3.64%	4 887 307	6.07%	2.00%
Expansion capital	638 463	23.78%	14 481 478	17.97%	4.41%
Rescue capital	-	-	254 799	0.32%	-
Replacement capital	24 624	0.92%	142 016	0.18%	17.34%
Management buy-out - MBO	1 885 972	70.24%	58 814 426	72.99%	3.21%
Total 2018	**2 684 943**	**100.00%**	**80 581 471**	**100.00%**	**3.33%**
Total 2017	3 532 018	-	75 324 705	-	4.69%

In table 1 we can clearly see that the ratio of venture capital investment in the Central and Eastern Europe countries to the total volume within Europe is only 3.33% in 2018 (the previous year 4.69%).

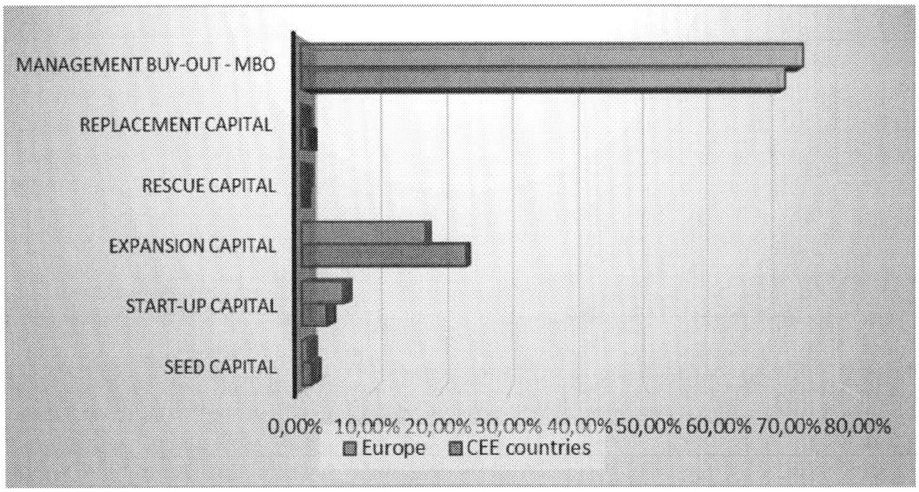

Fig. 1. Types of venture capital investments in CEE countries and in Europe in 2018

However, as regards the structure of the use of individual types of venture capital, this is approximately the same in both groups under review. Within individual types of venture capital, there is a clear prevalence of management buy-outs. MBO represents more than 70 % of its total volume of venture capital funding in CEE countries as well as in Europe as a whole. The second most commonly used type of venture capital is expansion capital (23.78% share in Central and Eastern European countries, 17.97% share in Europe as a whole).

In table 2 we can see the review of the volume of venture capital funding in Visegrad Group countries (Slovakia, Czech Republic, Poland and Hungary) in years 2010 – 2018.

Table 2. Venture capital investments in Visegrad Group countries (million EUR, years 2010 – 2018).

Country/Year	2010	2011	2012	2013	2014	2015	2016	2017	2018
Czech Republic	229	144	106	134	299	16	168	67	767
Poland	653	678	473	380	251	896	805	2505	850
Hungary	65	195	103	56	170	194	101	209	346
Slovakia	15	9	98	2	12	17	17	4	26
Total V4	**962**	**1026**	**780**	**572**	**732**	**1123**	**1091**	**2785**	**1989**
Total CEE	**1336**	**1247**	**1005**	**789**	**1311**	**1700**	**1703**	**3532**	**2685**
Visegrad Four´s percentage on a total CEE countries	72.01	82.28	77.61	72.50	55.84	66.06	64.06	78.85	74.08

As can be seen from the table, the share of V4 countries in the total volume of venture capital used in Central and Eastern Europe varies from year to year. The lowest share can be seen in 2014 (55.84%). On the contrary, we see the highest proportion in 2011. Up to 82.28% of the venture capital used in the countries of Central and Eastern Europe was in the Visegrad Group countries.

Development of risk capital in all countries during the period is considerably volatile. As we can see, the share of investment in V4 countries in the total volume of venture capital investments in CEE also fluctuates. In general, since 2010, the largest volume of venture capital within these countries has been used in Poland. In 2017, the volume of venture capital used in Poland was 2 505 million EUR. Venture capital investments in Poland represent more than 89 % of its total venture capital investments´ s volume in CEE countries in this year. On the contrary, the lowest values were reported by Poland in 2014 (251 million EUR). The table shows that venture capital is least used in Slovakia. While in 2012 the volume of venture capital utilized in Slovakia was 98 million EUR, in the last reference year it was only 26 million EUR.

Although the ratio of Slovakia, Czech Republic, Hungary and Poland in the total volume of venture capital used in Central and Eastern Europe is relatively high, the level of utilization of capital, particularly in Slovakia and the Czech Republic is low. What are the main reasons that may cause this? Marková and Balcová [18] present two main types of barriers to the development of venture capital in Slovakia. These are barriers on the demand side and barriers on the supply side. Among the main obstacles on the supply side, the authors cite insufficient information about the possibilities of using venture capital as well as its weak promotion. On the supply side, the most important barrier was identified as the low innovation capacity of small and medium-sized enterprises. Due to the weak innovation potential, investors are not interested in financing projects of small and medium-sized enterprises, as they are not attractive enough for them. In the case of the Czech Republic, the Czech Republic has several assumptions that could positively support the use of venture capital - characterized by a relatively high education level of the population, a high number of small and medium-sized enterprises which undercapitalization is good prerequisites for venture capital use [10]. Despite the above assumptions for increasing the use of venture capital in the Czech Republic, this level is relatively low. Rajchlová et al. [13] ranks among the obstacles to the development of venture capital in the Czech Republic include inadequate support for start-ups and companies in the development phase in the area of tax benefits, no support for venture capital financing and insufficient knowledge and misunderstanding of this form of financing by entrepreneurs.

A comprehensive assessment of risk capital in individual European countries was developed by Groh et al. [3]. In The Venture Capital & Private Equity Country Attractiveness Index, they assessed the attractiveness of countries for venture capitalists according to several criteria (economic activity, capital market level, taxation, investor / corporate governance, human and social environment and business culture and opportunities). The individual criteria examined are listed in Table 3.

Table 3. VCPE Country Attractiveness Index (V4)

Country	**Slovakia**		**Czechia**		**Poland**		**Hungary**	
VCPE Country Attractiveness Index	**score**	**rank**	**score**	**rank**	**score**	**rank**	**score**	**rank**
	50,5	**71**	**65,7**	**33**	**72,4**	**26**	**57,7**	**52**
Economic Activity	73,7	65	85,4	28	89	19	81,3	43
Depth of Capital Market	34,7	78	47,7	61	75,8	22	46,6	64
Taxation	103,7	43	102	48	107,8	24	101,5	49
Investor Protection and Corporate Governance	58,1	71	70,5	36	63	57	57,4	74
Human and Social Environment	44,3	69	72,5	18	60,8	34	52,6	48
Entrepreneurial Opportunities	58,6	46	72,9	26	66,8	34	59,9	45

From the table above, then the graph we can assess which factors are barriers to the development of the venture capital market in the countries of the Visegrad Group and conversely, what are the advantages of the business environment of the country, which could attract investors. Only the main evaluation criteria are listed in the table below. 125 countries were evaluated in 2018. Among the monitored countries, Poland was the best (26th place), followed by the Czech Republic (33rd place), Hungary was in 52nd place and Slovakia was the last (71st place).

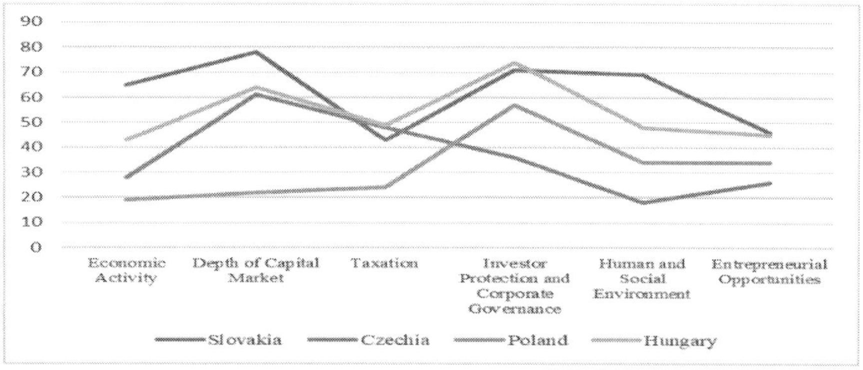

Fig. 2. Rank of Visegrad Group countries in 2018

In the case of Poland, taxation and economic activity were rated the best. Compared to other countries, Poland also has a high rating in the case of the depth of capital market. On the contrary, the weaknesses of the Polish market could be the investor protection and corporate governance. The main barriers of Slovakia as the worst rated country among the V4 countries can be seen especially in the case of the depth of capital market - weak IPO activities in the market, low trading volume and the size of the stock exchange received a weak rating. The investor protection was also assessed negatively in the case of Slovakia - the level of enforceability of the law is low, as well as the weak security of property rights. On the contrary, Slovakia managed to place better than the Czech Republic only in the criterion of taxation (43rd place, while the Czech Republic took 48th place). Entrepreneurial culture and the offer of opportunities were also evaluated relatively well in comparison with other factors. In the case of the Czech Republic, as the main barriers to the development of the use of risk capital, we can observe, as in the case of Slovakia, the level of the depth of capital market - the stock exchange is small, and the share of non-performing loans in total loans is not low. On the contrary, in the case of the Czech Republic, the human and social environment as well as economic activity are evaluated positively. Hungary has a relatively good score in terms of economic activity and entrepreneurial opportunities. On the contrary, Hungary achieved a weak assessment in terms of investor protection and depth of capital market.

4 Conclusion

In this paper, we analyzed the use of venture capital in CEE countries. We came to some interesting findings. The most used forms of venture capital in CEE countries are management buyouts, which account for about 70% of total investment. Although there are rapid differences in the volume of venture capital utilization in CEE and other European countries, the structure of the various forms of investment is similar.

We also found that venture capital investment in the CEE countries represent only a small fraction of the investment of capital across Europe (3.33%). The low level of use of risk capital is due to several factors that are interrelated. Authors Marková and Balcová [18]; Rajchlova et al. [13] consider the most important obstacles and barriers to venture capital development in particular:

- the low innovation capacity of SMEs, making them less attractive to venture capital investors,
- for some SMEs, their weaknesses are management and marketing skills,
- relatively low awareness among entrepreneurs about the possibilities of using venture capital,
- the tax and legal environment is not sufficiently supportive of the use of risk capital,
- the lack of legislative elements to promote the use of risk capital,
- underdeveloped capital markets (except Poland),
- poor support for R&D,
- excessive bureaucracy (eg in the case of Poland, when starting a business, the registration period is approximately 3 times longer than in the average OECD country and at the same time it is more expensive: 4 times more expensive than the OECD average).

For the above reasons, despite the fact that in our opinion there is scope and relatively good conditions for the use of risk capital in the CEE region, the use of capital still on a low level. This implies that the State should take part in promoting venture capital support through various programs and raise awareness of this form of SME financing.

The presented working paper is the output of the scientific grants VEGA n. 1/0007/19 "Asset allocation in a low interest rate environment in the financial and non-financial companies in the Slovak republic".

References

1. A . P. Groh, H. Liechtenstein, International allocation determinants for institutional investments in venture capital and private equity limited partnerships. *International Journal of Banking, Accounting and Finance,* **3**, 176-206 (2011)
2. A.P. Groh, H. Liechtenstein, Determinants for allocations to central eastern Europe venture capital and private quity limited partnerships. *Venture Capital - An International Journal of Entrepreneurial Finance,* **13**, 175-194 (2010)
3. A.P. Groh, H. Liechtenstein, K. Lieser, M. Biesinger, The Venture Capital & Private Equity Country Attractiveness Index. https://blog.iese.edu/vcpeindex
4. A.P. Groh, J. Wallmeroth, Determinants of venture capital investment in emerging markets. *Emerging Martets Review,* **29**, 104-132 (2016)
5. A.Soloma, The importance of the depth of the capital market on international venture capital and private equity allocation decisions: The case of Poland. *Economics and management,* **18**, 237-245 (2013)
6. C. Cabral-Cardoso, M. C. Cortez, L. Lopes, Venture capital and the financial crisis in Portugal : the VC managers' viewpoint. *Journal of Small Business and Enterprise Development,* **23**, 1032-1056 (2016)
7. C. M. Mason, R. T. Harrison, Business angel investment activity in the financial Crisis : UK evidence and policy implications. *SAGE Journal* **33**, 1 (2015)
8. E. Gomes Santana Félix, C. Pacheco Pires, M. A. Gulamhussen, The Determinanst of venture capital in Europe – Evidence across countries. *Journal of Financial Services Research,* **44**, 259-279 (2013)
9. G. Prelipean, M. Boscoianu, M. Lupan, C. E. Nastase, Innovative financing solutions based on venture capital and private equity to support the development of enterpreneurship in Romania. *Transformation in Business & Economics,* **13**, 331-347 (2015)
10. H. Doláková, Rizikový kapitál a ČR – antonyma nebo homonyma? *Ekonomika, financie a manažment podniku* (Fakulta podnikového manažmentu EU, Bratislava, 2009)
11. J. Block, P. Sandner, Venture capital funding in the middle of the year 2011 : are we back to pre-crisis boom levels ? *Startegic Change : Briefings in Entrepreneurial Finance,* **20**, 161-169 (2011)
12. J. Block, P. Sandner, What is the effect of the financial crisis on venture capital financing? Empirical evidence from US Internet start-ups. *Venture Capital - An International Journal of Entrepreneurial Finance,* **11**, 295-309 (2013)
13. J. Rajchlová, M. Baranyková, M. Polák, Z. Brož, Podmínky pro využití financování podniků formou private equity a venture kapitálu v České republice a vybraných evropských zemích - identifikace barièr, možnosti řešení. *Trendy ekonomiky a managementu,* **5**, 33-43 (2011)

14. J. Srpová, V. Řehoř, *Základy podnikání. Teoretické poznatky, příklady a zkušenosti českých podnikatelů* (Praha, Grada Publishing, a.s., 2010)
15. L. Jeng, P. C. Wells, The determinants of venture capital funding : evidence across countries. *Journal of Corporate Finance,* **6**, 241-289 (2000)
16. M. Kleinschmidt, *Venture capital, corporate governance and firm value* (Wiesbaden, GWV Fachverlage GmbH, 2007)
17. N. Bedu, M. Montalban, Analysing the uneven development of private equity in Europe: legal origins and diversity of capitalism. *Socio-Economic Review,* **12**, 33-70 (2014)
18. V. Marková, P. Balcová, Odstránenie prekážok vo využívaní rizikového kapitálu slovenskými malými a strednými podnikmi. *Ekonomika a spoločnosť,* **12**, 67-75 (2011)
19. Y. Ning, W. Wang, B. Yu, The driving forces of venture capital investment. *Small Business Economics,* **44**, 315-344 (2015)

Current measures to maintain the stability of workers in Slovak companies

Jana Kissová[1*], *Gabriela* Dubcová[2]

Abstract. Over a short period of time, individual countries in the world face a common problem that affects them and adversely affects the lives of individuals. In connection with the current emergency situation related to the corona virus pandemic, it is possible to notice fundamental changes and enormous impact in the social or economic dimension. The aim of the article is to provide an overview of the current situation in selected countries and to compare the system of measures in the Slovak Republic and the Czech Republic that were adopted in order to stabilize or retain workers or aimed at elimination of imminent damage.

1. Introduction

Over a short period of time, individual countries in the world face a common problem that affects them and adversely affects the lives of individuals. In connection with the current emergency situation related to the corona virus pandemic, it is possible to notice fundamental changes and enormous impact in the social or economic dimension. In the fight against the "invisible enemy", countries are taking different kinds of measures to protect the health of all their people. Procedures, steps, regulations or restrictions are adopted autonomously based on their own decisions and estimates of the development of the situation.

The aim of the article is to provide an overview of the current situation in selected countries and to compare the system of measures in the Slovak Republic and the Czech Republic that were adopted in order to stabilize or retain workers or aimed at elimination of imminent damage.

1.1 Theoretical bases related to the stabilization of workers

Stabilization of workers is the subject of human resources research - generally called human resources. For the sake of prosperity, the goal of any business is to retain and stabilize workers. For this purpose, each company has defined stabilization measures, which may vary depending on the decision and discretion of the company managers. [2, 13]

At the time of the current crisis, the set of measures to retain workers is fundamentally changing as it affects a large number of businesses and people. Therefore, whether they are kept in their positions or not in a given situation depends on the decisions of the state or the extent of the support the state will provide and how it will be implemented. [1, 3]

* Corresponding authors: jana.kissova@euba.sk, gabriela.dubcova@euba.sk

1.2 Overview of the current situation and measures taken in selected countries to limit the spread of coronavirus

Among the countries of Europe, Italy was the most affected by corona virus, gradually followed by other above mentioned countries. The way and extent of financing of the adopted measures is preliminary, as it may change in the light of developments.

An overview of the adopted measures in the selected countries is illustrated in Table 1 below, which sets out the scope and funding of the adopted measures. The common objective of the measures is to stop the spread of the virus as quickly as possible and to eliminate the damage.

Table 1. Overview and comparison of adopted measures in the most affected countries of Europe

Germany	Italy	Spain	France
The State Development Bank KfW provides loans. Total volume 500 bil. Eur.	Money to support hospitals (3.5 bil. Eur.).	Guarantees of 100 billion Eur. for loans to businesses.	Funds to support employment (estimated Eur. 45 billion)
600 bil. Eur. to help companies is under preparation	Subsidies for those who still have to work and are in high-risk employment environment.		
Aid from the government to exporters.	Support for companies (Eur. 10 billion).	Deferment of mortgage and loan repayments for people who cannot work.	Deferred payment of taxes for companies.
Help for companies to maintain employment while employees are at home through the Kurzarbeitergeld.	Support for parents who have to stay home with their children.	Some social aid benefits will be suspended and redistributed in favor of vulnerable population groups.	Support for parents who need to stay home with their children
Deferred taxes are planned for companies.	One-time support of EUR 500 for sole traders.	Introduction of so-called institute of temporary release from work.	300 billion Eur in the form of guarantees for the banking sector to help companies.

Source: Financial Times

2 Aim

The aim of the article is to provide an overview of the current situation related to the spread of coronavirus in individual countries. The priority is to identify and compare the adopted measures to support and maintain employment in the Slovak Republic and the Czech Republic.

3 Results and discussion

The subject of this section is to provide an overview of measures that are aimed to help with maintaining stability and providing support for employment in connection with the spread of coronavirus in Slovakia and to compare them with the measures adopted in the Czech Republic.

3.1 The system of adopted measures in Slovakia and in the Czech Republic related to the current situation

Due to the emergency situation, the Government of the Slovak Republic took vigorous measures almost at the beginning. Implementation of measures to prevent the spread of coronavirus was slower and more moderate in the Czech Republic, which was reflected in a faster increase in the number of infected patients and unfortunately also in the number of deaths. Currently, the measures taken in both countries are extreme. As the impact of these measures has a significant negative effect on the further functioning of the economy or the quality of life of the population, they consequently require a set of measures to mitigate these impacts.

General immediate and strict measures were taken mainly to protect the health of the citizens. Although they can be perceived as a major intervention in the way of life and quality - the priority is to prevent the spread of the virus in the shortest possible time and to ensure the protection of public health. Specific decisions were implemented on an ongoing basis in rapid succession as the situation evolved.

3.1.1 The system of measures to retain workers in the SR related to the current situation.

The strict measures concerning the compulsory closure of various types of services have undoubtedly and within a very short time period caused unfavorable and fundamental problems in the country's economy, industry or the business sector. According to experts and analysts, this situation will cause a temporary deep downturn and a high deficit, but a very rapid rebound after the crisis has subsided. The current situation represents "stopping or freezing" of almost all economic activities.

In addition to the economic impact, strict measures also have an impact on the overall level and quality of life of citizens as well as on the labor market and employment. In order to overcome the emergency situation and to maintain an acceptable standard of living for people and workers, the Slovak Republic is preparing legislative changes in the area of business and employment policy to help employers maintain the job positions for their employees. These changes will be incorporated into the amendment to the Social Insurance Act and also into the amendment to the Employment Services Act.

The changes will concern several groups in the labor market, such as employers (legal persons), employees or sole traders (self-employed) etc. They will be implemented with the aim to alleviate the unfavorable situation for these group, since they are not responsible for it. The following are the measures presented so far to mitigate the impact of the crisis for the three groups mentioned above in order to maintain and promote employment.

A. The following measures are prepared for employers: [6 - 8, 10 - 12]

- ***JobKeeper payment*** - it aims to prevent redundancies and to simplify the conditions for recruitment
- ***deferred payment of health and social aid contributions*** - deferment of the obligation to pay health and social contributions for the employer (not for the employee) for the

months of February, March and April 2020, without entry in the register of defaulters or other sanctions. Payments for May would be due without deferment (in June). The amount of the temporarily unpaid contributions will be subsequently calculated free of interest and divided into 18 monthly installments, which will be payable from July 2020 to December 2021

- ***cancellation of the obligation to pay corporate income tax advance payments*** - cancellation of the obligation to pay corporate income tax advance payments by June 30, 2020 (i.e. the next advance payments will be payable from October 2020). Interests on late payments or other sanctions will be waived and the missed advance payments will be reflected in the annual accounts.
- ***deferment of VAT for three months*** - deferment of VAT for February, March and April 2020. VAT for May would be payable in due time (25 June) without any sanctions. The amount of the temporarily unpaid VAT will then be alculated interest-free and divided into 18 monthly installments, which will be payable from July 2020 to December 2021. It applies only to retail and service establishments (according to measure No. OPL / 2596/2020 ÚVZ SR), as well as to specific justified cases (for example, companies that supply exclusively to retail or prove that they were directly affected by the coronavirus-related situation) and those operations that have been affected by the closure of schools (such as private nurseries)
- ***reimbursement of salary for an employee without allocated work (20-20-60) over a period of three months*** - if the employer is unable to allocate work to the employee during the months of April, May, June, due to the coronavirus situation, he may order him to take time off, paying him a salary of 80% of his average salary in proportion to: –20% will be paid by the employer, – 60% will be paid by the Social Insurance Agency
- ***wage for "home office"***- if the employee works from home, he / she is entitled to the wage in full
- ***priority completion of VAT inspections and refunds without delays*** - Priority completion of VAT inspections where a taxpayer has applied for a VAT refund. Allocataion of an amount of € 10,000 to be rejected or paid no later than 30.04. and the amount of € 50 thousand, which must be rejected or paid by 31.05. If the tax office did not find any disparities during the inspection within the above-mentioned deadlines for which VAT can be withheld, it must pay it back immediately. This does not concern the period for carrying out tax audits, which is one year.

B. The following measures are prepared **for employees:** [6 - 8, 10 - 12]

- ***entitlement to "pandemic" nursing care*** - in the case of quarantine or home isolation, it is proposed to extend the period of disbursement of nursing care for children up to 11 years of age or 18 years of age for children with long-term adverse health conditions for the whole period of quarantine and isolation, as well as during the entire closure of schools and pre-school facilities,
- ***adjustment of sick pay*** - the employer paid the amount of sick pay for the first 10 days up to now; from the first day of the quarantine period, it will be reimbursed by a social insurance company amounting to 55% of the daily assessment base
- ***filing an application for inclusion in the job seekers register by electronic means without a qualified electronic signature***

C. The following measures are prepared **for entrepreneurs**: [12]

- ***Possibility to defer the obligation to file tax declaration and the obligation to pay income tax till 30.6 for all entrepreneurs* -** tax declarations and tax payment defferal has already been approved
- ***Payments of minimum contributions will be forgiven for a period of three months.* -** for sole traders who shall pay more than minimum contributions, only the amount of minimum contributions would be forgiven.
- ***Suspension of trade for sole traders for at least one month or for an indefinite period* -** possibility to suspend trade for at least one month (currently 6 months), for an indefinite period but for a maximum of three years.

The most recent steps in the area of employment support include the specific measures especially for employees, smaller companies or sole traders, listed in Table 2:

Table 2. Program: "First Aid to Employees, Companies and Sole Traders"

Measure	Method of support
The state pays 80% of the employee's salary	➢ in companies whose operations have been compulsorily closed. Those employees who stayed at home will receive wages from the state. ➢ they will not cost anything for the employer. The payment will also include tax per employee and contributions per employee. The amount of the contribution will have its financial ceiling as well as other eligibility conditions.
Contributions for sole traders and employees	➢ in companies that have seen a drop in sales, it will depend on how much they were affected. Calculations for March will be halved. ➢ in the case of April: o *if sales drop by more than 20 percent, the state will contribute 180 euros to salaries.* o *with a drop of more than 40 percent, the state will contribute 300 euros.* o *with a drop of 60 percent, it will be 420 euros,* o *with a drop of more than 80%, the state will contribute 540 euros.*
Provision of bank guarantees worth half a billion euros	➢ for companies to be able to finance their business
Reimbursement of wages	➢ Employees in quarantine and parents taking care of a family member will be paid 55 percent from their gross wage. ➢ this measure was approved by the government and parliament during the week
Deferred payments of contributions	➢ when sales drop by more than 40 percent
Deferred income tax advance payments	➢ when sales drop by more than 40 percent ➢ Entrepreneurs will start to pay advance payments since October
Possibility to deduct losses from previous years	➢ companies will be allowed to deduct losses from the year 2014 (including) ➢ It is only applicable for companies who haven´t deducted the losses yet

Source: processed according to [9]

3.1.2 The system of measures to retain workers in the Czech Republic related to the current situation.

In comparison with Slovakia, there is a broader range of measures applied in the Czech Republic in the framework of "support programs", which specify the situations or the conditions under which they will be provided.

A. The following measures have been taken **for employers**: [5]

- ***Antivirus - Employment Promotion*** – employment protection program to help protect jobs, the state will use it to compensate for money paid. It is intended to help employers better cope with the current situation without having to make redundancies. To become eligible for the program depends on specific situations:

Table 3. Employment promotion program according to specific situation

Sitution	Method of support
Quarantine order for employees	➢ the employee receives a wage compensation of 60% of the average reduced earnings ➢ contribution for the employer 'amounting to the full salary compensation
Obstacles to work on the part of the employer due to the obligation to close operations due to the emergency situation	➢ the employee receives 100% of the average earnings ➢ contribution for the employer 'amounting to 80% of paid salary compensation
Obstacles to work on the part of the employer due to quarantine order or child care in significant proportion of employees (30%)	➢ the employee receives 100% of the average earnings ➢ Contribution for the employer 'amounting to 80% of paid salary compensation
Limited availability of inputs (raw material, products, services) necessary for operations	➢ the employee receives 80 % of the average earnings ➢ Contribution for the employer 'amounting to 50% of paid salary compensation
Reduction in the demand for services, manufacturing and other products of the company	➢ the employee receives 60 % of the average earnings ➢ Contribution for the employer 'amounting to 50% of paid salary compensation

Source: processed according to [5]

B. The following measures have been taken **for employees** [4]:

Within the framework of labor - law relations, the so - called. **"Labor Ten Commandments for handling Coronavirus in Czech Republic"**, have been issued which define in detail the material security of employees according to specific situations.

- ***Wage compensation*** – for the first 14 calendar days, the employee is entitled to get a wage compensation of 60% of reduced average income. From the 15th calendar day, the sickness pay is paid to **an employee who is in quarantine, to an employee who was acknowledged as temporarily incapable of work** due to illness, due to the temporary closure or restriction of operations as a result of emergency measures imposed by the public health authority,

- ***Nursing care*** – belongs to an employee who cares for a child up to the age of 13 or from the age of 13 with the first degree of dependence for the whole time of closing the school facility
- ***Wage compensation of 100% of average income*** - belongs to the remaining employees due to the absence of most employees (e.g due to quarantine, childcare...)
- ***Wage compensation of 80% of average income*** - belongs to employees as a result of restrictions in inputs (supplies of raw materials, services, components…)
- ***Wage compensation of 100% of average income*** - belongs to employees due to temporary closure or restriction of operations (e.g clothing stores) or because of a reduction in demand for services or products
- ***Home office wage (home office)*** - if the employee works from home, he / she is entitled to the full wage

C. The following measures have been taken **for sole traders:** [5]

- ***possibility to defer filing of the income and expenditure statement*** – entrepreneurs can submit it without sanctions until 03.08.2020. In case also the insurance premium has been paid till this date, the enterpreuner will not be penalized for late payment.
- ***forgiveness of social insurance*** – entrepreneurs will not have to pay the specified premiums from March to August, thus reducing their total premiums by the amount corresponding to the minimum pension prepayments for half a year. Sole traders will not have to pay anything at all. The period will be added to the pension entitlements.
- ***Suspension or termination of trade license, registration at the Labor Office of Czech Republic*** – To terminate a business activity, it is necessary to notify the relevant authority in writing (electronically) and to register with the Labor Office.
- ***deferred income tax declaration*** - Deferred income tax declarations till 1.07.2020 for all legal and natural persons.
- ***individual remission of a fine for late payment of all taxes*** - in cases where the taxpayer proves the reasons related to coronaviruses.
- ***final phase of EET deferment (deferment in the last phase of implementation of electronic sales registration)*** – the original date is postponed by 3 months until 1 August. During this time the Financial and Customs Administration will proceed with maximum tolerance and will operate only in advisory function.

4 Conclusion

The situation concerning the spreading of corona virus is constantly evolving. Individual countries are affected differently. From the current situation and development it is clear that countries that underestimated this disease at the beginning and took only mild and moderate measures are currently hit the most with a high death rate. It appears that strict measures that prioritize health over the prosperity of the economy are necessary to be able to manage and overcome the crisis.

The article provides an overview of measures taken to mitigate the crisis in some countries. It focuses more on the system of measures adopted for individual groups of the labor market in order to maintain jobs in the Slovak and Czech Republic. Based on the comparison, it can be stated that the Czech Republic has adopted a broader set of measures, while in the Slovak Republic the measures are formed at a slower pace due to the need for legislative changes, amendments to laws and the current change of government. Several of the measures mentioned to be taken in the Slovak Republic are still in the process of being negotiated, i.e they may not be adopted eventually. To conclude, it should be noted that these measures cannot be considered as definitive, they will be modified, amended or supplemented in the coming weeks, depending on the evolution of the situation.

Acknowledgement

This paper is an outcome of projects: „Stabilization on Human Capital in Slovak Enterprises as an Immanent Determinant of their Success, Competitiveness and Sustainable Development ", VEGA No. 1/0569/18 - funded by the University of Economics in Bratislava; led by doc. Ing. Mgr. Gabriela Dubcová, PhD. - project share is 100%.

References

1. H. Huettermann – H. Bruch (2019). Mutual profits? HRM in health, collective well-being and organizational performance. *Journal of Managemet Studies,* vol. 56, No.6 pp.1405 -1072
2. K. Burdett – C. C. Tudela – M. Coles (2020). The cost of Jobs Loss. *The Review of Economics Studies.* No. 1/2020 *ISSN* 0034-6527
3. M. Vooren at.al. (2018). The Effectiveness of active labor market polices, *Journal of Economics Survey,* vol. 33, No.1, pp. 125-149
4. Ministerstvo financií Českej republiky. Ministersvo financí přehledne ke koronaviru. (1903.2020) Dostupné na: https://www.mfcr.cz/cs/aktualne/koronavirus-covid-19/2020/ministerstvo-financi-prehledne-ke-korona-37925 [accessed 28.03.2020]
5. Ministerstvo práce a sociálních věcí Českej republiky. Informace ke koronaviru. Dostupné na: https://www.mpsv.cz/web/cz/zamestnanci-info [accessed 28.03.2020]
6. N. Seneši. Príspevok na podporu udržania pracovných miest v roku 2020. (23.3.2020) Dostupné na: https://www.podnikajte.sk/financie-na-podnikanie/prispevok-na-podporu-udrzania-pracovnych-miest-2020 [accessed 28.03.2020]
7. N. Seneši. Predĺži sa kvôli koronavírusu termín na podanie daňových priznaní? (13.03.2020) Dostupné na: https://www.podnikajte.sk/dan-z-prijmov/predlzi-sa-kvoli-koronavirusu-termin-na-podanie-danovych-priznani [accessed 28.03.2020]
8. Poslanci schválili zavedenie pandemického ošetrovného a poskytovanie príspevkov na udržanie pracovných miest. (25.032020) Dostupné na: https://denniks.sk/32208/poslanci-schvalili-zavedenie-pandemickeho-osetrovneho-a-poskytovanie-prispevkov-na-udrzanie-pracovnych-miest/ [accessed 28.03.2020]
9. Prvá pomoc pre zamestnancov či firmy. Pozrite si prehľad siedmich ekonomických opatrení (29.03.2020) Dostupné na: https://index.sme.sk/c/22370764/povedali-ako-chcu-pomoct-firmam-a-zivnostnikom-pozrite-si-prehlad.html?ref=FB [accessed 28.03.2020]
10. V. Folentová – M. Barcíková. Aktuálny zoznam návrhov na pomoc podnikom, o ktorom koalícia rokuje. (27.03.2020) Dostupné: https://e.dennikn.sk/1825413/aktualny-zoznam-navrhov-na-pomoc-podnikom-o-ktorom-koalicia-rokuje/?ref=side&_ga=2.101082945.1124802050.1585333719-841732987.1583937711 [accessed 28.03.2020]
11. Vláda schválila pandemické ošetrovné a ďalšie sociálne opatrenia. Pozrite si ich prehľad. (25.03.2020) Dostupné na: https://finweb.hnonline.sk/ekonomika/2119105-vlada-schvalila-pandemicke-osetrovne-a-dalsie-socialne-opatrenia-pozrite-si-ich-prehlad [accessed 28.03.2020]
12. Z. Kollárová. Vláda odobrila odloženie daní do konca júna, spísané sú aj ďalšie opatrenia (18.03.2020) Dostupné na: https://e.dennikn.sk/1808736/vlada-odobrila-odlozenie-dani-do-konca-juna-spisane-su-aj-dalsie-opatrenia/ [accessed 28.03.2020]

13. Z. K. Chulanova (2019). Profesional standards as factor of daptation of human resources to the industry 4.0 Aproaches to development and impelementation, *Journal of Human Resource Management*, vol. XXII, No.1/2019, 12–20 ISSN 2453-7683

Contact

Ing. Jana Kissová, PhD.
University of Economics in Bratislava, Faculty of Business Management
Main Street No. 1, 852 35 Bratislava, Slovakia
E-mail: jana.kissova@euba.sk
ORCID: 0000-0002-4140-0890

doc. Ing. Mgr. Gabriela Dubcová, PhD.
University of Economics in Bratislava, Faculty of Business Management
Main Street No. 1, 852 35 Bratislava, Slovakia
E-mail: gabriela.dubcova@euba.sk
ORCID: 0000-0001-5024-4335

How leverage can improve performance

Miroslav Kmeťko[1*], and *Eduard* Hyránek[2]

[1]University of Economics in Bratislava, Faculty of Business Management, Department of Business Finance, Dolnozemská cesta 1, 852 35 Bratislava, Slovakia
[2]University of Economics in Bratislava, Faculty of Business Management, Department of Business Finance, Dolnozemská cesta 1, 852 35 Bratislava, Slovakia

Abstract. One of the best-known Capital Asset Pricing Model (CAP/M) provides us with a methodology for measuring the relationship between the risk premium and the impact of leverage on expected returns. However, this model is not used only to value the cost of capital but also to evaluate the performance of managed portfolios. We will test how the expected return changes in percent by changing the debt-equity ratio and the tax rate based on following assumptions: market return 7%, risk-free rate of return 1% and beta 1.2. These assumptions will be constant and we will change the debt-equity ratio and tax rate. Based on these results, it is clear that the change in profitability varies, in relation to the change of the DE ratio by one tenth. As for changes I n tax rates, changes in expected profitability are not entirely in direct proportion to these changes.

1 Introduction

Leverage causes two problems: it increases the company's risk to shareholders and can also cause the company to go bankrupt in bad times. The total risk of the company can thus be divided into business risk (without debt) and financial risk (with debt).

2 Methods

We started with characteristics of the model. Subsequently, we used this model with variants by changing variables. These solutions are presented in tabular form. As companies in real life respond to the change in interest rates, we observed the four largest companies in the United States by market capitalization. We also compared the DE ratio with the ROE ratio to prove whether the theory holds.

* Corresponding author: miroslav.kmetko@euba.sk

3 Beta and leverage

One of the best-known model Capital Asset Pricing Model (CAPM) provide us with a methodology for measuring the relationship between the risk premium and the impact of leverage on expected returns. The model was created by Sharpe [1] and Lintner [2] to estimate the cost of capital for project selection.

A survey by Graham and Harvey [3] that up to 73.5% of CFOs use this model in decision making. However, this model is not used only to value the cost of capital but also to evaluate the performance of managed portfolios. In the literature we can find many studies that point to different returns using different strategies [4,5,6] and results for equally and value-weighted portfolios [7,8,9]. Other studies have shown [10,11,12] that the CAPM applies in January, a month when inflation is usually low, important macroeconomic indicators are announced, and also in months when investors can borrow cheaply. This model is presented as one of the most important model in the financial economy [4,5]. Even Berk and van Binbergen [13] found in their work that this model is closest to the models used by the investor in capital allocation decisions.

Some authors also included macroeconomic factors [14], or human capital [15], or growth in consumption [8].This model has the following basic formula:

$$R_s = R_f + Risk\ premium \quad (1)$$

The risk premium is calculated:

$$R_s = R_f + \beta(R_m - R_f) \quad (2)$$

If the company has no debt, then in this case the has unlevered beta β^U, so the company has only business risk, By adding debt to the balance sheet, the company has also financial risk and the beta is called levered beta β^L.

Then the relationship between the unlevered and levered beta is as follows:

$$\beta^L = \beta^U[1 + (1-t)D/E] \quad (3)$$

Or:

$$\beta^U = \frac{\beta^L}{[1+ (1-t)D/E]} \quad (4)$$

By modifying the original formula by unlevered and levered be, then we get the equation:

$$R_s = R_f + \beta^U(R_m - R_f) + \beta^U[(1-t)D/E](R_m - R_f) \quad (5)$$

4 Results

We will test how the expected return changes in percentage by changing the debt-equity ratio and the tax rate based on the following assumptions: market return 7%, risk-free rate of return 1%, beta 1.2. These assumptions will be constant and we will change the debt-equity ratio and tax rate.

Table 1. Changes in debt-equity ratio

	DE ratio 0,3	DE ratio 0,33	DE ratio 0,6
Change in DE	0	10 %	100 %
Beta	1,2	1,22	1,36
Return in %	8,2	8,29	9,14
Change in return in %	0	1 %	11 %

Based on these results, it is clear that the change in profitability varies, in relation to the change of DE ration by one tenth.

Table 2. Changes in tax

	Tax rate 25 %	Tax rate 30 %	Tax rate 50 %
Change in tax	0	20 %	100 %
DE ratio	0,3	0,3	0,3
Beta	1,2	1,2	1,2
Return in %	8,70	8,59	8,2
Change in return in %	0	-1,26 %	-5,74 %

As for the changes in tax rates, the changes in expected profitability are not entirely in direct proportions to these changes, as it was the case in the previous case.

5 Discussions

The largest company is characterized as a publicly traded company with the largest market capitalization. Data are as of 31.3.2020. To this date was Microsoft largest company (MSFT) with a market capitalization of 1.2 billion of US dollars. Its historical development of DE ratio for the previous 5 years was:

Table 3. Historical development of DE ratio of MSFT

	2015	2016	2017	2018	2019
DE ratio	0.44	0.74	1.04	0.99	0.77

If we compare these values with basic interest rate of the US Federal Reserve, we would conclude that the value of the DE ratio of the company changed significantly with the conditions on the credit market.

In 2018, the Fed´s key interest rate reached its maximum (within the period under review). If we monitor the interdependence of the DE ratio and ROE, then in 2017 was the value of the ROE indicator the second highest (after 2019).

The second most valuable company is Apple (AAPL) with a market capitalization of 1.11 billion US dollars. Compared to MSFT, the ratio is still increasing.

Table 4. Historical development of DE ratio of AAPL

	2015	2016	2017	2018	2019
DE ratio	0.54	0.68	0.86	1.07	1.19

This company thus not relate to changes in interest rates. On other hand, the increase in debt also leads to an increase in the ROE ratio, which was the highest in the last year.

The third largest company is Amazon (AMZN) with a market capitalization 0,97 billion US dollars. The development of the DE ratio is very similar to MSFT.

The development of the ratio is in the following table.

Table 5. Historical development of DE ratio of AMZN

	2015	2016	2017	2018	2019
DE ratio	1.06	0.79	1.37	0.76	1.02

The DE ratio was also the highest in 2017 and also rose in 2019 when the Fed began to reduce interest rates. However, it is in conflict with previous companies because in 2017 was the ROE indicator the second lowest (after 2015). The increase in indebtedness thus did not support the growth of the ROE ratio.

The latest company is Alphabet (GOOG or GOOGL) with a market capitalization of 0.8 Billion US dollars. Compared to the other previously mentioned companies, it has the lowest values of the DE ratio. This indicator has also not changed significantly.

Table 6. Historical development of DE ratio of GOOG

	2015	2016	2017	2018	2019
DE ratio	0.04	0.03	0.03	0.02	0.08

This company thus not take an advantage of debt financing. This is also confirmed by the ROE values, which are significantly lower than the previously mentioned companies.

Conclusions

Some companies seek to take advantage of low interest rates to increase shareholder value, other such as AAPL ignore this possibility. Other large companies are adjusting their DE ratio according to the current situation on the credit market. Another exception concerns Amazon, which has kept the DE ratio low for a long time. This was ultimately reflected in the value of the ROE ratio, which was the lowest of all examined companies.

The article is an partial output of research project VEGA MŠ SR No. VEGA 1/0462/19 entitled Implementation of key determinats into performance models as tools of financial decision-making of the company in the current conditions.

References

1. W. Sharpe, Capital asset prices: a theory of market equilibrium under conditions of risk. *Journal of Finance* **19**, 425-442, (1964)
2. J. Lintner, The valuation of risky assets and the selection of risky investments in stock portfolios and capital budgets. *Review of Economics and Statistics* **47**, 13-37, (1965)
3. J. R. Graham, C. Harvey, The theory and practice of corporate finance: evidence from the field. *Journal of Financial Economics* **60**, 187-243. (2001)
4. A. W. Lo, *Adaptive Markets: Financial Evolution at the Speed of Thought.* Princeton, NJ: Princeton University Press. (2017)
5. J. Y. Campbell, *Financial Decisions and Markets: A Course in Asset Pricing*. Princeton, NJ: Princeton University Press, (2018)
6. J. B. Berk, J. H. van Binsbergen, Assessing Asset Pricing Models Using Revealed Preference. *Journal of Financial Economics*, **119** (1), 1-23, (2016)
7. N. Chen, R. Roll, S. Ross. "Economic Forces and the Stock Market." *Journal of Business*, **59**, 383-403, (1986)
8. R. Jagannathan, Z. Wang. "The Conditional CAPM and the Cross-Section of Expected Returns." *Journal of Finance*, **51**, 3-53. (1996)
9. R. Jagannathan, Y. Wang. "Lazy Investors, Discretionary Consumption, and the Cross-Section of Stock Returns." *The Journal of Finance*, **62**, 1623-1661, (2007)
10. E. F. Fama, K. R. French, Dissecting Anomalies, *Journal of Finance*, 63, 1653-1678.
11. D. Li, E. X. N. Li, L. Zhang, Costly External Equity: Implications for Asset Pricing Anomalies, NBER Working Paper 14342.

12. T. G. Bali, N. Cakici, R. F. Whitelaw, Maxing out: Stocks as Lotteries and the CrossSection of Expected Returns, *Journal of Financial Economics*, **99**, 427-446, (2011)
13. J. B. Berk, J. H. van Binsbergen (2016). Assessing asset pricing models using revealed preference. Journal of Financial Economics **119** (1), 1-23, (2016)
14. P. Jylha, Margin requirements and the security market line. *The Journal of Finance* **73** (3), 1281-1321, (2018)
15. T. Hendershott, D. Livdan, D. Rosch Asset pricing: A tale of night and day. Working paper, (2018)

Are Supporting Sectors of the Creative Industries Inferior? Current study from Slovakia.

Mária Kmety Barteková[1,*], *Helena* Majdúchová[1]

[1]University of Economics, Faculty of Business Management, Department of Business Economy, Dolnozemská cesta 1, 852 35 Bratislava, Slovakia

Abstract. The aim of the contribution is to analyse the current state of the supporting sectors within the creative industries, to identify and quantify the role of the supporting sectors in the Slovak economy. In the European union, there are huge expectations associated with the creative industries. Based on the following findings, we can conclude that the low level of external investments is caused by the low possibility to get the loans. There is no relation between the regional density of supporting sectors' enterprises and the economic results reached by these regions.

Introduction

The well-known authors and researchers [1,2,3,4] tried to considerate a new aspect, that is, the multidimensional nature of creativity and the potential synergic and complementary effects of creative talents of different nature. The merging of talents is, indeed, a source of original, complex and synergic ideas which are at the basis of the positive relation between creativity and local development.

The objectives of the first companion of cultural economists—a few of them are even in our midst—were good. They wanted to prove all other economists that their mechanisms are applicable in a study of the arts, too, and looked for recognition for their exercise. At the same moment they expected to influence cultural policy, bringing in a much needed economic perspective [5,6]. Maybe some hoped to influence the world of economists as well.

The motivation for the value-based approach came about in the study of the world of the arts. As many of us have experienced, people in that world do not take easy to the standard economic approach. The value-based approach begins with what theater makers and others involved in the arts world find important, that is with their values, or with the qualities that they want to realize. Usually those values, or qualities, are artistic in kind; they concern theatrical or musical or multimedial qualities. Often they will mention social and societal values, like when they want to contribute to local communities, social cohesion, justice and such [7]. While using relevant economic policy tools, the state at national, regional and

* Corresponding author: maria.bartekova@euba.sk

local levels may help to rise creative industries in general by determining priorities for the socio economic development [2].

This is a useful starting point because it allows us to focus attention on the relational nature of processes of value allocation and cultural validation: cultural value does not operate and is not generated in a social, cultural and political vacuum, but is in fact shaped by the power relations predominant at any one time, and is a site for struggles over meaning, representation and recognition [8].

Creative industries are in a strategic position to promote smart, sustainable and inclusive growth in all EU regions and cities, and thus contribute fully to the Europe 2020 Strategy, which is the EU's growth strategy [9]. While some EU regions have been very good at tapping into this extraordinary potential as a way to promote socio-economic development - including through the use of EU Structural Funds -, it however appears that many others have not been making most of this potential.

2 Methods

The following chapter provides the information about research aim, the detail description of the research object and the methodology used in this paper.

The aim of the paper is to research the specifics of the supporting sectors of the creative industries. In the paper, we articulated the following research questions:

- Research question no. 1: Are supporting sectors of the creative industries inferior?
- Research question no. 2: Have the supporting sectors of the creative industries reported the weak results in the selected indicators (the debt level, the share of the bankrupted entities, etc.)?
- Research question no. 3: Are the clusters with the highest amount of supporting sectors' enterprises located in the regions with the highest GDP per capita?

Supporting sectors of the creative industries include the following business activities according to the numeral classification provided by the Statistical Office of the Slovak Republic:

- Other professional scientific and technical activities – SK NACE code 74.900
- Organization of congresses and business exhibitions – SK NACE code 82.300
- Artistic education – SK NACE code 85.520
- Stage - related supporting activities – SK NACE code 90.020.

We collected the information about all enterprises, which are active in Slovakia and belong under the selected SK NACE codes. SK NACE represents statistical classification of economic activities in the Slovak Republic.

The most valuable sources were founded in the scientific database Web of Science. We focused on the Q1 and Q2 - ranked journals in management, innovation, economics, business research and regional studies.

To fulfil the research aim, we collected the information about the enterprises belonging to the supporting sectors of the creative industries in the Slovak Republic. The data used in the research were provided by the Statistical Office of the Slovak Republic, Finstat and Eurostat. To find the answers for the research questions, we used the descriptive statistics.

3 Results and Discussion

Creative industries (supporting sectors included), in particular SMEs, encounter difficulties in accessing the funds they need to finance their activities, both in terms of credit and equity. While this is a common challenge for SMEs in general, the situation is significantly worse for those working in the creative sectors for the following reasons:

- unlike other industrial products, creative industries' products are generally not mass-produced;
- in most cases the financial requirements of creative industries' companies are not substantial enough for banks to find them commercially interesting enough to undertake the analyses of projects,
- creative industries also face specific challenges in achieving investment readiness, in particular at a time when they have to find innovative business models for the new digital environment.

One can see the summary of the financial results in the table 1. The debt ratio of the supporting sectors' enterprises reached in average 40%. As it was mentioned before, the supporting sectors of the creative industries are not substantial for banks. In most of the cases the banks reject the applications for the business loans.

Table 1. Financial characteristics of the supporting sectors' entities in Slovakia.

Variable	N	Mean	Std Dev	Minimum	Maximum
Assets	236	67 733.88	190 934.46	1 196.00	2 272 275.00
Equity	236	30 749.53	129 188.65	-92 412.00	1 828 241.00
Debt ratio	236	40.37 %	47.78 %	-4.24 %	350.66 %

Culture and creative-based development has relevance for regions in all stages of development - in the 'convergence' regions, where support for infrastructural development may continue to be needed, but also in both convergence and 'competitiveness' regions where the infrastructure and other local resources can be exploited to enhance the comparative advantages of the local economy and to stimulate creativity and enterprise [10].

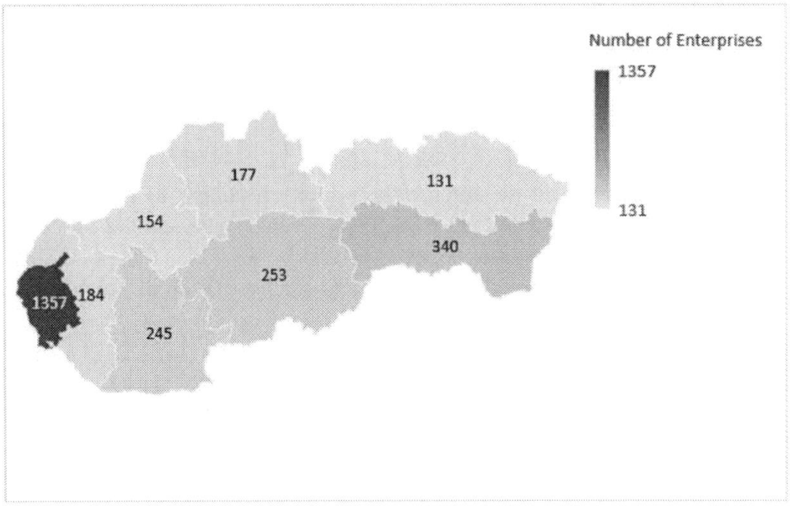

Fig. 1. Regional location of the supporting sectors' entities of the creative industries

Source: own processing, data extracted from Finstat database (2018).

Figure 1 represents the regional density of the supporting sectors' enterprises in Slovakia. The cluster with the highest density of the mentioned enterprises is the Bratislava region (1357 enterprises). The second one is the Košice region with the number of enterprises 340 [11].

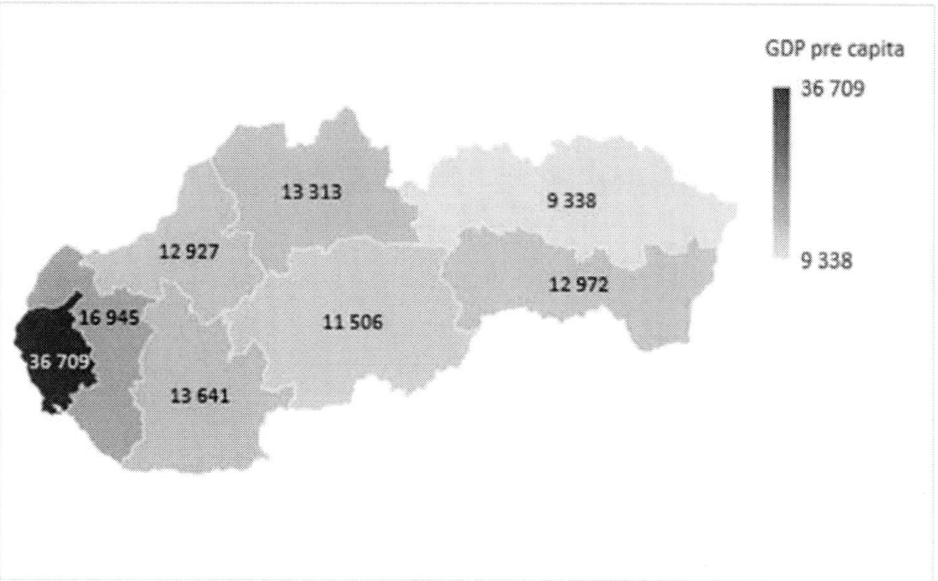

Fig. 2. Regional Differences in Slovakia measured by the indicator GDP per capita

Source: own processing, data extracted from the Statistical Office of the Slovak Republic database (2018).

The most creative regions are characterized by having more high-tech manufacturing industries than the rest of the regions although the number of low-tech manufacturing firms is similar. Lastly, the industrial structure of each region has a greater influence on regional wealth than the existence of industrial agglomerations [12].

In many regions the creative sector is rapidly growing, outperforming other more established sectors in terms of growth in new businesses, turnover and employment; in many instances the absolute size of the sector now exceeds that of a range of traditional industries. Many cities and regions have some form of creative sector strategy; the focus and detail of this strategy, however, vary considerably; in the most developed regions the promotion of the creative sector is part and parcel of a strategy for knowledge-based industries. As well as its inherent dynamism and growth potential, the creative sector has a number of other attractive qualities from a local development perspective; it makes use of a range of skills at a series of different levels, it tends to be socially responsible and inclusive and it usually involves 'clean', environmentally friendly processes. The region with the highest GDP is also a region with the highest density of the supporting sectors' enterprises – the Bratislava region. One can see that the rest of regions in Slovakia has the similar number of enterprises (figure 2). We can't prove the relation that the clusters with the highest number of the supporting sectors' enterprises are located in the regions with the highest GDP per capita.

In more technical terms, creative activities often generate positive externalities in the areas where they are located, their openness and interaction with other activities give rise to agglomeration and cluster effects and they tend to generate a high proportion of total value added locally. A longer-term and more strategic approach to culture-based development is required at a local and regional level. This should be mainstreamed into development strategies that are integrated and built on partnerships between public authorities, cultural organisations, the relevant business interests and representatives of civil society.

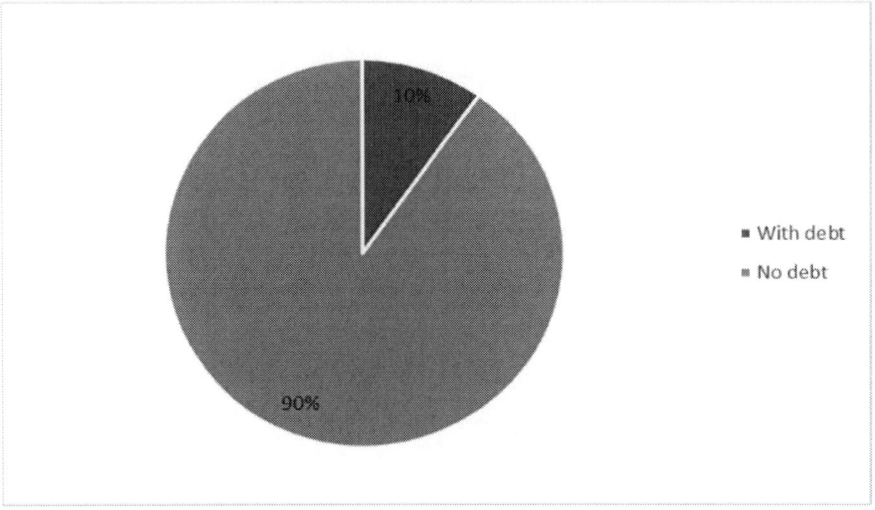

Fig. 3. External sources in the capital structure of the supporting sectors' enterprises.

Source: own processing, data extracted from Finstat database (2018).

The enterprises operating in the supporting sectors of the creative industries are using mainly the own financial sources as the own equity. One can see only 10% of the all enterprises used the external sources of financing their activities or projects (see Figure 3).

Before the corona crisis have begun, more than 99 % of enterprises were financially stable in 2018 (Figure 4). The supporting sectors of the creative industries consists of 2 841 enterprises in Slovakia [13]. Only a few enterprises operating the supporting sectors were abolished or under restructuring (0.45 %).

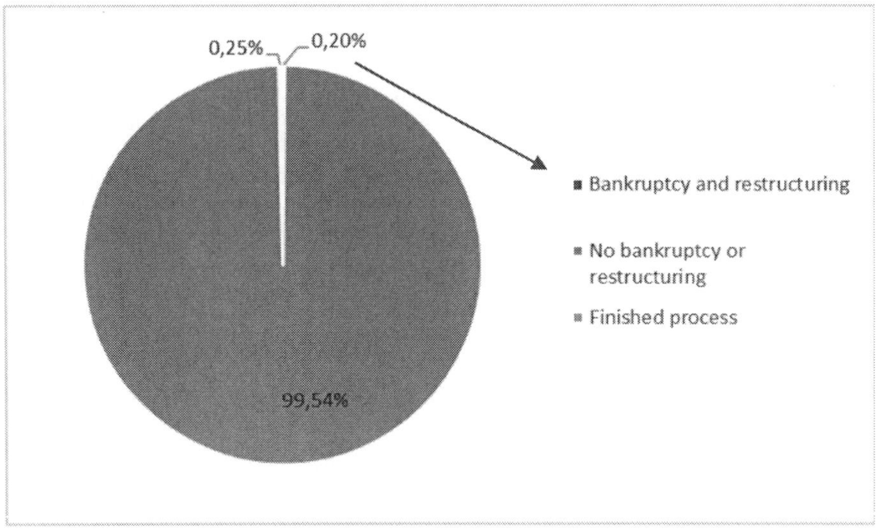

Fig. 4. The financial wealth of the enterprises in the supporting sectors.

Source: own processing, data extracted from Finstat database (2018).

Figure 5 shows the ownership structure of the supporting sectors' enterprises in the Slovak Republic. Most of the enterprises (around 89%) has the domestic private ownership.

The second biggest group of enterprises (the enterprises with the foreign ownership) reached 7%.

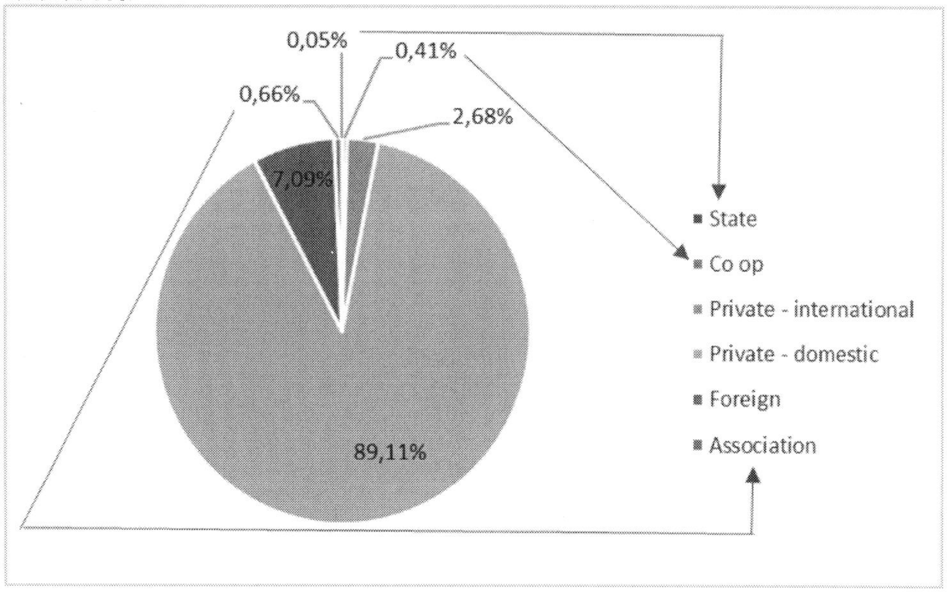

Fig. 5. The ownership structure of the supporting sectors' enterprises in Slovakia.

Source: own processing, data extracted from Finstat database (2018).

Against this background, there is a crucial role to be played by public authorities – particularly regional authorities – in stimulating private investment and promoting public private partnerships for the benefit of their Creative industries' SMEs, notably through guarantee mechanisms or venture capital funds [14,15].
In the last section of this paper, we have provided the main conclusions.

.4 Conclusion

Based on the finding stated above, we can conclude that the most of the enterprises operating in supporting sectors of the creative industries are own by the domestic owners. The reason is simple. In most cases the financial requirements of Creative industries' companies are not substantial enough for banks to find them commercially interesting enough to undertake the analyses of projects, then develop their expertise in a field that is still perceived as risky and build up specialised Creative industries departments able to understand the special business models of creative industries (supporting sectors included).

From this point of view, the supporting sectors' enterprises are inferior for the banks as the clients. The financial structure of these companies is influenced by the lack of external financial sources. The main source of the capital is their own equity. One can see that the assets of the supporting sectors' enterprises are lower than in the manufacturing industries.

This paper is the output of the scientific grant VEGA n. 1/0340/19 "The Entrepreneurial Dimension of Creative industries in the Context of Innovation and Smart Growth" (50 %) and the internal grant project of the University of Economics in Bratislava (Faculty of Business Management) no. I-20-101-00 called "Research of Creative Industry Enterprises with regard to Their Life Cycle in the Slovak Republic" (50 %).

References

1. S. Cerisola, A new perspective on the cultural heritage–development nexus: the role of creativity. *Journal of Cultural Economics*, **43**, 21-56 (2019)
2. A. Drab-Kurowska, State intervention in the context of creative industries. Management – Poland, **22**, 74-84 (2018)
3. S. Cerisola, Creativity and local economic development: the role of synergy among different talents. *Papers in Regional Science*, **97**, 199 – 209 (2018)
4. A. Lemmetyinen, Entrepreneurship in culture and creative industries: perspectives from companies and regions. *International Small Business Journal – Researching Entrepreneurship*, **37**, 417-418 (2019)
5. Y. Wijngaarden, E. Hitters, P.V. Bhansing, 'Innovation is a dirty word': contesting innovation in the creative industries. International Journal of Cultural Policy, **25**, 392-405 (2019)
6. W. Yu, Creative industries agglomeration and entrepreneurship in China: necessity or opportunity? Industry and Innovation, 27, 420-443 (2020)
7. XQ. Li, D. Gagliardi, I. Miles, Innovation in R&D service firms: evidence from the UK. *Technology Analysis & Strategic Management*, **31**, 732-748 (2019)
8. E. Belfiore, Whose cultural value? Representation, power and creative industries. *International Journal of Cultural Policy*, **26**, 383-397 (2020)
9. F. Angelini, M. Castellani, Cultural and economic value: a critical review. *Journal of Cultural Economics*, **43**, 173-188 (2019)
10. R. Boix, B. De-Miguel-Molina, J. L. Hervas-Oliver, Creative service business and regional performance: evidence for the European regions. *Service Business*, **7**, 381 – 398 (2013)
11. The Statistical Office of the Slovak Republic, (2020, 25 March). Enterprise births in the SR by NACE Rev. 2 and size class category by number of employees. Retrieved from http://datacube.statistics.sk/#!/view/sk/VBD_SLOVSTAT/og2005rs/v_og2005rs_00_00_00_sk.
12. Eurostat, (2020, 24 March). Value added and turnover of enterprises in the cultural sectors by NACE Rev. 2 activity. Retrieved from https://appsso.eurostat.ec.europa.eu/nui/submitViewTableAction.do.
13. Finstat, (2020, 1 April). Firmy s finančnými údajmi - databáza hospodárskych výsledkov slovenských firiem. Retrieved from https://finstat.sk/databaza-financnych-udajov.
14. T. Paris, S. Ben Mahmoud-Jouini, The process of creation in creative industries. *Creativity and Innovation Management*, **28**, 403-419 (2019)
15. S. Kaymas, Globalization, Neoliberalism and Creative Industries: A Note on the Alternative Governance Proposition. *Amme Idaresi Dergisi*, **52**, 79-114 (2019)

On the eve of tomorrow

Enikő Korcsmáros[1], *Bence* Csinger[2*]

[1]UJS Faculty of Economics and Informatics, Department of Economics, P.O.Box. 54, Komárno, Slovakia
[2]UJS, Faculty of Economics and Informatics, MSC student

Abstract: The main purpose of our research is to examine the role of social media in the marketing strategy of small and medium-sized enterprises. The reason for our choice of topic is that social media is extremely important nowadays, which also influences the marketing strategy of the sector under investigation. For this reason, we examined the relevance of this topic in the Central and Western Transdanubian region. In the first part of our research, we collacted information on social media concepts and their distribution internationally and domestically. As a next step, we focused on presenting the most common social media and the opportunities they offer. Then we provide an insight of the situation of the regions under study, and the author's hypotheses are formulated, followed by an examination of their relevance.

1 Introduction

Our research aims to examine the role of social media in the marketing strategy of small and medium-sized enterprises with the help of domestic and foreign literature.The significance and topicality of this topic is given by the fact that the role of social media has greatly increased in relation to the sector we are examining.The aim of the research is to provide the reader with an insight into the appreciation of the role of social media in the marketing strategy of small and medium-sized enterprises, especially in relation to the two regions of Hungary.Our research covers the Western Transdanubia region and the Central Transdanubia region.Our choice of this topic was also influenced by the answer to the question of how well companies operating in the small and medium-sized enterprise sector are aware of the opportunities provided by social media today and how much they use these opportunities to create or increase a competitive advantage.In our dissertation, we place great emphasis on the role of social media within the marketing strategy of small and medium enterprises.As a first step, our goal is to introduce the ever-changing, renewable conceptual definitions of social media.After that, with the help of secondary research, we provide a comprehensive picture of their development and spread, both internationally and domestically.Each of the social media that is most popular in the world today is presented.

*Corresponding author: csinger.bence@gmail.com

In the following chapters of our theoretical review, the authors provide an insight into the opportunities provided by social media and their situation today. With the help of primary data collection, we conducted a regional study of the two regions of Hungary.Our research covers the Western Transdanubia region and the Central Hungary region. Finally, we conducted a comparative analysis of the two regions we examined, as a result of which we would like to provide an insight into the importance and appreciation of the role of social media. As a result, the research goal of our dissertation is to provide a comprehensive picture of the situation of the topic we are examining and the opportunities it offers in the Western Transdanubia and Central Transdanubia region.

2 Domestic and international guidelines of the topic

In the conceptual definition of social media, the chapter presents to the reader the path of the development and spread of the researched topic in an international and domestic context. This is followed by an introduction to the most widespread social media and the opportunities they offer and their importance in today's world.

2.1 The path of social media development

As a first step in our research, we examine the conceptual definitions of social media:"Social media, or the commonly used English term' social media ', ["] is a set of media that users fill with content. Anyone can participate in social media, as it is a community where users engage in open dialogue with each other without control, mainly for the purpose of connecting. " [1]

Social media has been specifically developed to fill this online interface exclusively with users with content that they can connect with each other, thus generating additional content. Another advantage is that it is free for customers and cost-effective on the corporate side [1]. The path of its formation can be divided into five parts: Web 1.0, Web 2.0, Web 3.0, Web 4.0 and Web 5.0.

In Hungary, it can be said that social media appeared with a significant time lag. The first truly successful website was the iWiW platform created by Média és Kommunikációs Zrt., Which started operating on April 14, 2002. The name refers to the abbreviation International who is who. After almost 4 years, the website was acquired by Magyar Telekom in April 2006. As a result of the implementation, numerous new features have emerged. These include, for example, the fact that individual users receive e-mail notifications about messages and mark-ups on the platform, which has led to a significant increase in the number of new user registrations.Meanwhile, iViW's number one competitor, myVIP, appeared on the market in 2006. The uniqueness of the site was given by the fact that in addition to allowing the creation of different communities and groups, its users had an unlimited number of invitations, as well as permission to upload any number of images and videos. In addition, two other social media platforms are worth mentioning. The first is none other than baratikor.com, which, although it started operating in Latvia in 2004, became available to Hungary by 2006. Although the leaders did not place much emphasis on the Hungarian market, the services of this website were the cornerstone of Facebook. The second social site worth mentioning is mutasd.be, which started with really innovative solutions. Facebook, known today as the best-known social website, appeared in Hungary in 2007, but until October 2010, iWiW was the most popular social platform. After 2010, the website, which has been constantly evolving since its launch, has taken over in terms of traffic and number of users [2]..

2.2 The most widespread social media and the opportunities they offer

The next step is to review the most common social media today. The beginning of the history of Facebook can be traced back to February 2004. This year, the four founding members, Mark Zuckerber, Dustin Moskovitz, Chris Hughes and Eduardo Saverin, created a community platform that was still known as "thefacebook". Its expansion was extremely rapid. In March of the year of its foundation, it was launched on three more campuses (Yale, Columbia, Stanford), and a year later, 800 campuses already used the site, as the message board function was introduced by the founders.Its uniqueness was further enhanced by the fact that after a year the "mobile version" was also released, so the service became available to almost everyone. The year 2009 can also be considered as a turning point, as the so-called "like" function was introduced at that time [3].

Currently, the number of active users per month reaches 2.38 billion and it has 1.56 billion active users per day [4].

The beginning of Twitter's history can be traced back to 2006, when four friends, Jack Dorsey, Noah Glass, Evan Williams, and Biz Stone, sought answers to the question of how to create a kind of community mini-blog. Initially, the platform provoked extreme dissatisfaction. It was only through luck, namely a conference held in Texas in 2007, that the company "escaped" where guests could communicate with each other and organize their programs exclusively through Twitter messages. Despite the initial difficulties, what made the website really great and unique was that it became popular in political circles, which is still true today. As a result, Twitter has now grown into one of the most popular social platforms [5].

Instagram, as one of the most popular platforms today, started operating in October 2010 on Apple-sold iPhones. The app was created by Kevin Systrom and Mike Krieger. Its popularity has been and continues to be due to its simplicity as well as the uniqueness of its creative photographs. The service will be available on Android devices from April 2012 and on Windows Phone devices from November 2013. Its development can be said to be unbroken, with the update of the new version released on the market in 2013, each user has the opportunity to make a video of up to one minute, to which the user can add captions, location tags and sharing pages like Instagram [6].

Currently, the number of people who actively use Instagram is more than 1 billion per month, nearly half of whom are daily active users. With regard to the companies present on the platform, it can be said that more than 25 million companies have a company profile, which shows a constantly growing trend [7].

LinkedIn, the largest business social network in the social media market, was founded on May 5, 2003 by Allen Blue, Konstantin Guericke, Eric Ly, and Jean-Luc Vaillant. The company has now become the biggest dream of startups, and has played and continues to play a major role in the emergence of professional life paths on social media platforms. Compared to previous founders, the primary goal of the founders of LinkedIn was to show people new opportunities to do business. In ten years, it has grown into an international company with more than 3,700 employees [8,9]. The company currently has 590 million users, half of whom are active monthly users [9].

The importance of social media in business is growing at an extremely rapid rate. By having a business brand present on these websites, it generates and realizes more business, and is able to satisfy the needs of customers at a higher level, thanks to which it facilitates the company's own digital marketing activities. It is vital for companies to take advantage of websites like Facebook, Twitter, Instagram, LinkedIn in case they want to stay competitive or just build an advantage over their competitors.The next step is to examine the importance of social media in business by analyzing their benefits. The first notable benefit is the promotion of community advertising. As more and more businesses experiment with digital advertising, they are realizing that this type of strategy has enormous benefits. These include the fact that advertising costs are significantly low, the target audience is targeted, and performance analysis is real-time, meaning that the company is able to continuously monitor the performance of its advertising.The next big benefit of social media is that it proves to be an effective tool to increase brand awareness. At the same time, communication is an important aspect, as social media is about building conversations. In addition, it is important that their goal be to establish two-way communication [10].

In terms of social media, Hungary is characterized by two important figures. Based on the NRC's research, it can be said that 84% of Hungarian Internet users use the Internet every month, while 73% use the Internet every day, of course in different proportions by age group. The proportion of Internet users in the 15-34 age group is almost 100%, while the over-50 age group is significantly behind. Furthermore, the research prepared for NMHH shows that the average time spent in Hungary is 3.8 hours per day.97% of users are considered active social media users, and in terms of popularity, Facebook is the leader with 89% [11,12].

Currently, social media itself is the most significant marketing communication tool. The reason for this is that companies and users have the opportunity to use a large amount of social platforms where two-way communication is dominant. From a corporate point of view, it can be said that having a presence on each platform is more cost-effective, all marketing tools are available [11].

The next step is to examine the importance of social media in business by analyzing their benefits. The first notable benefit is the promotion of community advertising. As more and more businesses experiment with digital advertising, they are realizing that this type of advertising strategy has enormous benefits.These include significantly lower advertising costs, targeted audience reach, and real-time performance analysis, meaning you can continuously monitor your ad performance. The next big benefit of social media is that it proves to be an effective tool to increase brand awareness. But what tasks can be done to accomplish this step? The number one activity is that a given company needs to find its target audience.In addition, it is a key step to make the content it creates spectacular. At the same time, communication is important, as social media is about building conversations. However, it is important that their goal be to establish two-way communication [10].

The goals of a company's well-designed marketing campaigns will only be successful if its strategy focuses on triggering activity. By this we mean the process by which each user responds to a company listing, for example by posting or sharing. This is necessary because, on the one hand, it is the basis of the conversion, i.e., the consumer most often buys online by contacting the brand in advance, and on the other hand, the algorithms of each platform support content that encourages consumers to be active. Sweepstakes, polls and competitions, for example, make a major contribution to this.So-called influencers, who are popular online personalities today with a large number of followers, can be of great help in the stimulation. If the target audience of the company and the selected influencer coincide, then the mediated marketing message can be successful. A key factor is also for the company to offer a "story" to its consumers, as in today's world, the emphasis is on human relationships. However, if the company is able to transfer this kind of interaction and content to a branded environment, it can become a market leader.Most people would like to see posts related to various promotions or videos about a new product / service that are funny or instructive.While compliance may seem like a difficult task, if a business is able to weave a story into its strategy, it will meet these criteria [7].

3 The aim of the dissertation

The main goal of our research is to provide insight into the role of social media within the marketing strategy of small and medium enterprises. The choice of topic is related to the interest of the SME sector in the marketing strategy, the knowledge gained so far, and the relatively large amount of used literature. In our research, we would like to point out the important role of social media as a marketing communication tool in the life of small and public enterprises, and what opportunities it provides for them in terms of positive development.The aim of the dissertation is to provide the reader with a comprehensive picture of the role of social media in the marketing strategy of small and medium-sized enterprises by examining two regions of Hungary, Central Transdanubia and Western Transdanubia. With regard to research, the following sub-objectives can be formulated:

• Overview of the development path of social media in domestic and international context.

• Presentation of the four most common social media today

• Explain the benefits of social media.

• Examination of the current situation of the Western Transdanubia and Central Transdanubia regions from an economic point of view.

• Presentation of the economic situation of the studied regions by means of primary data collection.

• Examination of the correctness of the formulated hypotheses.

With the aim of achieving the main goal of the dissertation, five hypotheses were formulated in connection with the examined problem. The answers obtained during our primary data collection, which were made possible by the application of a quantitative, questionnaire method in the studied regions, greatly contribute to the achievement of our goals. With the help of these data, the authors highlight the appreciation of the role and importance of social media in Hungary as well. Finally, we aim to make proposals based on research plans that SMEs can fully turn to their own advantage.

4 Research methodology and method

In conducting their research, the authors examined the role of social media in the marketing strategy of small and medium-sized enterprises. As a first step, a research plan was developed, which included the method of examining the SME sector in the research topic, and the individual hypotheses were formulated. As a first step, the development, spread and role of social media in relation to enterprises were presented with the help of domestic and international literature. To perform our practical research, we performed primary data collection.The query lasted from November 2019 to February 2020, and the implementation was implemented with a direct mail strategy. With the help of a questionnaire, we examined the small and medium-sized enterprise sector for the two regions of Hungary. In the first part of our method of quantitative analysis, we were primarily interested in demographic data. We then looked for answers as to exactly what the company uses social media for, which website is most important to it, how important it is for it to be present on the platforms mentioned earlier, and whether they monitor their competitors in this regard.In the third part of our questionnaire research, we placed great emphasis on getting the reader an answer to the question of what percentage of their marketing activities and marketing costs are spent on social media presence and future plans. As the next step of our research plan, we examined the Western Hungary and the Central Transdanubia regions separately with the help of secondary data collection. As a final step, the evaluation process followed, in which we compared the information obtained during the primary and secondary data collection and performed a regional study for the two regions.In the processing of secondary data, the applied analysis methods were divided into groups of basically simple descriptive statistical methods and methods that allowed deeper statistical data analysis. The PSPP program we used was a great help to us in the evaluation process. During the examination of the formulated hypotheses, we used descriptive statistical methods, and within that, our assumptions were analyzed with the help of cross-tabulation analyzes.A significance level of 5% was determined during the evaluation. In the analysis of each variable, we performed a Chi-square test, within which Pearson's Chi-square Asym was taken into account. sig value, with the help of which we determined that the null hypothesis or the alternative hypothesis will be accepted taking into account the specified 5% significance level.

5 Research results

The chapter provides the reader with an insight into the importance of the role of social media for the two regions of Hungary through our primary data collection. The regions under study and the fact that our hypotheses are correct or not are presented.

5.1 Analysis of questionnaire query data

During our primary data collection, we conducted a questionnaire survey on the presence of small and medium-sized enterprises operating in the Western and Central Transdanubia region in social media. Our questionnaire contained 27 questions that covered not only the general characteristics of each company but also the importance and future of the role of social media. In the course of our study, 591 responses were returned and processed from the 3,000 questionnaires sent out, of which 61.1% came from the Western Transdanubia region and 38.9% from the Central Transdanubia region.In terms of the number of employees, it can be said that in the Central Transdanubia region 70% of the respondents are micro enterprises, 21.4% are small enterprises and 8.6% are medium-sized enterprises. According to the nature of the activity, it can be stated that the majority of micro, small and medium-sized enterprises operate in the service sector, followed by the commercial sector and then the manufacturing sector.Similar results were obtained in the Western Transdanubia region. 72.1% of businesses have more than 10 years of professional experience in the market. After analyzing the general characteristics, we move on to questions related to our research topic. Surprisingly, we found that 23.6% of the companies surveyed are not present on any social media platform. Graph 1 illustrates well the proportion of companies using each social media platform in terms of headcount. The size distribution of companies is shaped by favoring Facebook in all categories, followed by Twitter with the exception of medium-sized businesses. Nowadays, in the case of Instagram, which is popular among individual users, it is said that micro-enterprises use it most often in terms of the number of employees, and in connection with LinkedIn, it is clear that medium-sized enterprises prefer it to small enterprises.

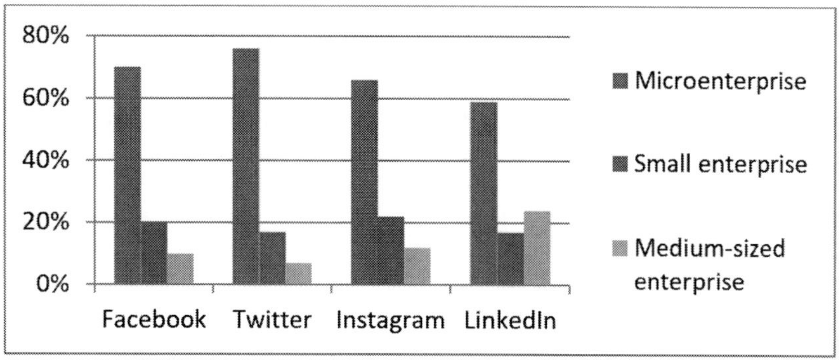

Fig. 1: Distribution of presence on each community platform in terms of size of enterprises

Our questionnaire shows that the primary goal for micro-enterprises is to build brand awareness, provide information and increase the number of target audiences. In the case of small businesses, this is as follows: building brand awareness for them and providing information is the primary mission, which is similar for medium-sized businesses. Breaking into new markets, creating a competitive advantage and increasing revenues are more important for micro-enterprises.In the next question, we were looking for the answer to whether companies employ a separate person, individuals, an outside company who deals with presence on social media. In this respect, 72.9% of firms answered in the negative and only 16.6% of them employ regularly. This is surprising as they do not put enough energy into social media presence based on this. Without the help of a person who does not specialize in this, they will not be able to reap the benefits of social media platforms.An important question in our research was the importance of presence in social media. The average value of all responses is 3.8, which means that the presence on each platform usually plays an important role for businesses. In our view, the value is not higher because organizations are only aware of a few benefits and do not have a comprehensive picture of how profitable this step can be for them overall and in the long run.When asked what percentage of their marketing activities and marketing costs their presence on social media accounted for, the largest number of respondents in the surveyed sector was only 0-25%, which is surprising as they suggest that they are probably unaware of social media opportunities..Graph 2 shows a comparison of the marketing costs of small and medium-sized enterprises and the distribution of enterprises by size, based on which it can be said that micro-enterprises spend the largest percentage of their marketing costs on social media presence. They are followed by small businesses and finally medium-sized businesses.

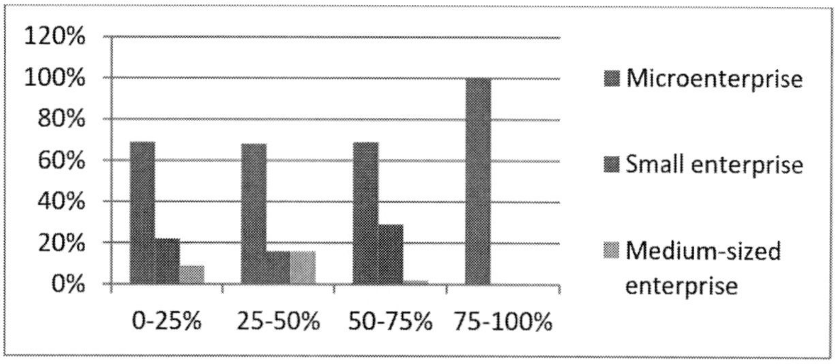

Fig. 2: Comparison of marketing costs by SME group

We then wondered whether there was a change in the lives of individual companies using the platforms mentioned earlier and, if so, in what direction. Only 0.7% of respondents felt the negative impact of this, 35.2% said there was no change, and 64.2% said it brought a clear positive change in the life of their company. The positive effect was seen in the increase in the number of addressed and persuaded individuals belonging to the target audience, as well as in the increase in revenues. Several opted for other options, in which they had the opportunity to provide their own response, such as the emergence of new business partners, increased commitment within the organization, easier recruitment, and increased brand awareness.We continued to place great emphasis on the future of the SME sector in question, in terms of whether there are plans to expand our presence in social media. Unexpected results were obtained, as 56.7% answered in the negative and 43.3% answered in the affirmative. The following questions concerned only those companies that answered yes. Of the 253 companies selected, 149 would be willing to spend more capital on this. In their view, the most obvious solution is to appear on new social platforms, followed by the launch of various media campaigns. As a next step we provides the reader with a comprehensive picture of how businesses that want to take the opportunity to expand their presence on social media want to do so. It can be clearly stated that in the case of micro-enterprises the involvement of influencers and youtubers, in the case of small enterprises the involvement of external companies specializing in social media, and in the case of medium-sized enterprises the involvement of well-known persons is the priority.By comparison, in both regions studied, the number of enterprises with 0-9 employees is the highest, and as mentioned earlier, Facebook proved to be the most frequently used social platform. The summary also shows that, regardless of the size of the business, the provision of information is a priority for all of them. In most cases, the appearance on each platform has made a positive difference in the lives of the companies that take advantage of the opportunities that come with it, but there are no plans to expand this presence outside the medium-sized companies of the Central Transdanubia region and small businesses in the Western Transdanubia region. However, their goal is to launch media campaigns using sponsored ads with more capital. The majority of our respondents have been in social media for 1-5 years in both regions, which, based on their feedback, plays an important role in their lives.

5.2 Hypothesis testing

After the preparation of the theoretical part of our dissertation, the next indispensable part of our research plan was the setting up of the hypotheses and the examination of their correctness. The hypotheses we set up were formulated in a deductive way, the meaning of which suggests that we started from theoretical theorems and formulated our statements for the practical application of these assumptions. Based on the topic under study, five hypotheses were formulated, which are the following:

1. The active presence of small and medium-sized enterprises in social media has an impact on the company's marketing activities.

For the first hypothesis, considering Pearson's Chi-square test, it can be concluded that the two variables are significant because the significance value is less than 0.05, which is illustrated in Table 2. Based on this, it can be said that there is a correlation between the two examined elements, i.e. in this case we reject the null hypothesis and the alternative hypothesis comes into force. From the value of Fi it can be concluded that the relationship between the two variables is one-way, and based on Cramer's V and Contingency coefficients it can be said that there is a stronger than average positive positive relationship between the variables. In ordinary language, it can be stated that the more active the company is on social media platforms, the greater the impact of this activity on the company's marketing strategy, ie, it places more emphasis on advertising, which increases revenue and increases the number of target audiences.Based on this, it can be said that the more significant a company is present on social media platforms, the greater the share of this activity among its marketing activities.

Table 2: Chi-square tests and symmertric measurements for Hypothesis 1.

		Chi-square rehearsals			
Statistics	**Value**	**df**	**Asszimp. szig. (2-tailed)**		
Pearson's Chi square	697,45	10	,000		
Likelihood rate	10,98	10	,000		
N/number of valid cases	588				
		Symmetrical measurements			
Category	**Statistics**	**Value**	**Aszimp.sztd.mistake**	**Approximate T.**	**Approximate szig.**
Nominal-nominal	Phi	1,09			
	Cramer's V	,77			
	Contingency coefficient	,74			
N/number of valid cases		588			

The questionnaire for the research shows that the majority of the small and public enterprise sector in the regions studied (70%) is present on a social media platform, however, the remaining 30% still do not have the opportunity to take advantage of social media. gave opportunities. In our opinion, the reason for this is that they are afraid of change, do not want to innovate, are satisfied with old, proven methods, and are unaware of the benefits provided by social media. For them, the authors suggest that they take advantage of these opportunities. Make the first impression in the hands of the business.

You can achieve this by creating a website, the amount of which is HUF 100,000 if it is done by an external company. . By moving to social platforms, which are today's dominant marketing communication tools, companies can share relevant articles, photos, videos, thereby maintaining contact with individual consumers.Having an online presence makes your business visible, makes your target audience easy to find, communicates with potential customers, and provides up-to-date information about changing consumer needs. Internet advertising contributes greatly to the growth of a company, but for this they need to constantly analyze the results, which they need to use. It is often heard that advertising is expensive, but in our opinion, advertising and online advertising are even more expensive in the long run, and that is nothing if the organization does not advertise. For them, the cost of advertising placement is about 150,000 HUF per month. In order to achieve continuous improvement, they need to set more and more goals. We also recommend companies to use the platforms on which they have the largest target audience. For young people, we recommend using Instagram. However, Linkedin is best suited for creating professionally relevant content.The authors 'critical remark extends to the fact that the SME sector employs only a very small percentage of individual / individuals / external companies to ensure a presence in social media. If they don't want to outsource this task, they have three options: educating, training their own employee, hiring a new employee, or hiring young, ambitious trainees with up-to-date information in the world of social media. In our opinion, this is essential, as the employment of a specialist specializing in this task would greatly contribute to increasing efficiency and speeding up the process. Based on payment data, today the salary of these professionals is around 200,000 HUF.Among the critical remarks, it is worth mentioning that, based on data collection, most companies do not monitor the presence of their competitors on social media at all. This is a problem because it often happens that a business could learn from the mistakes made by the competition, or even from their successes. In addition, if they get to know them well, the organization can get a comprehensive picture of what can give it its uniqueness, not to mention that the competitor encourages it to perform better day by day, and in terms of budget, this step does not involve capital expenditure. With regard to feedback from the target audience, we recommend that feedback be taken into account and used in all cases, as they need to adapt to ever-changing consumer needs if they want to be present in the market, and this step contributes to efficiency. takes into account and uses the suggestions, in the long run it will also have the opportunity to increase the existing consumer base.The authors concluded that businesses are not fully satisfied with the benefits of being present on social media platforms because they do not spend a sufficient percentage of their marketing costs or marketing activities on this factor. As a result, most of them believe that no level of change is taking place since they have been present in the online space. This short-term thinking suggests that most of them will have a negative opinion and thus will not reap the benefits, returning to the marketing and advertising tools previously used. nearly 50% of respondents do not plan to expand their presence on social media. We recommend that they also appear on other social platforms and, in terms of implementation, involve so-called influencers which are a great help, for example, in launching a targeted media campaign.

6 Conclusions

The main goal of our dissertation was to give the reader an insight into the role of social media in the life of small and medium-sized enterprises through the two regions of Hungary. As a first step, we described the exact definition of social media with the help of domestic and international literature, and then we examined its formation and development path in an international context. We have illustrated all this through the development of the Web. Our dissertation also covers when and how social media appeared in Hungary. It is important to mention that two social platforms have played a leading role in Hungary, iWiW and myVIP. One of the most well-known websites today, Facebook, was launched in 2007, one of the cornerstones of which was baratikor.com. As a next step, we put a lot of emphasis on introducing the most common social media, including Facebook, Twitter, a social mini-blog, Instagram for sharing creative photos, and Linkedin, which is responsible for the emergence of professional life paths on social media platforms. After reviewing the possibilities, our dissertation examined the importance of social media in the small and medium-sized enterprise sector. In this subsection, we provide help and information to the managers of individual companies on why it is essential to be present on the mentioned platforms in today's world, be it small, medium and large companies. We have dedicated a separate subchapter to reviewing the most significant communication tool, social media, nowadays, during which we examined the market position of the most popular social platforms. In addition, it is important to highlight that the trends present in social media are constantly changing, which can be said to be the only thing in the world of the internet that is constant. In addition, we examined the situation in Hungary on the basis of research conducted by the NRC and the NMHH. Turning to the practical part of our dissertation, the focus is on the examination of two regions of Hungary.The territorial, social and economic situation of the Central Transdanubia and the Western Transdanubia region was presented, which shows a kind of comprehensive picture of the location of these areas in Hungary from the point of view of small and medium-sized enterprises. As the next step of our research, we analyzed the data of our questionnaire query. Of the nearly 3,000 questionnaires sent out, which contained 27 questions, 591 responses were received from the two regions surveyed. Subsequently, the authors formulated five hypotheses on the topic, the validity of which was processed with the help of the PSPP program. The formulated assumptions were evaluated using descriptive statistical tools.For the study, we used a cross-tabulation analysis in which the Pearsons Chi-square helped to establish the validity of the hypotheses. In several cases, the authors supplemented the research with symmetric indicators (Fi, Cramer's V and Contingency coefficients), which provided a comprehensive picture of the strength of the correlations between the variables. In the case of the hypotheses, we thought to discover a relationship between four pieces, while in the case of one piece there was no correlation. This means that the null hypothesis was rejected in four cases and accepted in one case. In the chapter entitled Conclusion, the reader gets an insight into the well-founded conclusions of the authors, as well as their suggestions, critiques and opinions based on them.It is also important to mention future research directions related to the topic under study, as the current state due to the coronavirus is expected to have a major impact on the future situation. It can be said about Hungary that webshops operate freely and more and more companies are starting to apply this kind of strategy, in which it is of great help to them that courier services deliver orders without any problems. In today's world, the consumer hears from all communication channels to stay at home! But how does a 21st century customer react to this? Due to the restrictions currently in place, of course, you place orders online. However, not only the consumer but also the businesses need to adapt to this sudden change! In each case, demand will determine supply and in the current situation, demand for online orders will increase greatly.

References

1. Dr. I. Eszes, Digitális gazdaság. *Nemzeti Tankönyvkiadó RT (Budapest),*376. ISBN 978-963-19-7139-2 (2011)

 https://doi.org/10.1051/shsconf/20208301033

2. F. Novák. A közösségi média megjelenése, üzleti- és marketing szintű lehetőségei: szakdolgozat. Budapesti Gazdasági Főiskola. 11. (2010)
3. ThePitch. Facebook története. Retrieved from https://thepitch.hu/online-marketing-szotar/facebook-tortenete/ (2018)
4. S. Kith. 53 Incredible Facebook Statistics and Facts. Retrieved from https://www.brandwatch.com/blog/facebook-statistics/ (2019)
5. E. Balla. Kulisszatitkok a Twitter alapításáról és működéséről. Retrieved from http://valasz.hu/uzlet/kulisszatitkok-a-twitter-alapitasarol-es-mukodeserol-76862 (2014)
6. Instagram blog. Az Instagram története. Retrieved from https://instagram.blog.hu/2018/02/10/az_instagram_tortenete (2018)
7. Sz. Szatmári. A legfontosabb Instagram statisztikák. Retrieved from https://marketingprofesszorok.hu/instagram-statisztika-2019.html (2019)
8. D. Csoma. *Adatok* a Twitterről és a LinkedInről, amit sosem mertél megkérdezni. Retrieved from https://kozossegi-media-mindenkinek.blog.hu/2019/03/02/adatok_a_twitterrol_es_a_linkedinrol_amit_sosem_mertel_megkerdezni (2019)
9. S. Gyöngyösi. A hétvégén lett 10 éves a LinkedIn. Retrieved from https://marketingmorzsak.hu/a-hetvegen-lett-10-eves-a-linkedin/ (2013)
10. S. Keran. The Importance of Social Media in Business. Retrieved from https://www.lyfemarketing.com/blog/importance-social-media-business/ (2019)
11. Ryckposter.A közösségi média ereje. Retrieved from https://www.ryckposter.hu/a-kozossegi-media-ereje/
12. H. Lőcsei Ph.D., B. Bublik, A. Nyeste. Magyarország régióinak társadalmi-gazdasági profilja.Gazdaság- és Vállalkozáskutató Intézet. 142. (2013)

Strategic planning for creative enterprises

*Andrej Kovalev**

University of Economics in Bratislava, Faculty of business management, Department of Business Economy, Dolnozemská cesta 1/b, 852 35 Bratislava, Slovak Republic

Abstract. Strategic Vision and a clear definition of Business Success are vital elements of strategic planning for all creative businesses and cultural enterprises. A Vision sets out the aspirations for the future of a creative business or cultural enterprise. This Strategic Vision is the 'dream' of the future, a picture painted in words, which is intended to inspire people by appealing to the heart as well as the head. One of the leader's main functions is to inspire people towards a vision. This is 'Inspirational Leadership'. A Strategic Vision must be a Shared Vision, embraced by everyone in the business or organisation - not only the leader, the senior management team and key staff.

1 Vision and Success

Successful creative businesses need a combination of creative talent and strategic business management. Sometimes, one individual has both of these elements in good measure, but more often than not, the harmony of creativity and business is formed by two people, or a larger team. So, there must be a shared vision for the enterprise, even if the partners involved are very different characters with different skills. Matters of risk, growth, financial reward and lifestyle are all issues about which partners might have different views. If these are not in harmony, each person may have a valid, but different, business strategy in mind as their road map to different destinations. So the shared vision is in many ways the starting point. If people are working hard together but with different goals in mind, conflict – or at least stalemate – is likely to occur.

The process of clarifying the organisation's Vision, and definition of Success, is a crucial starting point for any business strategy or business plan. It is also useful in so far as it explores what is at the heart of the organisation, its Mission and its Values.

2 Internal Analysis

Classic 'SWOT Analysis' involves evaluating the Strengths and Weaknesses within a business or organisation, ie an 'Internal Analysis'. It's important to assess the strengths and weaknesses of our business, focusing particularly on strengths, especially strengths

* Corresponding author: andrej.kovalev@euba.sk

which competitors do not have. These relative strengths are an essential element of enterprises Competitive Advantage. Business strategy is about deploying enterprises characteristics in markets and other situations where they become strengths in relation to the competition. The PRIMEFACT Checklist below provides a useful structure for a comprehensive analysis. Checklist specifically for the creative and cultural industries and have used it successfully with a range of clients including creative businesses and cultural organisations.

The PRIMEFACT checklist:

- **People**. What are the strengths and weaknesses of our people? Employees, directors, members, associates, advisers and other stakeholders.
- **Reputation** (or Brand). What is our reputation with our target customers? What are the strengths – or weaknesses – of our brand or brands?
- **Intellectual** Property. What intellectual property do we have? How is it protected? How easily can it be turned into income streams?
- **Market Research**/ Marktet Information. What information do we have about market segments and market trends? What do we know about individual clients and their specific needs?
- **Ethos** (or Values or Culture). What is our ethos, our values and our organisational culture? Do all stakeholders subscribe to this same ethos?
- **Finances**, ie money. What is the current state of profitability, cashflow and assets? How much money do we have to invest or can we borrow?
- **Agility** (or nimbleness or change-ability). Are we agile enough to seize new opportunities. Are people prepared to change and ready for change? Or are we unable to change?
- Collaborators (Alliances, Partnerships and Networks). What are the strengths and weaknesses of our associations with other businesses and organisations (including government)?
- Talents (competencies and skills). What are our core competencies. What skills do we have available and what gaps are there? Are we able to learn new skills?

3 External Analysis

All businesses and organisations operate in a changing world and are subject to forces which are more powerful than they are, and which are beyond their control. Just as a ship at sea is subject to powerful natural forces of which it needs to be aware and deal with, organisations are influenced by forces in their external business environment.

Any business strategy needs to take account of all these forces so that opportunities and threats can be identified and the organisation can navigate its way to success by matching its internal strengths to external opportunities. As an aid to identifying all these external forces, a couple of acronyms come in handy.

A PEST analysis invites enterprises to list all the relevant external forces using four headings: Political, Economic, Sociological and Technological. These are useful headings; it doesn't matter that some items might be both political and economic.

However, to look in eight rather than four directions, use the ICEDRIPS checklist:

- **Innovation** including new technology and the Internet (of course) but other innovations too which may be particular to an industry.
- **Competitors.** Not only direct rivals but threats from substitute products, new entrants to the market, the changing power of suppliers and the changing power of customers.
- **Economic factors** such as inflation, exchange rates, downturns in the industry, public spending etc.
- **Demographics.** The relevant statistics of age, gender, geography, social class etc, and changes in these.
- **Regulatory environment,** ie laws, regulations, agreements and conventions.
- **Infrastructure** such as telecommunications networks, transport, public services and utilities.
- **Partners.** Strategic alliances with other companies or organisations.
- **Social trends,** including acceptance of technology, use of leisure time, fashions and changing beliefs.

Since the external environment in which the enterprise operates is constantly changing, the External Analysis needs to be repeated frequently and ideally constantly. In this way, the enterprise will be able to identify opportunities and threats more quickly.

4 Competitive Advantage

If enterprises are competing in a market place by offering the same thing as many other rivals, then enterprises can only compete on price: it's a buyer's market and prices will fall. This is the case for many creative businesses that do not succeed commercially. On the other hand, prices rise in a seller's market, with a monopoly being the strongest position to achieve in relation to customers. A profitable strategy shifts enterprises position from one of price competition towards a monopoly position. Enterprises can then raise enterprises prices. By focusing on a product or service that enterprises can deliver and rivals cannot. Yes, for that product there is a smaller demand, but enterprises are much more powerful in that particular market because there is even less supply. Our herd instinct is to follow others into bigger markets where there is more demand, but here there is also excess supply, so it's a buyer's market and prices fall.

Focusing on enterprises unique strengths and not following the crowd is counter-intuitive but strategically intelligent. Enterprises difference is enterprises strength! Focus on how enterprises are different and build enterprises business around this strength. Not every potential customer will want this, but those who do will love enterprises and pay the right price. It makes perfect sense to focus on the areas in which enterprises have competitive advantage – to do only those things at which enterprises can excel in relation to competitors. In this way enterprises are competing on enterprises own terms, choosing enterprises battles, and dealing with rivals from a position of strength.

This competitive advantage may be technical, for example, expertise in a particular business area that others do not have. On the other hand, competitive advantage may derive from infrastructure, systems, logistics, or economies of scale that cannot be replicated easily by rivals. For example, wide distribution networks, detailed information about sales trends, production capacity or rapid delivery can provide a competitive advantage over rivals for particular businesses.

Enterprises competitive advantage may be also derived from social connections or status. It may be that enterprises are well respected in a particular community, or have connections in all the right places in a certain industry, or can relate better than rivals to a particular group of customers because of cultural, language, or other affiliations. This is especially the case in businesses where personal relationships matter. We cannot recognise where we have competitive advantage unless we analyse our competitors. This is crucial because we need to base our marketing strategy on our competitive advantage. That's to say, if we want to focus on the things we can do better than our rivals, we need to know what our rivals are good at, not so good at, and how we can outmanoeuvre them.

If enterprises are one of many supplying the market with similar products, enterprises only option is to drop enterprises prices. On the other hand, if enterprises stand out from the crowd enterprises will be in demand, and able to increase prices. It's all about supply and demand and the relationship between these two things. A large market may seem attractive, but if there's an even larger supply, then it's a buyer's market. On the other hand, a smaller market but with even fewer suppliers works to the advantage of suppliers. It's better to be a big fish in a small pond than a small fish in a big pond: small fish get eaten by bigger fish, and businesses without a competitive advantage fail.

5 Strategic Marketing

Strategic Marketing relates to the 'big picture' and the big decisions, such as these:

- What business should we be in?
- What do customers really want?
- At what can we excel in a competitive marketplace?
- Which customers should we decide strategically not to serve?
- What is our Unique Business Formula?

Strategic marketing involves the recognition of competition, the identification of competitive advantage and as a result, serving markets enterprises can serve better than enterprises rivals. Consequently it involves deciding which markets not to serve, because rivals can do it better. This requires both an accurate understanding of market needs, and a willingness to shape the business and its products around the needs of selected customers. Strategic Marketing is at the heart of devising a feasible Business Formula because a Business Formula has to take into account an enterprise's competitive advantage and its choices to serve some markets and not others.

In short, Strategic Marketing requires a complete change of thinking, from: 'Let's do everything we can – and try to sell to everyone' to 'Let's do only what our rivals cannot do as well as us – and sell only to those markets that want what we excel at.' Only when the big issues of marketing strategy have been decided that we can turn to the details of operational marketing. Having decided which precise markets to focus on, we can tailor our messages to the precise needs of our chosen customers. Furthermore, we can listen to customers in our chosen markets, in order to understand more fully what they really want to buy. Marketing can be described as 'looking at the enterprise from the customer's point of view'.

In the end, Strategic Marketing is about deliberately aligning the whole enterprise to the changing needs of carefully chosen customers. This means that some markets are deliberately avoided in order to focus entirely on strategically selected markets, niches and customers. Consequently Strategic Marketing is about high level decisions about the direction of the enterprise as it manoeuvres successfully amongst competitors, uses its competitive advantage and selects particular customer groups to serve in order to fulfill its objectives and achieve its Vision.

Confirmation

This contribution is the result of the project VEGA 1/0340/19 „Podnikateľský rozmer subjektov kreatívneho priemyslu v kontexte inovácií a inteligentného rastu“

References

1. D. Parrish, *T-shirts and Suits: A Guide to the Business of Creativity,* Merseyside ACME; 2Rev Ed edition (2007)
2. D. Parrish, *Chase One Rabbit: Strategic Marketing for Business Success: 63 Tips, Techniques and Tales for Creative Entrepreneurs,* Wordscapes Ltd, (2014)
3. D. Hesmondhalgh, *The Cultural Industries*. London. SAGE Publication (2003)
4. J. Howkins, *Creative Economy*. Penguin Group Incorporated (2001)
5. R. Landry, M. Saïhi, Patenting and Spin-off Creation by Canadian Researchers in Engineering and Life Sciences. *Journal of Technology Transfer* **32**, *3* (2007)
6. I. Miles, S. Gee, Innovation and Creative Places. *Innovation Policy Challenges for the 21st Century*. 227-243 (2013)
7. E. Munro, *Building soft skills in the creative economy: Creative intermediaries, business support and the 'soft skills gap"*. University of Glasgow Centre Department of Theatre, Film and Television Studies, Gilmorehill Centre, Glasgow, (2017)
8. J. Potts, S. Cunningham, S. Hartey, P. Ormerod, Social network markets: A new definition of the creative industries. *Journal of Cultural Economic* **32**, *3,* 167-185 (2008)
9. D. Throsby, *Economics and Culture,* Cambridge University Press (2001)

https://orcid.org/0000-0001-8183-7053

Analysis of market concentration in creative industry

Mária Kozáková[1,*], and *Mária* Kmety Barteková[1]

[1] University of Economics in Bratislava, Faculty of Business Management, Department of Business Economy, Dolnozemská cesta1, 852 35 Bratislava, Slovak republic
[2] University of Economics in Bratislava, Faculty of Business Management, Department of Business Economy, Dolnozemská cesta1, 852 35 Bratislava, Slovak republic

Abstract. For the companies the situation in the industry is important in terms of identification opportunities and threats that occur in the industry. Given the increasing competition should be monitored as concentration develops in the industry in which it operates. This article examines the use of the Herfindahl-Hirschman index and Concentration ratio in selected sectors of the creative industry. The development and degree of concentration in the sector is an essential factor characterizing the level of the competitive environment and its exact definition is based on the system of quantitative characteristics that are part of the analytical materials of the institutions dealing with the enforcement of competition rules in most developed economies.

1 Introduction

In recent years, the importance of the cultural economy and of the cultural and creative industries has grown. The creative and cultural economy is said to be an important and growing part of the global world because of socio-economic potential, creativity, knowledge and information, job creation, wealth and cultural involvement. The basic condition for the existence of a company is competitiveness. One way to monitor competitiveness in the industry is measuring concentration in industry. The development and degree of concentration in the industry is an essential factor characterizing the level of the competitive environment. The paper focuses on the measurement of concentration in selected sectors of creative industry using the Herfindahl-Hirshman index and concentration ratios.

2 Creative industry

Creative industries can be defined as the area of overlap between culture, technology, science, and commerce. They involve the supply of goods and services that contain a substantial element of artistic and intellectual activities associated with a vital role in social

* Corresponding author: maria.kozakova@euba.sk

and human development. [13] Although we can find different definitions of the creative industries, in most research that sector has been investigated mainly based on specific characteristics [9].

Creative enterprises are particularly important for the economy as they have an impact on building the economic potential of countries, cities and regions. [1]. The creative industries involve the concretization of an image, through whatever medium for some form of economic return [2]. Due to creative industries major impact on the development of the economy and the society, cultural organizations and creative companies, which generate creativity, should attract more interest and support from state institutions [9]. Especially at the early stage of its development, the creative sector needs to be supported by the state. [11] According to Campell [8] there is much high-level assertion of creative industries role both for social cohesion and economic growth. The importance of creative industries is that the creative industries are a key contributor to innovation economies [7]. Ideas, processes, products and talent that are developed by the creative industries drive productivity in and outside these industries [14]. Games [6] states, that combining small and medium-sized enterprises (SMEs) and creative industries may offer an advantage of the economic development of the emerging economy. SMEs simply want to find a way to survive because they do not have alternatives, but this may not be the case in SMEs in the creative industries as they pursue business for innovation rather than merely survival. Creative industry is also important job creator. Combined with the expansion of the media, entertainment and design industries, the appeal of creative work in recent years has increased the number of people who do work and want to work in the creative industry [5]. Cnossen [5] also states that workers and entrepreneurs in the creative industry are differently motivated than workers in other sectors, because they are fuelled by "psychic rewards".

UNCTAD [15] makes a distinction between "upstream activities" (traditional cultural activities such as performing arts or visual arts) and "downstream activities" (much closer to the market, such as advertising, publishing or media-related activities) and argues that the second group derives its commercial value from low reproduction costs and easy transfer to other economic domains. The UNCTAD classification of creative industries is divided into nine subgroups, as presented in figure 1.[15]

Fig. 1. Classification of creative industries

2.1 Measuring market concentration in industry

Monitoring the situation in the sector is extremely important for the company in identifying opportunities and threats that exist in the industry. One of the possibilities is to monitor competitors, whether existing or potential, which could affect the industry but also the attractiveness of the industry itself. If the industry is attractive, prosperous, it is likely to also attract new competitors to enter it, which may affect the industry's competitive environment. Concentration by industry is a function of the number of companies operating and their respective market shares in total sales. According to Kavčáková [16] measuring of the market concentration of a given sector is particularly important because it is closely linked to the competitiveness of the companies in that sector.

The concentration ratio (CR) and Herfindahl-Hirschman index (HHI) have become the most popular indicators for measuring market concentration in empirical research due to the simplicity of their calculation. However, the latter indicator requires the market shares (or sales volumes) of all the companies in the market to be known. [4] The concentration ratio CR_m can be defined as the combined market share of the m largest firms within the market. The most common concentration ratios are the CR_4 and the CR_8, which means the market share of the four and the eight largest firms. The Hirfindahl-Hirschman Index (HHI) is a standard index used in analysing the degree of market concentration of a particular industry in a particular geographic market. It has been employed as a proxy for the level of competition in a market, measuring how close the market is either to a monopoly or to a perfect competition scenario. [3]

3 Methodology and data

Within the creative industry, we focused on measuring concentration in 3 sectors: Book publishing (NACE 58110), Publishing of newspapers (NACE 58130) and Publishing of journals and periodicals (NACE 58140). The results were processed on the available information on revenues of individual entities for 2017 and 2018. The data was obtained from database FinStat.

First, it was necessary to calculate the market share of each company in the industry, because the methods we use to evaluate the degree of concentration in the industry are based on the market share. Market share constitutes the share of firm in a given market or industry:

$$r_k = q_k/Q \tag{1}$$

where:
q_k – q_k indicates the volume of the i-th firm in the industry,
Q – represent the total volume of industry output.

Concentration Ratio (CR_m) is the most widely used index due to its simplicity and low requirements for input. It is sum of shares of n biggest companies on a market. [17]

$$CR_m = \sum_{i=1}^{m} s_i \qquad m \epsilon < 1; n > \tag{2}$$

where:
si – the market share of the i-th enterprise,
m – the number of enterprises analysed.

Value of the index ranges from 0 to 1. The higher the value is the higher is rate of market concentration on given market. To consider the degree of concentration according to the four most powerful companies in the industry (CR4) by the Federal Trade Commission in the USA the industry is [18]:

- unconcentrated - if the 4 strongest companies produce less than 25% of industry output ($CR_4 < 0,25$),
- moderately concentrated - if the four most powerful companies produce at least 25% and less han 50% of industry output ($0,25 \leq CR_4 < 0,5$),
- concentrated - if the 4 strongest companies in the industry producing at least 50% of industry output ($CR4 \geq 0,50$).

In our contribution we also calculate marginal concentration ratio. This latter measure is defined as the difference between CR_4 nad the eight-firm concentration ratio CR_8.

Herfindahlov-Hirschmanov index (HHI) as a tool used for the measurement of the market concentration is calculated as the sum of the squares of the market shares of the individual companies HHI is used in the two forms [19]:

- Value of HHI is left as a decimal,
- Value of HHI is multiplied by 10 000.

$$HHI = \sum_{i=1}^{n} \left(\frac{x_i}{\sum_{i=1}^{n} x_i} \right)^2 \quad (3)$$

where:
n – marks number of comapnies on a market,
xi – is share of i-th company on a market.

Value range of the index is from 0 to 10 000. If there is only one company in the industry, the index reaches the highest value of 10 000. Classification of the degree of concentration in the industry by the calculated index value, which indicates the Federal Trade Commission (FTC) in USA is [16]:

- unconcentrated - (HHI<1000)
- moderately concentrated - (HHI od 1000 - 1800)
- concentrated – (HHI > 1800).

4 Results

We analyze concentration in selected sectors on the basis of revenue. On the basis of this data the share of individual companies in sector was calculated. Concentration rates for m of the strongest firms in the industry are displayed in figures.

For year 2017, concentration ratios for sector Book publishing indicate concentrated market. Concentration ratio for the largest company in 2017 is 43,7 %, the strongest company has a market share of 43,7 %. In 2018 is concentration ratio 41,78 %. The combined market share of the four biggest companies is 59,76 % in 2017 and 61,81 % in 2018. According to Federal Trade Commission in the USA the industry is concentrated, because the 4 strongest companies in the industry producing more than 50% of industry output (CR4 ≥ 0,50). Marginal concentration ratio indicates that the difference between CR4 and the eight-firm concentration ratio CR8 is 11,09 % in 2017 and 8,98 % in 2018.

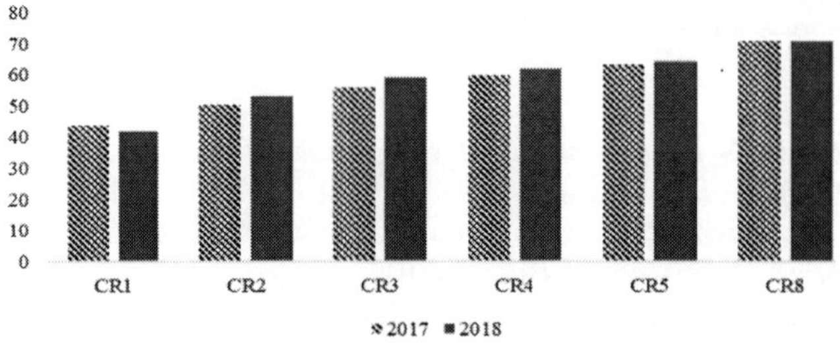

Fig. 2. Concentration ratio – Book publishing

Results of concentrations ratios in second sector – Publishing of newspapers are similar to first industry. Concentration ratio for the 4 largest companies in 2017 is 79,07 % and in 2018 80,24 %. We can conclude, that this industry is concentrated, because 4 strongest companies in the industry producing more than 50 % of industry output (CR4 ≥ 0,50). Marginal concentration ratio indicates that the difference between CR4 and the eight-firm concentration ratio CR8 is 9,54 % in 2017 and 10,14 % in 2018.

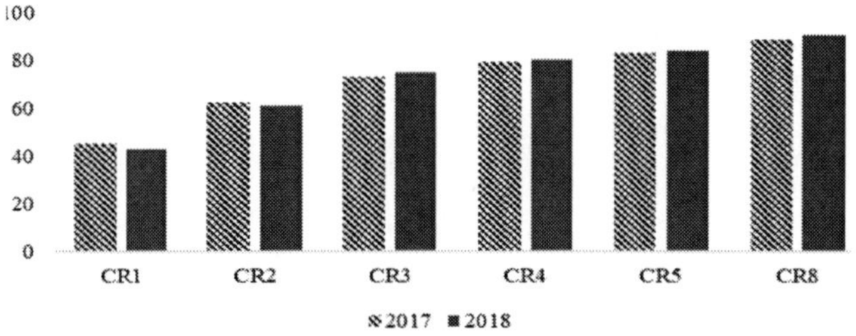

Fig. 3. Concentration ratio – Publishing of newspapers

Results of concentrations ratios in third sector – Publishing of journals and periodicals are also similar to first industry. Concentration ratio for the 4 largest companies in 2017 is 61,08 % and in 2018 59,96 %. We can conclude, that this industry is concentrated, because 4 strongest companies in the industry producing more than 50 % of industry output (CR4 ≥ 0,50). From all selected sectors is the highly concentrated sector- Publishing of newspapers. Marginal concentration ratio indicates that the difference between CR4 and the eight-firm concentration ratio CR8 is 11,98 % in 2017 and 11,12 % in 2018.

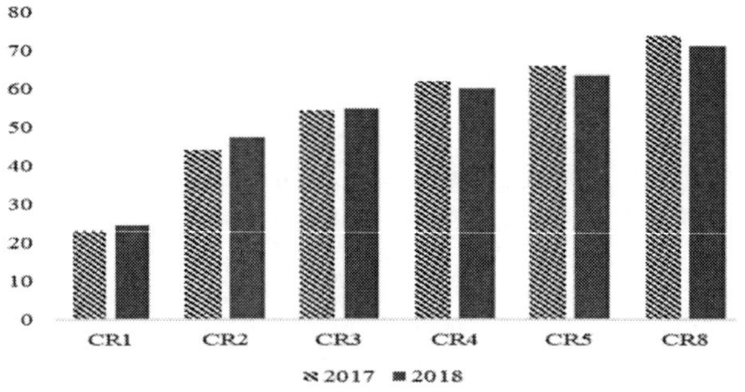

Fig. 4. Concentration ratio – Publishing of journals and periodicals

An analysis of computed results of the Herfindahl-Hirschman Index, which is based on the classification of the degree of concentration in the sector, shows that all of the calculated values of HHI are greater than 1800. Only HHI for sector Publishing of newspapers is from interval (1000 < HHI < 1800). This sector can be classified as moderately concentrated. The results of HHI are displeyad in figure 5.

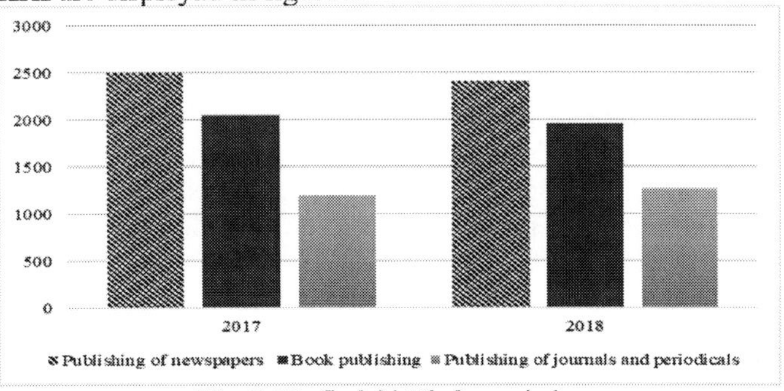

Fig. 5. Herfindahl-Hirshman index

We used Lorenz curve for illustration the situation in selected sectors. The Lorenz Curve is a tool used to represent income distributions as proposed by and it tells us which proportion of total income is in the hands of a given percentage of population.

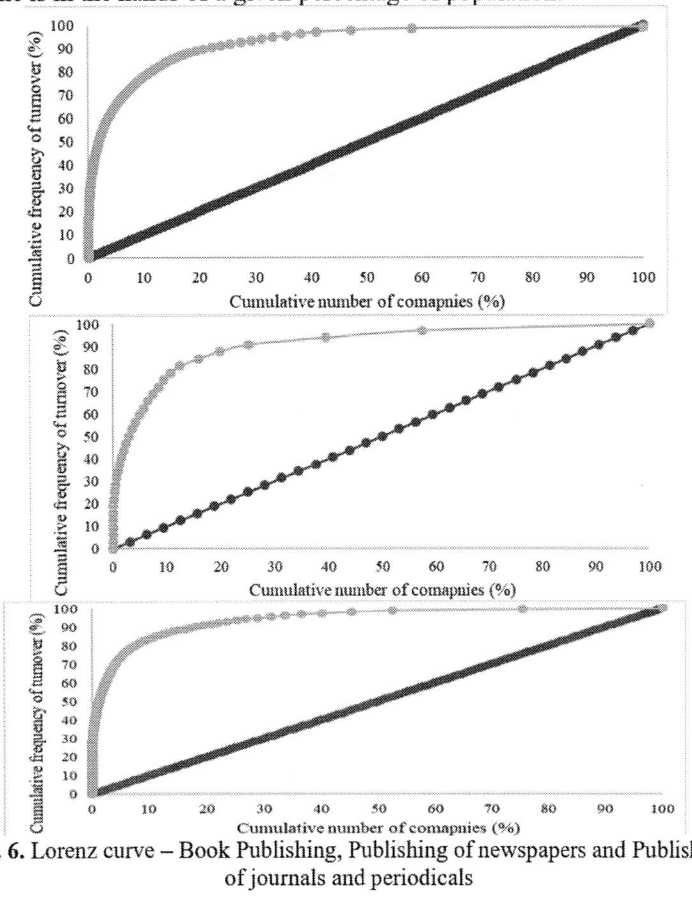

Fig. 6. Lorenz curve – Book Publishing, Publishing of newspapers and Publishing of journals and periodicals

5 Conclusions

Over the last years, the importance of the so-called cultural economy and of the cultural and creative industries has greatly increased. Today, cultural and creative industries are driving factors for economic growth and according to global demand, also stimulated by the new economy. [12] The basic condition for the existence of a company is competitiveness. One way to monitor competitiveness in the industry is measuring concentration in industry.

Market concentration analysis of the sectors - Book publishing, Publishing of newspaper and Publishing of journals and periodicals, based on indicators such as HHI and concentration ratio, that the selected sectors of creative industry are concentrated .The concentration rate for the sector Book Publishing confirms that the 4 strongest companies control from 59-62% of the industry's total sales (CR4 = 059-0.62), in the sector Publishing of newspapers 4 strongest companies control from 79-81% of the industry's total sales (CR4 = 0,79-0,81) and in sector Publishing of journals and periodicals 4 strongest companies control from 60-61% of the industry's total sales (CR4 = 0,6-0,61) . In drawing the conclusions, we use d the US FTC methodology, which states that the industry is highly concentrated if the share of the 4 largest companies in the industry is more than 50%. The results are also confirmed by the HHI. HHI for each sector was analyzed for year 2017 and 2018. According to the used classification of HHI, the selected sectors are concentrated. Only sector -Publishing of newspaper can be classified as moderately concentrated. Lorenz curve graphically displays the unequal distribution of the turnover on the selcted sectors of creative industry.

This paper is an outcome of the research project VEGA n. 1/0340/19 "The Entrepreneurial Dimension of Creative industries in the Context of Innovation and Smart Growth" in the range of 50% and Internal Grant I-20-101-00 Research of creative industry enterprises with regard to their life cycle in the Slovak Republic in the range of 50 %).

References

1. M. Ratalewska, The Role of Entrepreneurship and Business Consulting in Companies in Creative Industries. *Management Issues – Problemy ZarzÈdzania*, **16**, 1 (2018)

2. J. Cerneviciute, R. Strazdas, Teamwork management in creative industries: Factors influencing productivity. *Enterpreunership and Sustainbility Issues*, **6**, 503 – 516 (2018)

3. T. O. Kvålseth, Relationship between concentration ratio and Herfindahl-Hirschman index: A re-examination based on majorization theory. *Heliyon*, **4**, 1-24 (2018)

4. A. Krivka, On the Concept of Market Concentration, the Minimum Herfindahl-Hirschman Index, and Its Practical Application. *Panoeconomicus*, **63**, 525-540 (2016)

5. B. Cnossen, E. Loots, A. van Witteloostuijn, Individual motivation among entrepreneurs in the creative and cultural industries: A self-determination perspective. *Creativity and Innovation Management*, **28**, 389 – 402 (2019)

6. D. Games, R. P. Rendi, The effects of knowledge management and risk taking on SME financial performance in creative industries in an emerging market: the mediating effect of innovation outcomes. *Journal of Global Entrepreneurship Research*, **9**, 1-14 (2019)

7. Y. Wijngaarden, E. Hitters, P.V. Bhansing, 'Innovation is a dirty word': contesting innovation in the creative industries. *International Journal of Cultural Policy*, **25**, 392-405 (2019)

8. P. Campbell, D. O'Brien, M. Taylor, Cultural Engagement and the Economic Performance of the Cultural and Creative Industries: An Occupational Critique. Sociology – *The Journal of the British Sociological Association*, **53**, 347-367 (2019)

9. J. Radomska, P. Wolczek, L. Soloducho-Pelc, S. Silva, The Impact of Trust on the Approach to Management: A Case Study of Creative Industries. *Sustainability*, **11**, 816 (2019)

10. I. Hermanson, M. McKelvey, O. Zaring, The evolution and embeddedness of knowledge-intensive entrepreneurial firms in creative industries: contrasting experienced and non-experienced entrepreneurs in the Swedish fashion industry. European Planning Studies, 26, 2387-2406 (2018)

11. A. Drab-Kurowska, State intervention in the context of creative industries. Management – Poland, **22**, 74-84 (2018)

12. N. Boccella, I. Salerno, Creative Economy, Cultural Industries and Local Development. *Procedia - Social and Behavioral Sciences*, **223**, 291-296 (2016)

13. E. Martinaitytė, R. Kregždaitė, The factors of creative industries development in nowadays stage. *Economics and Sociology*, **8**, 1 (2015)

14. S. Cunningham, Hidden Innovation: Policy, Industry and the Creative Sector. Queensland: University of Queensland Press (2013)

15. UNCTAD. Creative Economy report, online at https://unctad.org/en/docs/ditc20082cer_en.pdf (2008)

16. M. Kavčáková, Hodnotenie koncentrácie odvetvia pomocou Herfindahlovho-Hirschmanovho indexu a jeho komparácia s ostatnými ukazovateľmi koncentrácie odvetvia. *Finančné trhy*, **4**, (2018).

17. P. Svoboda. Analysis of market concentration in selected sectors of public procurement. Acta Universitatis *Agriculturae et Silviculturae Mendelianae Brunensis*, **64**, 3 (2016)

18. I. Brezina, J. Pekár, Z. Čičková, Zmeny v koncentrácii slovenského bankového sektora. *Finančný manažér*. **12**, 3 (2012)

19. M. Šindelář, The analysis of market concentration of audit services in the Czech Republic. European Financial Systems 2015. Proceedings of the 12th International Scientific Conference, Brno: Masaryk University, 618-624, (2015)

Dynamic Models of the Creative Industries' Behaviour

Mária Kozáková[1,*], Kristína Krúpová[2]

[1]University of Economics, Faculty of Business Management, Department of Business Economy, Dolnozemská cesta 1, 852 35 Bratislava, Slovakia

[2]University of Economics, Faculty of International Relations, Department of International Political Relations, Dolnozemská cesta 1, 852 35 Bratislava, Slovakia

Abstract. The aim of the paper is to analyse the models describing the development of the Creative Industries in Slovakia. Creative industries are described as the industrial components of the economy in which creativity is an input and content or intellectual property is the output. The creative industries have therefore appeared to be newly represented as a significant and rapidly growing set of industries; an important sector, in other words, for policy consideration. Based on the following findings, we can conclude that the second model is precisely predicting the relationship between the growth in the creative industries and in the aggregate economy in Slovakia. With improved cultural statistics, also a more developed and theoretically better founded analysis would be possible. We therefore see our article primarily as a much-needed step towards developing statistical tools in empirical cultural policy on a consistent basis.

Introduction

According to the literature preview, there is a lack of models explaining the economic value and impact of the creative industries. Indeed, in times of financial and economic crisis, these industries become even more attractive as a source of potential employment and entrepreneurial endeavour. Output and value added, and clearly the main instrument of economic measurement for production is GDP. The contribution of culture to this global measure, i.e., the value added of culture industries, is therefore the main economic indicator often mentioned by the various authors [1-3].

In static models, the time factor is not explicitly taken into account. However, such models are best adapted for the solution of certain tasks. For example, in order to conduct a statics analysis, it is necessary to lay the foundation for the dynamic analysis [4,5]. A static state is such state to which the principles of statics are applicable. Static state is not the state of a "timeless" nature, and the static economy is not the "timeless" economy, but it will be an economy in which there is no "temporary order".

Static models possess property of zero lags. The last means that the value of the output parameter of a static model of t represents at any moment a certain function of value

* Corresponding author: maria.kozakova@euba.sk

of its output parameter in the same time point of t and does not depend on values of input parameter to t time point. Static models are set in the form of one-dimensional and multidimensional algebraic functions. The models which are directly considering time factor are usually called dynamic. In such models all variables of economic processes and systems are functions of time. Examples of dynamic models are equilibrium processes by Walras and the interaction of supply and demand by Marshall. Economic systems possess property of a lag effect [6,7]. This property of system is characterized by the speed of change of its output parameters in response to changes of its input parameters and parameters of its functioning, average time of receiving a result when modifying functioning parameters.

We start by describing the two possible fundamental relations between the creative industries and the rest of the economy. We interpret the relation of each model to different theoretical associations, what we should expect to observe if each model were true, and the appropriate policy framework in each case. Using various mapping documents, we then undertake an initial consideration of a set of data samples connected to the two models. However, this paper does not attempt a complete analysis. For that, a much more rigorous approach to modelling, data and statistical analysis would be required. What we aim to provide here is only a theory of the classes of models involved and an illustration of how different sorts of data might be applied to them. This seems to us a necessary first step (prior to a more rigorous approach) in developing the economics of the creative industries [8,9].

This is a useful starting point because it allows us to focus attention on the relational nature of processes of value allocation and cultural validation: cultural value does not operate and is not generated in a social, cultural and political vacuum, but is in fact shaped by the power relations predominant at any one time, and is a site for struggles over meaning, representation and recognition [10,11].

2 Methods

The chapter provides the information about research aim, the precise description of the research object and the methodology used in this paper.

2.1 Research Aim

The aim of the paper is to propose two dynamic models describing the behaviour represented by development and growth of the Creative Industries in Slovakia. The following research questions were articulated:

- Research question no. 1: What is the dynamic relation between the creative industries' prosperity and the prosperity created by the rest of the Slovak economy?
- Research question no. 2: What is the dynamic relation between the creative industries' growth and the rest of the Slovak economy growth?

2.2 Object of the Research

Sectors of the creative industries are formed by the business activities based on individual creativity, skill and talent. They are also the ones that have the potential for capital formation and job creation through the use of intellectual property. The sectors as advertising, architecture, art, antique shops, computer and video games, crafts, design, fashion design, film and video, music, performing arts, publishing, software, broadcasting channels and radio belong to the Creative Industries. These areas are mainly related to the high added value and significant impact on the quality of people's lives.

The notion of creative industry implies a number of areas (mainly the art of science and, last but not least, culture) and at the same time it affects the social demand for a change in the perception of the concept of industry, which is still understood only in terms of pure production characteristics. In this context, industry must be seen as an exploitation chain that is already created by creating the conditions for creative activity then includes creative activity, production, reproduction, presentation, representation, including export, storage, archiving and recovery.

2.3 Methodology

The economic value of the whole economy in Slovakia is defined as Y, and the economic value of the creative industries as CI.

In the first model (so called welfare model), the creative industries are hypothesized to have a net negative impact on the economy, such that they consume more resources than they produce. A dynamically equivalent statement is that the rate of total factor productivity (TFPCI) growth is less in the creative industries than in other sectors (TFPY). In this model, the creative industries are essentially a 'merit good' sector that produces cultural commodities that are welfare enhancing ($dU/dCI > 0$), but that are only economically viable with a transfer of resources from the rest of the economy ($dY/dCI < 0$).

The model is described by the two following formulas:

$$(dU/dCI) > 0 \tag{1}$$

$$(dY/dCI) < 0 \tag{2}$$

In first model, the creative industries are a net drain on the economy, although a net drain worth having, as the overall effect is welfare positive. This is due to the production of commodities of high cultural value ($dU/dCI > 0$) but low market value ($dY/dCI) < 0$), as production is inherently unprofitable because demand curves lies everywhere below cost curves. The economic justification for such restitution must ultimately then rest on a market failure argument, with policy appropriately calibrated to estimates of this nonmarket value.

Specifically, it means that growth in the creative industries comes at the cost of aggregate economic growth, as their growth is not what 'the market' wants but must be compelled to support it through transfers. Evidence for first model may therefore accrue in several ways. These include: high levels and rates of negative profit among creative industries firms; low total factor productivity (TFPCI < TFPY); persistently lower income to factors of production in creative industries compared to other industries; and other indications that the economic viability of activities organizations within the creative industries is critically dependent upon resource transfers from the rest of the economy to maintain prices, demand or supply.

If first model is true, we would expect to observe not just an economically stagnant or low-growth sector, but also one with lower performance levels (e.g. return on investment, incomes, etc). Such decomposition allows multiple opportunities for empirical assessment. This is an interesting empirical question to re-ask, for the implicit truth of first model is almost axiomatic in the field of cultural economics where few disputes the implicit assumption of below-average income or productivity growth.

Our starting axiom is that change in CI affects GDP (Y) in some way ($\Delta CI \leftrightarrow \Delta Y$). The two models proposed are the set of possible dynamic interrelations in which a change in CI activity has either: a negative (first model) or positive (second model) effect on the economy. This is plainly simplistic, yet we suggest it offers a useful starting point to orient both empirical analysis and policy discussion in order to be clear and explicit about this hypothesized relation and its evidential support.

For analytic convenience, we also assume that $d\mathrm{CI}/d\mathrm{Y}= 0$, meaning that economic growth affects the creative industries no differently to other industries, or technically that income elasticity is unitary. We strongly suspect this not to be the case, but rather that growth in income disproportionately effects demand for the output of the creative industries, but we shall set that aside here. Policy is analysed in terms of whether change in the creative industries changes aggregate utility welfare (or utility, U). Again, we presume $d\mathrm{U}/d\mathrm{CI}$ can increase, decrease or leave utility unaffected.

The second model explicitly proposes a positive economic relation between growth in the creative industries and growth in the aggregate economy, such that $d\mathrm{Y}/d\mathrm{CI} > 0$. In this model the creative industries are a growth 'driver' in the same way that agriculture was in the early 20th century, elaborately transformed manufacturing was in the 1950s–60s, and Information and Communication Technology was through the 1980s–90s. There are many possible explanations, but all are some variation upon either the notion that the creative industries introduce novel ideas into the economy that then percolate to other sectors (e.g. design-led innovation), or that the creative industries facilitate the adoption and retention of new ideas or technologies in other sectors.

Involvement of the creative industries in the growth of the economy can occur in two principle ways: supply-side and demand-side. The supply-side interpretation of this model emphasizes the export of new ideas from CI to Y. The demand side interpretation emphasizes how growth in Y causes a proportionate increase in demand for CI services. In practice, it is extremely difficult to separate these two forces without recourse to advanced micro-econometric techniques, which are not attempted here due to data limitations. This model may therefore be true, but with different policy implications depending upon whether causality runs predominantly from CI to Y – the supply-side growth driver model – or from Y to CI – the demand side induction model.

The second model is described by the two following formulas:

$$(d\mathrm{Y}/d\mathrm{CI}) > 0 \tag{3}$$

$$(d\mathrm{U}/d\mathrm{CI}) \geq 0 \tag{4}$$

If the second model is true, then there is a clear economic case for redirecting resources, not just for the benefit of the creative industries per se, but for the benefit of all. The creative industries, in this view, are clear winners to be backed. Evidence for this model would come from association of the creative industries with growth. This would accrue not just in jobs and commodities, but in new types of jobs and new sorts of commodities and services. The second model proposes the creative industries as growth drivers not because of operational expenditure multipliers, but due to their role in the adoption, retention and absorption of new ideas and technologies.

3 Results and Discussion

The creative industries would be therefore considered to create new industries and market niches and to sustain and strengthen extant industries. And specifically, without such advanced investment, aggregate economic growth would suffer. This is the contrary of model 1, in which economic growth suffers when there is such continued investment. Model 2 then argues that the creative industries are good for the economy because they introduce and process the new ideas that drive economic growth.

Figure 1 shows the development of the GDP as the main indicator of the economic growth in Slovakia. In 2014, GDP growth was 2.4% and in 2015 and 2016 Slovakia's economy grew 3.6% and 3.3% respectively. The highest GDP was reached in 2019.

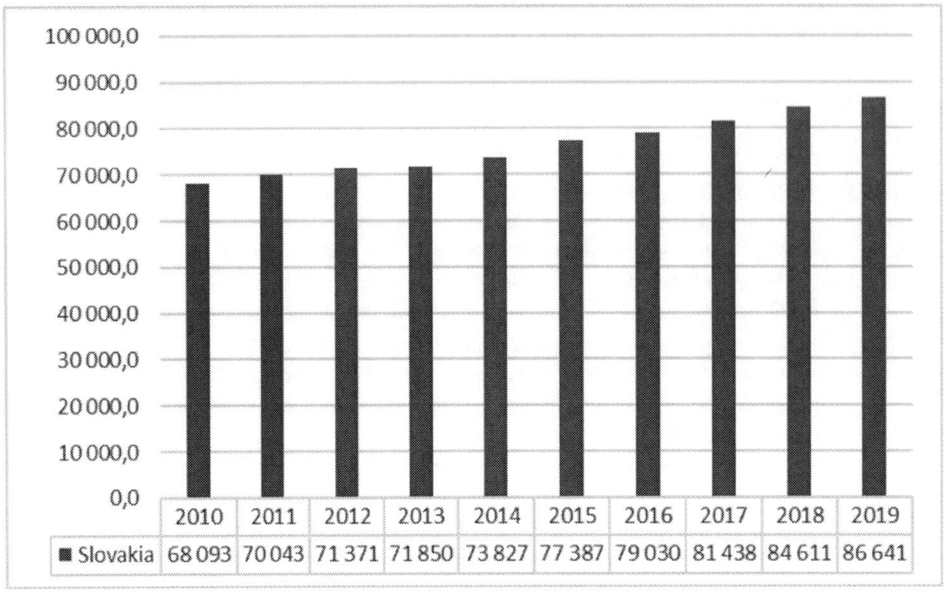

Fig. 1. GDP of Slovakia in chain linked volumes (2010), million euro

Table 1 contains the results of our analysis and data modelling [12, 13]. The first model related to the welfare, the creative industries are essentially a 'merit good' sector that produces cultural commodities that are welfare enhancing ($dU/dCI > 0$), but that are only economically viable with a transfer of resources from the rest of the economy ($dY/dCI < 0$). The first model is not valid in Slovakia. With improved and continual cultural statistics, also a more developed analysis would be possible.

Tab. 1. The comparison of the model 1 and the model 2

Year	Y	U	CI	*d*Y/*d*CI	*d*U/*d*CI
2010	68 093,0	418,3	9 777	0,0005127	0,031982942
2011	70 043,4	433,3	10 246	0,0007531	-0,119220056
2012	71 371,3	476,1	9 887	uncountable	uncountable
2013	71 850,1	469,2	unavailable	uncountable	uncountable
2014	73 827,8	457,4	unavailable	uncountable	uncountable
2015	77 387,7	452,6	11 690	0,0006086	-0,013011152
2016	79 030,9	438,6	12 766	0,0004154	0,072202591
2017	81 438,4	499,9	13 615	0,0003151	uncountable
2018	84 611,9	unavailable	13 789	0,0004928	uncountable
2019	86 641,3	unavailable	14 115	0,023303	uncountable

The second model is based on the premise that (dY/dCI) > 0 and at the same (dU/dCI) ≥ 0; and proposes a positive economic relation between growth in the creative industries and growth in the aggregate economy. Based on the results of our research, we can conclude that according to the premises of the second model, the second model is predicting the relationship between the growth in the creative industries and in the aggregate economy in Slovakia. In years 2011 and 2015, *d*U/*d*CI was almost the null, so we count it as the null for our purpose. The limitation of this research is in the missing data from few years.

.4 Conclusion

Policymakers must acknowledge the conceptual and empirical challenges in defining and measuring the creative economy. Analysts should reflect on the different approaches that have been adopted in different studies and develop a clear understanding of why transparent definition and measurement is important for policy-making purposes. There is no better way to engage with the issues than looking at some data.

The first model mentioned in the paper was not confirmed and it is not valid in Slovakia. The second model linked to the growth of the creative industries and the rest of the economy is predicting the dynamic behaviour of the creative industries' entities. Yet in both cases, policy should properly treat the creative industries as a 'special sector'[14,15]. This is not because it is economically significant, but because it powers the growth of other sectors. This may plausibly lead to intervention, but unlike model 1, the ostensible purpose of this is to invest in economic growth, or to invest in the development of capacity to meet growth in demand.

There are several issues open for further research. Firstly, some dimensions of culture that could be included are at present not included in the model 1 and 2, such as existing legislation in culture. Finally, the analysis should be broader in terms of time dimension and accuracy of data, but we are unfortunately limited in this aspect with existing cultural statistical data. With improved cultural statistics, also a more developed and theoretically better founded analysis would be possible. We therefore see our article primarily as a much-needed step towards developing statistical tools in empirical cultural policy on a consistent basis, hoping to stimulate research, including the verification of our findings.

This paper is the output of the scientific grant VEGA n. 1/0340/19 "The Entrepreneurial Dimension of Creative industries in the Context of Innovation and Smart Growth" (50 %) and the internal grant project of the University of Economics in Bratislava (Faculty of Business Management) no. I-20-101-00 called "Research of Creative Industry Enterprises with regard to Their Life Cycle in the Slovak Republic" (50 %).

References

1. D. Rybárová, Creative Industry as a Key Creative Component of the Slovak Economy. In *Globalization and its Socio-Economic Consequences 2019* (Rajecké Teplice, Slovakia, 1-8, 2019)
2. A. Srakar, V. Čopič, M. Verbič, European cultural statistics in a comparative perspective: index of economic and social condition of culture for the EU countries. *Journal of Econometrics*, **42**, 163-199 (2018)
3. S. Šagátová, The Impact of globalization on the development of creative industries businesses. In *Globalization and its Socio-Economic Consequences 2019* (Rajecké Teplice, Slovakia, 1-8, 2019)
4. M. Bertanha, Regression discontinuity design with many thresholds. *Journal of Econometrics*, **218**, 216-241 (2020)
5. L. Bauwens, E. Otranto, Nonlinearities and regimes in conditional correlations with different dynamics. *Journal of Econometrics*, **217**, 496-522 (2019)
6. S. Cerisola, Creativity and local economic development: the role of synergy among different talents. *Papers in Regional Science*, **97**, 199 – 209 (2018)

7. A. Lemmetyinen, Entrepreneurship in culture and creative industries: perspectives from companies and regions. *International Small Business Journal – Researching Entrepreneurship*, **37**, 417-418 (2019)
8. Y. Wijngaarden, E. Hitters, P.V. Bhansing, 'Innovation is a dirty word': contesting innovation in the creative industries. *International Journal of Cultural Policy*, **25**, 392-405 (2019)
9. W. Yu, Creative industries agglomeration and entrepreneurship in China: necessity or opportunity? *Industry and Innovation*, **27**, 420-443 (2020)
10. XQ. Li, D. Gagliardi, I. Miles, Innovation in R&D service firms: evidence from the UK. *Technology Analysis & Strategic Management*, **31**, 732-748 (2019)
11. E. Belfiore, Whose cultural value? Representation, power and creative industries. *International Journal of Cultural Policy*, **26**, 383-397 (2020)
12. Eurostat, (2020, 24 March). GDP and main components (output, expenditure and income). Retrieved from https://appsso.eurostat.ec.europa.eu/nui/show.do?dataset=nama_10_gdp&lang=en.
13. Eurostat, (2020, 24 March). Value added and turnover of enterprises in the cultural sectors by NACE Rev. 2 activity. Retrieved from https://appsso.eurostat.ec.europa.eu/nui/submitViewTableAction.do.
14. F. Angelini, M. Castellani, Cultural and economic value: a critical review. *Journal of Cultural Economics*, **43**, 173-188 (2019)
15. R. Boix, B. De-Miguel-Molina, J. L. Hervas-Oliver, Creative service business and regional performance: evidence for the European regions. *Service Business*, **7**, 381 – 398 (2013)

Analysis of Reporting Period Flexibility

Zuzana Kubaščíková[1,*], *Zuzana Juhászová*[2], and *Miloš Tumpach*[3]

[1,2,3] University of Economics in Bratislava, Faculty of Business Informatics, Department of Accounting and Auditing, Dolnozemská 1, Bratislava, Slovakia

Abstract. Entities take advantage of the possibility of changing the reporting period (to other than the calendar year) because such a change has a rational basis, usually regarding the specific type of business of a particular entity. They are therefore entities that are seasonal in nature respective to their economic activities. As part of the analysis, of the use of the alternative reporting period by the Slovak accounting entities, we present in addition to the number of accounting entities also their area of business to substantiate the reasons why the entity lead to change in the reporting period.

1 Introduction

The option of voluntary choice of the reporting period, as allowed by current accounting acts and accounting standards, allows entities to choose a financial year other than the calendar year as the reporting period that follows the production process of the entity and thus takes into account the specificities of individual industries and adapts various factors [1,2].

Farmers, foresters, mountain operators, guest houses, hotels, ski resorts, children's parks, indoor pools are largely dependent on a temperate climate zone and it may not always be economically profitable to end the reporting period on 31 December. Taking into account, for example, large shopping chains such as Kaufland, Hypernova and Tesco, for such entities that are busy during Christmas and are preparing their upcoming season in full swing, it is disadvantageous and impractical for them to close their accounts as of December 31, and thus in this hectic period they have to perform reconciliation and closing operations and all the tasks necessary to prepare and publish their financial statements.

One of the possible reasons for changing the reporting period to the financial year other than the calendar year may be the entity's efforts to save its cash for the external processing of financial statements and tax returns, as external entities providing such services are very busy at year-end and higher than in the second half of the year [3,4,5]. Entities may negotiate a more advantageous price with external audit entities because it is not the main business season of the audit entity, allowing the entity to calculate with better, more efficient services, including better staffing and external audit entity access, which immediately leads to saving entity's funds [6,7,8].

* Corresponding author: zuzana.kubascikova@euba.sk

2 Methods and Data

For testing and evaluating data from financial statements we used database available on web page www.registeruz.sk. Financial statements for years 2018-2011 were processed.

Table 1. Research sample

Year	No. of financial statements
2018	219173
2017	208228
2016	198414
2015	190486
2014	186348
2013	166269
2012	90451
2011	96535

Various scientific research methods were applied in writing this paper. Predominant were methods of analysis, synthesis and comparison. Based on the analysis, we have applied theoretical information to problems related to reporting periods from various sources. Then we focused on the target group – accounting entities with other reporting periods than a calendar year, the process of changing the reporting period, the types of accounting entities which use different reporting periods. We used the induction method and a generalization, to characterize the beneficiaries when changing the reporting period and to characterize the advantages and disadvantages in which the entity learned about the alternative choice for the reporting period.

3 Results and Discussion

The decision to change the reporting period is taken by economic entities because they can take advantage of such an alternative and the benefits of choosing a financial year as this reporting period outweigh the negatives for the accounting entity.

The following may be considered as advantages of the financial year:

- the financial year is used by those entities with a seasonal dimension to their activities,

- there is no need to prepare different financial statements for the parent company if parent company is using a reporting period other than a calendar year,

- more accurate and truthful business profit calculations for entities of a seasonal nature,

- an effective and fair assessment of the entity's overall business for entities with seasonal activities based on financial and operating information that is consistent with the entity's actual performance,

- lower fees for audit services,

- payment of dividends and shares that better correspond to the entity's actual performance (in seasonal accounting entities),

- fewer errors in the closing operations of the seasonal accounting entities,

- maintaining a true and fair view of the facts that are the subject of the entity's financial statements and providing truthful information about the entity to investors.

The following may be considered as disadvantages of the financial year:

- it is not so easy to compare financial indicators with other entities from different sectors,

- sometimes there are problems in changing settings in entity information systems.

Entities in the territory of the Slovak Republic may choose their reporting period - calendar year or financial year according to the Accounting Act. How many accounting entities choose the possibility of changing the reporting period we have found in the following graph: We have chosen the years 2011 – 2018 because all the data for 2019 were not yet available (it will only be available in the autumn 2020). When processing the data, we worked with a financial statement published on the website www.registeruz.sk.

Graph 1. Number of accounting entities with another reporting period than calendar year.

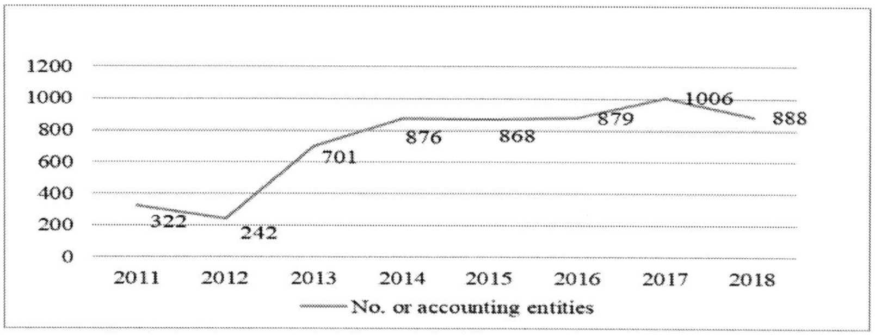

The share of accounting entities that use the fiscal year is extremely low in Slovakia, it is approximately 4.5% of accounting entities, in the years 2011-2012 it is only 2.5% -3.5% of entities.

The graph with the number of accounting entities using the fiscal year by type and ownership shows that the fiscal year is used more by international and foreign companies.

Graph 2. Number of accounting entities with another reporting period than calendar year according ownership.

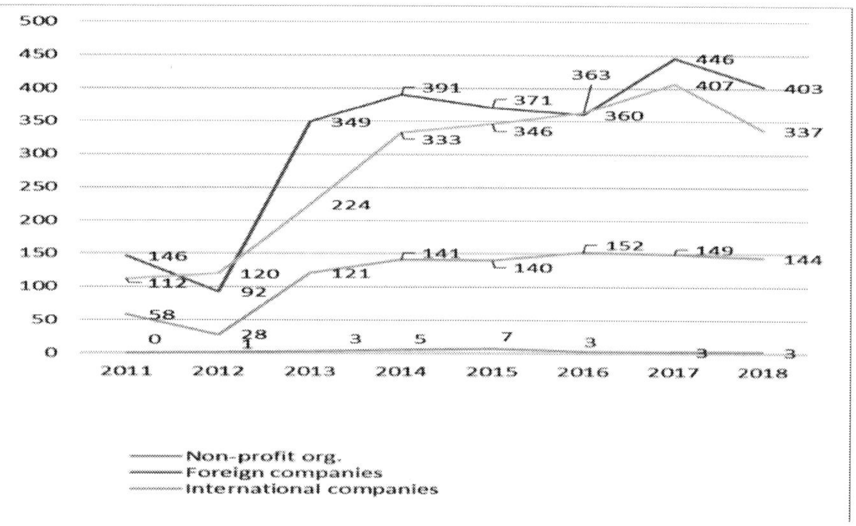

The following table shows the numbers of entities by industry for 2018, the results in individual years are similar. In each sector, the number of such entities is below 1% of the total. Only a slightly higher share is in the automotive industry (2.9%), mechanical engineering (1.26%), electrical engineering (1.81%), chemistry and plastics (1.06%).

Table 2. Use the alternatives of the reporting period by Slovak accounting entities by industry

Industry	No. of accounting entities	% from of all entities
Car industry	9	2,90
Tourism and gastronomy	21	0,16
Transport and logistics	19	0,21
Wood and paper	6	0,25
Electrical engineering	26	1,81
Energy and mining	4	0,31
Finance	28	0,55
Chemistry and plastics	12	1,06
Information technologies	64	2,88
Metal productinon and metallurgy	26	0,59
Retail	72	0,37
Media, publishing and culture	8	0,2
Properties	48	0,31
Clothing and footwear	3	0,24
Agriculture and forestry	6	0,11
Food industry	13	0,63
Law, consulting and accounting	114	0,56
Sale and maintenance of vehicles	13	0,33
Design and engineering	25	0,35
Advertisment	17	0,34
Services	35	0,23
Waste treatment	1	0,12
Brokerage services	25	0,27
Construction industry	56	0,23
Engineering	26	1,24
Education and training	5	0,19
Telecommunications	5	0,75
Wholesale industry	118	0,71
Other production	1	0,11
Development and testing	12	0,4
Health	70	0,71
Total	**888**	

One sector that may benefit from setting a reporting period other than the calendar year is the agricultural sector, whose overall activities, and thus profitability, is due to the rotation of the seasons. From a cost perspective, entities have higher costs during the sowing period, and then the harvest and selling of crops may arise in the following reporting period. The agricultural sector is a sector exposed to high risks and uncertainties that are not within the competence of an entity that is unable to accurately determine the evolution of the climate, weather conditions and also changes in prices of agricultural products and commodities on the market. The stages and phases of the industry are subject to a cycle of alternating winter, spring, summer and autumn and corresponding climatic conditions.

If we perceive this topic historically, the state (Slovak Republic) has strictly mandated the calendar year beginning on 1 January and ending on 31 December as the reporting year, but taking into account the alternation of seasons and their impact on the agricultural sector, such behavior is disadvantageous for the state because the economic indicator of gross domestic product does not contain all the appropriate and profitable data necessary to determine the desired level of the macroeconomic indicator of the state for the reporting period. In some economically undeveloped countries, economic production has the strongest impact on a country's economy, so such countries benefit from the flexibility of the reporting period in their favor to increase their gross domestic product.

The attempt to establish a financial year for the agricultural sector implements the requirement to 'count' the sales phase into the reporting year. If an entity wants to meet such a condition, its production process must be less than one year and the products can be resold relatively quickly, most often to animal production for further processing, but also to other industries. In such a specialized holding, it is appropriate to set the beginning of the reporting period at a time when stocks are at a minimum because the minimum stock level reduces the risk of, for example, incorrect inventory counting and inventory valuation. Based on the climatic conditions of our geographical environment the financial year should start on 1 July and end on 30 June. However, taking into account the smaller farmers' business as a private entrepreneur, the start of the 1 July reporting period is economically disadvantageous for them because of the lack of time to carry out the seasonal farmer's work. It is necessary to optimally determine the beginning of the reporting period for the period when the results of the production processes of the previous reporting period can be realistically predicted. For example, if an entity is engaged in the production of oilseed rape as one of the types of winter crops that are sown in the fall, it is convenient for the entity to begin its reporting period on 1 October and end on 30 September the following year.

If an entity chooses an optimal reporting period, it will not only eliminate the risk of miscalculation of its own inventory but if prices fall, the entity must take this into account adjusting its inventory to reduce its profit or loss. Furthermore, it eliminates the risk of unsettled receivables for which the company creates provisions, recognizing in its financial statements the economic benefits of such receivables, which leads to the sustainability of the entity's cash liquidity.

For entities deepening during economic recession, there is an effort to eliminate as many risks as possible from the macroeconomic but also the microeconomic environment. There are a wide variety of tools and methods for such elimination. One of the options is to eliminate risks through the use of accounting policy, by choosing the appropriate reporting period to eliminate accounting and tax risks.

The alternative choice or reporting period cannot be used by entities that are public entities operating under budgetary rules, due to the linking of their budgets and cash flow to the state budget, the budget of municipalities and higher territorial units, which are compiled according to the calendar period. Those that carry on business or perform other self-employed activity can change a reporting period but they should consider the reasons and consequences of that decision. Changing the reporting period may increase but also decrease the usefulness of the information provided by the financial statements. 'A change in the reporting period is justified only if the information, once carried out, leads to greater usefulness and a true and fair view of the business.'

The reasons for applying changes in the reporting period can vary from the need to adapt to the production cycle, through adherence to the accrual principle, to the reconciliation of the reporting period of the subsidiary with the parent company within the consolidated group. Such a change has the advantage of eliminating various risks in terms of the inaccuracy of the data provided in the consolidated financial statements of the consolidated group. Under Slovak legislation, companies that were part of the consolidated group until 2004 had to prepare their financial statements at the end of the calendar year, which however did not correspond to the reporting period of the whole group. Therefore, in many cases, two accounting systems were applied in one accounting entity, one for the purpose of act on accounting and the other for consolidated decision-making purposes. However, such measures greatly complicated the functioning of accounting entities.

In companies with a specific length of production process, there is a problem with allocating revenues to costs. For example, taking into account the construction industry, their costs outweigh the revenues that an entity achieves in a given year or in subsequent years or revenues are based on estimation.x

For entities that decide to change the reporting period it also impacts on the tax period, and therefore entities decide to change their reporting period based on the needs they want to meet.

However, a change in the reporting period can be used to the detriment of the state. Therefore, if an entity which hasmade high profits and is one of the largest entities and whose income tax accounts for a significant portion of its revenue to the state budget, decides to take advantage of the change in reporting period from calendar year to business year to manipulate the amount of income tax paid , and therefore part of the profits would be taxed at the 'old' rate of income tax, the State would reduce by a significant proportion of the income tax received and paid.

4 Conclusion

Any company should be aware of the consequences of the use of different reporting periods. This study aims to throw some light on the relevance of choosing a specific reporting period for accounting entities in a European country. However, our analysis indicates that there are some significant advantages when using different income reporting periods to the number of entities using the reporting period another than financial year in Slovakia is still low and no significant year-on-year increase or decrease was recorded.

This paper concludes therefore that, entities should choose a financial year as their reporting period rationally when it comes to entities that are characterized by seasonality – agriculture and construction sector, services, entertainment. In an effort to faithfully and truthfully present accounting data, accounting entities change the reporting period as one of the possible tools for eliminating the risks of failing a true and fair view.

Acknowledgments

This article is an output of the project APVV no. APVV-16-0602 "Enhancement of the relevance of the accounting data in the SR - from expenses to value".

References

1. Andrejovská, A., Zhodnotenie ekonomickej situácie v obciach pomocou pomerových a absolútnych ukazovateľov. *Verejná správa a spoločnosť : [vedecký časopis]*. - Košice : Fakulta verejnej správy Univerzity Pavla Jozefa Šafárika, roč. 13, č. 1, pp. 102-108 (2012)
2. Canto, O., Gradin, C., Del Rio, C., Poverty statics and dynamics: does the accounting period matter? *International journal of social welfare*, vol. 15, issue. 3, pp. 209-218 (2006)
3. Krišková, P., Užík, J., Management of audit risk and its impact on the audit of financial statements. *Managing and modelling of financial risks : proceedings : 8th international scientific conference : 5th - 6th september 2016, Ostrava, Czech Republic*, pp. 460-466 (2016)
4. Lehenchuk, S. F., Valinkevych, N. V., Vyhivska, Irina M., Khomenko, H. Y. The Significant Principles Of Development Of Accounting Support For Innovative Enterprise Financing. *International Journal of Advanced Science and Technology, 29*(8s), pp. 2282 – 2289 (2020)
5. Miller, H., Accounting period concept and its effect on management decisions – discussion. Journal of accounting research, vol. 4, supplement S, pp. 15-17 (1966)
6. Ondrušová, L., Management decisions leading to ownership transactions. *Strategic management : international journal of strategic management and decision support systems in strategic management,* vol. 16, no. 3, pp. 18-22 (2011)
7. Pakšiová, R., Janhuba, M., *Teória účtovníctva v kontexte svetového vývoja.* Bratislava: Iura Edition. (2012)
8. Parajka, B. Are information needs of financial entities served by financial statements in the Slovak Republic. *Financial management of firms and financial institutions*, pp. 950-957 (2015)

Characteristics of expert´s report as evidence

Milan Kubica[1*], *Nikola* Švejdová[2**]

[1] Department of Business Economy, Faculty of Business Management, University of Economics in Bratislava, Dolnozemská cesta 1/b, 852 35 Bratislava, Slovakia
[2] CORPORA, a.s., Záhradnícka 68, 821 08 Bratislava, Slovakia

Abstract. In recent years, there has been an increasing need for private expert´s report for judicial evidence. In practise, it appears that a well-developed expert´s report is an important bases for the court´s decision-making. It is no secret that the quality of the expert´s reports presented varies. Therefore, when expert drawing up his expert´s report, either directly for the court or for the parties, the expert should avoid mistakes and pay attention to the quality of the expert´s report, which should have certain characteristics that we address in this paper.

1 Introduction

In recent years, there has been an increasing need for private expertise for judicial evidence. In practice, it appears that a well-prepared expert report is an important basis for the court's decision-making. It is no secret that the quality of the expert opinions presented varies. Therefore, when drawing up his expert's report, whether drawn up directly for the court or for the parties, the expert should avoid errors [1] and pay attention to the quality of the expert's report [2], which should have certain characteristics that we address in this paper.

2 State of the problem at home and abroad

2.1 Expert as a perpetrator of a crime

The expert may be the perpetrator of the offense by giving **false expert opinion**. The facts of this offense are contained in § 347 of the Criminal Code.

If an expert before a court, prosecutor or police in criminal proceedings, before a court in civil or execution proceedings, or before a public authority, or before an arbitration tribunal, declares an untrue circumstance that is essential to the decision or such circumstance conceals, or causes minor damage to another, by misrepresenting circumstances that are material to the person to whom the expert report relates, or material to the decision underlying the expert report, or to conceal such circumstance, will be punished by imprisonment for one to five years.

* Corresponding author: mkubica@euba.sk
** Corresponding author: svejdova@corpora.sk

The expert will be punished by imprisonment for three to eight years if he commits the offense in a more serious manner or on a specific motive.

The expert shall be liable to a term of imprisonment of four to ten years if he commits the act in question and causes considerable damage or any other, extremely serious consequence.

The features of this offense will not be fulfilled not only when the expert in court, the prosecutor or the investigator in the criminal proceedings declares the facts which are essential for the decision, but even if they conceal such circumstances. The expert may commit this criminal offense both by action and omission of his obligation arising from his position in criminal proceedings.

The offender may only be an expert in connection with the performance of his duties. The perpetration of the offense is deliberate, and the perpetrator's intention must also apply to the fact that it is a circumstance which is essential to the decision. This factual nature of the offense protects the interest in the correct establishment of the facts as a basis for the legal decisions of courts, prosecutors, investigators or police authorities. [3]

2.2 Mistakes in expert´s acts

Article 146 of the Code of Criminal Procedure deals with the errors of a written expert act in criminal proceedings: "If doubts arise as to the accuracy of the expert report or if the expert report is unclear or incomplete, the expert must be asked to explain or complete the report. If this would not remove the doubts or uncertainties of the expert opinion or the completeness of the expert opinion, another expert shall be added."

Errors in expert activities can be divided into [4]:
- formally,
- methodological and content.

The most common **formal errors** in the submitted written expert opinion occur due to failure to comply with the formal requirements laid down in Act No. 382/2004 Coll. and in Decree no. 228/2018 Coll. These are:
- the individual pages of the expert report are not numbered or tied and stapled; the free ends of the line are not obscured by a sticker or not stamped with an official stamp,
- lacks any of the particulars of the title page,
- the composition is not observed (valid only in written expert opinion),
- missing expert's signature,
- missing official stamp or non-blue stamp,
- the date or place of the expert's work is not stated,
- there is no expert clause,
- the wording of the expert clause is not within the meaning of the legal standard,
- the expert clause is not signed by the expert or the person responsible for the expert activity (legal entities),
- the number of the expert act does not correspond to the number recorded in the expert's journal.

An expert act shall be considered to be **methodologically incorrect** if:

- is based on incomplete or technically unacceptable documents and the expert did not draw attention to this fact and did not ask the contracting authority to supplement the documents,
- the expert did not take into account all the facts that were relevant to the opinion, but selected only some and with others, e.g. he did not deal with contradictions,
- the expert unambiguously solved the expert tasks, although the range of input values or the level of scientific development of the given field does not allow such unambiguous solutions,
- the expert is unauthorized in dealing with legal issues or pre-settles legal issues unequivocally and fails to draw attention to the possibility of another alternative,
- the opinion is ambiguous or unclear, the exact opinion of the expert on the issue is not obvious,
- the report is incomplete - the expert did not provide answers to all questions in the expert's report; the answers to the questions are incomplete,
- the expert based on illegally obtained documents,
- when drawing up the report, the expert is based on knowledge that no longer corresponds to the given scientific level of knowledge at the time of submitting the report;
- in carrying out the task, the expert is based on an incorrect or invalid regulation for the date of the assessment;
- the expert did not approach the examined issue systemically and did not consider the properties of all elements and their interaction,
- the expert has prepared an report/supplement in the field/sector in which he is not registered, the expert has also prepared answers to those questions (tasks) that belong to a field/sector in which he is not registered,
- the expert in the expert act solved and provided data which he did not have in the task of expert.

2.3 Assessment of expert report as evidence

The objective of rating an expert report as evidence is to assess it in terms of factual, formal and content correctness [5].

Rate of ***factual accuracy*** of an expert report represents a rating of its quality, resp. truthfulness. The problem in this area may be, in particular, that the expert report is prepared by the expert in the matter, but is assessed by a layman (court or judge). That is why the expert should prepare his report not only in the statutory, but also in sufficiently comprehensible form, so that the review is written in a way that allows other stakeholders to understand its content.

The ***formal correctness*** of the expert report is also rated, i.e. whether it is at all an expert opinion from a formal point of view (whether the composition and particulars of the expert opinion are respected, whether it is signed, stamped, or missing an expert clause, etc.). [6] Subsequently, the ***procedural requirements*** of the report are also rated, whether it is even possible to authorize the expert report as evidence. Indeed, if the evidence for obtaining the expert's report were obtained in an unlawful manner, if the expert's person were biased, if the expert evaluated the legal aspect of the matter, etc., it would not be possible to use that expert's report as evidence.

In addition to factual and formal correctness, the assessment is also rated in terms of ***content correctness***, which mainly concerns the verification of input documents (especially whether these documents are sufficient to answer questions, whether they are true and logically acceptable) and the rating of the expert's work (rating whether the expert chose the right method, whether he used it correctly and so on).

In practice, there may be situations where the court's rating of an expert report raises doubts about the correctness of the expert opinion. Such a situation is dealt with in Section 146 of the Code of Criminal Procedure [7]: "If doubts arise as to the accuracy of the expert opinion or if the expert report is unclear or incomplete, the expert must be asked to explain or complete the expert opinion. If this would not remove the doubts or uncertainties of the expert opinion or the completeness of the expert opinion, another expert must be added." In the past, the equivalent provision corresponding to civil litigation was contained in § 127 of Act No. 99/1963 Coll. The Code of Civil Procedure, as amended, under which "an expert opinion may also be reviewed by another expert, scientific institute or other institution". However, since this legislation, the recodification of civil procedural law, by the adoption of Act no. 160/2015 Coll. Civil Procedure Code and Act No. 161/2015 Coll. The Civil Extrapolation Code has been dropped. In § 207 para. 3 of Act no. 160/2015 Coll. the Civil Procedure Code, as amended, provides for an indirect arrangement which allows such a procedure under strictly defined conditions. These conditions are (1) the fact that there is a clear discrepancy between the conclusions of the experts so far, and (2) the person conducting the audit report will be an expert institute. In these cases, so-called the ***review (control) expert opinion***, while the expert appointed for the preparation of such opinion is obliged to cope with the previous opinion.

For example, in German civil proceedings, the court may invite the parties to nominate an expert. If the parties agree on the expert, this decision shall be binding on the court. At the same time, the court will determine the extent to which the expert can maintain contact with the parties involved. The expert must be impartial and independent in the drawing up of the expert report, with the result that, according to the decision of German judicial practice, the expert report submitted by a party in civil proceedings has less weight. It is permitted in Germany that if the judge has the expertise necessary to consider the case, he may judge it himself. However, the parties must be informed of that procedure. In the field of expert evidence in the German legislation, we can register the difficult position of the expert in the preparation of a written expert opinion, because the German legislation does not contain any provisions on the formal and content requirements of the written expert opinion.

3 Research results – characteristics of expert´s report as evidence

It is often the case that an expert report is the basis for a final court decision. Consequently, the person of the expert, as well as the expert opinion, must fulfill all the attributes placed on them in order for the expert evidence to be considered reliable, true, high-quality and objective. The aim is that the court can base its decision on such evidence and has no doubt about it. The court decision thus issued also guarantees a fair trial.

Thus, in addition to the statutory requirements [8], an expert opinion should also meet the following requirements [9]:

a) complexity,
b) internal consistency,
c) independence and impartiality,
d) completeness,

e) repeatability,
f) the soundness and reasonableness of the methods used in drawing up the report,
g) mutual control of the methods and procedures used in the expert report,
h) the reliability and decisiveness of the data used in drawing up the expert report.

Ad a)

The **complexity** of the expert opinion must be perceived on two levels, namely the factual and temporal level. On the merits, an expert's report is complex if it takes into account all the legitimate interests, rights and obligations of all parties concerned independently, impartially, justifiably and justifiably. Complexity in the temporal plane is determined by taking into account the past, present and future state of the examined phenomenon. Consequently, an expert's report is considered to be complex if the expert has taken into account, independently and impartially, all the relevant facts which he should have had and could have taken into account in its preparation.

Ad b)

The expert opinion is **internally consistent**, unless the general assumptions, principles, methods used, etc. mutually contradict each other. Thus, the expert must use procedures and methods that are consistent with the general assumptions of their use and on which those methods and procedures are based, as well as with the assumptions explicitly set out in the expert opinion, when drawing up the expert opinion. If this principle were not respected, it would not be possible to determine the correct result.

Ad c)

The requirement of **independence and impartiality** is met if the expert is in no way interested and dependent on the outcome of the expert report or on the course and outcome of the court proceedings in which the expert report is to be used. Nor should the expert intentionally influence the outcome of the assessment to the benefit of any interested party by choosing certain methods, data or procedures used in the preparation of the opinion. [10]

Ad d)

The aim of the requirement for **completeness** of the expert opinion is, in particular, to ascertain the methods used, the assumptions used, the whole procedure by which the expert reached at the result given in the expert report, the information and data used. Compliance with this principle is a condition for compliance with the repeatability principle.

Ad e)

The **repeatability** of the expert report is associated in particular with the possibility of reviewing the examination carried out, calculations and procedures used by the beneficiary. The structure and form of the review should be clear, not misleading the recipient, etc. The expert opinion must be repeatable also from the perspective of another expert. This means that even an impartial expert must be able to repeat the procedure used by the expert in making the report and to produce similar results, using an average effort and based on information on procedures, methods and data sources.

Ad f)

The **soundness and reasonableness** of the methods used in the preparation of the opinion means that all the procedures and methods used in the preparation of the expert report were by the expert in his independent and impartial consideration of all the relevant facts and information he had and could have known in the opinion - that there was a good reason to use them. On the contrary, the concept of justification must be distinguished. It is justified to use certain methods and procedures in an expert opinion if the expert explicitly states directly in his expert opinion the reasons which led him to use them.

Ad g)

Compliance with the requirement of **mutual control of methods and procedures** is necessary, in particular, where at least two different procedures or methods have been used to obtain a set of output data, and those methods or procedures produce identical or insignificantly different results for identical or similar input data. Experts often use this principle, for example, in determining the rate of capitalization, growth rates, etc.

Ad h)

Data coming from **reliable** sources of information that are independent of both the expert and the stakeholders may be considered reliable, with each source used indicating its source. Failure to comply with this requirement shall also be considered if the expert prefers data from non-trusted sources over data from trusted sources.

Another view of the evaluation of expert opinions is given by J. Musil [11], who divided the criteria of evaluation of expert opinions into three groups:

1. Legal correctness of evidence

When rating the legal site, respectively the legal admissibility of an expert opinion as evidence shall be assessed whether:

- the expert report has been prepared by a person having the status of an expert, an expert organization or an expert institute and whether these entities have been properly involved in the process,
- all legal regulations were complied with in the preparation of the expert opinion,
- the expert has not exceeded the limits of his professional competence, in particular by: addressing legal issues, evaluating evidence, using expertise and methods belonging to a specialization other than that in which he is qualified; .
- the report has the content and formal requirements given by the legislation.

2. Evidence relevance of the expert opinion

In the opinion of J. Musil [11], in assessing the relevance of the expert opinion, it is assessed whether it is possible to deduce from the content any evidence on the subject of the evidence and what it is, or whether the expertise contributes to establishing the facts of the case. Legal relevance of expertise - consistency between the subject of the evidence and the subject of the expert's opinion must be considered by the law enforcement authority as soon as it decides to recruit the expert and formulates questions for the expert in order to avoid inefficient and uneconomical procedures.

3. Rating of the credibility of the expert report

The credibility of an expert report can be said if the truthfulness of knowledge results from the method and procedure we have come to know. [12] Thus, a testimonial which, by its content, guarantees a true outcome of knowledge is credible.

The criteria for assessing the credibility of an expert report are [11]:

- the ability of the expert to provide credible expert opinions (requirements for the expert in terms of professional qualifications, personality characteristics and objectivity);
- the completeness and error-free nature of the underlying materials, whether they were obtained legally, etc.,

- expert justification of the expert opinion, what is that the expertness of the expert examination and the professional justification of the conclusions of the expert opinion are rated.

Musil also emphasized the clarity of the expert report, which he expressed as follows: '"Experts seem to be unaware that their report has a specific function. It is to convey expertise not to scientists but to lawyers and procedural parties. In particular, the defendant must have a real opportunity to be fully involved in the process of taking evidence and in evaluating evidence, which, of course, means that they are understandable to him. Many experts believe that the authority of their judgment will be higher the more Latin words, English abbreviations, mysterious marks and symbols are in it." [11]

4 Conclusion

In many legal areas, expert activity is increasingly important. In some cases, for example, in investigating the mental state of persons, in traffic accidents or in determining the value of real estate, it would not be possible at all to successfully conclude legal proceedings (criminal, civil, etc.) without the assistance of experts. It is often the expert statements and opinions of experts that the state authorities and courts base their decisions in dispute resolution. The aim is that the court can base its decision on such evidence and has no doubt about it. The court decision thus issued also guarantees a fair trial. Therefore, the expert should, with due diligence and precision, draw up his expert opinion, process all the data and information that influence the conclusion of the expert report and consequently the fundamental impact on the court's decision on guilt, innocence or the length of the sentence. Consequently, the expert should avoid giving false or distorted and incomplete expert opinions, thereby committing a criminal offense.

This paper is an outcome of research project: "Systematic knowledge transfer in the field of valuation and economic forensic expertise", KEGA No. 025EU-4/2018.

References

1. K. A. Martire, B. Montgomery-Farrer. Judging experts: Australian magistrates´evaluations of expert opinion quality. *Psychiatry psychology nad law.* (2020)
2. M. Bugeja, RD. Rosa, T. Walter. Expert in Australian takeovers: Fees and quality. *ABACUS-A JOURNAL OF ACCOUNTING FINANCE AND BUSINESS STUDIES,* **41**, 307-322 (2005)
3. T. Wolf. On the quality of forensic expert opinions from a penal perspective. *Forensische psychiatrie psychologie kriminologie.* **6**. 235-242 (2012)
4. A. Bradáč, M. Kledus, P. Krejčíř. *Soudní znalectví.* 342 (2010)
5. C. Will. Experience with expert opinions (and experts). *MKG-chirurg.* **6**. 269-274 (2013)
6. E. Ludolph. Cardil errors in expert opinions for the statutory accident insurance. *Unfallchirurg.* **118**. 963-975 (2015)
7. Act no. 301/2005 Coll. Code of Criminal Procedure as amended
8. A. Kozierkiewicz-Hetmanska. The analysis of expert opinions´consensus quality. *Information fusion.* **34**. 80-86 (2017)

9. A. J. Bělohlávek, R. Hótová. *Znalci v mezinárodním prostředí – v soudním řízení a trestním, v rozhodčím řízení a investičních sporech.* 555 (2011)
10. E. Cunliffe. Independence, reliability and expert testimony in criminal trials. *Australian journal of forensic sciences.* **45**, 284-295 (2013)
11. J. Musil, J. (2010). *Hodnocení znaleckého posudku.* (2010)
12. G. Edmond. Expert evidence in reports and courts. *Australian journal of forensic sciences.* **45**. 248-262 (2013)

Cash Concentration and Liquidity Policy

Magdaléna Kubranová[1*]

[1] University of Economics in Bratislava, Faculty of Business Management, Department of Business Finance, Dolnozemská cesta 1/b, 852 35 Bratislava, Slovakia

Abstract. The purpose of this article is to present how to create an effective payment and collection policy in the corporation. The paper shows defining characteristics and analyses of chosen aspects of corporate payment and collection policy which are the important field in corporate liquidity policy. The rules on payment terms are one of the key factors which are controlled in most contracts involved in businesses. A key treasury task is to collect and concentrate incoming cash as quickly as possible, so that it is made available for operational and investment purposes. The methods used in the solving the problem include analysing the current solution in liquidity policy when many organizations have a large number of highly dispersed locations that collect or disburse cash. This article uses a combination of qualitative and deductive methods. A comprehensive analysis of liquidity policy is also related to cover the details of how to create such a forecast, the reliability of the source information used within it, and how to improve the document about liquidity policy on an ongoing basis.

1 Introduction

Liquidity policy should be an organized, repeatable practice or philosophy that a corporate adopts in order to control the risk it assumes when extending net payment terms to its customers. It can be as general or as specific as corporate would like it to be.

First and for a most corporate should have a credit policy in order to protect its accounts receivable. The corporate probably go to great pains to conceive design, manufacture, warehouse, market, sell and ship great product. Obviously, these are essential to businesses success. Credit policy which is included payment and collection policy will give stability, efficiency and effectiveness to company. A solid, repeatable policy allows all departments to understand company's values and work flows. Without it, there is no set way of doing things that pertain to cash flow. It also prevents any misunderstandings between the sales and credit departments which can sometimes cause unnecessary friction. Another benefit to having a set policy is that it will provide consistent customer care. By knowing how accounts will be cleared for credit and the manner in which collections will be conducted, customers will know exactly what to expect when dealing with company. The better company's focused is that customers will understand and follow requirements.

* Corresponding author: magdalena.kubranova@euba.sk

Payment and collection policy should be a central function of any corporate. Good collection policy balances conflicting goals. The company wants cordial relations with its customers. It also wants them to pay their bills on time. [1, 7] Cash must be made available in the correct amounts and with the proper timing to ensure that company functions are not impeded. It is impossible to manage effectively without an accurate cash forecast that is completely reliable.

2 Presentation of surveyed results

The type of payment method that a company uses to pay its suppliers employees has a considerable bearing on the timing of when cash departs a company´s bank accounts, which in turn impacts the cash planning of the treasury department. The current account will handle all the business´s receipt and payments so every transaction will pass through this account at some stage. In structuring transaction services there are several aspects to consider.

2.1 Payment transaction services

There are some types of transaction that are likely to pass through the current account: physical cash (deposits and withdrawals); paper (lodgements and cheques, this type of transaction is diminishing as means of settlement); electronic (receipts and payments) the most common way for business to transact with another; card (credit and debit card receipt and payments) the most common way for a business to transact with a individual consumer; foreign (exchanges to or from business´s trading currency).

The cost of running a current account normally consists of monthly account maintenance charge and transaction chargers that depend on type, with electronic transactions being cheaper than others.

In a world of sophisticated fraud, businesses need strict internal controls over account access, payment originations, authorisations, validations and reviews. [2, 10] These internal controls need to be operated in conjunction with bank to ensure there is rigorous application.

2.2 Subsidiary liquidity management

After accounting for all outflow and inflow payments, the subsidiary will find itself with either excess or deficient cash. It uses liquidity management to either invest its excess cash or borrow to cover its cash deficiencies. If it anticipates a cash deficiency, short-term financing is necessary. If it anticipates excess cash, it must determine how the excess cash should be used. Investing in foreign currencies can sometimes be attractive, but exchange rate risk makes the effective yield uncertain.

Liquidity management is a crucial component of a subsidiary´s working capital management. [6, 7, 9] Subsidiaries commonly have access to numerous lines of credit and overdraft facilities in various currencies. Therefore, they may maintain adequate liquidity without substantial cash balances. While liquidity is important for the overall multinational corporation, it cannot be properly measured by liquidity ratios. Potential access to funds is more relevant than cash on hand.

2.3 Centralized cash management

The management of working capital such as inventory, accounts receivable and payable - debtors and creditors and cash has a direct influence on the amount and timing of cash flows. Working capital management and the management of cash flow are integrated. We discuss these elements before focusing on cash management.

Each subsidiary should manage its working capital by simultaneously considering all of the points discussed thus far. Often, though, each subsidiary is more concerned with its own operations than with the overall operations of the multinational corporation. Thus, a **centralized cash management** group may need to monitor, and possibly manage, the parent-subsidiary and inter subsidiary cash flows. This role is critical since it can often benefit individual subsidiaries in need of funds or overly exposed to exchange rate risk. [3]

While each subsidiary is managing its working capital, there is a need to monitor and manage the cash flows between the parent and the subsidiaries, as well as between the individual subsidiaries. This task of international cash management should be delegated to a centralized cash management group. International cash management can be segmented into two functions: **optimizing cash flow movements** and **investing excess cash.**

The centralized cash management division of an MNC cannot always accurately forecast events that affect parent-subsidiary or inter subsidiary cash flows. It should, however, be ready to react to any event by considering: **any potential adverse impact on cash flows**, and **how to avoid such an adverse impact**. [3, 8, 10] If the cash flow situation between the parent and subsidiaries result in a cash squeeze on the parent, it should have sources of funds (credit lines) available. On the other hand, if it has excess cash after considering all outflow payments, it must consider where to invest funds. This decision is thoroughly examined shortly.

2.4 Techniques to optimize cash flows

Cash inflows can be optimized by the following techniques: accelerating cash inflows; minimizing currency conversion costs; managing blocked funds; managing inter subsidiary cash transfers. The first goal in international cash management is to **accelerate cash inflows**, since the more quickly the inflows are received, the more quickly they can be invested or used for other purposes. Several managerial practices are advocated for this endeavour, some of which may be implemented by the individual subsidiaries. First, a corporation may establish **lockboxes** around the world, which are post office boxes to which customers are instructed to send payment. When set up in appropriate locations, lockboxes can help reduce mailing time **(mail float).** A bank usually process incoming checks at a lockbox on a daily basis. [4, 5] Second cash inflows can be accelerated by using preauthorized payments, which allow a corporation to charge a customer´s bank account up to some limit. Both preauthorized payments and lockboxes are also used in a domestic setting. Because international transactions may have a relatively long mailing time, these methods of accelerating cash inflows can be quite valuable for multinational corporations.

Minimizing currency conversion costs. Another technique for optimizing cash flow movements, **netting**, can be implemented which the joint effort of subsidiaries or by the centralized cash management group. The technique optimizes cash flows by reducing the administrative and transaction costs that result from currency conversion. Over time, netting has become increasingly popular because it offers several key benefits. First, it reduces the number of cross-border transactions between subsidiaries, thereby reducing the overall administrative cost of such cash transfers. Second, it reduces the need for foreign exchange conversion since transactions occur less frequently, thereby reducing the transaction costs associated with foreign exchange conversion. Third, the netting process imposes tight control over information on transactions between subsidiaries. Thus, all subsidiaries engage in a more coordinated effort to accurately report and settle their various accounts. Finally, cash flow forecasting is easier since only net cash transfers are made at the end of each period, rather than individual cash transfers throughout the period. Improved cash flow forecasting can enhance financing and investment decisions.

A **bilateral netting system** involves transactions between two units: between the parent and a subsidiary, or between two subsidiaries. A **multilateral netting system** usually involves a more complex interchange among the parent and several subsidiaries. For most large multinational corporations, a multinational netting system would be necessary to effectively reduce administrative and currency conversion costs. Such a system is normally centralized so that all necessary information consolidated. From the consolidated cash flow information, net cash flow positions for each pair of units (subsidiaries, or whatever) are determined, and the actual reconciliation at the end of each period can be dictated. The centralized group may even maintain inventories of various currencies so that currency conversions for the end-of-period net payments can be completed with-out significant transaction costs. Multinational corporations commonly monitor the cash flows between their subsidiaries with the use of inter subsidiary payments matrix.

Managing blocked funds. Cash flows can also be affected by the host government´s blockage of funds, which might occur if the government requires all funds to remain within the country in order to create jobs and reduce unemployment. To deal with funds blockage, the multinational corporation may implement the same strategies used when a host country government imposes high taxes. To make efficient use of these funds, the multinational corporation may instruct the subsidiary to set up a research and development division, which incurs costs and possibly generates revenues for other subsidiaries.

Another strategy is to use transfer pricing in a manner that will increase the expenses incurred by the subsidiary. A host country government is likely to be more lenient on funds sent to cover expenses than on earnings remitted to the parent. Most tax treaties require that transfer pricing be conducted at market rates. There for a policy of artificially inflated transfer prices resulting in subsidiaries in developing countries paying very high prices for services or goods from head office is of dubious legality and ethics.

When subsidiaries are restricted from transferring funds to the parent, the parent may instruct the subsidiary to obtain financing from a local bank rather than from the parent. By borrowing through a local intermediary, the subsidiary is assured that its earnings can be distributed to pay off previous financing. Overall, most methods of managing blocked funds are intended to make efficient use of the funds by using them to cover expenses that are transferred to that country.

Managing inter subsidiary cash transfers. Proper management of cash flows can also be beneficial to a subsidiary in need of funds. The leading or lagging strategy can make efficient use of cash and thereby reduce debt. Some host governments prohibit the practice by requiring that a payment between subsidiaries occur at the time the goods are transferred. Thus, multinational corporations need to be aware of any laws that restrict the use of this strategy.

2.5 Complications in optimizing cash flow

In some cases, optimizing cash flow can become complicated due to characteristics of the multiplication corporation. If one of the subsidiaries delays payments to other subsidiaries for supplies received, the other subsidiaries may be forced to borrow until the payments arrive. A centralized approach that monitors all subsidiary payments should be able to minimize such problems.

The existence of **government restrictions** can disrupt a cash flow optimization policy. Some governments prohibit the use of a netting system, as noted earlier. In additions, some countries periodically prevent cash from leaving the country, thereby preventing net payments from being made. These problems can arise even for multinational corporations that do not experience any company-related problems.

The abilities of banks to facilitate cash transfers for multinational corporations vary among countries. Banks in the developed world are advanced in the field, but banks in some other countries do not offer services. Multinational corporations prefer some form of zero-balance account, where excess funds can be used to make payments but earn interest until they are used. In addition, a bank may not update the multinational corporation bank account information sufficiently or provide a detailed breakdown of fees for banking services. Without full use of banking resources and information, the effectiveness of international cash management is limited. In addition, multinational corporations with subsidiaries in, say, eight different countries will typically be dealing with eight different banking systems. Much progress has been made in foreign banking systems in recent years. As time passes and a more uniform global banking system emerges, such problems may be alleviated.

2.6 Centralized liquidity policy

The multiplication corporation short-term investing policy can either maintain separate investments for all subsidiaries or employ a centralized approach. Recall that the function of optimizing cash flows can be improved by a centralized approach, since all subsidiary cash positions can be monitored simultaneously. With regard to the investing function, centralization allows for more efficient usage of funds and possibly higher returns. Here the term centralized means that excess cash from each subsidiary is pooled until it is needed by a particular subsidiary.

Centralization when subsidiaries use the same currency. To understand the advantages of a centralized system, consider that the rates paid on short-term investments such as bank deposits are often higher for larger amounts. Thus, if two subsidiaries have excess cash of 50 000 € each for one month, the rates on their individual bank deposits may be lower than the rate they could obtain if they pooled their funds into a single 100 000 € bank deposit. In this manner, the centralized - pooling approach generates a higher rate of return on excess cash. The centralized approach can also facilitate the transfer of funds from subsidiaries with excess funds to those that need funds.

Centralized cash management of multiple currencies. Centralized cash management is more complicated when the multiplication corporation uses multiple currencies. All excess funds could be pooled and converted to a single currency for investment purposes. However, the advantage of pooling may be offset by the transaction costs incurred when converting to a single currency. Centralized cash management can still be valuable, though. The short-term cash available among subsidiaries can be pooled together so that there is a separate pool for each currency. Then excess cash in particular currency can still be used to satisfy other subsidiary deficiencies in that currency. In this way, funds can be transferred from one subsidiary to another without incurring transaction costs that banks charge for exchanging currencies. This strategy is especially feasible when all subsidiary funds are deposited in brunches of a single bank so that the funds can easily be transferred among subsidiaries.

Another possible function of centralized cash management is to invest funds in securities denominated in the foreign currencies that will be needed by the subsidiaries in the future. [8] Multinational corporations can use excess cash to invest cash in international money market instruments so that they can cover any payables positions in specific foreign currencies. If they have payables in foreign currencies that are expected to appreciate, they can cover such positions by creating short-term deposits in those currencies. The maturity of a deposit would ideally coincide with the date at which the funds are needed.

Impact of technology on centralized cash management. International cash management requires timely information across subsidiaries regarding each subsidiary's cash positions in each currency, along with interest rate information about each currency. A centralized cash management system needs a continual flow of information about currency positions so that it can determine whether one subsidiary's shortage of cash can be covered by another subsidiary's excess cash in that currency. Given the major improvements in online technology in recent years, all multinational corporations can easily and efficiently create a multinational communications network among their subsidiaries to ensure that information about cash positions in continually updated.

The electronic communications network may be more sophisticated than the one described here, but this description illustrates how easy it is for an multinational corporation parent to continuously monitor the cash balances of each subsidiary and communicate instructions among subsidiaries. The process of transferring funds among subsidiaries may be especially easy when all the multinational corporation's subsidiaries use branches of the same bank. [11, 12] A communications network allows the multinational corporation to make the best use of each subsidiary cash, which reduce the amount of external financing needed and reduce the multinational corporation's exchange rate risk.

3 Discussion

There are several cash aggregation services available from banks that can assist in this process. One approach, called cash sweeping, moves cash into a central concentration account; this is useful for centralized payment and investment systems. An alternative, notional pooling, aggregates cash from multiple accounts for investment purpose, without actually shifting the cash into a concentration account. Banks offer a wide range of liquidity management and optimisation services. Cash pooling is a product designed for corporation operating in interrelated groups. Main advantages are that all members of the group, whether large or small, get better interest rates, whether for deposits or loans, based on the overall position of the group; centralisation of balances and more efficient management of group's cash flows and more efficient use of group's own resources. Fund sharing within the group gives the corporate opportunity to decrease the cost of external resources.

Local effective cash pooling is efficient use of liquidity of the accounts in cash pooling and option of common drawdown of a credit line by all cash pooling members. Physical concentration of funds on a master account with or without return to individual accounts of the system and money flows which need to be posted really take place between the accounts included in the system in order to determine the consolidated status. There is option to make an individualised setting of balances, limits, and sharing of funds at individual Cash Pooling levels as required by the client and next option to calculate and keep records about the loans within the group along with the information concerning the interest rates applicable to these intra-group loans.

Fictitious cash pooling firms will have better interest rate resulting out of the consolidated balance of the group. No transactions take place between individual accounts of the system in order to ascertain the consolidated balance. Benefits resulting from inclusion of the account into the system are allocated to the master account or directly to individual accounts of the system. Firm does not have to post loans within the group since there is no real money flow.

Cross border cash pooling implementation may be connected with the implementation of the Real Cash Pooling. Automated transfers of balances at the end of each banking day between accounts of the firms kept in various banks of the bank group. There is option to set the transfer details (credit , debit, both credit and debit) and option to set the permitted daily debit balance on the subordinated account and its automated settlement at the end of each banking day from the master account kept in another bank of the bank group.

Conclusion

Irrespective of the type of cash concentration system being used, the treasury manager should periodically review a number of structural issues involving a company´s use of bank accounts. These issues are examined low-usage accounts, review the accounts of acquired businesses, mandate deposit cut off times and administration charges. The ongoing concentration of accounts has an additional benefit, which is that the smaller number o remaining banks will see that more cash is being stored in their accounts. The need for a cash concentration system depends upon the structure of business.

References

1. S. M. Bragg, Corporate Cash Management. A Treasurer´s Guide. Published by Accounting Tools, Inc., Centennial, Colorado. 245 p. (2014)
2. E.F. Brigham, P.R. Daves, Intermediate financial management. Mason: Thomson / South-Western Academic Resource Center USA,. 1040 p. (2007)
3. Ch. W. L. Hill. International business : competing in the global marketplace. New York : McGraw-Hill/Irwin, 738 s. (2009)
4. A. Penning, M. Fardon, Cash management. Published by Osborne Books Limited, Worcester. 2010. 190 p. (2010)
5. J. Tennent, Guide to cash management. How to avoid a business credit crunch. London: The economist in association with Profile Books Ltd. p. 211. (2012)
6. J. Kotlebová, Changes in the Holding of Liquidity during the Present Financial Crisis. In *European Financial Systems 2014. International Scientific Conference*. Masaryk University, 2014. pp. 322-329. http://is.muni.cz/do/econ/sborniky/2014/proceedings-EFS-2014.pdf (2014)
7. K. Belanová, Financial market imperfections and corporate investments. -. In *Challenges for financial sector of CEE countries in overcoming problems of economic integration in the EU.* International conference on currency, banking and international finance. Bratislava, Slovak Republic. pp. 17-23 (2017)
8. K. Belanová, The Impact of financial market imperfections on the investment – cash flow sensitivity in Slovak companies. *International scientific conference. European financial systems 2017.* proceedings of the 14th international scientific conference : Czech Republic. - Brno : Masaryk University. pp. 16-19 online (2017)
9. R. Pakšiová, The Critical analysis of profit for its allocation decision-making. *Scientific Annals of Economics and Business.* Alexandru Ioan Cuza University of Iasi, vol. 64, no. Special Issue, pp. 41-56 online.(2017)
10. R. Pakšiová, Z. Kubaščíková, Business property of company and investments. *Procedia economics and finance : 16th Annual conference on finance and Accounting, Prague,* vol. 25, pp. 70-78 online (2015)
11. Ł. Kurowski, P. Smaga, Monetary Policy and Cyclical Systemic Risk - Friends or Foes? *Prague Economic Papers: <A> Bimonthly Journal Of Economic Theory And Policy* **27,** 5, 522-540 (2018)

12. A. Belke, J. Klose, Forecasting ECB Policy Rates with Different Monetary Policy Rules. *Finance A Úvěr: Czech Journal Of Economics And Finance*, **69,** 3 .238-252 (2019)

SHS Web of Conferences **83**, 01040 (2020) https://doi.org/10.1051/shsconf/20208301040
Current Problems of the Corporate Sector 2020

Impact of the Covid-19 pandemic on the automotive industry in Slovakia and selected countries

Iveta Kufelová[1*], *Monika* Raková[2]

[1]University of Economics in Bratislava, Faculty of Business Management, Department of Business Economy, Dolnozemská cesta1, 852 35 Bratislava, Slovak republic

[2] University of Economics in Bratislava, Faculty of Business Management, Department of Business Economy, Dolnozemská cesta1, 852 35 Bratislava, Slovak republic

Abstract. The new coronavirus has far-reaching economic impacts all over the world and, due to the openness of the Slovak economy, it also has an impact on Slovak companies as well as car plants. Except that paralyzed European demand for vehicles, as well as the entire supply chain. The result was the fact that there was no one to make or something to make. And so the manufacturers, together with the suppliers, closed their productions for operational and hygienic reasons. However, this will have a very big impact on the economy as a whole, as the car industry is a fundamental pillar. The paper is pointed to highlight the current effects of the pandemic and measures in the automotive industry in Slovak republic and selected countries.

1 Introduction

The virus, which is faced by almost all countries of the world, has affected all countries and economies of Western Europe. Many of them have adopted restrictive measures that have slowed, paralyzed and even stopped the economy. As in European countries, the economies of the United Kingdom and the United States can be expected to weaken, especially due to the late response to the current situation. For this reason, a recession can be expected in 2020 not only in Europe but also in the USA. and in the euro area by 6% or more. The Italian economy is most likely to be affected. The Slovak economy could decline by approximately 11 to 12% (7.2% of GDP). The reasons for this decrease stem from the following situations. Due to the spread of the COVID-19 pandemic, all schools and school facilities in Slovakia were gradually closed. Subsequently, all retail outlets were closed, except for food and drugstores, which caused the household services sector to be completely shut down. (restaurant, cultural, recreational sports, etc.) This sector was the first to fully feel the effects of restrictive measures in its business. Gradually, self-employed people, who did not have the opportunity to buy various consumables, to close their business activities were closing down.

* Corresponding author: iveta.kufelova@euba.sk

After the closure of restaurants, shops and, consequently, manufacturing companies, there was a big drop in all statistics that capture economic life. Industrial production fell by 20-30%, retail sales by 40%, restaurants and hotels by 80%.
The following figure shows the added value according to the individual sectors of the national economy and points out which sectors will be most affected.

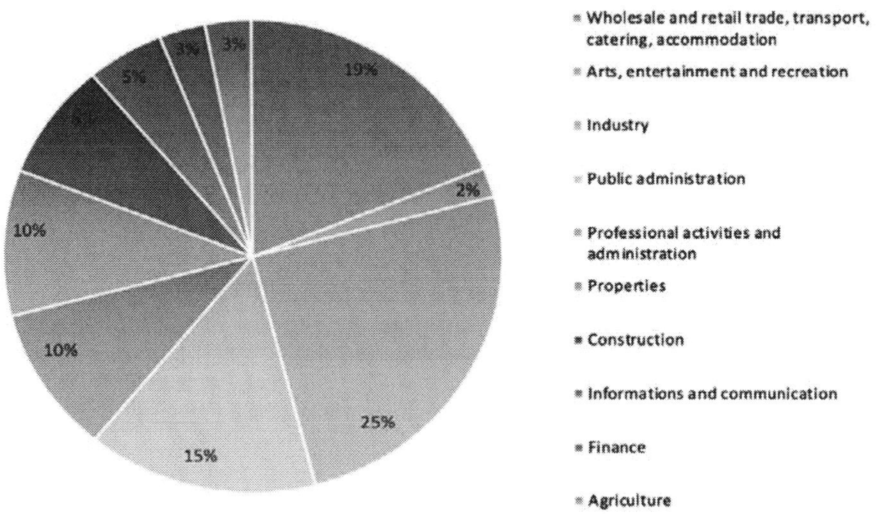

Figure 1. Added value of individual sectors of the national economy

From the above, it can be stated that the industry accounts for about 25 percent of total value added, followed by wholesale, retail, transport, catering, art, recreation contribute to the creation of added value by about 19 %. Of course, the suspension of business in this will also affect other sectors of the national economy.

2 Characteristics of automotive industry

According to history, manufacturing sectors is a key pillar of economic growth and development and until nowadays it becomes positively moderate for the economy.There are a few types of manufacturing industries which are textiles, food and beverages, aerospace and also automotive manufacturing. The automotive industry is a worldwide industry which has high competition all over the world. As the automotive sectors contribute high economic revenue for the country, the growth of the sectors should be intensified[2]. The role of the automotive industry in the economic structure and its impact on investment flows are assessed through social accounting matrices. Under the proposed methodology, the classification of sectors and the impact. [4] To boost up the sectors by incorporate strategies that help in increasing the production and quality, all issues that involve in the manufacturing should be solved. [3] Automobility is defined as self-organizing autopoetic, non-linear system that spreads worldwide, and includes cars, car-drivers, roads, petroleum supplies and many novel objects, technologies and signs. [1]

Automotive industry is looking for both competitive materials and production processes, in order to act on a possible mass reduction and fuel economy which in turn involve a real improvement in the use of environmentally friendly products. [2] Automotive manufacturing is a complex and energy-intense process which consumes a significant quantity of raw materials and water. Toremain competitive, automotive original equipment manufacturers have to strive for better product quality by continuously improving their production process and driving towards low-carbonemissions and enhanced sustainability [5]

The automotive industry is a key industrial sector and economic pillar of several Central countries and Eastern Europe. This sector is one of the main sources of FDI in the region for the last 20 years. Car manufacturers use educated, productive and relatively cheap labor and quality connections with Western European markets as well as a favorable position for exports to the east. Slovakia is currently one of the important centers of the global automotive industry, producing the highest number of passenger cars per capita in the world. It owes its position mainly to the presence of modern ones three car races: Volkswagen (Bratislava); PSA Peugeot Citroën (Trnava) and KIA Motors (Žilina)

and global supply companies. The development of the automotive industry in Slovakia and its integration into the world's automotive centers began in. In the 1990s, when the German car company Volkswagen AG launched a car plant near Bratislava. This important step helped the development of the Slovak economy after the demise of the former markets CMEA and the cessation of arms production. The entry of the automobile company Volkswagen into Slovakia also began to build a supply chain, which attracted further investment in the country. Slovakia so has embarked on a new path of development of industrial production in the sectors: automotive and mechanical engineering.

3 Methods

The main aim of this paper is to point out the change in the automotive industry brought by the current pandemic situation in the world. The first part of the paper focuses on the characteristics of the automotive industry in general and pointing out its main importance in the Slovak Republic. In the results and discussion, it is tried to demonstrate how Covid-19 caused the closure of operations, and thus a decline in sales and profits in the automotive industry in 2020 not only in the Slovak republic but also in selected countries. In this paper there are used general methods to process the paper, especially the method of analysis, synthesis and deduction.

4 Results and Discussions

In this chapter there is tried to map the current situation in the automotive industry affected by the pandemic covid 19 not only in the Slovak Republic, but also in selected countries of the world. The figure 2 shows the sales of cars in the years 2018 and 2019 in the world before pandemic situation cause by Covid-19. Part of figure is also comparison between shown years. As it could be seen, production of car was break down at all over the world also before Covid-19 come into people lives and paralyzed economy.

	2019[8]	2018	% change 19/18	% share 2019
EUROPE	**18,548,160**	**19,440,981**	**-4.6**	**25.0**
European Union[9]	15,769,041	16,644,609	-5.3	21.3
Russia	1,513,640	1,527,928	-0.9	2.0
Turkey	944,087	983,952	-4.1	1.3
Ukraine	6,367	5,659	12.5	0.0
Others: Europe[10]	315,025	278,833	13.0	0.4
NORTH AMERICA	**12,165,682**	**12,968,219**	**-6.2**	**16.4**
of which the United States	7,452,191	8,028,375	-7.2	10.1
SOUTH AMERICA	**2,676,738**	**2,754,002**	**-2.8**	**3.6**
of which Brazil	2,449,242	2,411,076	+1.6	3.3
ASIA	39,179,376	41,897,507	-6.5	52.9
China	20,675,662	22,726,556	-9.0	27.9
Japan	8,187,935	8,214,183	-0.3	11.0
India	3,571,298	3,987,912	-10.4	4.8
South Korea	3,629,252	3,692,685	-1.7	4.9
Indonesia	1,028,782	1,040,322	-1.1	1.4
Thailand	955,105	1,057,504	-9.7	1.3
Others: Asia[11]	1,131,342	1,178,345	-4.0	1.5
MIDDLE EAST/AFRICA	**1,537,412**	**2,065,538**	**-25.6**	**2.1**

Figure 2 The car production in the world in 2018 and 2019 and its comparison

Due to the new covid 19 virus, sales of new cars in **Russia** in April 2020 fell by 72.4% compared to the previous year. The Association of European Enterprises states that production has slowed down and sales have almost completely stopped. In April, only 38,922 new cars and light commercial vehicles were sold in Russia, which recorded the highest decline in sales. The most popular were Lada cars, which accounted for almost 9,395 vehicles in total sales. [6]

In April, car sales in **Germany** fell to the lowest monthly level since unification in 1990. Sales of new vehicles fell by 61 percent year-on-year to 120,840 units due to coronavirus restrictions. According to the automotive industry association VDA, car production in Germany almost stopped in April, falling by 97% year-on-year to 10,900 vehicles.

Sales of new passenger cars in **Britain** almost stopped in April due to coronavirus. It fell by about 97 percent to about 4,000 vehicles compared to last year when it was 161 thousand. It thus reached the lowest level since 1946. The car market in Britain has been hit hard by the temporary closure of car dealerships as part of measures against the spread of coronavirus.

The British Association of Car Manufacturers and Dealers predicts that car sales in Britain will fall to 1.68 million in 2020 from 2.31 million last years.

Japanese carmakers fell global sales by a third to 1.82 million vehicles in March. The decline in sales due to the corona crisis will probably continue in the coming months. Toyota sales in March fell by 22.6% year on year to more than 779 thousand units, Nissan's sales fell by 43% to more than 315,000 cars. Worldwide sales of 7 Japanese automobiles for the fiscal year to March (Toyota, Nissan, Mazda and four other manufacturers) fell by 7.3% to 26.5 million, at least in four years.

According to analysts, consumers will postpone the purchase of new cars. Global sales could fall by a third this year, falling by 11 percent at the height of the global financial crisis. [7]
In the **Czech Republic**, sales of new passenger cars fell by 53 percent to 1,0679 vehicles in April due to anti-coronavirus measures and lower overall demand. In four months, the market fell by 26.3% to 60,873 cars sold. The decline affected all segments and vehicle designs with the exception of electric and hybrid passenger cars. The registration of these vehicles increased year-on-year due to the pressure on car manufacturers through stricter emission limits. The biggest drop in sales was recorded by Volkswagen and Škoda vehicles. In the first four months, light commercial vehicles fell by 28 percent to 4,787 vehicles in the Czech Republic. Truck sales fell by 32 percent compared to the previous period to 2,278 vehicles. In April alone, it was a decrease of 52 percent. There was also a decrease in the sale of motorcycles by about 18%. The carmaker Škoda Auto (in Mladá Boleslav) interrupted production on March 20, 2020 and employees remained in quarantine at home. The cessation of Škoda production will also affect employment in supplier companies. The carmaker itself employs approximately 34,000 people in the Czech Republic and another approximately 3,000 people through employment agencies. On March 25, Toyota Peugeot Citroën Automobile (TPCA) in Cologne stopped production due to the spread of the virus. The interruption could last until April 17. TPCA currently employs approximately 2,500 employees and the company is one of the largest exporters in the Czech Republic and also the largest employer.
Another discontinued production of the Hyundai car plant in Nošovice in order to prevent the spread of the corona virus infection. Production was suspended for about three weeks. After strict hygienic measures, production at Automobilka Hyundai in Nošovice was 14.4. 2020 renewed. Initially, the company will work on two shifts and will cover up to 12,000 employees - 3,300 employees are employed by Hyundai in total and 8,700 are their direct suppliers. Employees stay at home for the time being from the night shifts and will receive 70% of the average earnings, or as substitutes in the day shift in the event of a failure. Toyota Peugeot Citroën Automobile (TPCA) has extended production shutdown due to a coronavirus pandemic. Production lines could start in Cologne in early May 2020.
Škoda Auto Mladá Boleslav has also not resumed production yet. [8] [9] [11] [13]

Automotive industry in Slovak republic

The Slovak economy was subsequently further affected by the situation in the automotive industry, when Volkswagen Slovakia announced that it was closing its operations in Bratislava and in all European cities. Production was also suspended by other car manufacturers (PSA in Trnava, Jaguar in Nitra). The production of car from 2006 to 2017 is in figure 2.

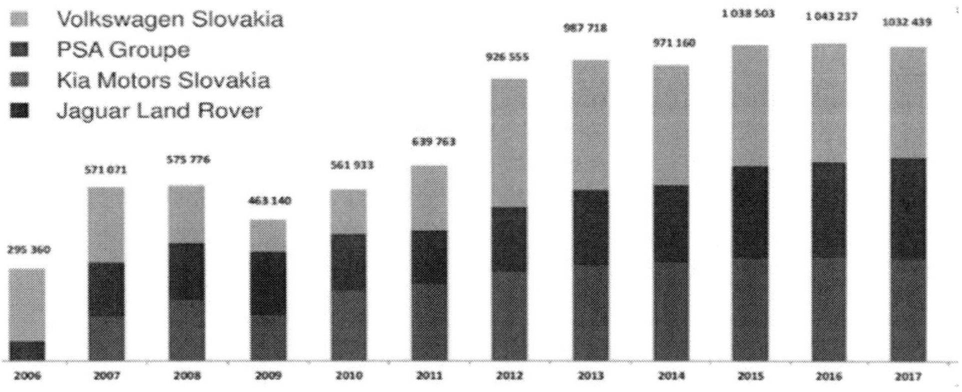

Figure 2 The car production from 2006 to 2017 in Slovak republic

Volkswagen Slovakia is a pillar of Slovak exports, one of the largest private employers and, together with its employees, also the largest contributor to public budgets. The company invested more than 4.5 billion euros during its operation in Slovakia. It currently employs 12,100 workers. It produces cars Volkswagen Touareg, Audi Q7, Audi Q8, Porsche Cayenne, Porsche Cayenne Coupé, Volkswagen up, SEAT Mii electric, ŠKODA CITIGOe iV. In addition to vehicles, it produces gearboxes, components for gearboxes and chassis, as well as equipment used in the manufacture of vehicles.

VW plans to start production in the Spanish region of Navarre from April 20, 2020. This is where the VW Group has its production plant, which has been up and running since mid-March 2020. After the necessary hygiene measures have been taken, only one change will start for the first four days at first, and two after the implementation of the new processes and compliance with the measures. [10] [11]

Production at Jaguar Land Rover was suspended on March 20, when only selected Land Rover Defender production teams work at the plant, and those designed to provide the necessary operational activities.

Lastly, it closed the KIA operation in Žilina, which subsequently started production in a two-shift operation on April 7, 2020 due to orders from the countries of the Middle East, Africa and Asia. The resumption of production is carried out in accordance with all hygienic preventive measures, which are recommended by the crisis staff of the Slovak Republic and the regional public health office. At present, the KIA plant in Teplička nad Váhom employs more than 3,800 people in Slovakia, making it the main employer in the region. Kia Ceed models, the XCeed crossover and the Sportage sports-utility model are produced here. Since the start of production, the company has produced more than 3.4 million vehicles and 5 million engines.

Carmakers in Slovakia currently employ more than 177 thousand people. At the same time, they make up half of the total economy of our country. With their closure, the Slovak economy will be paralyzed for some time. The closure was, on the one hand, a response to declining demand in vehicle markets and supply chain constraints, and, on the other hand, a measure to close operations was necessary to prevent further spread of the virus.

The first quarter-on-quarter decline in GDP will be visible as early as the first quarter of 2020 and can be expected at 1%, as January and February were still quite good. Subsequently, a decline of more than 10% can be expected in the second quarter.

From the third quarter, gradual economic growth can be expected not only in Slovakia but also in Western Europe. According to the above scenario, the recession should be V-shaped. [9] [10] [11] [12]

Conclusion

The pandemic of the new coronavirus causing COVID-19 poses a global health threat with a direct impact on the economy. As the viruses know no borders, the governments of the affected countries had to take tough measures to slow the spread of COVID-19. One of the restrictions brought by pandemic was the closure of car production plants both in the world and also in the Slovak republic. This cessation of production will have the main impact on the country's GDP as the automotive industry because it is a pillar of the Slovak economy. From the point of view of the future development of Slovakia, it is therefore important to take such measures that will economically improve public administration, create favorable conditions for the activities of business entities in Slovakia and at the same time increase the level of preparedness for crisis situations in the future.
Such measures may be included tax reduction, saving in public administration, rationalize the social system, create flexible conditions on the labor market. State funds should focus primarily on funding teaching and research at universities of technology, science and medicine, to adopt the principles of financial policy of public administration (state administration and self-government), which will create financial reserves in addition to debt reduction.

This contribution is the result of the project VEGA no. 1/0368/18 "Cenové stratégie v prostredí pôsobenia efektívnych regulačných mechanizmov na nadnárodných trhoch sieťových odvetví slovenskej ekonomiky" in extent 50 % and the project IG no. I-20-103-00 "in „Hodnotenie inovačného potenciálu v podmienkach nových manažérskych trendov na základe navrhnutých determinantov v automobilovom priemysle"extent 50 %.

References

[1] Wells, P., Wang, X., Wang, L., Liu, H., Orsato, R., 2020. More friends than foes? The impact of automobility-as-a-service on the incumbent automotive industry. Technological Forecasting and Social Change.. doi:10.1016/j.techfore.2020.119975
[2] Peter, I., Fracchia, E., Canale, I., Maiorano, R., 2019. Incremental sheet forming for prototyping automotive modules. Procedia Manufacturing.. doi:10.1016/j.promfg.2019.02.182
[3] Mohd Fazi, H.B., B Nik Mohamed, N.M.Z., Bin Basri, A.Q., 2019. Risks assessment at automotive manufacturing company and ergonomic working condition. IOP Conference Series Materials Science and Engineering.. doi:10.1088/1757-899x/469/1/012106
[4] Garcia-Remigio, C.M., Cardenete, M.A., Campoy-Muñoz, P., Venegas-Martínez, F., 2020. Valoración del impacto de la industria automotriz en la economía mexicana: una aproximación mediante matrices de contabilidad social. El Trimestre Económico.. doi:10.20430/ete.v87i346.852
[5] Giampieri, A., Ling-Chin, J., Ma, Z., Smallbone, A., Roskilly, A.P., 2020. A review of the current automotive manufacturing practice from an energy perspective. Applied Energy.. doi:10.1016/j.apenergy.2019.114074
[6] https://www.reuters.com/article/us-russia-autos-salesfigures/russia-car-sales-plunge-72-4-in-april-year-on-year-aeb-idUSKBN22O18C?il=0

[7] https://www.reuters.com/article/us-japan-automakers-sales/march-global-sales-slump-spells-trouble-ahead-for-japans-carmakers-idUSKCN22A14C
[8] https://www.lidovky.cz/byznys/auto/nosovicka-automobilka-hyundai-obnovila-vyrobu.A200414_070619_ln-auto_ele
[9] https://www.czso.cz/csu/czso/cri/miry-zamestnanosti-nezamestnanosti-a-ekonomicke-aktivity-unor-2020
[10] https://ec.europa.eu/eurostat/web/lfs
[11] https://www.mfcr.cz/cs/verejny-sektor/makroekonomika/makroekonomicka-predikce/2020/makroekonomicka-predikce-duben-2020-38089
[12] Sario. Automobilový priemysel. https:/www.sario.sk/sites/default/files/content/files/automobilovy_priemysel_0.pdf

The Economic and Social Consequences of Tax Havens in the World

Gizela Lénártová[1*]

[1] University of Economics in Bratislava, Faculty of Business management, Department of Corporate Finance, Dolnozemská 1, 851 04 Bratislava, Slovak Republic

Abstract. The tax havens in the world have become the global phenomenon related tax avoidance, tax fraud and evasion and money laundering. The aim of the paper is to analyze their scope and to assess economic and social consequences of their existence in the world society, world economy, international and national tax systems. Many analyzes of the current situation and reported cases show that tax havens are threatening the stable development of the world economy, causing negative consequences of the economic, social, security and humanitarian nature of the global scale. Combating tax avoidance, tax fraud and evasion through tax havens must be stronger and more effective all around the world.

1 Introduction

Globalization significantly affects all ongoing processes in the world economy, while its positive and negative consequences can be observed throughout society. Piketty [1] has analyzed the current problems of the world economy, examining the causes of inequality, and his former doctoral student Gabriel Zucman [2] has focused on tax havens in the world. Tax havens are used for a variety of purposes, in particular to achieve tax savings, exploit anonymity, and protect assets and hide wealth. Tax savings may or may not mean tax evasion. Sometimes it is about tax avoidance. There is no big difference between them. This depends on the specific jurisdiction, as the concepts of tax evasion and tax avoidance are distinguished in tax legislation and criminal law. Many authors defined a tax evasion as a tax reduction, which is a criminal offense. However, advanced tax jurisdictions consider not only tax evasion but also tax avoidance unacceptable. A series of transactions in which tax legislation is abused leads to unacceptable tax avoidance. Krištofík [3] states that tax savings are among the most common motives for using tax havens, while he further states that the category of taxes (international tax planning and international tax optimization) is often examined from multiple points of view and according to different criteria. According to Graham and Tucker [4], many tax legislative provisions provide incentives or discourage companies from making certain decisions. But some inadequate incentives like a direct support are similar to tax avoidance. According to Ozili [5] factors encouraging tax evasion are: high tax rates, tax evasion sophisticated technologies, weak tax enforcement strategies, inaccurate tax data/records, booming tax avoidance industry, corruption, inadequate

* gizela.lenartova@euba.sk

collection mechanism and non-transparency, self-employed income and other factors. Other factors that encourage tax evasion may include excessive tax burden, lack of honesty in the government, perceived unfairness, tax authorities' poor institutional infrastructure and responses, financial benefits of evading taxes, perceptions of inequality, low level of trust in tax authorities, perceived poor use of tax revenues, poor treatment of taxpayers, corruption in government, increase in banks' offshore activities with non-financial companies connected to banks, etc.

The term "tax haven" comes from the English term "tax haven", which refers to countries (territories) with a minimum, respectively no taxation. These are countries with non-standard, very advantageous tax regimes focused on attracting foreign capital to their territory. Tax havens are the target area of the so-called tax planning of many entrepreneurs - individuals or large business entities. Thus, a large proportion of the funds that are initially taxed in the country of the source of this income is spilled from the world. Domestic countries thus lose a large proportion of their tax revenues in their domestic government budgets. In addition to the term 'tax haven', the term 'the offshore financial center' (THOFC) is often used in the literature [6], means a country "off the coast" to designate smaller islands outside the territories - off the coasts of developed countries. Over time, these islands, especially from the United States and the United Kingdom, began offering various kinds of benefits to investors from developed countries around the world in order to attract foreign capital to their territory. Offshore centers are therefore countries that, in addition to a special tax regime, also offer other services and exceptional trading and business conditions for investors on their territory, which include quality infrastructure, a stable legal environment, a reliable banking system, a functioning judicial system, low administrative restrictions on the simple and fast establishment of companies, etc. These conditions and unmistakable advantages have enabled them to become major global financial centers. In the last decades of the 20th century, the offshore industry has developed and it is estimated that the volume of liquid capital flowing through these countries accounts for approximately 70-80 % of private liquid capital worldwide. With regard to offshore centers, serious problems of a global nature are also coming to the forefront. Used only as tax havens, these territories did not cause such concern to experts of international institutions (e.g. OECD) and did not pose such serious threats to international development as in the last decades of the 20th century. The point is that these territories (whether we call them tax havens or offshore centers) have been used by criminals for money laundering in addition to tax planning and these areas are soaked in by the dirty money economy of the world (drugs, prostitution, kidnapping, financial fraud, terrorism, trade in arms and strategic raw materials). This is reason why these teritories are a real threat for stability in the world.

2 The aims and methods

The aim of this paper is to characterize the nature of tax havens, to investigate their extent and effects on whole of society, next to evaluate the negative consequences of their existence in society and economy in the world and in national tax systems. The methods used are normative methods of analysis of tax legislation and legal measures and regulations of G20, OECD, EU and other international institutions relevant to the given topic, comparative methods, abstraction, inductive and deductive methods. Effectiveness of taken measures were evaluated by normative methods and negative consequences in national and international context were summarised on the base of methods of synthesis.

3 Results and Discussion

Tax havens have been the subject of research and studies by economists around the world for many decades. They examine issues of domestic tax legislation, international treaties to eliminate double taxation, the application OECD documents, EU directives and their impact on the decision making of individuals and companies to use tax havens.

Hong and Smart [7] are exploring the use of various tax structures - the use of preferential debt financing, royalties and dividend policy, intercompany trade. Desai, Foley and Hines [8] have shown in their studies that branches in countries with high tax rates use much more to finance their foreign capital investments than equity. They also discussed tax rate issues, finding that lower tax rates abroad stimulate US companies to use indirect ownership structures in their foreign direct investment. This is related to the increasing share of multinationals investing through entities in third countries. Studies suggest that pre-tax profitability negatively correlates with local tax rates, resulting in active tax avoidance and low connectivity, respectively none between the place where economic activity took place and the place where profits are allocated. In his regression analysis, Weyzig [9] confirmed that international double taxation treaties are key determinants of foreign direct investment directed through the Netherlands. Other studies [10] point out that some entities including MNE not only pursue tax benefits but also other criteria when setting up companies in advantageous jurisdictions.

Zucman [11] presents the results of research that focused on tax havens around the world and their role in deepening inequalities. His sharp criticism is mainly Switzerland, which hides more than 1/3 of the wealth of the world. The analysis of tax havens is based on data from central banks of different countries, comparing the financial liabilities of banks with financial assets and the difference between them is hidden in tax havens in various forms, e. g. in the form of mutual fund shares, in particular in Luxembourg, the Cayman Islands and Ireland.

Table 1. The World's Offshore Financial Wealth

	Offshore wealth ($ billions)	Share of financial wealth held offshore	Tax revenue loss ($ billions)
Europe	2,600	10 %	75
United States	1,200	4 %	36
Asia	1,300	4 %	35
Latin America	700	22%	21
Africa	500	30 %	15
Canada	300	9 %	6
Russia	200	50%	1
Gulf Countries	800	57 %	0
Total	7,600	8,0 %	190

Source: Zucman, G. *The Hidden Wealth of Nations. The Scourge of Tax Havens* (2015)
Notes : Author's computation: Offshore wealth includes financial assets only (equities, bonds, mutual fund shares, and bank deposits). Tax revenue losses only includ the evasion

of personal income taxes on investment income earned offshore as well as evasion of wealth, inheritance, and estate taxes.

According to Zucman, up to $ 7.6 trillion of privately owned world wealth was stored in the tax havens of the world in 2014, which represents about 8% of the world's net financial wealth. The author states that this is an estimate because the difference between financial liabilities and financial assets in banks does not include the value of assets in the form of works of art, jewellery and real estate. Out of $ 7.6 trillion, $ 1.5 trillion was declared, $ 6.1 trillion was undeclared, of which annual tax revenue loss was $ 190 billion. (approximately $ 125 billion for interest income, dividend and income tax, $ 55 billion for inheritance tax and $ 10 billion for wealth tax).

In 2014, the International Association of Investigative Journalists (ICIJ) revealed hundreds of multinationals that had collusion with some European governments. The cases revealed include: Apple in Ireland did not have to pay 12.5% tax; apply an effective tax rate of 0; Amazon in the UK reported minimal revenue as UK online store revenue flowed through a Luxembourg branch that reimbursed the British for the cost of distributing goods; Starbucks in the Netherlands has had unjustified tax benefits since 2008 - it has not paid any taxes for 4 years; Fiat in Luxembourg has unjustifiably reduced its tax liability by EUR 20-30 million since 2012 due to the tax advantage. McDonald's in Luxembourg has not paid corporate tax on European and Russian subsidiaries for more than 8 years ($ 500 million). The European Commission has assessed these practices as unlawful State aid.

In 2016, the International Association of Investigative Journalists (ICIJ) published, under the name Panama Papers [12], a database of the Panama consulting firm Mossack Fonseca with 11.5 million documents with 200,000 companies and individuals (presidents, diplomats, businessmen, athletes, artists, royal family members) focusing on the transfer of funds to tax havens. A portion of the company's clients in its 40-year history have used the company for tax avoidance, Money laundering, hiding money from corruption, illegal arms trading, diamonds. It has been estimated that since 1977 hundreds of billions of euros have "flowed" through this company. The world's leading economists as well as international institutions responded to the Panama Papers, such as OECD's BEPS initiative and the OECD Black List of non-cooperating countries. The EU has adopted ATAD Directive and the Directive on the automatic mandatory exchange of information for multinational companies in the EU Member States. Since 1 January 2016, multinationals have been reporting in the EU Member State where the Group's principal entity is located for tax purposes: amount of income, profit before tax, tax paid, declared capital, retained earnings, tangible assets, number of employees by country where they do business. Reports shall be provided to other Member States where the group is located.

In 2017, ICIJ published a Paradise Papers [13] - database of 13.4 mil. documents of two offshore companies Apple based in Bermuda and Estera based in Singapore. The records include emails, contracts, bank records of more than 25,000 people-related businesses in 180 countries around the world. They uncover increasingly sophisticated and accounting operations and methods of how huge multinational corporations benefit from tax havens, while revealing the links of personal wealth and large corporations to the worldleaders.

Paradise Papers contains 200 names of Slovaks. According to the authors Ištok (ed.) [14] between 2005 and 2014 there was a dramatic increase in the number of Slovak companies that moved their registered offices to jurisdictions in tax havens. In 2005 there were 1510 of these companies, in 2014 the number increased to 4362 Slovak enterprises, which represents an increase of 189%. The most used tax havens by entrepreneurs from Slovakia are in the Netherlands 1154, in the USA 1028, in Cyprus 750, in Luxembourg 393, in Seychelles 205, in the Isle of Man 155. The most common motives for the use of tax havens are tax savings.

3.1 Consequences of tax havens in national and international context

In response to these ICIJ publications, 300 world economists from more than 30 countries around the world have decclared that tax havens distort the functioning of the world economy. However, the negative consequences of their existence are much wider. These are not only economic, but also social and security implications in a national and international context, which overlap, interact, multiply their effects in a negative synergic effect.
As a results of study of tax havens in the world we can summarize consequences:

1. Economic consequences and threats: decrease in revenues to the domestic budgets; the lack of funds to finance the needs of the state – for education, health, (hospitals, prevention of diseases and epidemics, fight against pandemics); the lack of funds for the public sector, environmental protection, development programs, support for science, research and development, business support and job creation etc.; the lack of funds to redistribute wealth and mitigate inequalities; tax havens distort the functioning of the world economy and the national economy of the world; tax havens destroy the world economy and national economies and cause economic and financial instability.

2. Social consequences and threats: deepen inequalities between individuals and economic actors in the economic and social context, as well as increasing inequalities in opportunities to learn, live and work in dignity; the increase in poverty and the resulting consequences of low levels of education and disease prevention, increased frustration of the population, growth in crime in society; decrease of legal awareness of the population.

3. Security implications and threats: the loss of transparency in the financial sector increases the risk of using funds for crime with negative social consequences, for terrorism and for threats to human security.

4. Political and social consequences and threat: the lack of transparency of national and national public finances makes it possible to link business and investment funds (net money) with funds from crime (money laundering), linking with world politics threatens the democratic foundations of states in the world, linking financial power and political power and non-transparency allow to misuse and threaten the foundations of democracy and human civilization.

5. Humanitarian consequences - humanitarian disasters: according to the world's leading economists, the existence of tax havens is literally a tragedy for developing countries. ActionAid estimates that developing countries will lose about € 160 billion in tax havens. The Australian mining company has saved so much money on taxes in Malawi, that would be enough to pay 39,000 teachers or 8,500 doctors a year. According to the UN, the Democratic Republic of Congo is one of the least developed countries in the world, with nearly half of the children suffering from malnutrition, but Glencore (copper mines) and Gentler (mining tycoon - a close friend of the Congolese president) invested more than half a billion dollars in Gertler's offshore companies. Its profit is estimated at 67 mil. dollars. All these negative consequences mean a serious threat to our society. This is the main reason why action needs to be taken against them.

Fazio [15] wrote in the Corona-time crisis "It has become evident that those who do not pay their taxes are not only guilty of a crime, but guilty of the most serious crime – murder: if the beds and the respirators are not there they are partly to blame".

3.2 Evaluation of effectiveness and success of the fight against tax evasion through tax havens

The OECD uses four basic criteria to identify tax havens: 1. no or low tax burden, 2. ineffective exchange of tax information, 3. absence of transparency, 4. investment primarily on the basis of tax incentives and savings without real economic activity. In addition to these 4 criteria, an important criterion is used to identify a tax haven - the placement of jurisdiction in the FSI (Financial Secrecy Index) ranking, which assesses, among other things, the degree of confidentiality, respectively transparency and use of preferential tax regimes. OECD publishes black and grey lists of countries that are often updated. There are many paradoxes in the fight against tax havens [16], one of them is that some OECD member states, resp. the EU is ranked in negative FSI rankings (e.g. USA, Luxembourg or UK).

The G20 leaders declared the "end of banking secrecy" in the world at the April 2009 summit. It should be noted that the enforcement of this measure is difficult and in real life there has been only a slight restriction on banking secrecy. Persuasion is not enough. Many tax havens agree to cooperate, often only formally, and provide information sporadically. The result is the absence of a strong tool to break banking secrecy and force the exchange of information. The sanctions applied so far are not sufficiently effective. Weichenrieder and Xu [17] have derived positive welfare effects of a crackdown on tax haven secrecy.

Since 2010, the US has been conducting an automatic exchange of information between foreign banks and the IRS through the Foreign Account Tax Compliance Act (FATCA). This measure met with great criticism.

In 2013, the OECD adopted a Global Automatic Exchange of Information (GAEI) document to ensure global automatic exchange of information (GATCA). It should be implemented as a global standard by the end of 2020. Key tax havens have announced that they will participate in this system. Overcoming financial opacity is still ineffective. Most accounts in tax havens are hidden in shell companies, foundations, funds, which are intended to legally or illegally break connections and hide the real wealthy money owners. Only a small fraction of taxpayers who do not have access to shell companies are involved in the automatic exchange of information, and it is only a small fraction of the data provided in the automatic exchange of information. Tax institutions do not have the means to verify the whole system and to set up a system of control and verification on 'provocateurs and denunciators' is a weak strategy to fight.

In November 2016, the OECD approved the Multilateral Instrument (MLI), which allows rapid revision of bilateral double taxation treaties and the implementation of adopted BEPS (Base Erosion and Profit Shifting) measures. The multilateral instrument will help to respond better and faster to inappropriate practices in the field of aggressive tax planning of multinationals.

The EU has not taken significant and open action directly against tax havens in the last decades of the 20th century. Based on analyses, EU legislation has so far been aimed at combating tax evasion and fraud, particularly in the field of VAT, aggressive tax planning and tax avoiding. At the beginning of 21st Century on the basis of the BEPS initiative, the EU has taken some measures to combat tax avoidance, in particular ATAP and ATAD. The latest proposals of the European Commission are focused on: automatic exchange of information and new transparency rules (new Directive in force), a new proposal for a directive on tax intermediaries, the adoption of the Fifth Anti-Money Laundering Directive, the creation of a black list of tax havens (following the OECD list).

The increasing pressure of the EU in the fight against tax havens can be viewed positively. In the EU Member States, the automatic exchange of information in the area of taxation and Money laundering has been improved both in the legislative area and in the use of IT systems in tax practice. The creation of a black and grey lists of tax jurisdictions includes 4 phases: country selection, screening, listing and monitoring. In 2017, ministers of EU Member States agreed at a meeting of ECOFIN on a black list of tax havens. This was the result of an assessment of 92 tax jurisdictions out of a total of 213 countries according to three main criteria: transparency of the tax system, fair taxation, and engaging the country in international cooperation to combat tax fraud. Countries that did not meet the criteria during the past period have reached the black list [18]. These are 17 states that have the status of non-cooperating tax jurisdictions: American Samoa, Bahrain, Barbados, Grenada, Guam, South Korea, Macau, Marshal Islands, Mongolia, Namibia, Palau, Panama, Saint Lucia, Samoa, Trinidad and Tobago, Tunisia, United Arab Emirates (UAE). In addition, 47 jurisdictions have been defined in grey list which must improve transparency, eliminate harmful tax competition, introduce the required criteria and implement OECD BEPS, of which 8 are so-called 'hurricane countries' have longer time to implement the measures.

In 2019 in group of non-cooperating tax jurisdictions there were 8 countries: American Samoa, Fiji, Guam, Oman, Samoa, Trinidad and Tobago, US Virgin Islands, Vanuatu. In February 2020 four other states have been included in the black list: Cayman Islands, Palau, Panama and Seychelles.

Several EU Member States support strict measures against black and grey countries, including upcoming to limit financial support from the European Union, the World Bank or the IMF. Research studies [19] documented that tax reforms themselves cannot to solve the tax avoidance and tax evasion through tax havens. Many experts claim that without sanctions and strict measures, which include raising the withholding tax, the possibility of non-deducting costs for companies, various audits and more, tax havens around the world will not be reduced. In its report, the international non-profit organization Oxfam drew attention to negative practices in the process of blacklisting. We mean the presence of strong political and economic pressures in blacklisting. As a result, the lists do not include states that have been involved in revelations and tax scandals over the past five years. According to Oxfam, another 18 countries, including EU Member States, which are among the largest tax havens in the world, should be blacklisted. Oxfam included the Netherlands, Ireland, Luxembourg and Malta. After frequent updates black and grey list of tax havens the Oxfam international [20] in press release wrote that EU' tax havens blacklist "more like a whitewash". The question arises, what is the real willingness of world politicians to address the problem of tax havens in the world and what will be the political continuity of the measures?

4 Conclusion

Tax havens negatively affect the world economy. The aim of this paper was to characterize the nature and extent of tax havens, to identify the negative consequences of their existence in the world economy and in national tax systems. The analyses show that tax havens are causing major tax evasion, which are missing from national budgets. The uncovered cases of Lux Leaks, Panama Papers, Paradise Papers documents show that tax havens are a big threat to the stable development of the world economy, causing negative consequences of the economic, social, security and humanitarian nature of the global scale. Turner [21] wrote "If one industry deserves to collapse in this crisis, it is tax avoidance". The initiatives and measures taken so far by the OECD and other international institutions and economic integration clusters are numerous but not yet effective. Therefore, the fight against tax havens must be more vigorous, stronger and more effective for all competent professionals and politicians internationally and nationally.

This paper is a partial outcome of the project of VEGA MŠ SR VEGA 1/0007/2019 "Allocation of assets in the environment of low interest rates in financial and non-financial corporations in Slovakia" in the range of 100%.

References

1. T. Piketty, *Capital in the Twenty-first Century*. s.237. (2014)
2. G. Zucman: *The Hidden Wealth of Nations. The Scourge of Tax Havens*. (2015)
3. P. Krištofík, M. Ištok, G. Nedelová, Motívy zakladania onshore a offshore společností slovenskými podnikmi v roku. *Politická ekonomie* **65**, 2, 198-216 (2014)
4. J. R. Graham, A. L. Tucker, Tax Shelters and Corporate Debt Policy. *Journal of Financial Economics*, **81**, 3, 563-594 (2006)
5. P. K. Ozili. Tax evasion and financial instability. *Journal of Financial Crime* (2019) also in *MPRA Munich Personal RePEc Archive,* PRA Paper No. 88430 (2018)
6. T. Sigler, K. Martinus, I. Iacopini, B. Derudder, The role of tax havens and offshore financial centres in shaping corporate geographies: an industry sector perspective. *Journal Regional Studies. (*2019)
7. O. Hong, M. Smart, M. In Praise of Tax Havens: International Tax Planning and Foreign Direct Investment. *European Economic Review* **54,** 1, 82-95 (2010)
8. A. Desai, F. Foley, J. Hines, A Multinational Perspective on Capital Structure Choice and Internal Capital Markets. *The Journal of Finance* **59,** 6, 2451-87 (2004)
9. F. Weyzig, Tax Treaty Shopping: Structural Determinants of Foreign Direct Investment routed through the Netherlands. *International Tax and Public Finance* **20,** 6, 910-937 (2013)
10. N. Brueller, A. Carmeli, G. Markman, Linking merger and acquisition strategies to post-merger integration: A configurational perspective of human resource management. *Journal of Management,* **44,** 5, 1793-1818 (2018)
11. G. Zucman: *The Hidden Wealth of Nations. The Scourge of Tax Havens*. (2015)
12. F. Obermaier, B. Obermayer, U. Schäffer. Panama Papers, Secret of Dirty Money. *Süddeutsche Zeitung*, (3.4.2016) https://panamapapers.sueddeutsche.de/en/
13. F. Obermaier, B. Obermayer, Paradise Papers, *Süddeutsche Zeitung*, (6.11.2017) https://www.sueddeutsche.de/thema/Paradise_Papers
14. M. Istok, M. Kanderova, P. Kristofik, K. Scerba, International Corporate Structuring of Slovak Companies Using Tax Havens. *Journal of Competitiveness,* **11,** 1, 22-40 (2019)
15. F. Fazio, Le cose che sto. *La Repubblica,* 16 March 2020 https://www.repubblica.it/cronaca/2020/03/16/news/le_cose_che_ho_imparato-251384470/, [accessed 1.04.2020]
16. P. Krištofík, M. Ištok, G. Nedelová, Motívy zakladania onshore a offshore společností slovenskými podnikmi v roku. *Politická ekonomie* **65,** 2, 198-216 (2014)
17. A. J. Weichenrieder, F. Xu, Are Tax Havens good? Implication of the crackdown on secresy. *Journal of Economics*, **127**, 2147-160 (2019)
18. http: //europa.eu.rapid/press-release_MEMO-175122_en.htm [accessed 31.03.2020]
19. J. Nebus, Will tax reforms alone solve the tax avoidance and tax haven problems? *Journal of International Business Policy*, Palgrave Macmillan, **2,** 3, 258-271 (2019)
20. OXFAM International, EU's tax havens blacklist' 'more like a whitewash', Press releases 12.3.2019. https://www.oxfam.org/en/press-releases/eus-tax-havens-blacklist-more-whitewash [accessed 30.03.2020]
21. G. Turner, Tax avoiders will receive coronavirus bailouts – we must redress this injustice. *The Guardian* 27.3.2020 https://www.theguardian.com/world/commentisfree/2020/mar/27/tax-avoiders-coronavirus-bailouts [accessed 1.4.2020]

Knowledge Management – situation in Poland and Slovakia

Renáta Machová[1,*], *Zoltán* Šeben[1], *Tibor* Zsigmond[1] and *Annamária* Bakulár[1]

[1]Univerzita J. Selyeho, Fakulta ekonómie a informatiky, Katedra manažmentu, Bratislavská cesta 3322, Komárno, 94501, Slovenská Republika

Abstract. The aim of the study is to present the situation of knowledge management at Polish and Slovak companies. The first part contains the introduction. We describe the importance of human knowledge which is one of the main resources in companies' life. In the theoretical part define the meaning of the knowledge and describe knowledge management. In the next part we describe the objectives of this study and the methods used in our research. Our study is based on secondary data. In the practical part we present the state of Polish and Slovak companies in terms of KM. The last part contains the conclusions. Some similarities can be found between the two countries. Large companies are more engaged in knowledge at some level. The biggest obstacles to KM is that the management and the employees have less time and information about the KM processes. In order to improve in the future, it is worth raising the awareness of companies about the usefulness of KM.

Introduction

In recent decades companies have realized that one of the most important sources of competitive advantage are human resources. Within this knowledge plays a key role, much of which lies in the people themselves. Not only is it important to acquire the knowledge for organizations, it must be ensured that it is accessible to everyone in the organization who needs it. Furthermore, it is important that the already acquired knowledge should be reused in the future. On the other hand, enterprises have to act socially responsible while acquiring the knowledge. There are methods like copying or stealing the knowledge, which are't the most ethical ways for knowledge acquisition. Enterprises have to minimize the use of these methods to be socially accepted. These processes (knowledge acquisition, sharing, storing, etc.) require a certain degree of competence and care, which is why the term Knowledge Management has been created. The KM summarizes the activities defined above. Companies that do not manage knowledge properly may have great difficulties. Knowledge "doesn't manage itself." So, due to untold knowledge, it may be hidden inside the company to those who need it, but aren't aware of its presence. In this case, the necessary knowledge must be re-acquired in some way. This will definitely cost the organization. These costs include monetary, time and energy costs. Of course, there are also

* Corresponding author: machovar@ujs.sk

costs associated with managing knowledge, but these are considered as a return on investment over time, to the benefit of companies. The aim of our paper was to learn about the spread of knowledge management among Polish and Slovakian companies.

In the first part of our study we review domestic and international literature and journals in more detail. The knowledge itself is presented, defined by the opinions of various internationally recognized professionals. In addition, researchers' definitions from the studied countries are included too. In the theoretical review, we also introduce the process of knowledge management.

1 Theoretical background

1.1 What does knowledge mean?

There is no unified definition that describes what knowledge means. The term has already been interpreted by several researchers based on different approaches. The work of Penrose [1] should be highlighted. According to her, some of the resources are physical while the other depends on people. The author also emphasized the growing importance of knowledge, but also added that it is a "vague" concept that cannot be treated as a solid factor.

At the end of the 20th century, further theories and studies were developed on the subject. It is important to mention the work of Davenport and Prusak [2]. In their book, *"Working Knowledge: How to Organize What They Know"*, knowledge is a coherent unit of experience, values, contextual information, and expert knowledge that helps evaluate and incorporate new experiences and information. [3-5]

According to Chaudhry, Li and Ren [6] in the business and organizational sense, knowledge is the so-called "Know-how". So it is a mixture of insight, perception, experience and foresight. In other words, it is a unique blend of intellect and intuition that enables people to be aware of what and how to act.

It is also important to highlight the definition of Peter Drucker [7]. In his view, the basic significant economic resource nowadays is not capital, natural resources and labour but knowledge. Its definition reads as follows: "Traditional factors of production - land (natural factors), labour and capital - have not disappeared but have become secondary. They are easy to obtain when we have the right knowledge. Knowledge in the new sense means knowledge as a utility and as a means of achieving social and economic results." [7, 8]

According to the interpretation of the Polish Wielka Encyklopedia [9], knowledge is the result of a cognitive act, but at the same time a reciprocal relationship between knowledge and cognition can be discovered. Knowledge is a consequence of cognition, and on the other hand, cognition largely presupposes knowledge.

Hittmár [10], a prominent figure in Slovak literature, also defines the meaning of knowledge in his book. According to his interpretation, knowledge is completely unstructured and is used inside the human mind. In its opinion, knowledge cannot be processed by information and communication technologies, so it depends on the conscious activity of its owner. This means that the owner cannot be anything other than human. Unlike data and information, it not only describes reality but also causes explanations and answers. This helps human to respond to the problems.

1.2 Managing the knowledge - KM

Companies need to manage their internal knowledge properly in order to be successful. Economically successful companies are aware of their goals. To achieve these goals, they need some tools. Knowledge is one of the main tools for achieving goals. It is another question how conscious this company is in managing this knowledge. In our opinion there are companies, especially smaller ones that do not consciously manage knowledge, but their goals "float in front of their eyes" and therefore do their best to achieve them – among other things they do everything to acquire and preserve the important knowledge. [3, 11, 12]

There are many definitions of knowledge management, but in essence, each term is similarly characterized. Bencsik [3] describes KM referring to Davenport and Prusak [2] and Gholami et al [13] as the management of all activities which aims mapping, gathering, systematizing, sharing, developing, leveraging internally accumulated and documented implicit knowledge, expertise and experience inside the organisation, etc. [4, 14]

2 Aims and methodology

We conducted a secondary research in our study. The results of the research show the situation of knowledge management in Polish and Slovakian enterprises. We studied the research of KPMG's local unit [15], Siuta-Tokarska [16] and Chmielewska-Muciek, Zinczuk [17]. In the case of Slovakia, our opportunities were relatively limited, as we found less previous researches on the subject. Finally, Gabris et al. [18] and Sujanova et al. [19] were the researches that we searched through. The authors examined the knowledge management of Slovak industrial companies. The disadvantage of these researches is the low number of items (based on 82 filled questionnaires). Another Slovak research is related to Tóbiás Kosár, Gódány and Szabó [20]. This study had a larger element count (360 respondents), but the research also examined other factors related to knowledge that aren't part of this study, so we could only use very little data from it.

3 The situation of the KM

3.1 Polish enterprises

3.1.1 The state of knowledge management at polish corporations

Firstly, we would like to present the results of a survey conducted by KPMG Polska – a local office of one of the world's leading audit, tax and business consulting firms. The research was conducted in 2004. 121 companies were surveyed, all of which were large companies, employing more than 250 people and with annual revenues in excess of € 40 million. [15],

Only 15 percent of the companies participating in the research indicated that they had already implemented some kind of knowledge management programme, while another 31 percent stated that the implementation of knowledge management programme was being developed. More than 50 percent of the filling companies (26 + 28 percent) indicated that they have not yet implemented any knowledge management program in the company and even 26 percent of them did not consider introducing a similar system at all. [15, 16]

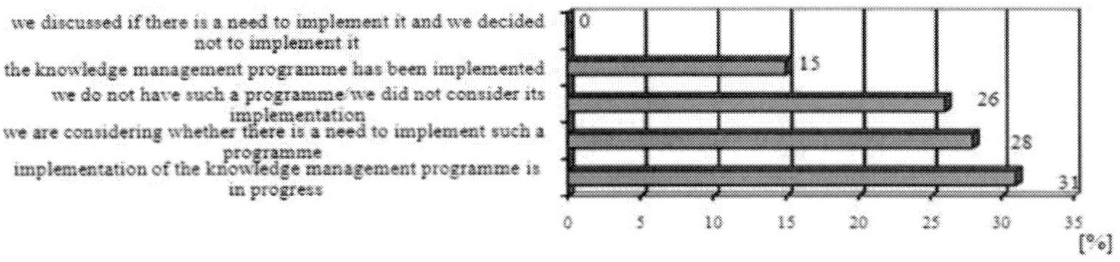

Fig. 1. The state of knowledge management at polish corporations. [15, 16]

On a positive note, 28 percent of companies planned to introduce some kind of knowledge management program. Furthermore, the percentage of companies that had already discussed the introduction of knowledge management but still decided not to implement it did not reach a single percentage. [15, 16]

Examining these data we have two observations – one positive and one negative. On the positive side, companies that have already discussed the issue of knowledge management have made a positive decision after the discussion. This can be assumed based on the first theorem ("we discussed if there is a need to implement it and we decided not to implement it"). Based on this, 28 percent of companies that indicated they are currently in the discussion phase are likely to introduce some kind of knowledge management program in their organization in the future. On the negative side – as explained above – the 26 percent of companies did not consider the implementation of such a programme. Overall, we assume that a quarter of corporations did not have adequate knowledge in terms of knowledge management. However, companies that were aware of the importance of knowledge management have already introduced it (15 percent), are in the process of implementing it (31 percent) or are considering implementing it (28 percent). [15, 16]

In its research, KPMG [15] also addressed the specific problems of large companies that oppose the introduction of knowledge management. The figure below clearly shows the main causes.

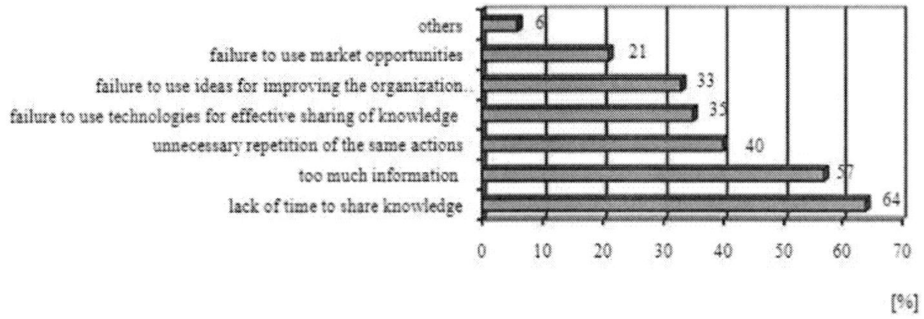

Fig. 2. Main factors motivating entrepreneurs to manage knowledge in the company. [15, 16]

Based on this, it can be seen that in the case of corporations, the lack of time is what prevents the sharing and the transfer of knowledge, and thus also the proper development of knowledge management.

The research also revealed that in 18 percent of the companies surveyed, inadequate treatment of knowledge caused 5 percent loss in income. For another 7 percent, income fell by 5-10 percent. For the remaining 65 percent losses are difficult to estimate. The loss can be assumed to be greatest for companies that do not employ any form of knowledge management at all. [15, 16]

The results show that for the majority of corporations operating in Poland, only the board of directors (42 per cent) and senior management (21 per cent) are involved in the knowledge management process. 7 percent of middle management is involved, while only 1 percent of additional workers are involved in similar processes. [15, 16]

3.1.2 The state of knowledge management at polish SMEs

According to the classification used in the European Union, all companies with less than 250 employees belong to one of the micro, small or medium-sized enterprises. These three types of companies are categorized into one group for most research. More than 99 percent of companies in Poland are SMEs. [16, 21]

According to a study [22] 25 percent of information sources for SMEs come from internal sources. Information is then obtained from clients (21 per cent), various exhibitions (12 per cent) and competitors (10 per cent).

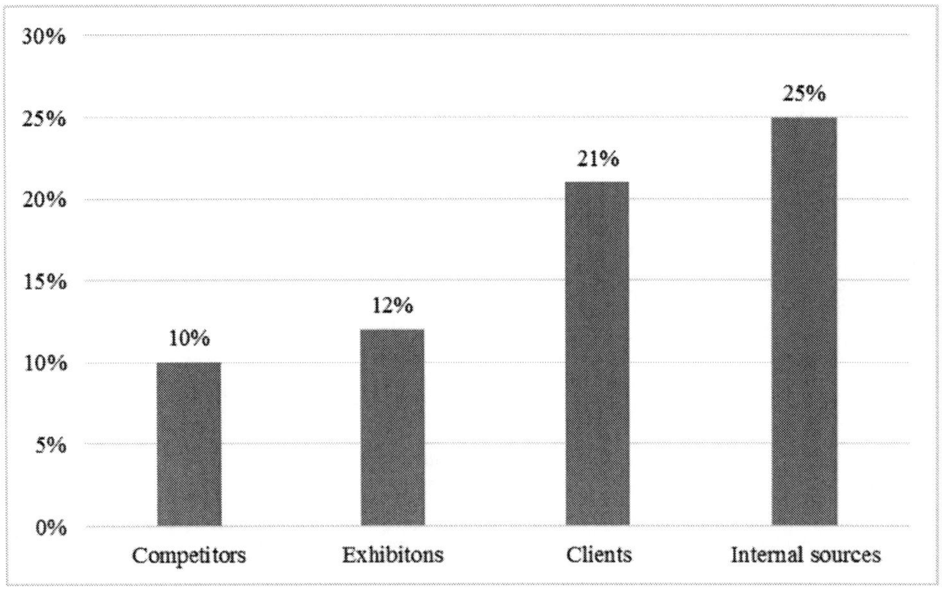

Fig. 3. Source of information for SMEs. Source: [22]

Wolański [23] points out that a group of SMEs in Poland is characterized by outstanding innovation, but most of them are medium-sized companies. Micro and small businesses tend to be less innovative.

Chmielewska-Muciek and Zinczuk [17] conducted studies among employees of 101 companies. The fillers of the questionnaire were from 1 large- 10 medium- and 90 small enterprises.

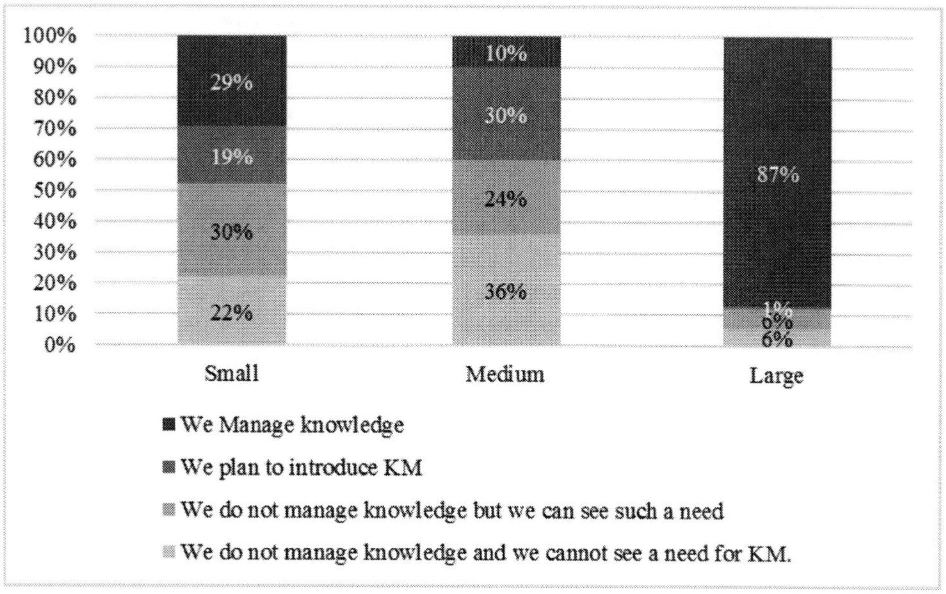

Fig. 4. Distribution of respondents' answers concerning the level of knowledge management. [17]

Based on the figure, it can be clearly seen that in large companies, employees found knowledge management to be very important. In the case of medium and small companies, the proportion of respondents who thought that knowledge management was important proved to be much lower. For these groups of companies, there was a much greater lack of interest in knowledge management. [17]

The research also revealed how proper knowledge management can help companies according to the respondents. The responses received were examined separately in the 3 company types. These results are summarized in the following table:

Table 1. The advatages of knowledge management for corporations and SMEs. [17]

Advantages / Type of business	Large	Medium	Small
Improved quality of services	66%	n/a	71%
Increased comeptitiveness	63%	n/a	51%
Improved customer satisfaction	60%	75%	54%
Cost reduction	57%	60%	n/a
Improved effectiveness	n/a	78%	56%
Time savings	n/a	71%	56%
Improved management	n/a	60%	57%

The table shows that employees working in different type of companies mentioned different benefits. Nevertheless, it is interesting that those working at large or small companies mentioned the "improved quality of services" in the first place. In comparison, employees working at medium-sized companies mentioned completely different advantages compared to those working in the other 2 types of companies. [17]

This research also examined what are the barriers that most hinder the proper management of knowledge according to employees. Employees working at large companies believe that the lack of financial resources (60 percent) is the biggest obstacle to knowledge management. 57 percent say ineffective communication, 51 percent say the fluctuation of specialists, 43 percent say lack of measurable benefits from knowledge management implementation and 42 percent say the wrong management style is what can prevent the introduction of knowledge management. 96 percent of respondents believe that knowledge could best be acquired through external training. Another 94 percent say an external – the so-called "third party expert", while 82 percent think that market research can also help. Previous experience was indicated by 73 percent, while negotiations (meetings) between managers and employees were marked by 72 percent. [17]

75 percent of employees at medium sized companies believe that inefficient communication is the main problem. 71 percent say the wrong management style and 61 percent say the lack of measurable benefits from knowledge management implementation is the biggest barrier to knowledge sharing. According to 86 percent of employees mark that the primary source of knowledge is the Internet. 75 percent say external training, while 73 percent think that professional literature can also help gain the important knowledge. According to the respondents, sources of knowledge can also be internal databases (63 percent), previous experience (52 percent), consumer opinions and feedback (51 percent), loyal employees who have been working for the company for a long time (46 percent), and unique knowledge and skills (46 percent). [17]

According to small business representatives, the biggest problems are ineffective communication (49 percent) and poor management style (47 percent). These barriers are followed by a lack of financial resources (40 per cent) and a lack of measurable benefits from knowledge management (37 percent). 63 percent of employees working in small businesses believe that databases can be the main source of knowledge. This source was followed by personal experience (55 percent), the Internet (54 percent), external training (43 percent) and literature review (41 percent). [17]

3.2 Slovak enterprises

In the case of Slovakia, the situation of knowledge management is relatively less researched. In this subchapter, we present the results of three studies that can be highlighted in some respects. One of them is related to the name of Tóbiás Kosár, Gódány and Szabó [20]. The authors examined 360 companies from three Slovakia districts (Komárno, Dunajská Streda, Nové Zámky). 73 percent of the enterprises participating in the research are micro and 11 percent small. 8% were medium-sized and 8% were large enterprises.

The answers show that 73 percent of respondents said that knowledge management plays an important role in corporate strategy. All large companies consider knowledge to be a tool of strategy. The research reveals that the majority of Slovak entrepreneurs have already recognized the importance of knowledge management, but the appropriate financial resources are't available for implementation. Most micro-sized enterprises do not have their own statements showing how much they spend to properly manage their knowledge. For companies that still have such a statement, only 0-2 percent of the total cost is spent on knowledge management. The situation is similar for small businesses. 45 percent of medium-sized companies spend 5-6 percent of their total cost on knowledge management. 43 percent of the surveyed large-sized companies spend 10 percent of their total spend on knowledge management. [20]

Gabris et al. [18] and Sujanova et al. [19] examined the maturity level of knowledge management in industrial companies. Their research was based on 86 completed questionnaires. The study reveals how companies relate to the implementation of knowledge management.

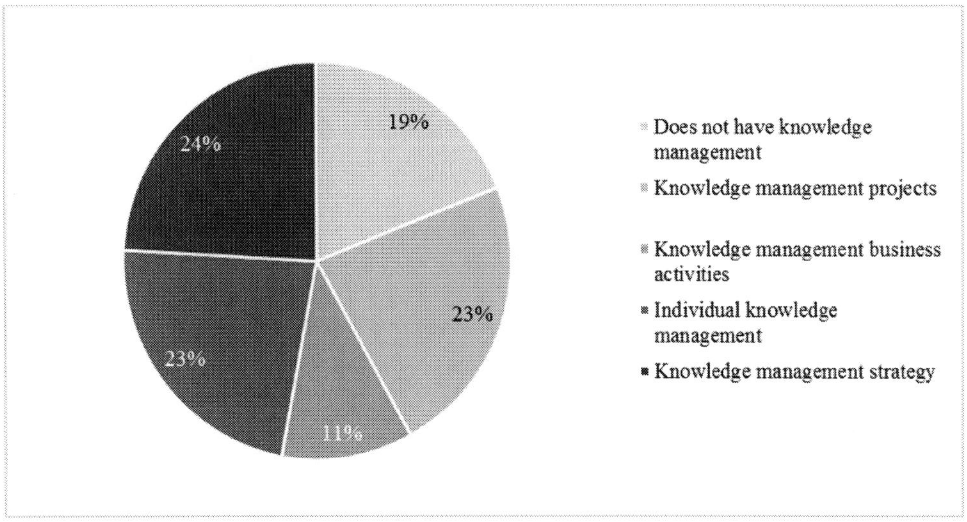

Fig. 5. The state of KM at Slovak industrial enterprises. [18, 19]

24% of participating companies have a knowledge management strategy. Another 23 percent have knowledge management projects, while 23 percent said that knowledge management works in an individual way within the company, so it is not "centrally" organized. In our opinion, this is a somewhat surprising result, since the right knowledge requires the right person to be guided and planned. It cannot "break the road" on its own. 19% of filling companies do not perform any form of knowledge management. We believe this is a relatively high rate. 11% of respondents indicated that their business was based on knowledge management. [18, 19]

The researchers also looked at employees' access to knowledge. According to the results, 57 percent of those who completed the survey believe that their organization has relatively easy access to the knowledge they need. In contrast, the proportion of those who think they have less access to the necessary knowledge is 15 percent. 28% of respondents indicated a moderate opportunity. Based on these answers, it is certainly positive that such a high percentage of people have access to the knowledge they need at any given time. The answers to this question are detailed below. [18, 19]

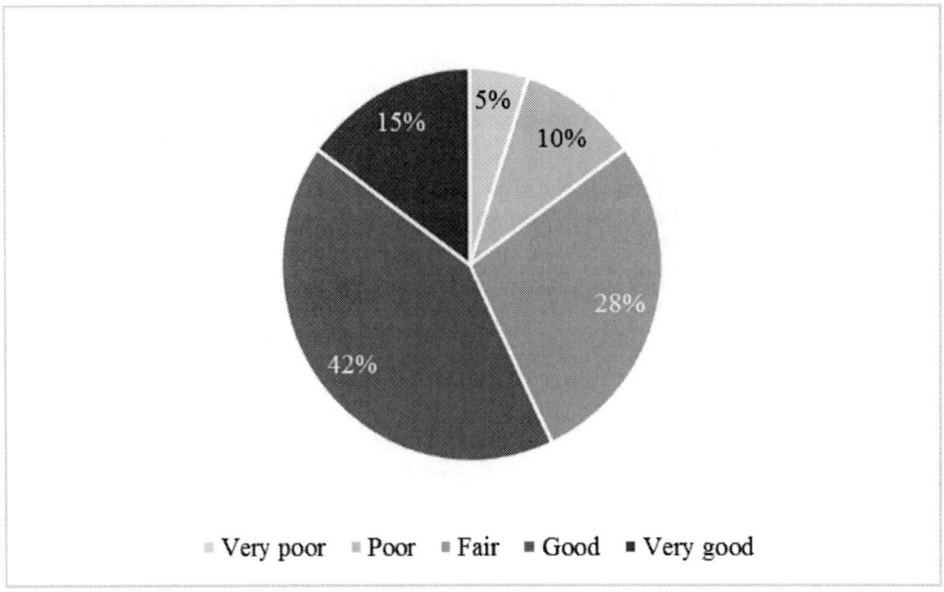

Fig. 6. Access to the company's knowledge. [18, 19]

Conclusions

In our study, we examined the relationship between knowledge management in companies in 2 countries using secondary data. Since our data came from several different studies, it wasn't possible to fully compare their results. Some studies have focused merely on certain areas of competition, while others have focused only on certain company sizes. Occasionally, some studies have investigated other factors that our research doesn't address. Because of that the results of these studies were of limited use. Comparison is also complicated by the fact that some studies have worked with relatively lower number of items.

Nevertheless, some similarities can still be found between the 2 countries. When we look at size, it is usually large companies that are more engaged in knowledge at some level. These companies have more sophisticated assets. One of the biggest obstacles to knowledge management is that the management and the employees have less time and information about the management of knowledge. In order to improve in the future, it is worth raising the awareness of companies about the usefulness of knowledge management. We believe that knowledge will become more and more important in the future, so we need young employees who are aware of it and are open to share their own knowledge in order to make the enterprises more successful.

Acknowledgement

This work was supported by the Collegium Talentum 2019/20 Programme of Hungary.

References

1. Penrose, E. *The theory of the growth of the firm*. Oxford: Oxford University Press (1959)

2. Davenport, T., Prusak, L. *Working knowledge: how organizations manage what they know*. Cambridge: Harvard Business School Press (1998)
3. Bencsik, A. Miért nem működik? Tudásmenedzsment magyar módra. *Tudásmenedzsment* **15**, 1 (2014)
4. Bencsik, A., Juhasz, T., Mura, L., Csanadi, A. Formal and informal knowledge sharing in organisations from Slovakia and Hungary. *Enterpreneurial Business and Economics Review* **7**, 3 (2019)
5. Mura, L., Gontkovicova, B., Dulova Spisakova, E., Hajduova, Z. Position of employee benefits in remuneration structure. *Transformations in Business & Economics* **18**, 2 (2019)
6. Chaudhry S.S., Li Y., Ren N. An enterprise knowledge management system based on the use case model. *Research and Practical Issues of Enterprise Information Systems II. IFIP International Federation for Information Processing* **255** (2008)
7. Drucker, P. F. *Post-capitalist society*. New York: Routledge (2011)
8. Hitka, M., Lorincova, S., Bartakova, G. P., Lizbetinova, L., Starchon, P., Li, C., Zaborova, E., Markova, T., Schmidtova, J., Mura, L. Strategic Tool of Human Resource Management for Operation of smes in the Wood-processing Industry. *Bioresources* **13**, 2 (2018)
9. *Wielka Encyklopedia*. Warsawa: Wydawnictwo Naukowe PWN (2005)
10. Hittmár, Š. Manažment. Žilina: EDIS (2006)
11. Vlacseková, D. Differences of motivation in public and private organizations. *Aktuálne problémy podnikovej sféry 2018 - Zborník vedeckých prác*, 1077-1085 (2018)
12. Peracek, T., Noskova, M., Mucha, B. Selected issues of Slovak business environment. *Economic and Social Development (ESD): Managerial Issues in Modern Business*, 254-259 (2017)
13. Gholami, M. H., Asli, M. N., Nazari-Shirkouhi, S., Noruzy, A. Investigating the influence of knowledge management practices on organizational performance: an empirical study. *Acta Polytechnica Hungarica* **10**, 2 (2013)
14. Bencsik, A., Juhasz, T., Seben, Z. The importance of informal knowledge sharing (workplace gossip) in organizations. Environment. *Economic and Social Development (ESD)*, 46-57 (2019)
15. KPMG Knowledge Advisory Services. *Insights from KPMG's European KPMG Polska. (2004). Zarządzanie wiedzą w Polsce 2004*. Warsawa: KPMG Polska (2004)
16. Siuta-Tokarska, B. Challenges and problems of knowledge management in enterprises in Poland. *China-USA Business Review* **12**, 2, (2013)
17. Chmielewska-Muciek, D., Zinczuk, Z. Knowledge management in it companies in Poland. *Human Capital without Borders: Knowledge and Learning for Quality of Life: Proceedings of the Management, Knowledge and Learning International Conference 2014*, 279-288 (2014)
18. Gabris, P., Bielik Marettová, M., Pavlenda, P., Ličko, M., Šujanová, J. Knowledge management maturity aspects in industrial enterprises. *8th International DAAAM Baltic Conference Industrial Engineering*, 454-457. (2012)
19. Sujanova, J., Gabris, P., Licko, M. Aspects of knowledge management in Slovak industrial enterprises. *Proceedings of The 13th European Conference on Knowledge Management, Vols 1 and 2*, 1135-1144 (2012)

20. Tóbiás Kosár, Sz., Gódány, Zs., Szabó, T. Culture and knowledge management in organizations in Slovakia. 5th *Central European Conference in Regional Science (CERS, 2014),* 1063-1073 (2014)
21. Jaki, A., Siuta-Tokarska, B. New imperative of corporate value creation in face of the challenges of sustainable development. *Enterpreneurial Business and Economics Review* **7**, 2 (2019)
22. *Raport o stanie sektora małych i średnich przedsiębiorstw w Polsce w latach 2008-2009.* Warszawa: Polska Agencja Rozwoju Przedsiębiorczości, PolishAgency for Enterprise Development (2010)
23. Wolański, R. *Wiedza i innowacje w małych i średnich przedsiębiorstwach—Postęp czy stagnacja*. Warsawa: University of Warsaw (2007)

Methods used in personnel audit in companies operating in the Slovak Republic in the stage of Industrial Revolution 4.0

Natália Matkovčíková[1]

[1]Faculty of Business Management, Department of Management, University of Economics in Bratislava, Dolnozemská cesta 1/b,852 35 Bratislava, Slovak republic

Abstract. The present paper contains the results of a research carried out in the framework of the VEGA project 1/309/18. Its aim was to identify and classify methods and approaches used in personnel audit in companies operating in the Slovak Republic in the stage of Industrial Revolution 4.0. to obtain objective information on compliance with legal standards and established standards for personnel work in the company in two areas: human resources and human resources management. The paper contains an analysis and description of the changes that have occurred in the companies with the onset of the Industrial Revolution 4.0 and the progress that these changes brought about in the repertoire of methods and approaches used in personnel audit. Key words: personnel audit, stage of industrial revolution 4.0, methods of personnel audit.

1 Introduction

The onset of industrial revolution 4.0 is the result of three interrelated and conditional processes. On the one hand, it is a process of globalization whose deepening breaks the regional borders of the labor markets and affects the symmetry on the labor market in favor of the demand for quality and prepared employees, thus starting the process of labor migration. On the other hand, it is a deepening process of internationalization that aims to create an international culture through the elimination of cultural and human differences, the labor force that enters the labor market and companies and brings an element of cultural heterogeneity into the working environment. Last but not least, it is mainly a massive entry of information and communication technologies into social processes and thus also in-house environment, which besides opening the space for changes in in-house communication in all directions, vertical and horizontal. This significantly influences the processes of management and leading of employees in the company, they changes the content of work and thus also the requirements for physical and mental potential/preparedness of employees for work performance (performance potential) [1]. These three processes also significantly affect the approaches and methods used in the personnel audit. In the period prior to the onset of the Industrial Revolution 4.0, the personnel audit mainly used simple mathematical statistical methods aimed at assessing

[1] Corresponding author: natalia.matkovcikova@gmail.com

administrative processes taking place in the field of personnel work [2,3,4], at the stage of the fourth industrial revolution these methods are no longer sufficient to respond to the needs of companies expecting a personnel audit. Progress in the content of personnel work from administrative to operational [5] caused the need to extend the repertoire of used methods in personnel audit. Companies that have entered the fourth stage of the Industrial Revolution expect the personnel audit to answer questions that can no longer be answered on the basis of information obtained by basic statistical mathematical methods. They need information on streamlining the management of personnel processes in the company, information on the selection of quality employees for the company and also issues related to retaining employees in the company. These requirements could only be met by using additional methods in the personnel audit. In addition to the basic methods used to obtain static knowledge of employees, their age, gender and practice, psychodiagnostic methods began to be used in the personnel audit during this period, in particular methods to identify the employee's working ability and to identify his/her ability to learn new technology, to acquire new skills resulting from the introduction of information and communication technologies into the work process. These methods were and are primarily oriented on the evaluation of the human capital quality of the company [6]. Personnel audit is understood as a tool for streamlining the management of personnel processes in the company and also as a tool for determining the value of the company that can enter into the business valuation processes in the case of prepared fusions and the like [7]. During this period, personnel audits in various industries are starting to use mainly methods using electronic and digital media as well as non-standard methods to respond to the identification of new phenomena in the workforce behavior caused by these changes in the work environment and work process. The basic methods, which have not ceased to be used in the personnel audit, are also being transferred to the electronic form, but on the contrary they have increased their share in the method repertoire at this stage. The methodological approach and the process of personnel audit thus combine the basic and psychodiagnostic methods, which meets the needs of companies that close themselves during this period and focus on strengthening the personnel management operation [8, 9]. "Personnel work in this period is characterized by the motto: the right people in the right place. It is the beginning of a new phase of personnel work, which in theory is called the phase of personnel management. Its content is "personnel insight into the company interior", focusing on operational management of personnel processes within the company. It is a stage that is characterized by the establishment of personnel departments and personnel managers, which are expected to stabilize/retain staff in the company, minimal personnel rotation, adherence to age automation in the evaluation of employees. This motto is also reflected in the basic objectives of personnel audit and also determines in the personnel audit the selection and application of methods and procedures" [3, p.63].

The main task of the personnel audit at this stage is to identify and characterize the mental and physical potential of the staff, which is referred to as "human capital" [10, 5, 14, 15]. As a result of several personnel audits carried out in companies during this period, it was found that psychodiagnostic methods were used the most [20] and were oriented to the determination of flexibility of the adaptation mechanism of employees, the ability to adapt their work stereotypes, work habits and attitudes to change [16, 17], triggering an increasingly massive entry of ICT into work processes. Concentrating on the use of these methods in personnel auditing has essentially been divided into a workforce that accepts and adapts to the changes these technologies bring to the work process and a workforce that is passive to rejection of ICT and is not able to adapt to this change. The preference of using these methods in the personnel audit led, according to some authors, for example J. Arnold et al.,[6], to some reduction of its goals and tasks and consequently to meeting the needs of companies. At the same time, some authors drew attention to the importance of the choice of methods and their possible combination in relation to the objectives of personnel audit and at the same time to the development stage in which the company and the personnel work carried out there are. In this context, for example M. Vetráková, [5] J. Anglim, V. Sojo, L. J. Ashford et al. [11] and other authors draw attention to the fact that personnel administrative activities are still in some percentage part of all activities in the field of personnel work and therefore the personnel audit cannot ignore the methods that quantify them. On the contrary, the results achieved by these methods point to reserves and sometimes also to the developmental stage of personnel work and indicate the need for changes [18, 19] in the company's personnel policy.

2 Results of the Research and Discussion

Based on the above, the basic objective of the research was formulated: to identify and characterize the methods that were used in personnel audit in companies operating in the Slovak Republic. The basis of the research was the main assumption that the main focus of the activities of personnel managers, personnel units, human resources units in companies operating in the Slovak Republic is mainly operational personnel activity, which corresponds to the developmental stage of personnel work in the stage of industrial revolution 4.0. At the core of operational work of personnel work should be three basic processes: recruitment and selection of employees, training of employees and stabilization/retention of employees in the company. Quality assurance of these processes depends on the systematic collection and evaluation of information and knowledge about their course and the reuse of the acquired information to improve it and the removal of reserves and deficiencies for which the personnel audit serves. Therefore, the main objective of secondary research was to systematize current knowledge of methods used in personnel audit and to find out in primary research what methods and procedures in personnel audit are used in companies operating in the Slovak Republic in which the personnel audit was performed.

The object of the research was a personnel audit in small, medium and large companies operating in the Slovak Republic. The subject of the research was the methods used in the past and now by companies in the personnel audit. The research sample consisted of 112 randomly selected respondents who performed a personnel audit in 445 companies operating in the Slovak Republic within 1 - 3 years. The research included only external and internal auditors who performed a personnel audit in the interviewed companies and had the necessary certificates (accreditation) that authorized them to carry out the personnel audit and also disposed the skills and competencies required to perform the personnel audit.

A questionnaire containing 30 combined questions was used to collect data and information. The questionnaire was distributed electronically. When sorting and processing the information and data obtained was used funnel-shaped procedure[2]. All 112 respondents responded to the questionnaire. The relevant answered questionnaires were 108. The structured interview method was also used to fill in the missing and inaccurate information, resulting in a 100% return. In addition to the above methods, methods of content analysis, synthesis, classification, comparison as well as statistical mathematical methods were used. These methods were also used to process the sets of methods used by the auditors in the personnel audits carried out by the companies surveyed by us. Standard quantitative and

[2] Note: The funnel procedure was based on our experience that respondents would respond positively to questions even if they were reluctant to disclose the methods they used in the personnel audit (copyright contradiction, recruitment agency know-how, etc.). This assumption has been confirmed.

qualitative methods were used to collect, process, evaluate and interpret the data. The research was conducted in October - December 2018.

Whereas previous research and case studies to identify methods and procedures used in the personnel audit have drawn attention to the fact that auditors and personnel auditing personnel managers in the personnel audit use mainly non-standard methods in addition to the standard methods and select them so that confirmed the predetermined expectations, we assumed that this approach would be confirmed in our research as well. However, the results obtained did not confirm this assumption. The interviewed respondents used the entire repertoire of basic methods in the personnel audit, including exploratory and statistical methods (Figure 1). All the methods used were standardized and auditors were able to handle them. In conducting personnel audits in subsidiaries of multinational corporations, the auditors used methods and procedures developed by the parent companies to varying degrees to adapt to the domestic/Slovak legislative environment and cultural behavior patterns. In particular, the basic methods and procedures used by the auditors in the personnel audit were based on standard methods or procedures respectively these were modified standardized methods. For example, in some test batteries the extend was adjusted, questions that contained a cultural conflict or contradicted the customs of the country in which the subsidiary operated were omitted. In the overall set of methods used in the personnel audit, standard basic methods were used, mainly in paper form, which were combined according to the needs of the personnel audit client by standard psychodiagnostic methods. Only 28 respondents in 68 companies used a combination of basic and psychodiagnostic methods in the personnel audit. The psychodiagnostic methods were used mainly test batteries focused on the detection of intelligence, memory and inventory of personality traits. All test batteries were used with the participation of a psychologist. Only 6 respondents stated that they also used non-standard diagnostic methods mainly in connection with obtaining information from social networks in order to assess one of the components of the intellectual capital of the company - social capital. The results of the research show that the methods used in the personnel audit reflect the nature of the development stage, which is characterized by the concentration of personnel work on "human capital", exclusively on one of the components of intellectual capital and not on the intellectual capital of the company as a whole. This point is pointed out by several authors (J. Dušek,[12], M. Vetráková, [5] and others).

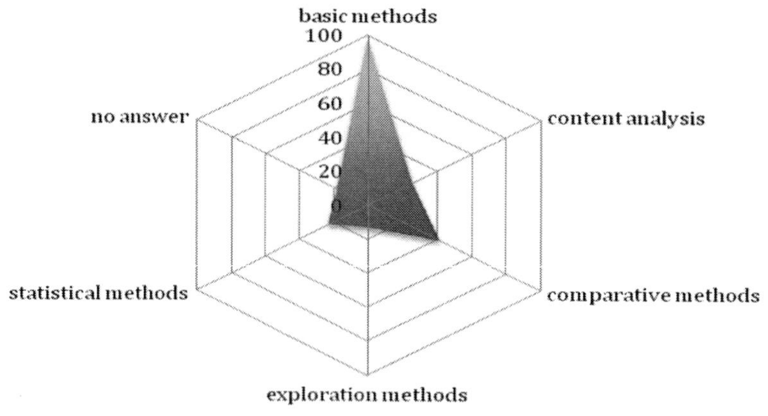

Fig. 1. Methods used in the personnel audit (Source: own elaboration)

3 Conclusion

Inventories of methods and methodologies used in personnel audit have evolved. In essence, at each stage of their development, they reflected the needs of personnel work and personnel audit on the one hand and on the other hand they reflected the knowledge peer of individual scientific disciplines, their methodological apparatus and the methods and procedures developed by them, methods and procedures applicable to personnel audits in companies. This finding was largely confirmed by the results obtained in our realized research. For example, it was confirmed that the companies we examined, which were already at the fourth industrial revolution stage in terms of the developmental stage of personnel work, used in the personnel audit mostly a combination of basic and diagnostic methods, with an emphasis on collecting information and their evaluation using information and communication technologies. According to respondents' answers, this requirement was made with the maximum effort to objectify the personnel audit process and to limit the subjective errors of auditors in the data collection process as well as in the interpretation. The research also pointed out that companies showing elements of the Fourth Industrial Revolution in personnel auditing used a wider repertoire of mathematical statistical and predictive methods, for example to identify and express the human capital potential of a company, to identify the training needs of employees and also, for example to identify the "hidden personnel reserves" of a company, such as company in which management persisted in old approaches to personnel work and employees. Although some of the information obtained did not reach the level of representativeness and the frequency of some responses was not statistically significant enough to allow generalization, the overall empirical research has yielded many original results and findings, and has also outlined the development of methods and approaches for personnel audit. The research predicted that in the new stage, in the industrial revolution stage 5.0, which companies are facing, the importance of personnel audit will increase and at the same time the repertoire of methods used to diagnose managers' soft skills as well as non-standard methods using information communication and digital technologies. Already, the first swallows of this process, which are associated with exploiting the information potential of social networks for the needs of personnel audits, are already noticeable.

This article is a partial output of the VEGA project no. 1/0447/19.

References

1. M. Bartekova, H. Majdúchová. Performance Potential and Development of Creative Industry: The Case of Slovakia. In 33rd International-Business-Information-Management-Association (IBIMA) Conference (Granada, Spain, 5895-5902, 2019)(2019)
2. L. Meňovský. Controlling ako nástroj personálneho manažmentu. Personálny manažment. Bratislava, Raabe (2005)
3. M. Szarková, M. Andrejčák, N. Matkovčíková. Personnel audit as a function of personnel marketing and personnel management. Brno: Tribun EU, 2014. ISBN 978-80-263-0809-6 (2014)
4. M. Szarková. Psychológia pre manažérov. Praha: Wolters Kluwer, 2016. 260 p. ISBN978-80-7552-175-0 (2016)
5. M. Vetráková. Riadenie ľudských zdrojov v ubytovacích zariadeniach hotelového typu. Bratislava: Wolters Kluwer. ISBN 978-80-8168-612-2 (2017)
6. J. Arnold et.al. Work Psychology: Understanding Human Behaviour in the Workplace. Harlow : Pearson Education Limited. 2016, 765 p. ISBN 978-1-2920-6340-1 (2016)
7. R.K.Yadav, N. Dabhade. Human Resource Planning and Audit–a Case Study of HEG Limited. Human Resource Management, 4:5 (2014)
8. J. Dušek et al. Nezaměstnanost a determinanty trhu prace v podminkach modernich evropskych ekonomik. Česke Budějovice: Vysoka škola evropskych a regionalnich studii (2017)
9. V. Bolek. Digitalizácia podnikových procesov v kontexte priemyslu 5.0. Bratislava: Fakulta podnikového manažmentu EU. ISBN978-80-225-4683-6 (2019)
10. J. Kintler. Eliminácia rizík spojených s investovaním do ľudského kapitálu podniku In: Personálny marketing a personálny manažment v organizácii. Bratislava: EKONÓM. Ed. ISBN 978-80-225-3843-5 (2013)
11. J. Anglim, V. Sojo, L.J. Ashford et al. Predicting employee attitudes to workplace diversity from personality, values, and cognitive ability. Journal of Research in Personality 83 (2019)
12. J. Dušek. Sociální kapitál a jeho vztah k rozvoji regionu. In.: Pána, J., Valášek, L., et.al.: Sociologie politiky – minulost, současnost, budúcnost. České Budejovice: VŠERS, o. p. s. 2012, p. 242-249. ISBN 978-80-87472-33.0 (2012)
13. Audit reports focused on personnel audit of examined companies from 2012 to 2014 and 2016 - 2018. Internal materials.
14. J. Kintler, Valuation of the company human capital. In 8th International Scientific Conference on Financial Management of Firms and Financial Institutions, Ostrava, Czech Republic, 342-348, (2013)
15. J. Kintler, Influence of Knowledge in the process of valuing company human capital. In International Conference on Current Problems of the Corporate Sector, Bratislava, Slovak Republic, 191-196, (2014)
16. G. W. Florkowski, R. S. Schuler, Auditing Human Resource Management in the Global Environment. Human Resource Management Journal, 5, 4 (2006)

17. S. L. Albrecht, A. Marty, Personality, self-efficacy and job resources and their associations with employee engagement, affective commitment and turnover intentions. The International Journal of Human Resource Management 31, 5 (2020)
18. O. Lundy, From personnel management to strategic human resource management. The International Journal of Human Resource Management 5, 3 (2006)
19. I. Roper, P. Higgins, P. James, Shaping the bargaining agenda. The Audit Commission and public service reform in British local government. International Journal of Human Resource Management 18, 9 (2007)
20. M. F.R. Kets de Vries et al., The development of the Personality Audit: a psychodynamic multiple feedback assessment instrument, The International Journal of Human Resource Management 17, 5 (2007)

Creation of Personal Career Path: Basis for the Development of Educational Products of Further Education

Marta Matulčíková[1,*] and *Daniela* Breveníková[2]

[1] University of Economics in Bratislava, Faculty of Management, 852 35 Bratislava, Slovakia
[2] University of Economics in Bratislava, Faculty of Applied Languages, 852 35 Bratislava, Slovakia

Abstract. The aim of the paper is, based on theoretical starting-points and empirical research, to describe types of career paths. Diversity of educational needs is reflected in that of career paths. Calculations of the correlation of job position and completed education of the respondent, respondent's mother and father and the correlation dependence of career path and completed respondent's education are presented in tables. Collected information was statistically processed with manual and automated programs: acquired statistics were collected and processed in MS Excel program and GNU PSPP. Based on empirical surveys, it can be concluded that there is a significant correlation between the employment status of both parents. Jobs of both mother and father are dependent on education. It can be stated that education is not such an important factor for the job as it used to be; other factors, such as material resources of the family, prospects of developing their own business, where the respondent can act either in the employer position or in that of self-employed, are important.

1 Introduction

Career is one of the most important aims in human life; it acts as a driving force of its activities. It is a projection of one's own future from working, material and psychic perspectives. Success or failure in one's career is a subjective concept: Everybody has some other ambitions and objectives in one's career, and what some people may consider an achievement, may be a disappointment for somebody else. In practice, we can differentiate between subjective and objective career failures:

– Subjective failure is based on a negative assessment of one's own career by a person himself. One assesses one's own career in relation to own desires, ambitions, interests, ideals as unsatisfactory, although people surrounding this person or his employer may view his career as successful.
– An objective failure in one's career is based on a negative assessment of the career by an employee's environment. The decisive influence is mainly that of the degree to which the employee fulfils the employer's expectations and requirements. The worker may or need not identify with this assessment.
– A combination of preceding possibilities does not mean that the carer is evaluated both by the employee as well as the employee environment.

* Corresponding author: marta.matulcikova@gmail.com

The failure in one's career may have several causes, which could be in employees themselves, their environment or in an employer entity. However, there is a more useful approach than analysing reasons for failure is searching for sources of success of individuals. We may come across with specific sources of personal success, which are based on practical experience of individuals; thus they can be very specific and in some cases can influence the entire career path. There are also joint factors, for instance, the need for permanent activity, creativity, perseverance, flexibility, ability of coping with stress and working under pressure, the aversion to being subordinate to somebody else, etc. These factors shape ambitions and affect a scenario of a particular type of career path.

2 Type of career path and career preferences

The problems of personality typology and career paths have been the subject of intensive attention throughout the entire development of mankind, and not only in academic works but also in practice. Typologies of personality are on the threshold of two scientific disciplines namely psychology and characterology. The first foundations were laid by Hippocrates [1], and further developed by Eysenck (personality questionnaire), Kretschmer [2], W.H. Sheldon in his Personality Theory, Pavlov, and others. Sheldon's physical-psychical typology was further developed by B. Janssen and H.T.A.Whiting (1984) in terms of the body type and extraversion/introversion personality dimension [3].
In more recent periods, we have to mention Holland's theory of career choice known as Holland's Hexagon [4] and Carl Gustav Jung's analytical psychology, which influenced developments in the explorations of personality typology [5]. Morgan, de Bruin, and de Bruin (2015), apply J. Holland's circular/hexagonal theory of vocational personality types in measuring vocational interests [6].

Career has different meaning for different people and may involve not only vertical but also lateral movements. Becker & Strauss (1956) define this concept as "an actor's movements through a social structure over time" [7, p.5]. Career shifts presented in career movement are analysed in works by Edgar H. Schein [8] and E. E. Jennings, who monitored the changes during managers' career [9].

The choice of career path is influenced by attitudes, traditions, as well as by family habits, which have been described by various names, which figuratively capture the essence of career paths, for instance, triangle, ladder, travelling, monkey paradise, and roller coaster [10], but are also referred to in a more specific way, namely expert career path, linear career path, spiral career path, and transition career path [1]. Interactions of personality traits and work behaviour and employee performance have been explored from various aspects, for instance [12, 13, 14, 15]. Other authors point out connections between negative personality traits as for example narcism, machiavellism and behaviour disorders to corruption inclination [16] Soodeh Hajhosseini, Zadallah Fathi, Hossein Shafiei, 2019). Yen, Cheng, Hsu, & Yen, (2019) focus on the interaction of of career adaptability, career satisfaction, person–job fit, and job uncertainty [17]. Biron and Ravid point out trends in preferred career path in the corporate sector, which is manifested in the offer of alternative career paths to their prospective employees [18] Perception of career success has been explored also in the context of cultural values [19].

Based on analyses and empirical researches carried out under the research project of the Ministry of Education, Science and Sports, Slovak Republic, VEGA No. 1/0309/18 Social network in human resource management, we describe career paths which are accompanied by achievement or failures or limitations in personal career, as follows:

- Linear (ascending) career path – at the beginning of their productive life, an individual chooses a certain field and gradually pursues a career whose level of achievement depends on the results of their work based on their efforts and skill. The aim of each individual is to rise in their career, which can be achieved within the same institution or can be linked to the transition from one company to another. This way they eliminate the stereotype that an employee perceives in the old job environment. The stereotype acts as an obstacle to their career growth or results in the failure to meet the required performance criteria, the failure to observe work discipline, difficult relationships between co-workers and management, and other consequences.

- Spiral career path – an individual develops in a given area during a certain period of time and then moves to a related or new field within a certain cyclical movement. This movement is associated with enriching one's work and gaining new experience, which can be achieved within one or more organizations.

- Professional career path – the employee chooses a profession for a lifetime, eager to become a recognised expert. He/ she acquires knowledge and skills connected with performing several work activities, gains social skills, is able to be at the head of meeting specific tasks, and has clear visions and desire for their realization. This kind of person can be a team leader and inspires the team to perform the necessary tasks. The employee usually stays in one job and his or her development opportunities are connected with self-improvement issues.

- Variable career path – the emphasis is on the 'work only' principle. It is typical of individuals who randomly move from one job to another without having the ambition to build their careers. They do not have a defined type of work or field of working; this applies to employees for whom the primary goal is to work to receive remuneration. This type of career may be long-term, but it can be also one stage in the career development. The aim is the volume of income and room for non-work activities, while the reconciliation of work and talent, or the overlap of work with the interests and needs of the individual are secondary. This may be due to the lack of interest of individuals in developing their experience, but it may also be a substitute solution in order to keep one's job while taking any job just to work. Often, this type of career path is pursued by less qualified employees who change jobs to improve their wages or salaries, or face problems with work discipline and performance. These people often leave their jobs under pressure of circumstances even on a voluntary basis.

- Independent career path – pursued by individuals who expect a great deal of freedom. They do not want to be overly structured or controlled by their superiors and are not satisfied with just performing their work activities and performing their work obligations.

- Multiple career paths – suitable for individuals who are interested in more career areas and expect to achieve self-realization by diversity in their working life. The individual acquires knowledge and skills associated with performing multiple work activities. The advantage of this orientation is reducing the risk of losing one's job. The disadvantage, however, is the time constraints when one engages in more than one job at a time, compared to one who only performs one job only.

- Discontinuous career path – is associated with individuals who see their self-fulfilment and satisfaction of their needs not only in work activities but also in a meaningful utilisation of their leisure time. They alternate employment with leisure activities. They are willing to work with different employers and find it important to satisfy their interests and self-fulfilment needs. They choose their employers with regard to earnings; their fluctuation is high, and they often interrupt their work activities due to travelling to other regions.

- Descending career path – is linked to economic and social changes that have affected the labour market when some categories of people have become secondary human resources. For some, the current situation significantly contrasts with the previous stable and even preferred working position. They associate this situation with disappointment, disillusion, and distrust, while the emergence of mental disorders may be manifested by depression, disinterest and low assertive behaviour. Pursuing a descending career is often associated with temporary or more permanent unemployment or a situation when training for a formal system is replaced by unemployment. Career preferences are not only relevant to the individual's knowledge and skills, abilities, attitudes and character traits, but also to family and social background. According to P. A. Sorokin, intergenerational mobility is also essential in addressing career paths [20], which ascertains social rise ascension or descent in the social hierarchy between generations of grandfathers, fathers, and sons. Career of each individual begins with the choice of education, which has its foundations in a family environment and affects the entire course of the career path.

3 Research design

Research in career preparation and its development has been more extensive, and it is not possible to present here its full results and reviews of opinions and the resulting conclusions. This paper deals with part of our research, which focused on the influence of family on the choice of formal education and career development, via presentation of respondents' opinions.

Respondents were human resources with complete secondary education concluded with a school-leaving examination and university education (all three levels), who are employed in selected divisions of section C - Industrial production, according to the Statistical Classification of Economic Activities SK NACE Rev. 2. 2 in accordance with Decree 306/2007 Coll. [21] Respondents were chosen by random selection; specific criteria include:

- Completed education: complete secondary education closed with a maturity examination (approx. GCE) and higher education;
- Currently employed and performing jobs ranging from auxiliary and preparatory jobs up to creative jobs on managerial level.

The research was carried out by means of the interview method and questionnaire method. The interview was used for the purpose of better familiarisation with the opinions and experiences of respondents. Primary data collection carried out by means of interview, was based on questions prepared in advance.

For the purpose of research, a relevant number of respondents was calculated on the basis of the number of employees of explored divisions of section C – Industrial production, amounting to 267 respondents.

Each testing of hypothesis focuses on the verification of the null hypothesis – H_0. Apart from null hypotheses we determined alternative hypotheses H_1.

The following hypotheses were determined:

Hypothesis 1:

H_0: Education of parents does not influence parents' job positions.

H_1: Education of parents influences parents' job positions.
Hypothesis 2:
H_0: Education of parents does not influence each other.
H_1: Education of parents influences each other.
Hypothesis 3:
H_0: Education of parents does not influence respondent's education.
H_1: Education of parents influences respondent's education.

Two kinds of methods were used to statistically process the information collected:
- Manual processing: bar code method was applied in some types of questions.
- Automated programs: acquired statistics were collected and processed in MS Excel program and GNU PSPP Statistical Analysis Software Release 0.8.2-gad9374.

Results of statistical processing are presented in statistic surveys in tables, while the findings acquired are described in words.

4 Research results

The working careers of individuals are associated with value preferences, interests, abilities and internal needs of the individual. The development of an individual's career depends on what opportunities actually exist and to what extent the individual is able to adapt to existing conditions and opportunities and use the opportunities available to build a career that meets certain needs, ambitions, and goals. The family environment has a significant impact in formulating needs, ambitions, and goals. The starting point of the research analysis was to ascertain the correlation dependences of the respondent, mother's and father's job positions at the level of the respondent, mother and father.

Table 1. Correlation dependence of job position and completed education of a respondent, respondent's mother and father

Correla-tions		**Respon-dent's job posi-tion**	**Job position of respondent's mother**	**Job position of respon-dent's father**	**Respon-dent's completed education**	**Completed education of respondent's mother**	**Comple-ted educa-tion of respon-dent's father**
Respond-ent's job position	Pearson Correlation	1	0,23	0,31	-0,28	-0,17	- 0,19
	Sig. (2-tailed)		0	0	0	0,01	0
	Number	267	267	267	267	267	267
Job position of respondent's mother	Pearson Correlation	0,23	1	0,51	-0,12	-0,52	-0,35
	Sig. (2-tailed)	0		0	0,05	0	0
	Number	267	267	267	267	267	267
Respondent father's job position	Pearson Correlation	0,31	0,51	1	-0,23	-0,29	-0,51

	Sig. (2-tailed)	0	0		0	0	0
	Number	267	267	267	267	267	267
Respondent's completed education	Pearson Correlation	-0,28	-0,12	-0,23	1	0,21	0,34
	Sig. (2-tailed)	0	0,05	0		0	0
	Number	267	267	267	267	267	267
Respondent's mother's completed education	Pearson Correlation	-0,17	-0,52	-0,29	0,21	1	0,42
	Sig. (2-tailed)	0,01	0	0	0		0
	Number	267	267	267	267	267	267
Respondent's father's completed education	Pearson Correlation	-0,19	-0,35	-0,51	0,34	0,42	1
	Sig. (2-tailed)	0	0	0	0	0	
	Number	267	267	267	267	267	267

Source: own processing by means of GNU PSPP Statistical Analysis Software Release0.8.2-gad9374.

Based on empirical surveys (Table 1), it can be concluded that there is a significant correlation between the employment status of both parents (0.51). The job is dependent on education, both that of mother (-0.52) and father (-0.51). Father' education also affects the mother's job position (-0.35), but mother's education does not significantly affect father's job position, although there is some dependence (-0.29). There is also high correlation dependence between mother's and father's education (0.42). The level of education completed by the respondent is also influenced by the completed education of the respondent (0,34). The correlation between mother's education and that of the respondent is low, based on the calculated correlation coefficient. The respondent's job position depends on the job position of the respondent (0.31); in the case of the mother, the dependence is low (0.23). Although the respondent's job position is to some extent correlated with education (-0.28), this dependency is more significant than in the case of his parents. Based on this, it can be stated that education is not such an important factor for the job as it used to be; other factors, for instance, material resources of the family, the possibility of developing their own business, where the respondent can act either in the employer position or in that of self-employed (entrepreneur without employees) are important. The possibility of starting one's own business is not connected with the level of education completed. It is complete secondary education graduates who are most often engaged in business and pursue an independent career (Table 2).

Table 2. Correlation dependence of career path and completed respondent's education

Types of career paths	**Respondent's completed education**				
	Complete general comprehensive	Complete secondary education	University bachelor (35 resp.)	University Master's, Engineering,	University postgrad. (doctoral)

	and follow-up at secondary vocational school (33 resp.)		(96 resp.)				Doctoral. (84 resp.)		(19 resp.)	
	Grad.	%	Grad.	%	Grad.	%	Grad.	%	Grad.	%
Linear	4	12.1	17	17.7	6	17.1	25	29.8	-	-
Spiral	-	-	11	11.5	-	-	12	14.3	-	-
Professional	19	57.6	21	21.9	19	54.4	23	27.4	12	63.2
Variable	-	-	12	12.5	-	-	6	7.1	-	-
Independent	4	12.1	16	16.7	4	11.4	7	8.3	2	10.5
Multiple career	1	3.1	7	7.2	6	17.1	9	10.7	5	26.3
Discontinuous	-	-	3	3.1	-	-	-	-	-	-
Descending	5	15.1	9	9.4	-	-	2	2.4	-	-

Source: own processing based on empirical research

The majority of graduates, with the exception of higher education of the second level, pursue a professional career. For university-educated respondents in master's and engineering studies, the mode is a linear career path and the subsequent performance of managerial functions. This number of respondents was the smallest, as there are few graduates in the enterprises monitored, and they were usually respondents in lower age categories, who are still pursuing their professional careers. The advantage of this orientation is the reduction of the risk of loss of work, the amount of experience, and contacts that are gained in performing one job and can be used in another job. According to respondents' opinion, multicultural orientation brings a significant synergy effect, even though it is very time-consuming. Since our respondents were employees of enterprises, we could observe that there was a group of employees in our research who have experienced a descending career, which was accompanied also with a period of unemployment.

5 Conclusion

The career paths of individuals should be the starting point for the product creation of further professional education and training institutions. Their product policy not only has to be in line with the needs of the labour market, but also with the needs of individuals who want to change their career paths, either voluntarily, due to subjective or objective failure, but also in order to maintain their lifelong employability. The decision-making bodies must also contribute to this trend in shaping educational policy by creating conditions for the establishment and operation of further professional education and training institutions.

This paper is an outcome of the research project "Social Networks of Human Resource Management"– funded by the VEGA no. 1/0309/18 primary research grant of the Ministry of Education, Science and Sports of the Slovak Republic, led by Prof. PhDr. Miroslava Szarková, PhD.; project share is 100%.

References

1. M. Szarková, *Psychológia pre manažérov a podnikateľov* (Sprint dva, Bratislava, 2009)
2. J. Boroš, *Úvod do psychológie* (IRIS, Bratislava, 2002)
3. B. Janssen, B., H. T.A. Whiting, Sheldon's physical-psychical typology revisited. *Journal of Research in Personality*, **18**, 4, 432-441 (December 1984)
4. T. Kollárik, E. Letovancová, J. Výrost, *Psychológia práce a organizácie.* (Vydavateľstvo UK, Bratislava, 2011)
5. M. Čakrt, *Typologie osobnosti* (Management Press, Praha, 2010)
6. B. Morgan, Gideon P. de Bruin, and K. de Bruin, Constructing Holland's Hexagon in South Africa: Development and Initial Validation of the South African Career Interest Inventory. *Journal of Career Assessment* **23,** 3, 493-511· (August 2015) 10.1177/1069072714547615. jca.sagepub.com
7. S. Matisa, Career paths for HR professionals within the shared service model-A comparative study of two cases. Master Thesis in Strategic HRM and Labour Relations 30 higher education credits. Supervisor: Julia Brandl Semester: Spring 2012 (University of Gothenburg, Sweden, 2012)
8. E.H. Schein, *The individual, the organization, and the career: A conceptual scheme. Scholar Select* (Amazon 2018)
9. G. Schanz, *Personal wirtschaftslehre: Lebendige Arbeit in verhaltens-wissenschaftlicher Perspektive* (Vahlen, München, 1993)
10. F. Bělohlávek, *Osobní kariéra* (Grada Publishing, a. s, Praha, 1994)
11. D. Lepeňová, M. Hargašová, *Kariéra v meniacom sa svete. Karierová výchova a kariérové poradenstvo pre žiakov* (Metodicko-pedagogické centrum, Bratislava, 2012)
12. R. Williamson Smith, Michael M. DeNunzio, Examining personality–job characteristic interactions in explaining work outcomes. *Journal of Research in Personality.* **84** (February 2020), 10.1016/j.jrp.2019.103884 103884
13. E. M. David, M.K. Shoss, Lars U.L. Johnson, A. Witt, Emotions running high: Examining the effects of supervisor and subordinate emotional stability on emotional exhaustion. *Journal of Research in Personality* **84** (February 2020), Elsevier, 10.1016/j.jrp.2019.103885
14. Jiang Y. Zhou, I. Nielsen, Thriving and career outcomes: The roles of achievement orientation and resilience. *Human Resource Management Journal* (First published: 06 March 2020) 10.1111/1748-8583.12287
15. A. Imran, Personality traits, individual innovativeness and satisfaction with life. *Journal of Innovation and Knowledge* **4**,1 (January–March 2019). 10.1016/j.jik.2017.11.002
16. S. Hajhosseini; Zadallah Fathi; H. Shafiei. Are those with darker personality traits more willing to corrupt when they feel anxious? *Iranian Journal of Management Studies* **12**, 3, 451-479 (2019). 10.22059/ijms.2019.266704.673335. Available from: Web of Science, EBSCO*host*, & DOAJ
17. H. Yen, J. Cheng, C. Hsu, & K. Yen, How career adaptability can enhance career satisfaction: Exploring the mediating role of person–job fit. *Journal of Management & Organization* (November 2018) 1-18. doi:10.1017/jmo.2019.75

18. M.M. Biron, R. Eshed, Gaps between actual and preferred career paths among professional employees: implications for performance and burnout. *Journal of Career Development* **44,** 3, 224-238 (2017). 10.1177/0894845316645664 Available at: journals.sagepub.com/doi/abs/10.1177/0894845316645664
19. George S. Benson, Cheryl K. McIntosh, Maritza Salazar, Hoda Vaziri. Cultural values and definitions of career success. *Human Resource Management Journal.* Special Issue (March 2020), 103884. 10.1111/1748-8583.12266
20. P.A. Sorokin, *On the practice of sociology* (The University of Chicago Press, London, 1998).
21. Statistical Classification of Economic Activities SK NACE Rev. 2. 2 in accordance with Decree 306/2007 Coll.

Distributional innovations of copyrighted works under conditions of sustainability

Martin Matušovič[1,*]

[1]University of Economics in Bratislava, Faculty of Business Management, Department of Business Economy, Dolnozemská cesta 1, 852 35 Bratislava, Slovak Republic

Abstract. Our research in the field of intellectual property has focused on selected innovative forms of distribution and protection. In examining classical innovations, we found different perspectives on their nature. Recently, we have witnessed social challenges and phenomena, which theory, practice and respondents understand as "social" innovations. The results of research into robotics, artificial intelligence, the Internet of Things and communication technologies need to be radically improved. The current epidemiological situation also significantly affects intellectual property in relation to sustainability. In terms of intellectual property records in society, our research is based primarily on generally binding internal standards, in particular our own and, in part, external resources in the field of intellectual property and innovation. Our research on boars also focused on the distribution of intangible products in pandemic conditions. We used the methodology of quantitative survey and quantitative statistical methods. Data were collected as metadata and also by direct collection. Subsequently analysed by statistical program by correlation and regression analysis. The results show that companies are aware of the importance of all kinds of innovation, intellectual property products and a significant change in distribution channels.

1 Introduction

The classical distribution of intellectual property has become unsustainable in pandemic conditions. In the past, we have encountered new forms of distribution of copyrighted works, intangible products using industrial property rights. Some have been less widely used in practice, others have been more popular and convenient in ways that allow us to distribute intellectual property anywhere, anytime, and all we need is internet access. The distribution of films from filmmakers to consumers through cinemas, classical theatrical

* Corresponding author: martin.matusovic@euba.sk

productions and concerts must change, adapt or innovate existing forms of distribution into new ones that are in line with the needs of practice. The classic established method of production and distribution of intellectual property products is too risky in the conditions of the epidemic. [1] Innovation and creativity are becoming a crucial factor in the sustainability of economic activity. [2] Technical progress leads to innovation in the form of irregular waves of creativity. Pulses and wave intensity are based on current practice. Waves stand in the way of a new current phenomenon, although it is usually not explicitly linked to it. Our knowledge from more than a decade of research suggests that the intensity of waves in the field of intellectual property can be measured / indicated / by the extent / development / cost / investment / research. Studies performed to evaluate the task

The EU, played by innovation in the sustainability of EU small and medium-sized enterprises (SMEs), in the manufacturing sector in China ... focused on the following phenomena: to explore the types of innovation that take place in manufacturing SMEs; identify factors that have influenced innovation in the productive sustainability of SME operations; identify the link between innovation and the sustainability of SME operations. The manufacturing sector has the skills needed to carry out innovation processes as well as to design innovative practices and activities that SMEs should put into practice.

The survey was used as a research proposal. Manufacturing SMEs were grouped into sectors, layers were selected and a sample of 30 SMEs from all sectors was selected. Questionnaires and interviews were used to obtain relevant data. The collected data were presented and analysed using tables, columns, graphs of their / individually obtained and extracted from statistical packages for the social sciences. Based on the findings, innovation has been identified as one of the main attributes that help SMEs remain competitive. [3] Today we are also witnessing a test of their flexibility and creativity, as they are connected to the final player with other actors in the distribution channels. [4] The findings also point to a strong link between intellectual property innovation and the sustainability of operations. Other research findings define the characteristics of management and the ability to use and protect intellectual property as the most important factor that positively affects the activities of companies. [5] On the contrary, environmental factors were considered insufficient.

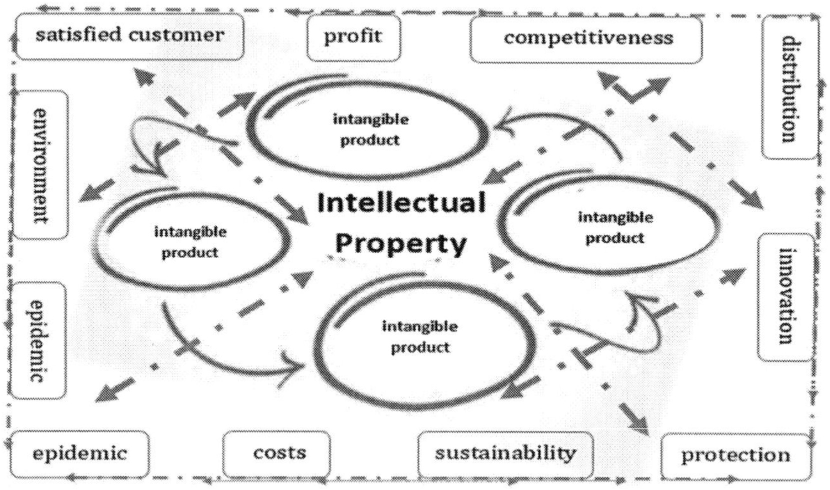

Fig 1. Selected indicators of intellectual property distribution.

Today, however, the situation is completely different. Probably research spending no longer indicates the intensity of intellectual property use. [6] This assumption can be understood as our hypothesis, which we have decided to confirm or refute our research by researche.

2 Present state of knowledge

The task of distribution is to realize the transfer of goods and services from the producer to the final consumer using distribution channels. It is a set of operations ensuring the delivery of goods, services in the right quantity, quality, price and time and contributed to satisfying the needs of the consumer. The delivery process needs to be managed effectively. Distribution management is / the management of resources and processes / used to deliver the product to the point of delivery from production to the point of sale, including storage, and including wholesalers and retailers. Distribution management also involves determining the optimal amount of product for a destination. [7] Distribution management can be defined as part of order management, production, marketing, sales and finance. It brings synergies to all activities. Distribution management is the provision of the movement of goods and services through various combinations of distribution channels, transport, planning and control of warehousing / delivery /, using an information system to monitor the processing of orders. [8] Physical distribution identifies the existing situation, monitors the environment, identifies potential opportunities and threats. Some manufacturers are also distributors. However, most manufacturers do not supply their products directly to the final consumer and use the services of intermediaries / distribution intermediaries, distribution assistants. Wholesalers, retailers, publishers and other entities together form a distribution channel. The decision to choose a distribution channel depends on various factors. Depending on the nature of the product and processes, intermediaries perform different roles in the distribution process. [8] The Typology of Distribution Channels can be derived from this. Direct distribution channel / manufacturer distribution itself directly to the consumer. Performs all distribution activities at its own expense and risk. In the case of copyrighted works, this situation depends on the product. [9] The indirect distribution channel / the manufacturer involves one or more entities in the realization of his products /. In part or in full, these entities assume production and distribution activities / costs, risk, and often participation in copyright with the exception of personal rights. The choice of distribution channel is determined by several factors: size of the target market, specificity of the target market, product characteristics, production and distribution costs, economic potential of the author, producer, distributor consequences of the pandemic, sustainability, new technologies, movement of copyright works in digital space, author's position, new protection requirements and others.

Slovak, European and world legislation regulates the rights of the entities involved and the relations between them. Emphasis is placed on: a high level of protection, settlement of rights, creation of a framework for the use of the work, other objects of protection, stimulation of innovation, creativity, investment, production, distribution, including in the digital environment, new business models, new entities, access to content, access to research, innovation, education and heritage conservation. Consideration is given to introducing / extending / mandatory and optional exemptions for the use of text and data mining technologies, for teaching purposes and for the preservation of cultural heritage in the digital environment. Existing exceptions and limitations in Union law should continue to apply as long as they do not limit the scope of the mandatory exceptions. Exceptions and limitations are intended to ensure a fair balance between the rights and interests of authors and other rightholders on the one hand and users on the other. They may be applied only in cases where they do not conflict with the normal use of the works and do not unreasonably prejudice the legitimate interests of rightholders.

The copyright law of each country regulates the producer of an audiovisual work, resp. the person who started or ensured his final production is considered to be the manufacturer. Unless otherwise agreed. [10] The exercise of proprietary rights depends on the authors and owners of the rights to the works. The original owner / author of the work / decides on the distribution of the work by exercising his personal rights. In the case of dissemination decisions, it significantly affects the distribution process. This decision is conditioned by many factors. The distributor chooses a direct or indirect distribution path. Indirect distribution is associated with the transfer of part of the property rights to third parties. The producer, seller, distributor, broadcaster of the work have partial and temporary rights to use the rights to the work in the form of licenses or sublicenses. A license may be granted for one or more uses, so the specific use of the work in the license agreement must be well defined. "The author may grant the user permission to sublicense third parties if the author agrees to the further use of the work in the license agreement. License fees for the use of the work may be agreed on a flat rate. They may also be determined as a proportion on an agreed basis. Stakeholders may also agree on a combination of these options. [11] The Copyright Act also regulates the representation of authors by a collective management organization. [12] This is the representation of authors or heirs and other owners in certain copyrights. The indirect form of distribution can be associated with a complex system of entities and relationships. The involvement of new digital / intelligent / works distribution platforms causes this system to be more complex and opaque. [13]

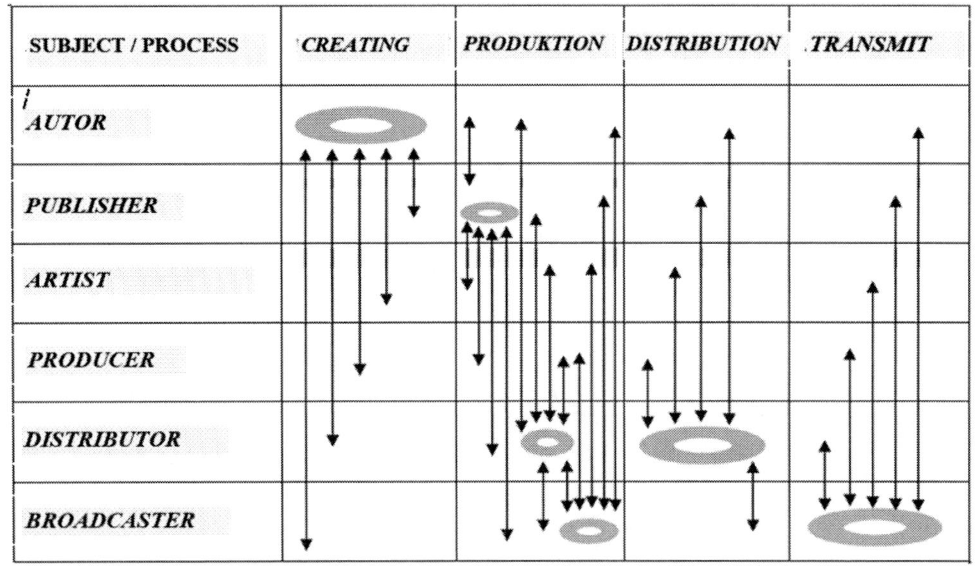

Fig 2. Selected subjects and process of intellectual property distribution.

Direct distribution of works is possible today mainly due to the rapid development of various technological platforms. However, the use of these platforms is associated with several pitfalls in securing distribution channels and new ways of licensing the provision of property rights to third parties. [14]

3 Methodology and research design

We used recognized general scientific methods in our research. We obtained information from existing literature. We also used empirical methods / observation, structured interviews /. We also examined existing metadata. When examining the data, a system approach / methodology of examining the spread of phenomena was applied without disturbing the overall context and links. We used other general methods of scientific research / analysis, synthesis /. The analysis was used to determine the bonds between the parts of the whole. The synthesis allowed us to formulate the conclusions of the research. We researched the content of selected concepts, current knowledge, models, selected processes in the real practice of sustainable distribution of copyrighted works. In the context of the most comprehensive review of the current state of innovation. We have defined selected indicators and problems of distribution process management in all phases of the life cycle. We focused our research goals on innovation and sustainable distribution of works: - we assume that in terms of innovation there is an ambiguous understanding and use of selected terms in terms of content, literature on the content of social innovation is inconsistent - we assume that more than half of respondents think they do not know or they do not know exactly what social innovation is. We assume that the distribution of works needs to be modified procedures. The formulated research goals were met by standard methodological steps. We also did our own research. We selected suitable research tools and methods, we identified a specific target object and subjects. We created our own database and processed the data and translated it into measures. The research used the method of purposeful analysis (questionnaire survey), which was focused on a selected sample. In the questionnaire, we formulated several questions in order to obtain evidence to achieve research goals. The questions were formulated in order to characterize the respondents, to characterize the individual subjects of the survey and to state the real state of innovations in the business sphere. First, we tried to look for data that someone in front of us had collected and analyzed. We had to supplement the existing data. For these reasons, we have collected our own data. We also examined the preferences of Slovak viewers for various forms of distribution of copyrighted works. A total of up to 100 respondents aged 15 and over participated in the survey; the upper age limit was not limited, more than half of the respondents were women and the remaining were men with different education. The questionnaire was distributed electronically and contained closed questions. Given the number of respondents involved, this is not a representative sample. The data obtained can at least partially serve to create an idea of the preferences and attitudes of the final consumer in the field of sustainability of the distribution of works.

4 Research results

We evaluated 250 opinions of respondents, mostly limited liability companies and sole traders. More than two thirds of companies are based in the Bratislava region with an average age of 15 years. The main activity of companies is mainly focused on services. We processed the collected data and gradually analyzed them. For some questions, respondents were offered possible answers, while others were asked to add specific information or express the respondent's opinion. Correspondents are relatively young people of working age. Externally cooperating respondents make up 18%. A small proportion of respondents also work as an accountant or economist. The structure of respondents is dominated by men aged approximately 35 years. Mostly these are the positions of owner, manager and manager. We briefly characterize the analyzed companies, the structure of legal forms, capital, including the extent of foreign participation, registered office, age, main activity, industry structure and size. We characterized the sample in more detail and took all the facts into account in the analysis and conclusions. The questions asked offer a choice of one or more legal forms of the business offered and at the same time provided respondents with the opportunity to add another institution dealing with intellectual property. When formulating the question, we assumed that the respondents would add other institutions, not just companies. It also happened in the range of 7%. However, respondents did not provide details of the institutions involved. We would also like to emphasize that 1% of respondents answered the questions asked in the questionnaire as a private person. Respondents' answers to equity documented more than half the dominance of domestic capital alone. Less than half of respondents declared a share of foreign capital of over 51%. We were also interested in the development of innovative activities. An increase in innovation activity was recorded in Bulgaria, Denmark, ahead of the Netherlands, Estonia, Ireland, Latvia, Lithuania, Belgium and the USA. The Slovak Republic and the Czech Republic recorded an increase in innovation activity at the same average level within Europe. Innovation activity was below the EU average was Finland, Germany, Luxembourg, Austria, France and Italy. The decline in activity was in several countries: Portugal, Malta, Greece. The most successful innovators grew by an average of 0.15%. The group of less successful innovators recorded an average decrease in innovation activity of 1.8%. The weakest group of innovators fell by 4.3%. Slovakia was a medium innovator. in 2010 and 2013 it also decreased. The performance of the Slovak Republic in relation to the EU has larger fluctuations. The EU peaked in 2012 (to 69% of the EU average), but also fell in 2014 (to 64% of the EU average). Slovakia maintained its position below the EU average in all monitored indicators. Recent findings at European level from 2019 show that innovation performance has been steadily improving in recent years. Europe has performed better in terms of innovation than the United States. Sweden performed best in terms of innovation, followed by Finland, Denmark and the Netherlands. The United Kingdom and Luxembourg have fallen from the top group of innovation leaders to strong innovators. Europe's innovation has improved in most indicators. The increase in indicators recorded for trademarks, patent applications, licensing, expenditure on research, development, innovation and others contributed significantly to the success. Despite the fact that the Slovak Republic experienced a shift in the period under review, its position within the Community did not improve much. It is still below the EU average. It is clear that the methodology for calculating the innovation of countries will not withstand frequent corrections. It would certainly be to the detriment of the comparability of historical data. In our view, it is possible to object to global comparability even now. Surprisingly, it is difficult to correctly examine and compare the state of intellectual property and its impact on the innovation of companies, employees and especially individuals. With certain reservations and commitment, it is possible to design and test a model involving industrial property. From our point of view, such an analysis in relation to the works of the authors is not feasible. Substantial innovations / especially in the field of technology / in relation to works are actually taking place in the environment. These innovations and related changes generate significant challenges for copyright entities and relationships. So far, we have not noticed any steps by the regulator in this area. New phenomena, subjects and relationships are emerging. The position of authors in securing their rights is constantly deteriorating. This situation is unsustainable in the future.

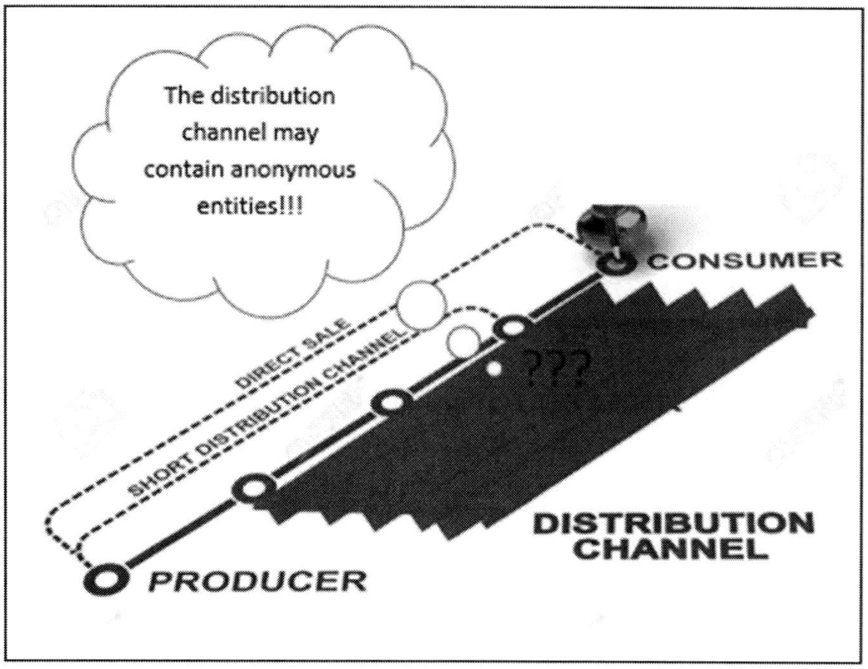

Fig 3. Distribution channel of intellectual property in internet.

5 Summary

Almost three quarters of respondents think that they do not know exactly how to define the essence of all kinds of innovation. Experts believe on the nature of social innovation. Some speak of a clear definition, others question the definition. Intellectual property research is based on legislation. The potential of intellectual property is not always the property of the company. Part of the rights / especially personal / are non-transferable. Therefore, the author, the originator is not always directly connected with the company and we only partially record intangible assets. leased, Externally purchased intangible products are problematic to keep track of records. Inaccurate data were obtained from the respondents' own survey. The questions were to refine the existing real data to achieve the research goal. We do not consider the obtained data set to be statistically significant, which may affect the reality of the results. The average value of the scope of research activities is M = 7,492; standard deviation SD = 10.9. Minimum value Min = 0, maximum Max = 50. The result of the research confirmed the inconsistency of records with reality. We also examined the correlation of selected parameters in order to identify the right indicators to measure the effects of the research results. We found a correlation between research spending, employment, turnover, wages and bonuses. The current situation has affected practically the whole world. Innovation has necessarily focused on digital technologies. They are being radically promoted in all areas of business and life. They have caused radical innovations in relation to the distribution of works. These innovations and related changes generate significant challenges for copyright entities and relationships. So far, we have not noticed any steps by the regulator in this area. New phenomena, subjects and relationships are emerging. The position of authors in securing their rights is constantly deteriorating. This situation is unsustainable in the future.

The paper is an output of the science project VEGA MŠ SR no. 1/0708/20 "Socio-economic determinants of sustainable consumption and production in terms of impact on the performance and competitiveness of enterprises"

References

1. G. Dubcová, K. Grančičová, D. Hrušovská, Customer Satisfaction and Loyaltytowards Companies in the Slovak Republic. *Business Economics, Management and Marketing 2018*. Proceedings of the International Scientific Conference, Brno : Masaryk University, 61-68, (2018)
2. G. R. Rughuram, L. Zingales, *Saving Capitalism from the Capitalists Crow Business*, 486 p., (2003)
3. M. Piko, *Autorské práva nie sú zadarmo ani bez byrokracie*, online at https://www.podnikajte.sk/pravo-a-legislativa/c/750/category/dusevne-vlastnictvo/article/autorske-prava-nie-su-zadarmo-ani-bez-byrokracie.xhtml (2012)
4. A. Jaffe, *Analysis of public research, industrial, and commercial innovation*, online at https://motu.nz/our-work/productivity-and-innovation/science-and-innovation-policy/analysis-of-public-research-industrial-r-and-d-and-commercial-innovation/ (2011)
5. D. Rybárová, Creative Industry as a Key Creative Component of the Slovak Economy. *Globalization and its Socio-Economic Consequences 2019*, Sustainability in the Global-Knowledge Economy. - Zilina : University of Zilina, Faculty of Operation and Economics Transport and Communications, (2019)
6. H. Majdúchová, B. Siváková, J. Kintler, Systematic risk determinants. *International scientific conference of business economics, management and marketing. ISCOBEMM 2017*. Proceedings of the [2nd] international scientific conference of business economics, management and marketing, Brno : Masaryk University, 60-170, (2019)
7. P. Kansal, S. Kapoor, *Basic of Distribution Management*, A Logistical Approach, 257 (2005)
8. F. Daňo, P. Kita, *Distribučný manažment. Bratislava* 2009, Daniel Netri. 235p., (2009)
9. H. Majdúchová, M. Kmerty Barteková. Innovations in the Creative Industry Entities, *Globalization and its Socio-Economic Consequences 2019*. International Scientific Conference. Sustainability in the Global-Knowledge Economy. - Zilina : University of Zilina, Faculty of Operation and Economics Transport and Communications, (2019)
10. T. Pacek, A. Mittelman, The current trends in the development of the right of intellectual property (not only) in the conditions of SR. *Social Sciences and Arts SGEM 2015*, Albena, Bulgaria, 715-721, (2015)
11. T. Gengyan, Analysis on the Position of Copyright Trade in the Sino–US Trade Frictio, *Publishing Research Quarterly* **36**, 284–295, (2020)
12. B. P. Krages, Pirates and Publishers: A Social History of Copyright in Modern China, *International journal of communication systems*, (2020)
13. J. Kissová. Problematika marží v obchode. *Aktuálne problémy podnikovej sféry 2015* Proceedings of the International Scientific Conference, Bratislava 2015, EKONÓM, 292-297, (2015)M. Matušovičová, D. Matušovič Recovery of the European Asset Management Ten Years After the Financial Crisis, *Croatian Economic Association, 2019*, **70**, no. 5, 782-801, (2019)

14. M. Matušovičová, D. Matušovič. Recovery of the European Asset Management Ten Years After the Financial Crisis. Croatian Economic Association, 2019, vol. 70, no. 5, pp. 782-801, (2019)

Startups and reasons for their failure

Lenka Mikle

University of Economics, Faculty of Business Management, Department of Management, 85235 Dolnozemská cesta 1 Bratislava, Slovakia
lenka.mikle@euba.sk

Abstract. The world is one big agreement. When we are going to school, when we are going to work or shopping. We can meet some type of agreement totally everywhere. In shop, we cannot take a thing without remittance. In school, we cannot take the best grades without knowledge and in work, we cannot take a financial reward without any added services. Business environment is very complicated, especially in these days. All trade activities are affected by a lot of factors, which are unpredictable. For customer is the most important right solution for their needs and for company is the most important find to way how to satisfy customer's need. The main purpose of this article is offering a view on small, beginning enterprise trying to provide customers something special through innovative product or used revolutionary technologies – the startup. On the other hand, as we know, minimum amount of startups can survive the first four years. Why is it so?

1 Introduction

Human individuality provides countless views of the world. Every person sees him differently. The physicist sees it as a set of laws that secure things around us. The doctor sees it as a mixture of human bodies, each one being identical and original. The mathematician sees it as a set of ubiquitous numbers. But what is the world really like? Dynamic, weird, unique and authentic. No action will take place again under the same conditions.

The current global economy is being commanded by large multinationals that set up smaller subsidiaries in different countries. The purpose of this business can be to gain a power position in the market or diversify risk. Small and medium-sized enterprises, which have a large representation in each country, are also helping to develop the economy. They are the driving force of any economy, bringing with them a number of positives. The last decade in the economic world has brought many innovations, new technologies and startups. The word, which refers to a small starting business with an innovative idea, began to spread from American Silicon Valley. This valley was the cradle for today's famous businesses, including Apple, eBay, NVIDIA Corporation, Intel, Google and Yahoo!.

Startups are among the greatest pioneers of business endeavour in the world economy. They bring innovative products and new approaches to customers, grow exponentially, become global enterprises.

1.1 Attribute of a startup

An American entrepreneur and author of many startups, Eric Ries, is undoubtedly one of the best-known authors on startups. Eric Ries has been involved in the creation of several startups, including IMVU. He advised business and product strategies for venture capital firms. Eric Ries [1] describes startup as an *attractive form of business that gives space for expectations to grow but can also be disappointing. He sees startup as an institution made up of people whose main goal is to create a new product or service in truly extreme and unstable conditions. According to Eric Ries, startup is an absolutely human business, not a business based only on a product, technology or innovative idea.*

Another expert for startup is Steve Blank [2], who has revolutionized business practice in business and innovation. Customer preference was a key factor in the search for a successful business model that came up with the topic of startups. Steve Blank and bob Dorf have worked on a book publication called The Startup Founder Handbook [3], in which they clearly and categorize startup as *a temporary organization looking for a scalable, repeatable and profitable business model.* He found that startups are not small versions of large companies, because large companies operate according to business models, startups are still looking for them. Startups on their business path have adopted product development methodology, product launches, life cycle phases and identical processes taught in entrepreneurship schools. Blank and Dorf examined in detail all too much the same reasons for startup failures. As a result, they came to the conclusion that startups need their own business tools, different from those used to manage large, existing businesses, because each startup moves in unknown and unexplored waters. The vision of startup is a series of untested hypotheses that require customer response. Information or data should be refreshed on a regular basis in days or weeks and not in months or years. The important thing is to keep it up-to-date and not waste the time spent unnecessarily improving a product that customers don't really want.

Also interesting is the view of the American businessman Paul Graham, who is a co-founder od Y Combinator. In 1995, together with Robert Morris, he founded Viaweb, the first SaaS company to become Yahoo!. In 1998 Y Combinator is a startup venture capital company. The main goal is to help startups bridge the first, most difficult phase of their existence and then introduce startups to larger investors or businesses interested in acquisition [4]. Paul Graham [5] from his own experience describes startup as a business designated for rapid growth. *Starting a business by itself does not mean setting up a startup. There is also no need for startups to start working on technology or take advantage of venture capital funding or have a fancy exit plan. The only important thing about startup is growth. Everything else that we associate with startups results from growth.*

The answer to the question "what is startup" can be provided by a person who has experienced the pitfalls of this business. Wil Schroter [6] founder of a web portal that gathers news from the startup environment, defines startup as the embodiment of the dreams of each founder. According to author, startup is the way from concept to reality, turning dreams into reality, which will be useful not only for the founder but also for the whole world.

The review of current literature has provided several points of view of startup and its main attributes. In addition, we can summarize that *startup is a small business in unstable or extreme conditions that tries to reach consumers by providing a product that is not yet provided by any other competing business or was the founder of an innovative idea in industry.* The business of each startup should be distinguished by its originality, which the startup can achieve through the use of original technology or creating a completely new need. In terms of startups, creating a new need is crucial, giving startup a competitive advantage in the industry.

1.2 Differences between startup and small or medium-sized enterprises

Startup as a business unit could be compared to small, respectively to medium-sized enterprises. But how can we tell which business is startup and which is simply a small business? Are there clear differences between them or will we have to rely on our assumptions?

According to the new definition, small and medium-sized enterprises are defined mainly by the number of workers. Micro enterprises are defined as enterprises employing fewer than 10 persons and whose annual turnover or annual balance sheet amount does not exceed 2 million Eur. Small enterprises are defined as enterprises employing fewer than 5é persons and whose annual turnover or annual balance sheet amount does not exceed 10 million Eur [7]

According to the Slovak Business Agency (SBA), in 2018 small and medium-sized enterprises (SMEs) made up 99.9% of the total number of business entities. Of this figure, up to 73.2% of SMEs contributed to employment in the business economy and 54.6% of SMEs contributed to total value added.

According to the Statistical Office, 560 521 business entities were active in 2018, of which up to 559 841 were small and medium-sized enterprises [8].

The question arises, how to distinguish startup from small business? Number of employees? Annual turnover or annual balance sheet total? Let's take a look at the differences that are undoubtedly between these business entities. Every founder, whether we are talking about startup or small business, strives to maximize profits and minimize costs. We know startups that also compress the employment curve as low as possible and contribute to overall added value.

According to Pope [9] the difference between a startup and a small business is often subjective in terms of growth goals and revenue forecasts. Startups are primarily aimed at "market distortions" and seek to increase revenues at a rapid pace. On the other hand, small businesses often set long-term goals and stable growth in the existing market. The author also highlights the high risk of startups. Every startup tries to find out if their product or service will find its place in the market. Investing a lot of time and money in startup is a huge risk.

The difference between a startup and a small business can also be seen from investor's perspective – investors have different expectations when providing funds. In the case of investing in a small business, the investor expects a slow repayment of the debt incurred. A small business poses less risk to the investor's capital in the long run but needs to be aware of a lower profit than the startup can generate. If the investor decides to put his money into the startup, he uses the portfolio investment that enables risk diversification [10].

Investing in every single type of business carries a certain amount of uncertainty. However, putting funds into startups is a bit riskier because the founder has only a business idea, no specific product or service. In the long run, a small business seeks to stay on the market, generate slow-growing profits and still be on the positive side. The return on funds from this type of business is more secure but should be expected in the longer term. Providing startup money is riskier because of having only an idea but if a business idea is taken up and successful in the market, the repayment of the deposit can be in a much shorter time frame. In the long run, however, it is more interesting for a startup to rocket up and make unimaginable profits or to be "bought" by a large, established company that can replace one of its departments, such as Research and Development.

Another difference between a small business and a startup can be seen in the question of the legal form companies can choose. Small and medium-sized enterprises have the largest representation in the form of a limited liability company. In 2017, the Commercial Code of the Slovak Republic was amended to a new legal form, which allows startups to use the legal possibilities that the state allow them.

1.3 Reasons of startups failure

The founders are very happy to share their advices, tips and experiences about the success of startups. Otherwise, if the startup does not make such profound profits? The American portal CB insights, which collects and analyses a huge amount of data using algorithms and data visualization, was engaged in this reverse side. The main task of the portal is to provide answers to important strategic questions and thus to provide data that enable easier, not only strategic, decision-making.

Based on the information gathered, CB Insingths [11] compiled a ranking of the most common factors of failure. The survey was conducted on 101 startups, which unfortunately could not resist the pitfalls of the startup environment.

The most common cause of failure is the no need of product in up to 42% of cases. The founder of the startup is firmly convinced of the originality and innovativeness of the product and if it comes to production without prior analysis of customer segments, the product may not find the right feedback from consumers. Startup will not help great technology, excellent data on customer buying behaviour, expertise of team members if it does not address the current market need.

Another serious reason for failure is the lack of capital. Money and time are limited not only in startups, but their prudent distribution is important. Up to 29% of startups cited the misapplication of capital as a source of their failure.

The third most important reason for a startup failure is to build the wrong team. A diverse team with different experiences has often been cited as critical to overall success. The founders are aware of the missing or inadequately assembled team. When starting a business, the founder is confident and do not feel the need to share their initial enthusiasm with other team members, but over time, with increasing activities, tasks and roles will be misaligned. Team question was crucial for up to 23% of startups.

Poor marketing may also be an interesting factor. Knowing your customer segment, reaching out to it and delivering the right product is one of the most important skills of a successful startup. Know-how to attract attention and turn non-customers into customers. The cause of the failure occurred when startup was unable to launch its product. The founders did not use the idea of proper product promotion.

Burnout is often a serious problem. The work-life balance cannot be overlooked indefinitely. This was cited by 8% of startups as the cause of the failure. Being able to respond correctly in a dead end, reorienting your mind from loss to profits is considered equally important for success or prevention of burnout. It is in this situation that we can draw attention to the necessity of the tam, given the possible division of competences, which would largely prevent burnout.

The first online publication on startups in Europe can provide an unusual look at the failure factors. The authors have provided an overview of all startups since October 2010. The authors of the articles are particularly interested in technological startups., provide readers with analyses, interviews with the founders of individual startups, or news from the startup environment. Startups are broken down by country. They also provide important information and updates for potential startup founder. The author Arnaud [12] created a ranking of the ten most important reasons for startup failures, many of the causes are the same as those provided by CB Insights by the public, but a few are different and meaningful. The author advises companies in commodity trading and has established several companies during his business life.

An important cause of failure can be attributed to founders who are looking into their product and unwillingly receive product feedback and criticism. It is essential to realize who the product is, whether it is the founder or the consumer. Without accepting customer feedback, startup can hardly think of success in the industry. The founder may not be willing to publish a prototype for several reasons, for example is not sufficiently prepared or is concerned about competition. But without first introducing the product, the founder will never know the response and feedback from customers.

The last most common factor of startup failure may be incorrect timing. Some startups launch products when the right technology is not yet available. It may be fatal for a startup to come up with a product that is not ready for the market or customers. It is essential to capture the right moment to launch the product.

2 Methods

The aim of the title is to provide a view of startup as a business entity, which is extremely important for the economic environment, but largely threatened by factors that will often not allow it to survive on the market.

The main goal of this article is to provide an overview of the most common and most important factors affecting the life of a startup. The main objective of the article is supported by a several intermediate goals:

- an overview of current opinions on startup from several experts who act in the startup environment as an investor, co-founders of startup or scientists,
- the differences between a startup and a small business, given that in some case it may be difficult to distinguish between these two business entities.

This article serves as a literature search for readers by providing the most up-to-date opinions of current experts on the issue being addressed and allowing further questions to be addressed.

3 Results

As we surely know, startup is a small, beginning enterprise existing in extremely unstable conditions. The main idea of startup is providing innovative product or service. Every startup should try to open a blue ocean – segment of customers who are not served by existing business in the market. The best situation would be to create a new need that customers could appreciate.

In the case of startup, we should be aware of the differences that distinguish it from a small business. The first main difference is innovation because a small business does not require innovation as the same level as startup, for which this feature is key. Another difference is the rate of profit, every entrepreneur is trying to maximize profits and minimize costs, but startup is expected to achieve exponential revenue growth. The vision can make the difference between a startup and a small business. The small business strives to be active in the market for as long as possible, to generate constant profits and to maintain its customer segment. Startup aims to create a boom, gain exponential growth, sells as many products as possible and then be bought by a large company.

Based on the research created by the CB insights web portal, it can be argued that the most common causes of startup failures are lack of experience, insufficient marketing, lack of funds, competition and the founder's unwillingness to receive feedback.

4 Discussion

The biggest problem for small and medium-sized enterprises is access to capital in the initial phase of business, which can be fatal for startups. For a more efficient functioning of the national economy, attention must also be paid to smaller business entities, startups, which however, rarely survive without help. Provide support and accept their market survival efforts.

Microenterprises, small and medium-sized enterprises contribute greatly to the proper functioning of each country's economy They create jobs, thereby reducing the level of unemployment and contributing to GDP. All of these benefits also apply to startups, which ranks among microenterprises by the number of employees.

At the end of this article we can express the importance of startups, small and medium-sized enterprises, support all government institutions for increased support of these enterprises.

References

1. RIES, E. The *Lean Start-up. How Today's Entrepreneurs Use Continuous Innovation to Create Radically Successful Businesses.* 1. edit. New York: Crown Business. 2011. 468s. ISBN 978-0-307-88791-7
2. BLANK, S. *About Steve.* 2015. [online]. Available on internet https://steveblank.com/about/
3. BLANK, S. – DORF, B. *The Startup Owner's Manual : The Step-by-Step Guide for Building a Great Company.* 1. vyd. Pescadero : K&S Ranch, Inc. 2012. 182s. ISBN 978-0-9849993-7-8
4. GRAHAM, P. *About Y Combinator.* 2014. [online]. Available on internet https://www.ycombinator.com/about/
5. PAUL GRAHAM. Want to start a startup ? 2012. [online]. Available on internet http://www.paulgraham.com/growth.html
6. McGOWAN, E. (2018). What is a Startup Company, Anyway? Available on internet <https://www.startups.com/library/expert-advice/what-is-a-startup-company>
7. EUROPEAN COMMISSION. (2006). New Definition of small and medium-sized enterprises : User Manual and Model Statement. Online: <https://www.unms.sk/swift_data/source/dokumenty/technicka_normalizacia/msp/def.%20podnikov-sme_user_guide_sk.pdf
8. SLOVAK BUSINESS AGENCY. (2019). Správa o stave malého a stredného podnikania v Slovenskej republike v roku 2018. Available on internet <http://www.sbagency.sk/sites/default/files/sprava_o_stave_msp_2018_aktualizovane.pdf>
9. POPE, E. K. (2020). Startup vs Small Business : What's the Real Difference ? Available on internet <https://www.fundera.com/blog/startup-vs-small-business>
10. STEPHENS, H. (2018). What's the Difference Between a Small Business and a Startup? Available on internet <https://www.smartcompany.com.au/startupsmart/advice/difference-between-small-business-startup/>
11. ARNAUD, B. *The 10 Most Common Reasons why Startups Fail.* 2018. [online]. Available on internet https://www.eu-startups.com/2018/09/the-10-most-common-reasons-why-startups-fail/

EVOLUTIONARY MODEL FAMILY BUSINESS - concept of special model

Zdeněk Mikoláš[1,*], *Iveta* Vozňáková[2]

[1]VŠPP Praha, Spálená 76/14, 110 00 Praha - Nové Město, Czech Republic
[2]VŠPP Praha, Spálená 76/14, 110 00 Praha - Nové Město, Czech Republic

Abstract. The authors of the article present the practical and theoretical knowledge which they have achieved in the framework of the entrepreneurship research over the past 10 years. The practical and research activities have focused in the last period, in particular, on the synergy of the links between the family business, innovative dynamism and competitiveness of the companies. The paper offers a generalization of the findings to the theory of synergy, economics of the time, the family business, especially in the innovation and dynamism of the evolutionary family business potential in the process of succession. The findings are also based on empirical research in more than 230 small and medium-sized family companies in the Czech Republic and Poland. Presented results deserves a broad discussion and verification in practice, depending on the cultural differences in manifestations of globalization in various parts of the world and under the influence of the industrial revolution 4:0

1 Teoretical background

The practical and research activities have focused in the last period (over the past 10 years), in particular, on the synergy of the links between the family business, innovative dynamism and competitiveness of the companies. The paper offers a generalization of the findings to the theory of the family business, especially in the dynamism of the evolutionary family business potential in the process of succession. The findings are also based on empirical research in more than 230 small and medium-sized family companies in the Czech Republic and Poland. Presented results deserves a broad discussion and verification in practice, depending on the cultural differences in manifestations of globalization in various parts of the world and under the influence of the industrial revolution 4:0. The following text is based on the theoretical conclusions of previous research projects [4,5,6,7,8,9,10,11,12,13].

* Corresponding author: zdenek.mikolas@vspp.cz, iveta.voznakova@vspp.cz

1.1 Key terms of the problem

Family business defines the synergies (unification) with 4 subsystems (circles) [5,7,8,11,12]: family, business (family business), environment and outputs.

Entrepreneurship is a process of the formation of something else, which belongs to a value, by spending the necessary time and effort, taking accompanying financial, psychological and social risks, and the acquisition of the resulting rewards in a form of funds and personal satisfaction .

Family is, in the sociology, a solidarity group of persons related to each other by marriage, relationships or adoption, that they live together in the long term and whose adult members are responsible for the upbringing of children.

Potential is the difference between what it is and what it can be (results from knowledge) or it must be (results from natural and legal laws).

1.2 The concept of evolutionary business modeling

The concept of evolutionary business modeling is based on business diagnostics and therapeutics, on a specific approach of scientific research known as "empiriology", based on experience, comparing knowledge with available theoretical knowledge, then formulating new theories and validating them empirically (experiments) in practice.

The author of the model (Z. Mikoláš) combines the Sarasvathy [15] effectuation and causal logic. Together, these principles lead to an effective way of reasoning and acting, drawing on the potential of previous generations.

2 Evolutionary model family business

2.1 The morphology of the potentials

The total potential of the family business has this basic morphology of P = A + B + C, where A is the effective potential (changed with the effect/E outputs), B is the spending potential (changed with the amount of spent resources M to achieve effects E, it also has the potential to eliminate the negative emissions and losses), and C is the stabilized potential (deferred for the next generation), in greater details [8,9,10,11,12,13].

2.2 Spontaneous order of network synergy potentials

The above mentioned morphology of the potential for the family business is dynamically changing by the spontaneous order of objective and subjective factors. It manifests itself in the form of a combination synergy of objective and subjective order effects and spontaneity in the form of threats, opportunities, strengths and weaknesses of the family business.

The following text contains generalized research findings of the authors that describe the "ideal theoretical" variant of a family business that is carried out in the framework of the objective of "natural" order and the rational subjective (human) behaviour (negentropy) by accepting the objective and the subjective spontaneity (entropy) [8,9,10,11,12,13].

2.3 Evolutionary dynamism

The ideal theoretical variant of the family business relies on the validity of the two basic natural and social evolutionary principles (it follows: laws of nature, instincts, reflexes):

$P_{i+1} > P_i$ a $(A_{i+1} + B_{i+1}) > (A_1 + B_1)$ two consecutive generations (i, i+ 1) of family business.

Therefore, under such conditions of the basic evolutionary (reproductive) legality for the two consecutive generations of the productive system, it is possible to write down in the following two variants:

a) $\chi_{i+1} = P_{i+1} : P_i > 2 - (C_i : P_i)$,

b) $\chi_{i+1} = 1 + ((A_i + B_i) : P_i)$.

Family business is characterized by these two propensities:

a) propensity to consumption $1 < \alpha_i = (P_i - C_i) : P_i > 0$,

b) propensity to save $\beta_i = 1 : ((P_i - C_i) : C_i) > 0$. Be notified of "a new fuzzy" philosophy of the concept of reproduction (evolution), which offers not only standard an analytical view, so-called, what proportion has the part in whole -see a) equation propensity to consumption, but it also searches for the answer to the question what is the future of all (i + 1) in the old part (i), from which it draws its foundation - see b) the equation of propensity to save. Propensity for consumption and savings are indirectly commensurate with the relationship.

It follows derivation of the formulas is indicated in sources the perfect growth potential (P) of family business the two consecutive generations (i, i + 1) is equal to the number $\varphi = P_{i+1} : P_i$. This growth dynamism is given by the ratio of stabilized (for the next generation of delay) potential (C) to the total potential (P) of the previous generation (i), that is, $\varphi = 2 - (C_i : P_i)$.

Finding the balance between the two propensities (for consumption and savings) we come to several optimal values:

a) $\gamma = C_i : P_i = 2 - \varphi$, approx. $\gamma = 0{,}382$,

b) $\omega = (A_i + B_i) : P_i = \varphi - 1$, approx. $\omega = 0{,}618$,

c) $\vartheta = C_i : (A_i + B_i) = \varphi - 1$, approx. $\vartheta = 0{,}618$.

Therefore, if the default (i) generation lays down approximately 38.2% of your potential for the future generation, then the following generation (i+1) will increase the total potential of approximately 61.8% compared to the total potential of the previous generation (i). Therefore, the total consumption (A+B) of default generation (i) is approximately 61.8% of the total potential (P), but also the proportion of deferred (stabilized) potential (C) on consumption (A+B) is also around 61.8%.

At the same time we can deduce that $\varphi = 1 + \sum_{i=1}^{n} (1 : \varphi^i)^2$, for **n** approaching asymptotically to (+ ∞).

In other words, the ideal dynamics of the growth potential of the family business is defined by a number of evolution potentials from (I) the present ($P_i = 1$) up to the past (asymptotically in the endless series of "ancestors", II, III, IV, ...). The referred argument we can graphically demonstrate.

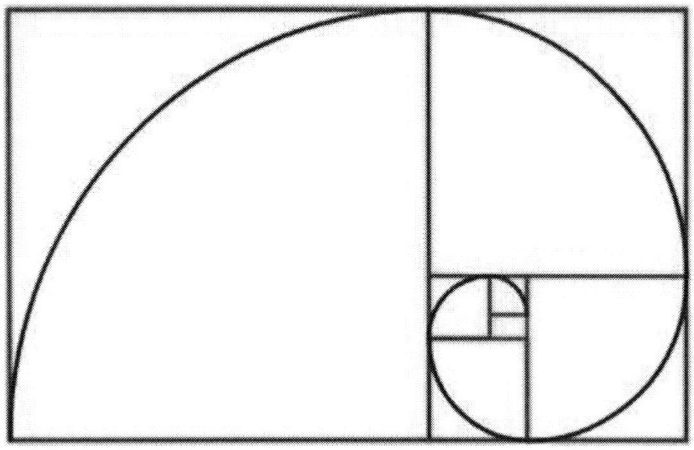

Fig. 1. Evolutionary series of potentials in business Source: authors

2.4 Optimization proportions of the golden ratio

Everything what is in nature composed so that the result was "appealing" harmony, is related to the golden ratio. Such objects and phenomena are not only more beautiful, but also healthier, effective and work better. This phenomenon has its deep causes that express themselves in many different areas of mathematics (such as the proportions of Platonic solids), computer science (e.g. the theory of heaps) through biology (shapes of the shells, horns, the proportions of DNA), music, modern and ancient architecture to astronomy (such as parameters or conditions of the orbits) or physics, which won its specific name golden physics . It is obvious that even most of the proportions of nature are directed by the golden ratio, and if it does not control, it is not such harmonious proportions.

Due to the fact that economics is an optimization science, it should be reflected also in this area. The task of an economist is in collaboration with experts in the area to find the most effective solution to achieve the given objective with regards to thrifty treatment of rare resources. So we find that the phyllotaxis, i.e. the spiral-shaped layout of the leaves on the stem in the proportions of the golden ratio allows to make the best use of solar radiation. Optimum use of space represents so-called the golden tree , whose branches are becoming shorter in proportion of the golden ratio (e.g. lungs).

This article is mostly implicating the issue of optimizing the development of the family business. Development of economic (commerce) systems in different hierarchical levels is rarely uniform, but almost it always shows a tendency. Here you can view the features in the chart with the Cartesian coordinates, where the x axis is time, or in polar coordinates, where the time is represented by evenly increasing angle (α) and development is given by extending of radius vector (r). Development is then modelled by the spiral. If the radius vector grows in proportion to the angle (a parameter of proportionality is (a)) that is the Archimedes' spiral, which corresponds to a clear extensive development , i.e. the development without innovation achieved by simply expanding of a production range. Therefore, $r = a*\alpha$.

Clear intense development, which is supported by the only innovation, is given by the logarithmic spiral and in a real life is virtually unattainable, therefore $r = a*e^{b.\alpha}$. The so-called "Golden spiral" represents a harmonious development, which is sustainably innovative.

2.5 The evolutionary law of economics time in family business

For the total potential of family business P = A + B + C results from economic theory (economics of time) relationship A = B*v. Where A is the effective potential (changing into effect E), B is spending (changing into resources spent on mass M), whereas v is the speed of productive transformation of the effective potential B upon the spending potential A. In economic theory and practice, these general characteristic is referred differently: labour productivity, profitability, yield, efficiency, etc.

Descriptive analysis (Figure 2) of behaviour in family companies we have come to the conclusion that the following two generation of family business innovate a productive transformation with increasing speed, i.e., $v_{i+1} > v_i$. Ideal acceleration can be described by circles k (circumscribed square ABCD) and l (inscribed in a square ABCD). Radius of circle k (abscissa u) is equal to the speed of the transformation of the new generation v_{i+1}. Radius of the inscribed circle l (abscissa w) is equal to the speed of the transformation of the previous generation v_i. The derivation of that relationship is done in resources [11,12].

It follows that the optimal pace of accelerating with innovation cycles (dynamism of generations change) of productive transformation is when $v_{i+1} = v_i*\sqrt{2}$, or approximately $v_{i+1} = v_i *1{,}414$. At the same time that the optimal pace of contraction of relative time in innovation cycles ("business" time) of productive transformation is t_{i+1} = 1: ($t_i*\sqrt{2}_i$) or approximately $t_{i+1} = t_i *0{,}7071$.

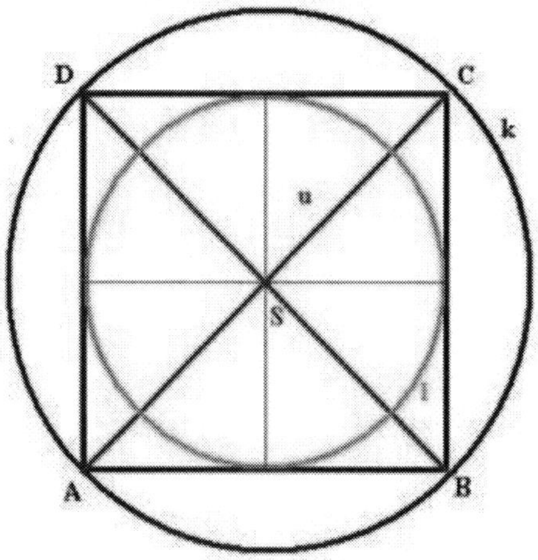

Fig. 2. Evolutionary series of productive transformation speed Source: authors

Summing up the above findings, then we can write an equation of $B_{i+1} = (\varphi - 1) : (v_{i+1} + 1)$, when an ideal evolutionary growth is true relationship $B_{i+1} = (\varphi - 1) : ((v_i*\sqrt{2}) + 1)$, for i = 0, 1, …., n (+ ∞). Alternatively, we can write $v_{i+1} = v_i*\sqrt{2}$, for i = 0, 1, …., n (+ ∞). Therefore, the speed in the following cycles in the ideal order increases in multiples of √2 and time of productive transformation (the existence of consecutive cycles) is shortened in multiples of 1: √2 [11,12].

If the speed of transformation vi is the basic unit (scale) for changes in the family business for the next two generations, we can be regarded as the unit metrics vi = 1, than $v_{i+1} = \sqrt{2}$. Therefore, $\xi = v_{i+1} : v_i = \sqrt{2}$. And it can be used for the default (basic) generation of a productive system to infer the approximate ratio of $\delta_i = B_i/P_i$ =0,309. And for the next generation of the productive system is: $\delta_{i+1} = B_{i+1}/P_{i+1}$ =0,256. It follows that the proportion of the mass of resources on the overall potential is reduced, $\delta_{i+1} < \delta_i$.

The optimum ratio of the speed of the two following (generations) transformations is derived from optima of positive synergies "coexistence and competition" of two generations, if $\xi = v_{i+1} : v_i = \sqrt{2}$, then roughly $\lambda_{i+1} = B_{i+1} : B_i = 1{,}340$ and $\mu_{i+1} = A_{i+1} : A_i = \xi^* \lambda_{i+1}$, or $\mu_{i+1} = 1{,}896$. The new generation (i+1) achieves higher speed in the reproductive cycle of productive transformations of 41.4%, than the previous generation cycle (i). It follows that the effective potential A of new generation (i+1) is about 89.6% higher than potential A, in the previous cycle (i) and at the same time spent potential B (resources) intergenerational rise about 34%.

The result of the phenomena described above is:
a) physical extension of family business (the family of entrepreneurs) $t_{fi+1} = (\varphi : \xi) * t_{fi}$, thus approximately $t_{fi+1} = 1{,}144 * t_{fi}$,
b) shortening of the business (working) time $t_{ei+1} = (1 : \xi) * t_{ei}$, $t_{ei+1} = 0{,}707 * t_{ei}$.
As a result of productivity growth of the family business is shortened productive (business) time between the following time cycles (generations) of around 29.3%, but at the same time increase of the total potential of the productive system causes the physical renewal of time of existence of the following generation of approximately 14.4%. There is therefore a paradox (the dilemma) of time: the new generation lives in business shorter time t_{ei+1}, its business (innovation) life cycle is shorter compared to the original generation of (i), and at the same time, its physical existence time t_{fi+1} is extended. Disparity or loss (excess) between the physical time and business between the following generations of the family business. Disparity (loss) in the following evolutionary cycles increases, provided that there are ideal conditions of spontaneous order of family business. Described phenomena bring into the family business a variety of other economic, social, ethical and other disparities. [4,5,6,11,12].

3 Discussion and simulation business

Based on the above theory we can make the following estimates of the evolution of the potential business in the researched family companies.

The researched generation of entrepreneurs (i) will be in business (as owners) in diameter $t_{ei} = 30$ years, as members of the business family will live approximately $t_{fi} = 80$ years.

For the successors, based on the simulation of the evolution of family business, we could estimate that

a) the physical extension of family business (the family of entrepreneurs) of new generation (i+1) will be approximately over 90 years of age, or $t_{fi+1} = 1{,}144 * t_{fi} = 1{,}144 * 80 = 91.5$ years,

b) and reaching the shortening of the business (working) time, which will exceed 20 years, so that $t_{ei+1} = 0{,}707 * t_{ei} = 0{,}707 * 30 = 21.2$ years.

Based on questioning in family companies it shows that the successor generation (i+1) begins to do business later (as owners) than their parents (between 30-35 years) and active business activity ends by transferring of the family business by their successors (i+2) at about 55 years of their age.

This simulation describes the ideal state, ie the development of family business with maximum intensity and efficiency. The applied research of the authors of this article provides two key features:

a) family businesses in the Czech Republic are conservative and are more prone to savings and investments with less risk, so overall potential growth is usually slower than the ideal value $\varphi = 1{,}618$ [11,12],

b) innovation dynamics in family firms in the Czech Republic usually do not reach the ideal value $\xi = v_{i+1} : v_i = \sqrt{2}$; only 3% of family companies included in the research sample were prepared for the Industrial Revolution 4.0 [9,13], ie for higher-order qualitative innovations - according to the scale of F. Valenta 6th and higher order [17].

Summarizing the above research findings, we can define the real parameters of family business in the Czech Republic: ,

a) $\varphi^r \in (1; 1{,}618)$

b) $\xi^r \in (1;\sqrt{2})$.

Based on the processing of research data, we conclude that the mean value of the parameters of contemporary family business is $\varphi^s = 1{,}3$ a $\xi^s = 1{,}2$. Then we can modify the previous simulation as follows:

1) The researched generation of entrepreneurs (i) will be in business (as owners) in diameter $t_{ei} = 30$ years, as members of the business family will live approximately $t_{fi} = 80$ years.

2) For the successors, based on the simulation of the evolution of family business, we could estimate that

a) the physical extension of family business (the family of entrepreneurs) of new generation (i+1) will be approximately $t^s_{fi+1} = 1{,}083 * t_{fi} = 1{,}083 * 80 = 86.6$ years,

b) and reaching the shortening of the business (working) time, which will $t^s_{ei+1} = 0{,}833 * t_{ei} = 0{,}833 * 30 = 25$ years.

The table shows the baseline data (tf0 and te0) and simulation data for the ideal variant (t_{fi+1} and t_{ei+1}) and the intermediate variant (t_{sfi+1} and t_{sei+1}).

Table 1. Simulation of family business dynamics in the Czech Republic

Time	t_{f0}	t_{e0}	t_{fi+1}	t_{sfi+1}	t_{ei+1}	t_{sei+1}
Years	80	30	91,5	86,6	21,2	25

These simulations are presented in the following graph.

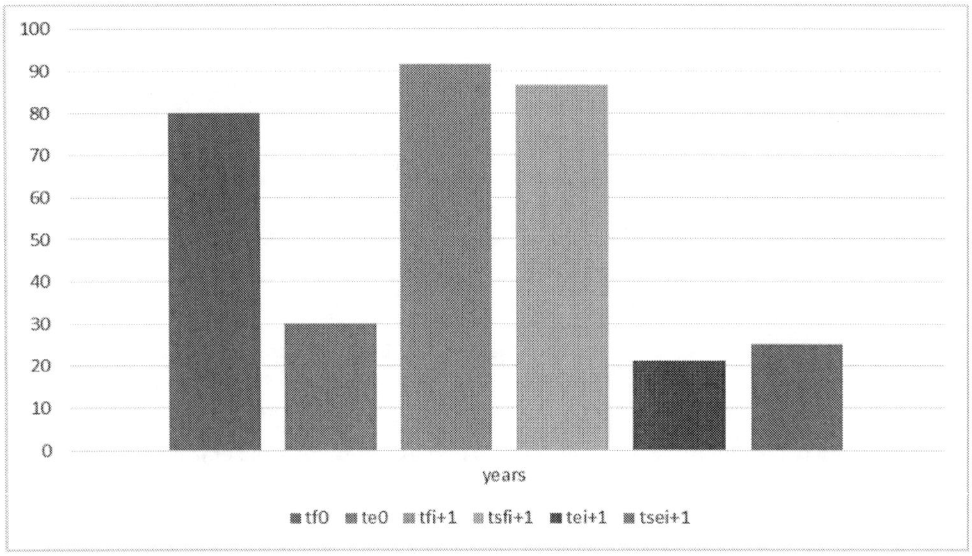

Graph 1 Simulation of family business dynamics in the Czech Republic

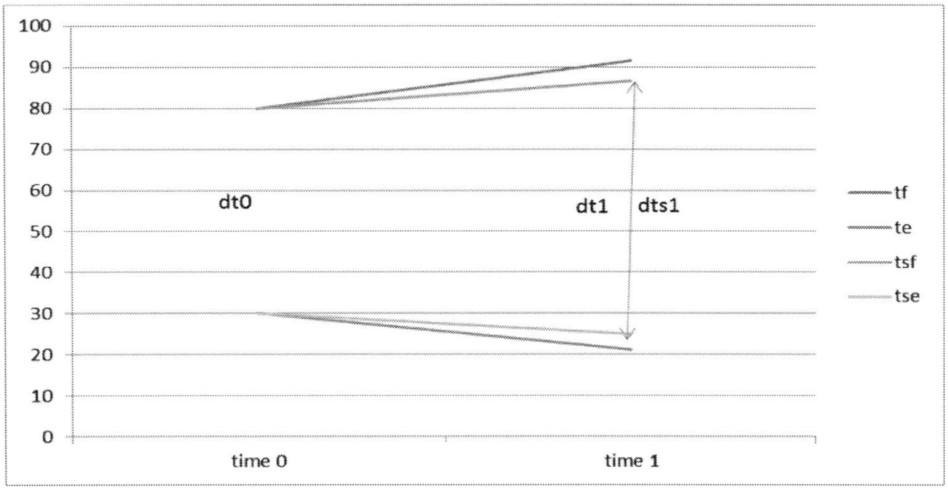

Graph 2 Disparity of time evolution

The disparity of time dynamics is shown in the graph and shows the variant fan between the optimal and medium trend:

a) default disparity is $d_{t0} = (t_{f0} - t_{e0}):t_{f0} = (80 - 30):80 = 50:80 = 0,625$

b) optimal disparity trend is $d_{t1} = (t_{f1} - t_{e1}):t_{f1} = (91,5 - 21,2):91,5 = 70,3:91,5 = 0,768$

c) the mean disparity trend is $d_{ts1} = (t_{sf1} - t_{se1}):t_{sf1} = (86,6 - 25):86,6 = 61,6:86,6 = 0,711$.

The conclusion is obvious from these results. The disparity of time is increasing under the influence of natural, technical, economic, social and psychological factors. The scissors between t_f and t_e open and pose considerable challenges. Pressure can be expected on qualitative changes in family development, entrepreneurship and in the priorities of society and the environment.

This means that there will be changes in the concept and functioning of families, changes in the governance of family businesses etc. under the influence of the industrial revolution (0.4 and following). We must also be aware of the historical context of family business research in the Czech Republic and the "new" EU members:

a) In post-socialist countries, especially in the Czech Republic, the family was renewed after about 50 years around 1990. The period of World War II completely suppressed the business of Jewish families and significantly reduced the business of the Czech population if it did not collaborate with fascists. The revival of family business after World War II was fully suppressed after 1948. As a result, before privatization (before 1990), the private sector in the Czech Republic had only 1.2% of property value and 2% of labor force in employment. From the above calculations and simulations follows the contemporary paradox. Entrepreneurship (incl. Family) has been developing in the Czech Republic again after 1990, therefore there are not enough families with a complete reproductive cycle of the next three generations, which is necessary for revealing the nature ("DNA") of family entrepreneurship in specific cultural, social, economic etc. state (eg Czech Republic). In this situation, the validity of the above model cannot be statistically verified at present. It is therefore a theoretical model derived from empirical knowledge and formulated on research principles other than statistical data analysis.

b) The advent of the Industrial Revolution 4.0, new technologies, socio-economic, cultural and other factors of the contemporary concept of entrepreneurship and social life raises the need to create new theoretical models of family business.

From this it can be inferred that these results come from the preservation of conservative values in families and national cultures of inter-generational conflicts, from the selective impact of globalization on social classes and business entities, etc. All referred findings it is possible to compare with the findings of other authors listed in the list of sources for this article.
It follows from the above text, the family business model and the available sources that there is a time when it will be necessary to reformulate the definition of basic paradigms, such as family definition, entrepreneurship, social evolution, etc.

4 Summary of findings

The evolutionary theory of the family business and related practical experiment introduces several final findings.Family firms represent the typical features of the business subject to objective and subjective spontaneous order. In the ideal (harmonic) variant very few companies work (these companies have already existed for a series of successive generations, even centuries). According to experts, a very risky is the third business generation (i+2), which only represents a few percent (2-3%) of companies founded by the first generation (i). This fact is not possible to verify in the Czech Republic, because at present, the transmission of the family companies from the first (i) on the second (i+1) generation is in progress.

It is confirmed the formation of evolutionary networks of family business (both in terms of the dynamics of the potentials, and in terms of the economics of time). By means of the comparative analysis we came to the conclusion that it can be applied successfully in research of spontaneous order within the theory of innovation by Professor František Valenta [17].

The group of family companies, which move beyond the ideal variant of development of spontaneous order, we can mean of rationalisation innovation. F. Valenta distinguishes four systems of rationalisation innovations, whereas the basic rule for the recognition of rationalisation innovations with different order is a finding what can sustain on existing economic status process (technical, social, economic, etc.), and what is going to be changed by innovation. This idea raises additional research orientation of the family business.

While every qualitative change remains something of the original initial element and something qualitatively changes. Then we are looking at the qualitative change from the perspective of what has changed, and this is something new, or we are looking at in terms of what persists, then it is an enhancement of what persists in a new quality [14].

Advanced the ideas of F. Valenta [17]., we see that innovation of the 6. order is innovation at the level of the new generation, which represents the start of production (transformation, manufacture or application) of the new economic subject (product, the production factor, technology, in our case the family company), which is from the closest relatives' other subjects diversified with a completely new structural (design or otherwise defined) layout (i.e. a solution to all its property, work, control functions, etc.) relevant to the user (owner).

If we convert this concept on family business, then we would understand for the fundamental qualitative innovation (6. order) in the family business the intergenerational succession of the family business by its successors. Therefore, it is a family business innovation on the level of the new generation of successors that represents business in undertaken or started a new business derived from the closest business relatives (genetically or adoptive), established with significantly different (new) concept, style, new business issues (or potential) and solution of business functions, relevant for the family, a business entity (company) and the relevant business environment.

We may declare our opinion, that the social and "industrial revolution" 5.0 is getting near and other global economic, social, cultural, ecological and other contemporary turbulences, built on the "microcosm optimization" of living matter, having two extreme marginal variants for the next (evolutionary) development of family business [13].

The present research has come to the fragmentary conclusion. The key factor for business development at the present time is the innovation process in the area of technology, often completely new and hitherto virtually unknown. Family businesses to absorb this trend, they should proceed to internal changes in paradigms and the dogmas of the behaviour of the families. There should be the innovation in behaviour of the families. It is essential that the successors should be trained from childhood to entrepreneurship, to contemporary technological mind-set, to maintain family traditions and values in the new historical and social context. Our conclusions are also confirmed by the latest medical knowledge, especially psychology, psychiatry and genetics.

The submitted theory deserves a broad discussion and further verification in practice, depending on the cultural differences in manifestations of globalization in various parts of the world and under the influence of the industrial revolution 4.0 or 5.0 ingoing, the migration of nations, environmental changes, etc. We should be aware of the current world is turned, without we exactly know what it will follow. Therefore, each expert opinion and scientific discussion are completely necessary and useful.

Roughly 30 years of renewed entrepreneurship and especially family entrepreneurship in post-socialist countries provide a number of incentives for theoretical research that do not have a knowledge base in a world where entrepreneurship has not been discontinued due to a different social order than in post-socialist European countries

References

1. B. de GROOT, H.P. Frances. *Cycles in basic innovations*. Technological Forecasting & Social Change, vol.76. (2009)
2. D. Filson, *Product and process innovations in the life cycle of an industry.* "Journal of Economic Behavior & Organization", Vol. 49. (2002).
3. M.Horooka. *Nonlinear dynamism of innovation and business cycles*. Journal of evolutionary economics, vol. 13, no. 2003. (2003).
4. A.Ključnikov, M. Civelek, P-Čech, J.Kloudová, *Entrepreneurial orientation of SMEs' executives in the comparative perspective for Czechia and Turkey*. Oeconomia Copernicana, 10(4), 773–795. doi: 10.24136/oc.2019.035, (2019).
5. A. Ključnikov, A. *Uncover SMEs finance through the impact of the specific factors. Evidence from Slovakia*. Transformations in Business & Economics, Vol. 15, 2B, pp.741-754, ISSN 1648 – 4460. (2016).
6. A. Ključnikov, A., J. Belás, J., L. Smrčka, L. (. The Role Of Risk-Taking And Competitive Aggressiveness In Management of SMEs. Polish Journal of Management Studies, 14 (1). ISSN 2081-7452, DOI: 10.17512/pjms.2016.14.1.12 (2016)
7. Z. Mikoláš, V.Fialová,. *The Dilemmas of the Power of Market Economy.* Studia Ekonomiczne. Zeszyty Naukowe, Nr 303, s. 9 - 23. Uniwersytet Ekonomiczny w Katowicach, (2016).
8. Z. Mikoláš, J. Peterková, *Dilemas of Innovative Dynamism and Social Responsility.* In. Interantional Journal of Investment Management and Finacial Innovations. Volume 2, Issus 1/2015, p. 35-40, (2015).
9. Z. Mikoláš, Z.Wozniaková, *Synergic Time Effect of Intelligent Enterprise.* In. Interantional Journal of Latest Research in Science And Technology. Volume 3, Issus 3, June 2014, (2014).

10. Z. Mikoláš, *Potential Theory and Innovation Dynamics of Enterprise.* Studia Ekonomiczne 2014, nr 183, s. 197 – 204, Uniwersytet Ekonomiczny w Katowicach, Katowice, (2014).
11. Z. Mikoláš, *Potenciál rodinného podnikání.* S. Majtán (red.), Aktualne problemy podnikovej sfery 2016: zborník vedeckých prác, Bratislava, Vydavatelstvo EKONÓM, s. 750-757, (2016).
12. Z. Mikoláš,J. Karpeta, *Family business – the Potential for futher Business Development in the Czech Rupublic.* International Journal of development Research. Volume 5, Issue 06, June 2015, (2015).
13. Z. Mikoláš, Z.Wozniaková, *Soudobá paradigmata rodinného podnikání v České republice na počátku průmyslové revoluce 4.0.* In. Štefan Majtán (ed.) a kolektív; Atuálne problémy podnikovej sféry 2017. Zborník vedeckých prác. Bratislava: EKONÓM, s. 735 – 744, (2017).
14. J. Peterková, *Využití konceptů inovací v průmyslovém podniku.* VŠB-TU Ostrava: Series on Advanced Economic Issues, Faculty of Economics, (2017).
15. S.D.Sarasvathy, *Effectuation. Elements of Entrepreneurial Expertise.* Cheltenham: Edward Elgar, (2008).
16. O.Scott, *Záhadný zlatý řez.* Praha: Dokořán, (2013).
17. F.Valenta, *Tvůrčí aktivita, inovace, efekty*, Praha:Svoboda (1969)

Convergence of two controlling terms. Do we face a common future between controlling and management accounting?

Juraj Mišún[1,*], and *Ivana* Mišúnová Hudáková[2]

[1]University of Economics in Bratislava, Faculty of Business Management, Department of Management, Dolnozemská cesta 1/b, 85235 Bratislava, Slovak Republic
[2]University of Economics in Bratislava, Faculty of Business Management, Department of Management, Dolnozemská cesta 1/b, 85235 Bratislava, Slovak Republic

Abstract. Every single theory is based on some fundamental terms, which can also be used by other, unrelated theories. However, in our case, two related theories use the same basic term. "Controlling" terms are used in management (one of the main management functions) and management accounting (a management subsystem that helps managers to make qualified decisions). This semi-continental problem begun to emerge in the last years, since English has become the main language of scientists, even the coexistence of two controlling terms has lasted for almost half a century. The German controlling term needed an appropriate translation back into English. While the German-speaking authors quietly found other terms for translating, many authors from Central and Eastern Europe (CEE) mistakenly assume that English and German "controllings" are equivalents, causing an increase of confusing literature. To solve this problem, we need a clear separation of the two theories and a qualitative prognosis of possible development. In this paper, we used a relatively simple but reliable method for the prognosis, which includes five different scenarios of development. Ultimately, management or managerial accounting is the most widely used translation. Therefore, we assume that in the long term, this term will gain the major position.

Introduction

Controlling – a powerful word causing feelings of great power or helplessness. Someone is controlling something or even somebody. Although there may be no doubt, what is meant, the opposite is true. These doubts stem from different countries and languages. The word controlling is used in English, German, in many CEE languages, and in management and business economics. Schwarz argues [21, p.VI]: "science is always world science. There is no German controlling, just as there is no German physics." Yet, here we are – saying one English word and meaning two different concepts. Why is it interesting to solve this problem? The main threat is an increase of confusing literature when German knowledge is translated into English and collides with Anglo-American knowledge.

We would not have to be in the current situation at all, if the:

a) discussion between Goodwin [8] and Urwick [24] about control and controlling did not end with the relative victory of Urwick;

* Corresponding author: juraj.misun@euba.sk

b) management literature would sufficiently deal with the manager-controller relationship;
c) German naming in the early seventies would be based on a sufficient examination of the use of this term;
d) German professional word "Kontrolle" gained the meaning it already has in common life.

1 Theoretical background

In this part, we explain the development of the word "control" in different languages, and look closely at the use of this term in German business economics and Anglo-American management. Our intention is also to determine the relationship between these terms.

1.1 Problem with the word "control"

When examining the issue, we immediately came to the root of the problem – the very word "control". Its meaning, its connotations, create two different directions that have caused different development with the same used term. Even in English, we cannot find a clear definition of the word and it has many meanings. There are different variations of differences, ranging from prohibiting to manipulating. According to Emmanuel et al. [6] there are two major themes in these variations a) control as domination and; b) control as regulation. However, in both, we see the power to influence the development in the way the one who controls (controlling entity) desires.

The real problem occurs actually, when we look at the word control in terms of German or French language. This issue is well described by Schwarz [21]. The word "control" has two almost opposite meanings in the German, French and English languages. The German word "Kontrolle" was borrowed from the French "contrôle" in 18th century. However, the original French meaning "counter register for confirming the information from an original register" acquired the German meaning "Supervision, (re)examination, oversight". However, we consider Latin as the origin of "control."

A different development took place in English. Based on Goodwin's research [9], the French word was adopted into English in 1475 and later obtained the meaning of regulation, domination and direction. A similar meaning as in English was adopted in Germany in the 20th century [20], but the history of "control" caused problems in the current use of the term controlling.

The comparatively large problem is that German literature still suggests that control is somewhat inferior and bad. Such expressions arise when comparing "Kontrolle" with "Controlling". Jung [12, Abb.78, p.1162], states in a scheme that "Kontrolle" is "backward-looking", "identifying errors, searching for culprits", "accusing, punishing". On the other hand, the knight in shining armor "Controlling", is "forward-looking", "helping, guiding".

1.2 Controlling in German business economics

To bypass the limitations resulting from the German meaning of the word "Kontrolle" Albrecht Deyhle introduced at the beginning of the 1970s a new term – controlling. "Well, we got the word or the term Controller from the US. The word controlling is not as common in the US as far as I know. ... It was probably myself. I did it as a twin sister to marketing. [2, p.100]" In addition, some authors argue that the term controlling is currently virtually meaningless in English science. "Rather, it is referred to as 'management accounting' [14, p.13]."

One of the definitions states, it "means the target-related support of management tasks serving the system-based procurement of information and the information processing for the drawing up of plans, co-ordination and control; it is therefore a systematology improving the quality of decisions on each management level of the company" [19, p.12]. However, many authors point out that there is still no consistent definition.

After studying nearly 70 German controlling publications on the issue, we found only one that describes our problem. Stoffel [22, p.9] states: "While in the Anglo-American-speaking world the term Controllership is used as comprehensive term for the specialization and centralization of certain tasks with the controller and the tasks of the controller are called Controller Functions and Controllership Functions respectively, the term Controlling is used in the US-American management literature quite predominant in the meaning of management process."

Another major problem is that while the meaning of the word Kontrolle has changed in the German language in the 20th century, the most German-speaking authors do not reflect this fact in their works. In many German publications, great emphasis is placed on the claim that control is not controlling (Controlling ist nicht Kontrolle). In English-written literature, we do not find such a discussion and controlling is simply a regular exercise of control.

A special feature of German controlling development lies in the international isolation, which lies in language barriers and separation of international developments chosen by the respective scientists themselves. Many scientists published long time only in German language [7]. However, the theory is not completely isolated and we certainly do not mean the other German speaking countries.

At the beginning of the 1990s, German controlling knowledge base was relatively large. When the Iron Curtain fell, it affected research at CEE universities. Language skills of CEE scientists included mainly Russian, but also German to some extent. Only relatively few scientists could speak well in English. It was a very fertile ground for any German theory, German foreign investments were also an important factor. The relatively large area of the former Warsaw bloc means, we can call the German controlling theory a semi-continental one. Yet, controlling research has considerably transformed in German speaking countries. For long, German speaking students, professors and researchers of controlling have been rather isolated, but in the last years, it increasingly changed [20]. Now, it would be appropriate to unify the basic concepts of terms controlling and management accounting in German literature to avoid incorrect explanation and interpretation of these concepts [15].

1.3 Controlling in management

The term controlling is in management used for describing the management function of control, as suggested by Henri Fayol and later accepted by many American and English speaking authors. Since then, control is a solid part of the division of management into functions, and almost all management authors respect this fact. "The management textbook by Koontz and O'Donnel (*Principles of Management: An Analysis of Managerial Functions*, McGraw-Hill, New York, 1955) contributed to the widespread use of 'controlling'." [20, p.1] Currently, controlling is used for describing the managerial function in many bestselling management textbooks. This management function is critical for organizational success [18] and is always purposeful – regulating a system according to someone's purposes [10]. Its essence is the regulation of work of subordinates [23] and can be considered as a convergence point between accounting, business policy and management theory [4]. Yet, even without this discussion, the literature that deals with the management function of control is full of terms that are used interchangeably [3].

Many definitions in the German literature focus on the comparative aspect of control [5]. It is often understood as a comparison of two variables (e.g. is/should be, norm/fact). It is surprising, because the common German phrase "Alles unter/ausser Kontrolle" does not mean "everything in/out of comparison". Management theory clearly states, that control process ends with corrective action. "The German-language controlling literature seems to have a rather disturbed relationship to 'Kontrolle'" [20, p.47]. The use of English words in Germany and the subsequent effort to translate into English can cause great confusion.

1.3 Relationship of the two controlling(s)

Particularly interesting is the question of mutual position of controlling in both meanings. We can conclude there is a relatively broad consensus: although a manager should be in a higher position, a controller is not subordinated. In German literature, a controller is often described as a helmsman/navigator on a ship, while manager represents the captain.

The literature on German controlling and global management accounting is heavily concerned with the relationship between the manager and the controller. Management publications do not come close to this issue at all and mostly ignore the fact that in practice businesses already have to deal with this relationship.

2 Aim, methodology and limitations

The aim of this paper is to separate the two controlling theories, and to outline possible scenarios of further development due to the confusing current state. Our goal is far from identifying which theory is correct, better or should be superior to the other. However, we would like to encourage a debate about the future.

In the case of the scenarios of future development, we chose a simple and relatively fast method. From the website WHU – Otto Beisheim School of Management [25], we gathered a list of institutes, chairs, departments and professorships of German controlling in Germany, Austria (A) and Switzerland (CH). Then, we then searched for official translations of their names into the English language.

We proceeded as follows: first, we tried to find the official site of the department in English; if the page could not be found, we were looking for an official translation within the organizational structure of the university; if this structure was not available, we looked for authors from the department in English written papers. Unfortunately, in some cases we were not successful, as many authors still publish their works exclusively in German language.

The first such survey was conducted in 2018, and this year we updated the list of departments/chairs as many of them have added their English website, or their members are publishing more in English-written publications.

Based on names, we have assigned individual chairs to individual scenarios, as their names should reflect their current and future focus. In the case of the four chairs, we have also made an in-depth analysis; since their focus has not been sufficiently clear (Chair of Business Taxation and Auditing; Chair of Finance and Control; Chair of Controlling/Performance Management; Chair of Internal Control/Internal Audit). Overall, we appointed 84 of 94 chairs to the scenarios. The remaining 10 did not have an official translation on their website or left their title in a German original, as justified earlier.

In this section, we would also like to draw attention to a certain limitation based on our position. Both authors are active in the field of management rather than the business economics, which can affect our findings. Even if we try to maximize our objectivity, we both study the management function of control. To avoid this limitation, we tried to study thoroughly the literature of the business economics field, especially in the German language.

3 Results and discussion

In this part, we focus on the foreseeable future. What are the possible scenarios for further development of this issue? In our opinion: a) coexistence with limitations, b) slow disappearance, c) fight for the term, d) the golden mean, and e) one future.

3.1 Coexistence with limitations

The first scenario expects no changes. The German/CEE authors will use the term controlling to describe the business-economics theory, the English-speaking authors the management function. Restrictions will start only when the language switches to English or the location changes to geographic CEE. However, these limitations can be eliminated by clearly set rules. One of these rules could be a unified translation or an explicit statement of meant meaning in the abstract of a paper. Due to compliance with the rules, there would be less confusion. Although it may currently seem useless to make any rules, it is necessary to realize that more scientists are interested in publishing in renowned journals in English. German journals will probably change their language to English to increase their global impact and increase their inclusion in prestigious scientific indexes. Examples for such rules might be publications by Luther, Jones & Saxl [16] or Horvath [11]. On the other hand, once we have discovered this issue, we started using the term "control(-ling)" in titles of our papers.

It is very interesting to see that some German and associate chairs are even using names, which lead more to management than to the management accounting. When translating the word controlling from German into English in names of German chairs, this applies to **eight** of the 84 cases (table 1).

Table 1. Translations of chair's names from controlling to management control

University	German name	English translation
ESCP Europe Business School	Lehrstuhl für Unternehmensethik und Controlling	Chair of Business Ethics and Management Control
Georg-August-Universität Göttingen	Professur für Management und Controlling	Chair of Management and Control
Julius-Maximilians-Universität Würzburg	Lehrstuhl für BWL, insbesondere Controlling und Interne Unternehmensrechnung	Chair of Business Management, Management Accounting and Control
Universität Innsbruck (A)	Institut für Organisation und Lernen - Bereich Controlling	Department of Organisation and Learning – Management Control
Alpen-Adria Universität Klagenfurt/Wien/Graz (A)	Institut für Unternehmensführung, Abteilung Controlling und Strategische Unternehmensführung	Department of Management Control and Strategic Management
Johannes Kepler University Linz (A)	Institut für Controlling & Consulting	Institute for Management Control & Consulting
Universität Wien (A)	Lehrstuhl für Controlling	Department of Business Administration
Wirtschaftsuniversität Wien (A)	Institut für Unternehmensführung, Controlling und Beratung	Institute for Strategic Management and Management Control

3.2 Slow disappearance

Another scenario suggests that German controlling authors will not accept the aforementioned restrictions. Many will want to continue to use the term controlling, but the next and the next generation will get used to the English terms management/managerial accounting. “Controlling” will gradually remain in management only, also because English publishing authors will still not be aware that this term is being used for something else. The conservative generation will thus have to follow the fact that the term is becoming less apparent in scientific work, and after some decades, it almost erodes. Reviewers of books and papers will require using English terms what will even more accelerate such erosion.

Of the 84 departments/chairs, we would assign **40** to this scenario, as they a) mostly highlight management/managerial accounting/overall accounting in the translation; b) have a clear financial focus on auditing, internal audit and internal control. In addition to the terms management/managerial accounting, we also find related terms in the names of these university departments/chairs: RWTH Aachen University, Freie Universität Berlin, Universität Bielefeld, Ruhr-Universität Bochum, Brandenburgusche Technische Universität

Cottbus – Senftenberg, Technische Universität Darmstadt, Heinrich Heine Universität Düsseldorf, Universität Duisburg Essen, Katholische Universität Eichstätt-Ingolstadt (2), Goethe Universität Frankfurt am Main, Europa Universität Viadrina Frankfurt, Justus-Liebig Universität Giessen, FernUniversität in Hagen, Technische Universität Hamburg-Harburg, Leibniz Universität Hannover, Friedrich-Schiller-Universität Jena (2), Technische Universität Kaiserslautern, University of Cologne, Leipzig Graduate School of Management, Johannes Gutenberg-Universität Mainz, Universität Mannheim, Philipps-Universität Marburg, Technische Universität München, Universität der Bundeswehr München, Universität Osnabrück, Paderborn University, Universität Potsdam, Universität des Saarlandes, Eberhard Karls Universität Tübingen, Universität Innsbruck, Universität Basel (2), Universität St. Gallen (4), Universität Zürich (2).

3.3 Fight for the term

The German controlling theory can still become global theory from the current semi-continental one. How could this be achieved? There would be a need for a greater number of publications in English in renowned journals that would adequately distinguish controlling and management accounting. There is still a chance to make this scenario happen, since there is not an absolutely compliance in using the term controlling for describing the managerial function. Another opportunity hides the fact that not so many scientists are exclusively devoted to the management function of controlling. This leaves it as a small part of something bigger, in our case management. Authors of general management publications are unlikely to oppose too much if they have to rename one of their chapters in later editions. However, they must learn that a second controlling term even exists.

We would assign **ten** departments/chairs to this scenario. The reason is simple: they translate the German term controlling into English as controlling (Table 2).

Table 2. Translations of chair's names from controlling to controlling

University	German name	English translation
Universität Augsburg	Lehrstuhl für Wirtschaftsprüfung und Controlling	Chair of Accounting & Controlling
Universität Augsburg	Lehrstuhl für Betriebswirtschaftslehre mit Schwerpunkt Controlling	Chair of Controlling
Technische Universität Berlin	Institut für Betriebswirtschaftslehre, Fachgebiet Unternehmensrechnung und Controlling	Institute of Business Administration Controlling and Accounting
Friedrich-Alexander Universität Erlangen-Nürnberg	Das Institut für Finance, Auditing, Controlling, Taxation	Institute of Finance, Auditing, Controlling, Taxation
		Finance, Controlling & Entrepreneurship
Universität Hohenheim	Lehrstuhl Controlling	Dept. of Business Administration: Controlling
Universität Regensburg	Lehrstuhl für Controlling und Logistik	Chair of Controlling and Logistics
Universität des Saarlandes	Lehrstuhl für Betriebswirtschaftslehre, insbesondere Controlling	Chair of Business Administration – Controlling
Universität Bern (CH)	Institut für Unternehmensrechnung und Controlling	Institute for Accounting and Controlling
Universität St.Gallen (CH)	Institut für Accounting, Controlling und Auditing, Lehrstuhl für Controlling / Performance Management	Institute of Accounting, Control and Auditing, Chair of Controlling / Performance Management

At first glance, this scenario is unrealistic with these low figures. However, what is more interesting about German controlling is its use in practice (e.g. SAP CO [Controlling] Module). This means that many large German companies also enter the battle for German meaning of controlling. Not only they export the application, but they also call organizational units controlling units.

3.4 Golden mean

The second and third scenario may seem too radical in the short term. The fourth assumes a slight adjustment of the term of controlling to controllership. This scenario would ultimately lead again to the abandonment of the term to the favor of management. In fact, we already have some authors who meet this scenario. However, many German authors point out that controllership is the activity of the controller but does not describe the whole issue of (German) controlling. None of the 84 examined departments/chairs in German-speaking countries uses the term controllership in the English translation of their title.

3.5 One future

During the examination of this issue, we also found several views of German authors calling for unity. This would require a rapprochement of both theories and cooperation as both business economics and management. Ultimately, there would be a theory created that further combines the knowledge of both areas. The result of such an approximation is shown in a figure by Becker & Heinzelmann [1, p.89], which displays controlling as the intersection of the manager's and controller's work when the manager manages the business and is responsible for the result and the controller has the economic withdrawal and is responsible for transparency of results.

What does such a future require? In any case, better communication between the supporters of both camps, which may also contain critic of the other. In the next step, cooperation with each other and the common workplace is the last important step.

In the names of German and associated chairs, we discovered **26** cases where their English name could symbolize the combination of two theories into one, as official translations combine management/managerial accounting and control. Without an in-depth study of publications and teaching, we cannot conclude whether they actually fulfill it. These chairs are part of the following universities: Technische Universität Berlin, Technische Universität Braunschweig, Universität Bremen, Technische Universität Chemnitz, Technische Universität Dortmund, Technische Universität Dresden, Friedrich-Alexander Universität Erlangen-Nürnberg, Europa Universität Viadrina Frankfurt, Georg-August-Universität Göttingen, Helmut Schnidt Universität, Technische Universität Ilmenau, Otto Von Guericke Universität Magdeburg, Ludwig Maximilians Universität München (2), Oestrich-Winkel EBS Business School, Carl von Ossietzky Universität Oldenburg, Universität Rostock, Universität Siegen, Universität Stuttgart, Otto Beisheim School of Management (3), Uni Witten/Herdecke, Bergische Universität Wuppertal, Karl-Franzens-Universität Graz (A), Universität Freiburg (CH).

We certainly admit that managers in large and medium-sized enterprises may exhibit cognitive limitations for processing the current amount of information, so controller's expert advice is appropriate when performing control function [Meyer].

Conclusions

In studying the issue, we have found that a major problem is that authors using this term do not have sufficient knowledge about the use of the term by others from another area. While German-speaking and associated CEE authors dealing with business economics have missed the use in management, English-using researchers in management have missed the use of the term in German business economics. More than 70 studied German publications do not mention that the term controlling is also used in a different meaning than the one described in the work and refers to the business economics. However, we cannot imagine the fact that all these authors have not found the meaning in management.

Often, in the arguments, we read in German works that there is a huge difference between the German verb "kontrollieren" and the English verb "to control" as the meaning of the second one is "to steer" or "to direct". However, how near is such a meaning to the phrase "to manage"? As mentioned at the beginning, this problem may really only concern a very small number of people involved. Yet, when one of the scenarios is fulfilled, it can go into really interesting dimensions.

From the point of view of management, we must really express the critic that literature in the field is inadequately researching the relationship between the manager and the controller. In addition, it also fails to mention the (increasing) use of the term controlling in German written literature.

For our own research, this issue is challenging as we operate in a region where the term controlling is predominantly used in the German meaning. This is in view of a) the teaching and research on universities; and b) practice of German companies that have made a significant contribution with their foreign investment in CEE markets. In the long run, we will see whether the meaning of the word will be adapted to the global management theory, or it will keep the semi-continental meaning. However, this will also very depend on the German authors who can make a significant contribution to fulfilling one of the presented scenarios, by directly pointing out in their publications that German controlling is not equivalent to English controlling.

As part of our research, we continue to explore the issue by gathering expert opinions on the issue through the ResearchGate scientific social network.

References

1. A. Becker, R. Heinzelmann, Management accounting und controlling: Ausbildungstraditionen in Großbritannien und im deutschsprachigen Raum [Management accounting and controlling: Teaching traditions in the UK and German-speaking countries]. In G. Seicht (Hrsg.). *Jahrbuch für Controlling und Rechnungswesen 2013 [Yearbook for Controlling and Accounting 2013],* 73-99. (LexisNexis Orac, Wien, 2013).
2. Ch. Binder, *Die Entwicklung des Controllings als Teildisziplin der Betriebswirtschaftslehre [The development of controlling as a sub-discipline of business economics].* (Deutscher Universitäts-Verlag, Wiesbaden, 2006).
3. R.H. Chenhall, Management control systems design within its organizational context: findings from contingency-based research and directions for the future. *Accounting, organizations and society,* **28**, 2-3, 127-168 (2003).
4. R.L. Daft, N.B. Macintosh, The nature and use of formal control systems for management control and strategy implementation. *Journal of management,* **10**, 1, 43-66 (1984).
5. C. Ebeling, *Erfolgsfaktoren einer wertorientierten Unternehmensführung [Success factors of a value-oriented management].* (Deutscher Universitäts-Verlag, Wiesbaden, 2006).
6. C.R. Emmanuel, D.T. Otley, K.A. Merchant, *Accounting for Management Control.* (2nd ed.). (Springer, London, 2013).
7. Ch. Endenich, *Comparative Management Accounting.* (Gabler, Wiesbaden, 2012).
8. E.S.L. Goodwin, A Further Sally on "Control". *Michigan Business Review,* **13**, 9-18 (1960).

9. E.S.L. Goodwin, Control: A Brief Excursion on the Meaning of a Word. *Michigan Business Review*, **12**, 13-28 (1960).
10. S.G. Green, M.A. Welsh, Cybernetics and dependence: Reframing the control concept. *Academy of Management Review*, **13** , 2, 287-301 (1988).
11. P. Horvath, What is "German controlling"? *Journal of cost management*, **23**, 2, 16-19 (2009)
12. H. Jung, *Allgemeine Betriebswirtschaftslehre [General business studies]* (12th ed.). (Oldenbourgh, München, 2010).
13. H. Koontz, C. O'Donnell, *Principles of Management: An Analysis of Managerial Functions*. (McGraw-Hill, New York, 1955).
14. U. Kück, *Schnelleinstieg Controlling [Quick Start Controlling]* (3rd ed.). (Rudolf Haufe, München, 2009).
15. J. Link, *Führungssysteme: Strategische Herausforderung für Organisation, Controlling und Personalwesen [Management Systems: Strategic Challenge for Organization, Controlling and Human Resources]*. (6th ed.). (Vahlen, München, 2011)
16. R. Luther, T.C. Jones, A. Saxl, *Experiencing Change in German Controlling: Management Accounting in a Globalizing World*. (CIMA Publishing, Burlington, 2010).
17. M. Meyer, Controlling und begrenzte kognitive Fähigkeiten: Ausgangspunkte und Grundfragen [Controlling and limited cognitive abilities: starting points and basic questions]. In M. Meyer, J. Weber (Hrsg.). *Controlling und begrenzte kognitive Fähigkeiten [Controlling and limited cognitive skills]*. (Gabler, Wiesbaden, 2011).
18. Merchant, K.A. The control function of management. *Sloan management review*, **23**, 4, 43-55 (1982)
19. T. Reichmann, *Controlling. Concepts of Management Control, Controllership, and Ratios*. (Springer, Berlin, 1997).
20. U. Schäffer, Management accounting research in Germany: From splendid isolation to being part of the international community. *Journal of Management Control*, 23, 4, 291-309 (2013)
21. R. Schwarz, *Controlling Systeme. Eine Einführung in Grundlagen, Komponenten und Methoden des Controlling [Controlling systems. An introduction to the basics, components and methods of controlling]*. (Gabler, Wiesbaden, 2002).
22. K. Stoffel, *Controllership im internationalen Vergleich [Controllership in international comparison]*. (Gabler, Wiesbaden, 1995)
23. T.R. Tyler, S.L. Blader, Can Business Effectively Regulate Employee Conduct? The Antecedents of Rule Following in Work Settings. *Academy of Management Journal* **48**, 6, 1143-1158 (2005).
24. L.F. Urwick, The meaning of control. *Michigan Business Review*, **12**, 5, 9-13 (1960).
25. WHU – Otto Beisheim School of Management, *Chairs – WHU – Otto Beisheim School of Management*. (2020) https://www.whu.edu/en/faculty-research/management-group/institute-of-management-accounting-and-control/students/links/chairs/.

Influence of competitive advantage on competitive positioning of Silesian companies in 2019

Olaf Flak[1,*] and *Grzegorz* Głód[2]

[1]University of Silesia in Katowice, ul. Bankowa 12, 40-007 Katowice, Poland
[2]University of Economics in Katowice, ul. 1 Maja 50, 40-287 Katowice, Poland

Abstract. Company competitiveness is still an important issue in management studies. It can be measured in several ways including competitive advantage and competitive positioning. The aim of the paper is to present the result of the annual research on company competitiveness conducted in Poland. The research question in this paper is: is there a relation between a certain type of competitive advantage and characteristics of a company competitive positioning? In the paper there is a description of the influence of competitive advantage on competitive positioning in six dimensions. Respondents in the research answered 45 questions in online survey at sensorium24.com (Company Competitiveness Barometer). 8 of the questions concerned the competitive advantage and 6 concerned the competitive positioning. There are a theoretical background of company competitiveness and phenomena which belongs to this general term – a competitive advantage and a competitive positioning, a methodology of research and results as well as conclusions.

1 Introduction

Competitive advantage and competitive positioning are used to describe an ability of a company to compete with one another. Dynamic and unexpected environment of a business needs from companies to be ready to use innovative and efficient competitive strategy in order to build a competitive advantage in the market [1]. Than the achieved competitive advantage influences on a competitive positioning which a company can be describe in terms of its profitability [2]. All of these phenomena belongs to a general term describing an ability of a company to compete on a market [3].

The aim of the paper is to present the result of the annual research on company competitiveness conducted in Poland. This project is called Company Competitiveness Barometer and its last edition was conducted form 09.01.2020 to 09.04.2020. Respondents from a few market sectors answered 45 questions in online survey at sensorium24.com. 8 of the questions concerned the competitive advantage and 6 concerned the competitive positioning. The research question in this paper is: is there a relation between a certain type of competitive advantage and characteristics of a company competitive positioning?

* Corresponding author: olaf.flak@us.edu.pl

Section 2 contains a theoretical background of company competitiveness and phenomena which belongs to this general term – a competitive advantage and a competitive positioning. Section 3 describes a methodology of research. The results were presented in Section 4 and in Section 5 there are conclusions.

2 Theoretical background of company competitiveness

Company competitiveness is defined in many ways which depend on the perspective of main factor of competition on a certain market. A wide overview of the approaches to this issue led to design the Company Competitiveness Integrated Model in which there are 5 elements influencing a strength of a company on a market.

Firstly, the source of any actions is a competitive potential. This means any resources, which the company has or should have to be able to use them to build, maintain and strengthen its competitiveness. Secondly, the way of the use of the competitive potential is a competitive strategy which is an adopted program of action aiming to achieve a competitive advantage [4]. Thirdly, the effect of the competitive strategy is a competitive advantage defined as the ability of the company to deliver the tangible and intangible assets to the buyer through the market [5]. The fourth element is a competitive positioning, understood as the synthetic economic results of the company, resulting from the ability of the enterprise to compete now and in the future [6]. The last element of the Company Competitiveness Integrated Model is an environment of a company. It is called the platform of competition and means a set of macro- and microenvironmental features in which the company operates in the sector [7].

Authors after a few years of company competitiveness research formulated their own definition of the company competitiveness which covers all these five elements of the Company Competitive Integrated Model. Competitiveness can be understood as "an multidimensional attribute of the company, resulting from both the internal features and the ability to cope with external circumstances. Competitiveness is relative, that means, there is no absolute scale for measuring competitiveness, (...) competitiveness can be used to describe the mutual relations of enterprises in the market." [5] The Company Competitiveness Integrated Model is shown in Fig. 1.

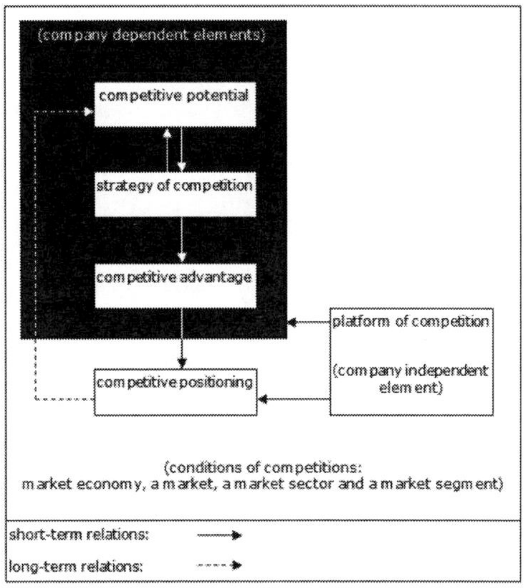

Fig. 1. The Company Competitiveness Integrated Model
Source: [5]

The competitive position of an organization is described as the position of an organization compared to its competitors in the same market or industry. Knowledge of competitive positions allows the enterprises to make tactical plans to maintain or improve their current positions or possibly withdraw from the market. Therefore, the knowledge of the competitive position of an organization and its rivals is critical. Since rivals are defined as organizations that are able to obstruct a company's market goals, and important moderators of a company's performance. They are considered as the most crutial elements in competitive strategies [8].

Competitive advantage is whatever value a business provides that motivates its customers (or end users) to purchase its products or services rather than those of its competitors and that poses impediments to imitation by actual or potential direct competitors. With a clearer understanding of competitive advantage, more systematic methods for gathering customer and competitor information, and careful analysis of the changes in both, businesses can do better job assessing, building, communicating, maintaining and enhancing competitive advantage [9].

Generating superior financial performance by exploiting sources of sustainable competitive advantage is an everlasting quest for commercial enterprises around the world [10].

To the extent that superior performance is not absolutely a function of competitive advantage, all firms (both these that have developed competitive advantage and these that have not) seem to have an opportunity to achieve superior performance. Practicing managers ought to bear in mind that even if they have ben able to develop competitive advantage, their firm maight underperform in case the benefit from the competitive advantage is lower compared to the cost incurred to develop it [11].

It should be mentioned, that quality as an element of competitive advantage next to the cost advantage seems to be the basic source of its construction [12].

The marketing mix comprises of four decisions which should be considered before launching a product. Firms ought to plan targeted approach on these four different components and they are product, price, and place & promotion. All the four variables are very helpfull the the firm in formulating strategic decisions necessary for competitive advantage [13].

Research on the impact of competitive advantage on the effectiveness of companies has been conducted for many years [14]. In many cases, the competitive advantage achieved is identified with the financial results of enterprises [15, 16].

3 Methodology of the research

The research was conducted in a period of 09.01.2020-09.04.2020 by the survey implemented in Sensorium24.com. The survey contained 45 questions of the Company Competitiveness Barometer [17]. 8 of them concerned a competitive advantage and 6 of them concerned a competitive positioning. Competitive advantage question were grouped in two according to four elements of marketing-mix which was a classic base of a competitive advantage. The description and literature study which are presented above show strength relationship between marketing mix and competitive advantage [18].

In the research 210 companies from Silesian Region took part from several different market sectors and different size, however, 89% of them were small and medium size. Companies were divided into 4 groups depending on the dominant source of a competitive advantage. There were 26 companies price oriented, 66 companies which used a place as a source of a competitive advantage, 72 companies focused in products as a competitive advantage and 46 companies using a promotion to create a competitive advantage. In the

Section 4 there are results presented in percentage of a share of every type of a competitive advantage source.

4 Results of the research

The results of the research contains relations of four different competitive advantages based on different sources coming from a marketing-mix (a price, a place, a product and a promotion) and six indicators of a competitive positioning.

The first indicator of a competitive positioning is a level of a cash flow. Fig 2 presents the trajectory of this level depending on a source of a competitive advantage. The exact percentage of companies there are in Table 1. The most number companies which had a source as a promotion had high level of a cash flow (45,65%). The companies using a place as a competitive advantage in 42,42% answered "very high". None of companies which used a product as a source had very low level of a cash flow, however, whose which used a price answered "very low" in 7,69%.

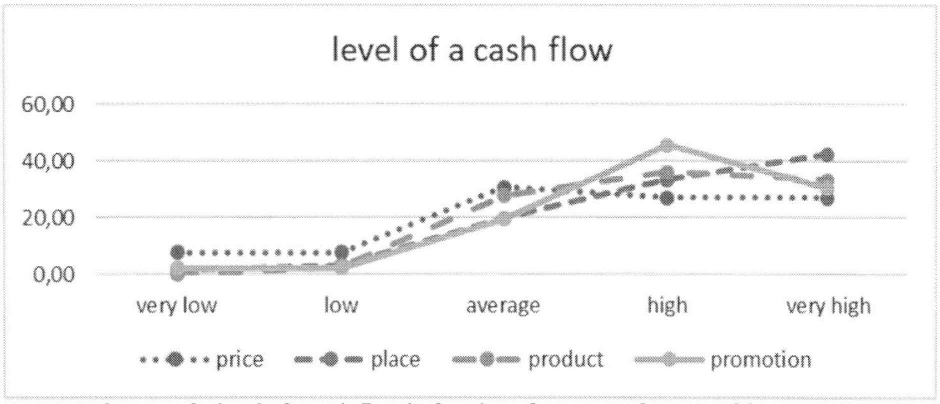

Fig .2. Trajectory of a level of a cash flow in function of a source of a competitive advantage

Table 3. Percentage of a level of a cash flow and a source of a competitive advantage relation

source of a competitive advantage	**indicator of a competitive positioning**				
	level of a cash flow				
	very low	low	average	high	very high
price	7,69	7,69	30,77	26,92	26,92
place	1,52	3,03	19,70	33,33	42,42
product	0,00	2,78	27,78	36,11	33,33
promotion	2,17	2,17	19,57	45,65	30,43

The second indicator of a competitive positioning is a level of external debts. Fig 3 presents the trajectory of this level depending on a source of a competitive advantage. The exact percentage of companies there are in Table 2. As it can be seen on the Fig 3, the highest numbers of answers (54,35% and 53,03%) of "very low" level of debts gave companies using a competitive advantage based on a promotion and a place, respectively. What is really astonishing, none of a company had very high level of external debts, no matter which source of a competitive advantage they used.

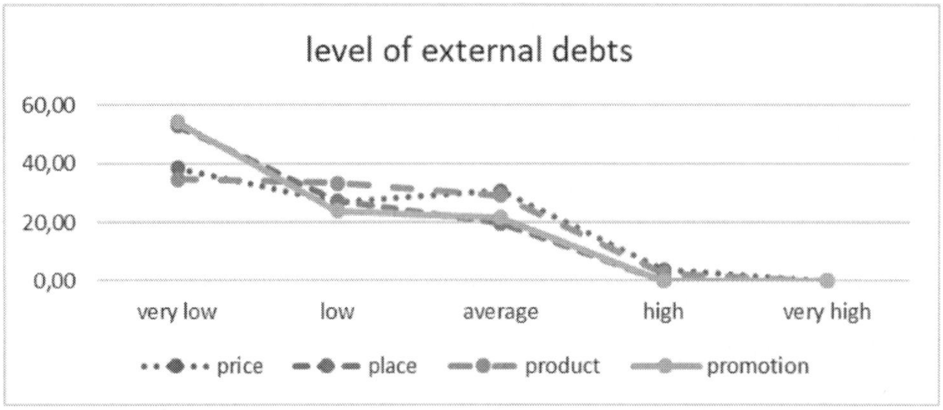

Fig .3. Trajectory of a level of external debts in function of a source of a competitive advantage

Table 2. Percentage of a level of external debts and a source of a competitive advantage relation

source of a competitive advantage	indicator of a competitive positioning				
	level of external debts				
	very low	low	average	high	very high
price	38,46	26,92	30,77	3,85	0,00
place	53,03	27,27	19,70	0,00	0,00
product	34,72	33,33	29,17	2,78	0,00
promotion	54,35	23,91	21,74	0,00	0,00

The next indicator of a competitive positioning is a level of market share. As it can be seen on the Fig 4 in Table 3, mostly all companies using different sources of their competitive advantages assessed their market shares as average without an impact of the source of a competitive advantage. This is quite interesting result which could tell us that the most of companies used adequate sources of their competitive advantages. However, there were 19,23% of companies which used a price as a source which noticed a very low market share. In the same time, 8,7% companies using a promotion claimed that they had a very high market share. So, on the one hand, companies can use a proper source of a company advantage. On the other hand, the source really influence on a market share.

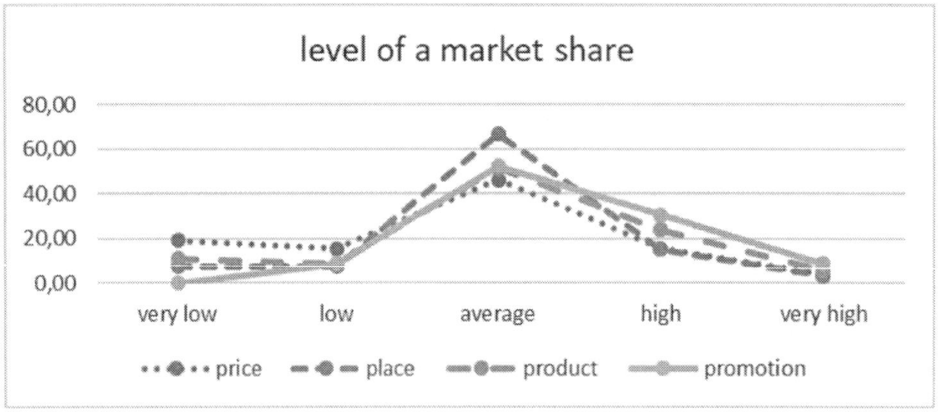

Fig .4. Trajectory of a level of a market share in function of a source of a competitive advantage

Table 3. Percentage of a level of a market share and a source of a competitive advantage relation

source of a competitive advantage	indicator of a competitive positioning				
	level of a market share				
	very low	low	average	high	very high
price	19,23	15,38	46,15	15,38	3,85
place	7,58	7,58	66,67	15,15	3,03
product	11,11	8,33	51,39	23,61	5,56
promotion	0,00	8,70	52,17	30,43	8,70

The fourth indicator of a competitive positioning is a level of profitability. As it can be seen on the Fig 5 and in Table 4, the highest number of respondents (54,55%) who used a place as a source of competitive advantage claimed that they achieved high profitability. Nearly the same number (53,85%) used a source of a price. What is really interesting, there were very few companies which noticed very low profitability in any types of a competitive advantage. The biggest difference in profitability was among companies which had average profitability – only 19,23% of "price" users and 45,83% of "product" users.

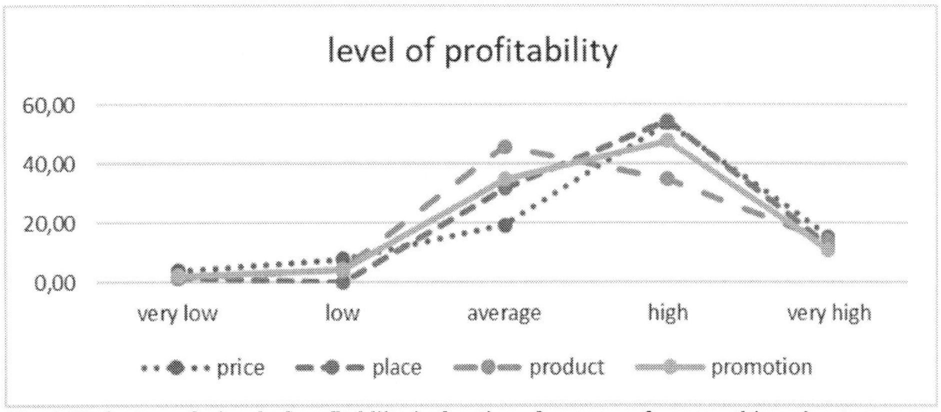

Fig .5. Trajectory of a level of profitability in function of a source of a competitive advantage

Table 4. Percentage of profitability and a source of a competitive advantage relation

source of a competitive advantage	indicator of a competitive positioning				
	level of profitability				
	very low	low	average	high	very high
price	3,85	7,69	19,23	53,85	15,38
place	1,52	0,00	31,82	54,55	12,12
product	1,39	4,17	45,83	34,72	13,89
promotion	2,17	4,35	34,78	47,83	10,87

The next indicator of a competitive positioning is a level of sale increase measured year by year. As it can be seen on the Fig 6 and in Table 5, most of companies said that they increased their sale on average. However, the companies using a product as a source of a competitive advantage were in 55,56%, and there were only 39,13% of companies used a promotion. What is astonishing, a quite low number of companies had very low and low sale increase – no matter which source they used to build their competitive advantage. However, the highest number of companies which had very high increase used a promotion as a source (15,22%).

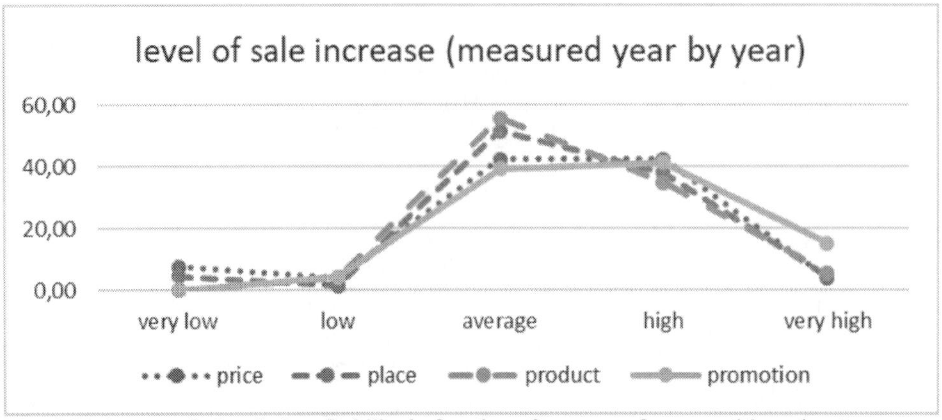

Fig .6. Trajectory of a level of profitability in function of a source of a competitive advantage

Table 5. Percentage of profitability and a source of a competitive advantage relation

source of a competitive advantage	indicator of a competitive positioning				
	level of sale increase (measured year by year)				
	very low	low	average	high	very high
price	7,69	3,85	42,31	42,31	3,85
place	4,55	1,52	51,52	37,88	4,55
product	0,00	4,17	55,56	34,72	5,56
promotion	0,00	4,35	39,13	41,30	15,22

The last indicator of a competitive positioning is equity of profitability. As it can be seen on the Fig 7 and in Table 6, none of the companies answered that this indicator is very low. Also very few had this indicator at a low level. It is possible to claim that this indicator did not depend much on a source of a competitive advantage – most of answers were similar.

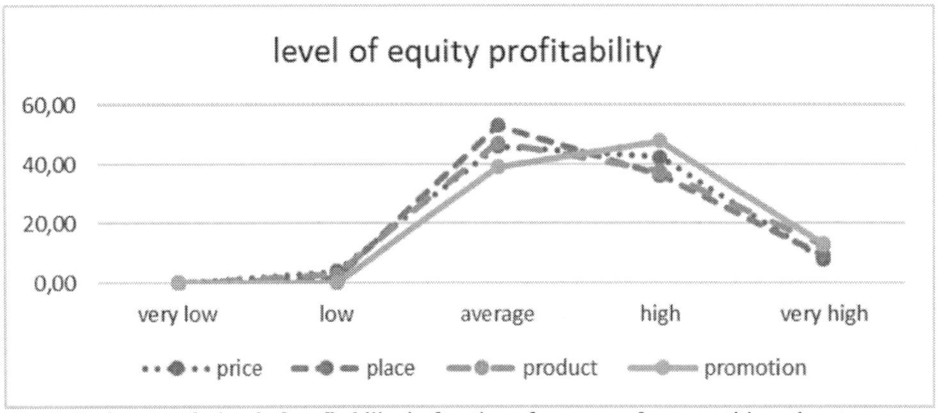

Fig .7. Trajectory of a level of profitability in function of a source of a competitive advantage

Table 6. Percentage of profitability and a source of a competitive advantage relation

source of a competitive advantage	indicator of a competitive positioning				
	level of equity profitability				
	very low	low	average	high	very high
price	0,00	3,85	46,15	42,31	7,69
place	0,00	1,52	53,03	36,36	9,09
product	0,00	2,78	47,22	37,50	12,50

promotion	0,00	0,00	39,13	47,83	13,04

5 Conclusions

The presented analysis of empirical research results shows the impact of individual competitive advantages on the competitive position, which was expressed through the financial indicators of the surveyed enterprises.

The analysis indicates the need for further search for sources of competitive advantage and research in the field of measuring the effects of achieved competitive advantage. Due to the cyclically conducted research using the Company Competitive Integrated Model, subsequent studies will be an opportunity to develop research in this area in the future. As a result, it will be possible to continue research in an in-depth range from those presented in this article.

References

[1] S. Farhikhteh, A. Kazemi, A. Shahin, A., M.M. Shafiee, How competitiveness factors propel SMEs to achieve competitive advantage? *Competitiveness Review: An International Business Journal* **1**, 13-21 (2020)

[2] M. Kharub, R. Sharma, An integrated structural model of QMPs, QMS and firm's performance for competitive positioning in MSMEs. *Total Quality Management & Business Excellence* **31**, 3-4, 312-241 (2020)

[3] E. Akben-Selcuk, Factors affecting firm competitiveness: Evidence from an emerging market. *International Journal of Financial Studies* **4**, 2, 1-10 (2016)

[4] O. Flak, G. Głód, G., Concept Research of the Competitiveness of Enterprises in Selected Countries in the European Union. S. Majtan (eds.) *Aktuálne problémy podnikovej sféry*. Bratislava, Vydavateľstvo EKONÓM, 86-92, (2014)

[5] O. Flak, G. Głód, *Konkurencyjni przetrwają*. Difin, Warszawa (2012)

[6] J.W. Bossak, W. Bieńkowski, W., *Międzynarodowa zdolność konkurencyjna kraju i przedsiębiorstw. Wyzwania dla Polski na progu XXI wieku*. Wyd. SGH, Warszawa (2004)

[7] O. Flak, G. Głód, *Konkurencyjność przedsiębiorstwa. Pojęcia, definicje, modele. Cz. 1*. Wydawnictwo Akademii Ekonomicznej w Katowicach, Katowice (2009)

[8] V.L. Dang, G.T. Yeo, A competitive strategic position analysis of major container ports in Southeast Asia. *The Asian Journal of Shipping and Logistics* **33**, 1, 19-25 (2017)

[9] H.K. Christensen, Defining customer value as the driver of competitive advantage. *Strategy & Leadership* **38**, 20-25 (2010)

[10] M. Holm, V. Kumar, T. Plenborg, An investigation of customer accounting systems as a source of sustainable competitive advantage. *Advances in accounting* **32**, 18-30 (2016)

[11] C. Sigalas, V.M. Papadakis, Empirical investigation of relationship patterns between competitive advantage and superior performance. *Journal of Strategy and Management* **11**, 1, 1-32 (2018)

[12] I.A. Elshaer, M.M. Augustyn, Direct effects of quality management on competitive advantage. *International Journal of Quality & Reliability Management* **29**, 3, 265-283 (2016)

[13] M. Singh, Marketing mix of 4P's for competitive advantage. *IOSR Journal of Business and Management* **3**, 6, 40-45 (2012)

[14] H. Ma, Competitive advantage and firm performance. *Competitiveness Review* **10**, 2, 15-32 (2000)

[15] S. Cantele, A. Zardini, Is sustainability a competitive advantage for small businesses? An empirical analysis of possible mediators in the sustainability–financial performance relationship. *Journal of Cleaner Production* **182**, 166-176 (2018)

[16] A. Agus, R.M. Sagir, The structural relationships between total quality management, competitive advantage and bottom line financial performance: An empirical study of Malaysian manufacturing companies. *Total Quality Management* **12**, 7-8, 1018-1024 (2001)

[17] O. Flak, A Method of Estimating the Determinant of Enterprise Competetiveness. A. Nalepka, A. Ujwary (eds.), *Business and Non-profit Organizations Fading Increased Competition and Growing Customers' Demands*, Wyższa Szkoła Biznesu, Nowy Sącz, 113-128 (2014)

[18] E. Chumaidiyah, The marketing mix strategy in influence to the competitive advantage. *Proceedings of the 2014 International Conference on Industrial Engineering and Operations Management* **7**, 9, 818-826 (2014)

Promoting process approach to management

Emília Papulova[1*]

[1]University of Economics in Bratislava, Faculty of Business Management, Department of Management, 852 35 Dolnozemská 1, Bratislava, Slovakia

Abstract. Increasing environmental dynamics and deepening competition are forcing businesses to seek effective responses and, within them, appropriate management approaches to be competitive and succeed in the market. The process approach, which in recent years has become an integral part of managerial practice in the company, emphasizes the business processes. Focusing on processes and improving them in companies is an effort to better respond to changing customer requirements. This paper deals with the possibilities of applying and using methods and techniques to support the business process management in companies and deals with the essence and benefits of individual methods to support the performance and competitiveness of a company.

1 Introduction

Business process management (BPM) based on process orientation of companies is not a new approach however many companies are still struggling with it application. BPM is often presented as a systematic managerial approach enables organization's workflow to be more effective, more efficient and more capable to adapt to an ever-changing environment.[1] In relation to the changing nature of the environment brings advantages in relation to the functional approach. Process approach offers new opportunities for companies from the view of increasing the performance and competitiveness of businesses. The advent of a process approach raises a shift in the concept of thinking, which represents a shift from division of labour, specialization and functional orientation towards the integration and business orientation of processes.

The goal of the article is to introduce and evaluate individual methods and techniques in terms of their application in the process management of a company and to point out their benefits for increasing the competitiveness and success of companies operating in a changing environment. As it is confirmed by researchers [2], praxis of the companies suggests that BPM methods indirectly contribute to process performance by establishing a BPM culture and that the correct application of methods yields direct performance benefits.

Methodology of the work is based on the analysis of theoretical knowledge and starting points in the field of process management, evaluation of individual methods and approaches usable in the management of process-oriented organization and based on the results of

* Corresponding author: papulova@euba.sk

existing research in this field. applicability of individual methods in the process management of the organization.

2 Process approach to management

2.1 Theoretical frameworks of knowledge about BPM

Today we are seeing a shift from an industrial society to a knowledge society. This shift has also led to changes in the way of thinking and approaches to business management.

Enterprises in an industrial company were characterized by their focus on achieving a competitive advantage, through the rapid introduction of new technology and technology, investment and improved asset management. Organizations in the knowledge-based economy are characterized by a different concept of thinking, a concept in which success depends more on the ability to mobilize and exploit intangible assets.

In order to ensure success over the long term, today's businesses are trying to meet the customer's requirements in the best and fastest possible way and to adapt their processes. Focusing on processes, their rational course and improvement with the aim of flexibly responding to changing customer requirements is the basic idea on which the process approach is built as a new concept of BPM. BPM is an approach used to manage business processes by facilitating the modelling, the execution as well as the improvement of a processes of organization and is considered as the most appropriate approach to continuously improving processes. [3]

Process-oriented company emphasizes in particular: [4,5]

- customer orientation, monitoring and adaptation to changing customer requirements, creating an immediate relationship with the customer and his loyalty
- focus on processes that are dependent on the customer and his needs
- product and service innovation, providing high quality customer-oriented products at affordable prices
- teamwork, quality, engagement and dedication of people in the team for continuous improvement of processes, continuous learning
- business process management systems using methods and approaches to support process efficiency with IT support for process management.
- a corporate culture that fosters interest in continuous process improvement, responsibility sharing.

The purpose of the transition from a functional hierarchy to a process-oriented organization is to support customer orientation and the organization's performance and competitiveness. The essence of the change is to replace the focus on functional units in the company system by focusing on business processes. The goal is to break down barriers between functional areas to promote organizational flexibility and value for the customer. Process organization is built on the principle of integration of activities into integrated processes provided by process teams.

2.2 Current methods and approaches to support the process approach to management

The focus of the process approach is on the business processes. The process approach is focused on the preparation and design of processes in the company so that they run efficiently and create maximum value for the customer. Several methods and techniques can be used to support the process approach:

- in the area of process improvement and product quality can be used methods from Total Quality Management through Lean Management to Six Sigma.
- in the field of process optimization and modelling, building business process architectures and defining business success metrics can be used Porter value chain methods, Business Process Reengineering, Business Process Redesign, Balanced Scorecard, Activity Based Management.

BPM requires a deeper understanding of the possibilities of using different tools and methods. Selection of methods should be connected to the organization's requirements and conceptions, but also to the skills and abilities to use the tool. [5] IT-supported methods and techniques create significant potential for improving process and management of the whole company.

2.2.1 Total Quality Management

Total quality management (TQM) is a comprehensive quality management approach that aims to ensure high product and service quality for maximum customer satisfaction.

The method assumes that the company is process oriented [4] and is based on philosophy of continuous process improvement, which begins with prevention, aimed at preparing quality inputs and production and management systems, which according to the author of this method, E. Deming, supports the idea of "quality needs to be built into the product." [6] It means constantly looking for ways to improve product quality and ensure minimum or zero error rates. Employees and their commitment and involvement throughout the process play an important role in the whole approach.

2.2.2 Lean Production and Lean Management

Lean production is considered to be a business philosophy that aims to reduce time between customer and supplier by eliminating wastage across the enterprise supply chain. It is not just about reducing costs but, above all, maximizing the value of the product for the customer. Slimming is the way to produce more, have lower overheads, use our areas and production resources more efficiently. Lean production cannot work without a close link to product development and production engineering, logistics and business management. [7]

Lean management is an approach where the concept of lean presents an effort to eliminate any losses and waste and reduce the costs of the company. The principles of slimming have gradually expanded from production to other areas, including the whole company. Some authors considered the main of Lean Management has become the reduction of all types of waste, like in transport, inventory, traffic, waiting, processing, etc. [8]

Lean management is a comprehensive approach aimed at maximizing customer value. The fulfillment of this goal is based on process-oriented production based on customer requirements, on the integration of suppliers and customers into the value creation chain and teamwork - process teams that are guided and motivated to create the highest value for the customer.

2.2.3 Six Sigma method

Six Sigma is focused on continually innovating and improving processes, reducing deviations and improving product quality and business profit growth by approaching all processes in the business to be defined, measured, analysed, improved and controlled.

In practice, the Six Sigma method is developed into the DMAIC model, which represents methodology to implement the Six Sigma method for improving processes in the company. [7] This methodology includes the following basic stages:
- D (Define) - the stage of defining a problem, which present aspects of the process or product that negatively affect customer satisfaction and business performance
- M (Measure) - stage of measurement and acquisition of objective data, which express the current state and achieved results
- A (Analyze) - stage of data analysis, identification of key causes of defined problems and based on analysis to identify possibilities of improvement
- I (Improve) - a stage in which the process is modified to provide key solution parameters to be accepted by the customer
- C (Control) - the aim of the stage is to evaluate the results of implemented solutions, whether they have a lasting effect and there are no other negative impacts.

The Six Sigma method is a metric method based on the analysis and using empirical data, quantitative measures for evaluating business objectives, where the basic measure is the number of defects, or deviations, resulting in lower customer satisfaction, increased costs and reduced business profit. Statistical evaluation and standard deviation are used statistical average and statistical deviation. The quality standard is set at a level that the business process must not produce more than 3.4 defects per million cases. [9]

2.2.4 Value chain approach

The perception of the transformation process in the company in its physical form is based on the perception of its material flow in the company, the movement of labour and capacity resources to ensure the course of activities so as to create products that can be placed on the market. Each of these activities, within the involvement of the defined resources, contributes to the creation of costs and participates in the creation of values and added value in the company. And here we come to perceive the transformation process as a value-creating process.

In an illustrative way, the process of value creation in the company and the share of individual activities in the creation of added value is expressed by M.E.Porter using the "value chain". The value chain encompasses a set of all related activities, from raw materials and materials processing to handing over the product to the customer and expresses how individual business activities contribute to value creation.

When analysing the value chain, we distinguish two basic groups of business activities, namely primary and secondary activities. The primary (basic) activities are arranged in the direction of material flow in the company. The performance and results of primary activities in the company depend on the course and involvement of secondary (support) activities. [10]

Value Chain Management is the process of managing the flow of integrated activities and information in a product flow with the aim of achieving efficiency, or greater value for the customer. [9]

2.2.5 Business Process Reengineering

Business process reenginiering is defined by Hammer and Champy as a fundamental reassessment and radical reconstruction of business processes so that dramatic improvements in critical performance measures such as cost, quality, service and speed can be achieved.

Business process reenginiering does not mean incremental changes and improvements in business but represents an endeavor to abandon existing practices and to look at new perspectives on the processes necessary to create products and services to meet customer needs. The idea is to find new and better solutions and ways of working. Business process reenginiering encompasses a range of changes, abandoning existing and established practices and changing work in terms of content and organization. It is important that changes are initiated from the customer and their needs, hence the new process design. Not only top executives are involved in this challenging plan, but it is important that other employees participate in it with their commitment, interest, creativity and willingness to succeed. [9]

The main expectations of fulfilling this goal include the ability to flexibly respond to customer requests, reduce under-sales, eliminate sources of errors and deficiencies, and improve quality, eliminate non-value-generating activities, reduce time and reduce overall business costs.

2.2.6 Balanced Scorecard

The Balanced Scorecard concept, commonly referred to as a strategic system for managing and measuring business performance, can also be used to support process management. This concept is used in practice to provide managers at all levels of business management with information about the activators, or drivers of business performance, and their statuses that affect business success. It is considered as a management system that provides a ground for businesses to clarify the vision and strategic plans and convert them into action with sufficient feedback about both internal processes and external business outcomes. [11]

The BSC approach structures business management issues into four areas or perspectives, namely: [7]

- financial perspective - through financial indicators that give a view of the company from the perspective of its owners
- Customer perspective - using indicators that present the customer's expectations and perspective
- process-organizational perspective - defines processes that are critical to customer satisfaction, focuses on their improvement, evaluation and measurement of their effectiveness
- Learning and growth perspectives - through the characteristics that a business must have for sustaining strategic goals in response to changes in the external and internal environment.

Since BSC is focused on managing the business according to the set goals and objectives, it is then important for strategic goals of the business to identify critical success factors that are further elaborated from the perspective of these four perspectives and define the links between them as well as evaluation of success.

The BSC monitors the interrelations and connections between the basic perspectives - perspectives, thus ensuring the link between strategy and business processes, as well as linking to the financial, customer, learning and growth perspectives. A prerequisite for the implementation of the BSC is a clear understanding of the organization's vision and strategy. The goal of organizational integration of Balanced Scorecard is its integration with management and reporting. [4]

The BSC concept thus provides business managers with the necessary information and insights to better understand goals, results, relationships, cause-effect relationships, activators in relation to financial and non-financial indicators.

Correctly understood and implemented Balanced Scorecard method brings to businesses numerous benefits. It serves as a support for implementing business strategy, it strengthens cooperation between business units, business departments, human resources of a businesses and it serves as a communication tool in a business. Another important benefit is orientation of managers on key customers and partners and improvement of relations with business stakeholders. [12]

2.2.7 Activity Based Management

Activity Based Management (ABM) was created by the improvement of the activity-based costing (ABC) method. Research proved that ABC and process modelling provide a good starting point in heading toward more cost-conscious design and help organization to learn the relationships between the activities performed in the organization and their associated costs [13]

ABM is a comprehensive cost management system that provides information from a wide range of business activities. It is an approach aimed at managing and evaluating processes in a company and identifying their strengths and weaknesses. This methodology is a comprehensive cost management system that provides information from a wide range of business activities thus becomes a source of information that has hitherto been hidden, creating room for better decision-making in the organization. [4]

The ABM methodology is based on a customer-oriented philosophy of process thinking and includes: [7]

- Identification of business activities from perspective of business processes and their hierarchy
- Measurement of performance, calculation of costs for individual activities
- Looking for ways to improve the overall value chain from product design, through production, sales to direct debit, continuous process improvement.

The essence of this approach is thus to constantly search for activities that need to be eliminated because they are unproductive and activities that need to be strengthened because they can contribute to increasing the profitability of business.

3 Further routing in the field of process approach

Increasing competition leads businesses to seek new innovative approaches, implement new equipment and workflows where people's work is being replaced by machine-driven computers. Integration of information technology into managerial work promotes closer interconnection of the company and its processes with the customer, the ability of managers to make quick and high quality decisions, ensure synchronization of business processes and thus reduce costs and increase efficiency, to ensure timely delivery of products and services that will be of value to the customer. [9]

Artificial intelligence is a science that seeks to simulate intellectual thinking using information technology. At the interface of informatics and biology, it offers tools that simulate various aspects of the world of living organisms by means of computer technology, such as. brain function. By means of these procedures it is possible to find solutions to various optimization or decision problems. [14]

The benefits of introducing process management in companies can be seen from operational, market and strategic aspects. Operational and strategic benefits are associated with benefits for an organization that has decided to look at its operation from a procedural perspective and actively approach process management. In the area of market benefits the positive aspects of process management on the part of the customer are included.

Process approach and BPM concepts have matured over the years and new technology, concepts, standards, and solutions appear. Recent trends of development in the context of process approach to management developments are in particular: [4,16,17]

- focus on development of process specialists and creation of qualified process teams,

- focusing more on process efficiency and innovation (moving away from focus on efficiency);
- creation of suitable organizational culture suitable for application of process management,
- integrating business applications and focusing on a user-friendly, user-friendly page
- use of social networks and mobile interfaces to support process management, using not only external social networks (such as Facebook and Twitter), but also internal networks such as SharePoint, etc. (Social & Mobile BPM Solutions)
- use of intelligent systems that enable the company to proactively adapt to changing market trends (iBPM - smart BPM solutions),
- use of cloud solutions (Cloud BPM solutions),
- the use of Gamification of BPM concepts and techniques.
- robotic process automation and opportunities connected to Industry 4.0.

New technologies and concepts such as automation, system integration, autonomous robotics, digitization, Internet of Things, artificial intelligence systems are also implemented in approaches and methods of BPM.

4 Conclusion

The concept of process organization and process management puts the integration of work at the center of managers' attention to bring the expected synergy effects and thus support the organization's performance and competitiveness. However, this concept also imposes new requirements on managerial work, the mastery of process philosophy, clear customer orientation and knowledge management supported by continuous learning of employees at all levels of management.

There are various BPM methods that contribute to process performance by establishing a BPM culture and lead to performance benefits.

Trends of the last phase of process management development are mainly focused on software solutions and IT support of application of process management and trends connected to Industry 4.0.

This article is a partial output of research project VEGA no. 1/0017/20 entitled Changes in the application of managerial functions in the context of the fourth industrial revolution and the adaptation processes of enterprises in Slovakia.

References

1. A. Gažová, Z. Papulová, J. Papula. The application of concepts and methods based on process approach to increase business process efficiency, *Procedia Economics and Finance*, **39**, 197-205 (2016) https://doi.org/10.1016/S2212-5671(16)30284-2
2. T. Schmiedela, J. Reckerb, J. Brockec. The relation between BPM culture, BPM methods, and process performance: Evidence from quantitative field studies. *Information & Management,* **57**, 103-175 (2020) https://doi.org/10.1016/j.im.2019.103175
3. K. Mehdouania, N. Missaoui, S. AyachiGhannouchia. An approach for Business Process Improvement Based on Simulation Technique, *Procedia Computer Science,* **16**, 225-232 (2019) https://doi.org/10.1016/j.procs.2019.12.176
4. Z. Papulová, A. Oborilová, J. Papula. *Procesný manažment ucelený pohľad na koncepciu procesného manažmentu* (Kartprint, Bratislava, 2014)

5. A. Oborilová, Z. Papulová, Business process management and its implementation in Slovak enterprises, *Comenius Management Review*, **5**, 4-13 (2011)
6. Š. MARSINA et al. *Management* (EKONÓM, Bratislava, 2007).
7. J. Řezáč, *Moderní management*. (Computer Press, Brno, 2009. 397)
8. R. J. Schonberger, The disintegration of lean manufacturing and lean management, Business Horizons, **62**, 359-371 (2019) https://doi.org/10.1016/j.bushor.2019.01.004
9. S.P. Robbins, M. Coulter, *Management*. (Pearson Education Limited, London, 2016
10. J. Papula, Z. Papulová, J. Papula, *Strategický manažment. Aktuálny concept pre rýchlo sa približujúcu budúcnosť*, (Wolters Kluwer, Praha 2019).
11. R. S. Kaplan, D. P. Norton, D. P. Balanced Scorecard (Management Press, Praha, 2007)
12. Ľ. Lesáková, K. Dubcová. Knowledge and Use of the Balanced Scorecard Method in the Businesses in the Slovak Republic, Procedia - Social and Behavioral Sciences, **230**, 39-48 (2016) https://doi.org/10.1016/j.sbspro.2016.09.006
13. K.Tornberg, M. Jämsen, J. Paranko. Activity-based costing and process modeling for cost-conscious product design: A case study in a manufacturing company, International Journal of Production Economics, **79**, 75-82 (2002) https://doi.org/10.1016/S0925-5273(00)00179-1
14. M. Jakubéci, *Využitie prostriedkov umelej inteligencie na akciových trhoch*. (Dizertačná práca. Bratislava, Fakulta managementu UK, 2015)
15. M. von Rosing et al., Business Process Trends, *The Complete Business Process Handbook*, Elsevier, **1**, 187-216 (2015)
16. T. Grisold et al. Exploring Explorative BPM - Setting the Ground for Future Research, *BPM 2019: Business Process Management*, *Springer* **11675** 23-3 (2019)
17. J. Papula et al. Industry 4.0: preparation of Slovak companies, the comparative study. *EAI/Springer Innovations in Communication and Computing*, **1**, 103-114 (2019)

SHS Web of Conferences **83**, 01051 (2020) https://doi.org/10.1051/shsconf/20208301051
Current Problems of the Corporate Sector 2020

Synergistic effect of joining process and its measurement

Anna Poledňáková[1]

[1]EU v Bratislave, Department of Corporate Finance, 85235 Bratislava, Slovakia.

Abstract. Rapid technological advances, globalization is leading to intensified global competition. In order to remain competitive and strengthen their market position, companies often opt for an external growth strategy through M&A. The most frequently mentioned reason for merging companies to achieve a synergistic effect. For this reason, it is essential that the buyer assesses the potential concentration in order to create value for shareholders. This paper focuses on analyses and characteristics and definition of synergy effect, analysis of main sources of synergy, definition of key variables in assessment of synergies.

1 Introduction

The technological advances, globalization is leading to intensified global competition. In order to maintain their competitiveness and strengthen their market position, companies often choose a strategy for external growth through mergers and acquisitions.

Mergers and Acquisitions [1] is a broad and challenging subject, as evidenced by a number of professional publications, studies conducted in business practice and the existence of a large number of consulting, financial and investment firms engaged in merger consultancy.

By merging with another business, they gain quick access to new technologies, markets, qualified human resources, built distribution channels, loyal customers and many other sources of future growth. The result of these processes is the emergence of financially strong enterprises. By contrast, companies in financial difficulty are trying to avoid bankruptcy by merging with another company.

It can be said that there is no industry in the world economy where there would be no mergers and acquisitions. [2] The synergy effect can be simply expressed as the fact that the combined effect of two or more activities is greater than their sum.

Methodology. Several scientific methods were used to process the present paper, both from the group of general logical methods and from the group of empirical and exact methods. The method of analysis was used in the examination of theoretical knowledge and opinions on the issue of the synergy effect of merging companies. We used the synthesis method to formulate conclusions synthesizing knowledge from the literature. Other methods used were induction, deduction and comparison. We used the abstraction method to define the essential characteristics of the examined phenomena - synergistic effects. Generalization

[1] Corresponding author: anna.polednakova@euba.sk

is a method of defining the context of real subjects and has been used to formulate conclusions. The paper was also based on the results of research by both foreign and domestic authors, especially in the area of merging companies and their motives.

2 Definition of synergy

Merger and acquisition synergies are an effect that adds value to two [3] or more companies when merged over the separate values of those companies. The concept of synergy is defined, according to the Cambridge dictionary [4], as follows: "Synergy is the combined power of a group of things which, when working together, achieve a higher overall strength than that achieved by the individual components separately"

The term synergy is usually associated with physical science as an economy. It refers to the type of reaction that occurred when combining two factors gives a higher effect than when the sum of two operational independent units is like the sum. The anticipated existence of synergies allows companies to incur the cost of the acquisition process and can still afford a premium for their target shareholders. Synergy is an increase in the performance of the combined company over what is already expected of the two companies or needs to be achieved as independent companies. Achieving synergy begins with a clear understanding of what it is. Defining synergies as "a combination of enterprises of which two plus two equals five" or "the amazing benefits of integration from combined strategies and economies of scale" is inaccurate and misleading.

Mark L. Sirower [5] in his book "Synergy Trap" gives the following definition of synergy. If the acquirers can achieve the performance that is already expected from the objective, the net present value (NPV) of the acquisition strategy is then clearly expressed by the following formula: NPV = Synergy - Premium "

Mergers and acquisitions synergies represent an effect that increases the value of two or more companies when merged over the separate values of these companies. Synergies can result from mergers and acquisitions through cost savings due to improved operational efficiency and / or revenue gains due to increased productivity in the use of company assets. The main objective of merger and acquisition companies is to achieve a synergistic effect. This effect is most objective and is assessed by increasing the value of the investment. The value of the company after the transaction is increased in case of synergies, whether profit or loss. Better financial results and higher business performance are achieved through proper investment selection and proper cost management.

Thus, the buyer usually approaches the assessment of the potential merger and acquisition target in a similar way to the choice of investment or investment project. The Company usually analyses the potential added value of the target company, which in practice uses primarily the net present value (NPV) of future projected cash flows, which are discounted by a discount factor. In practice, inputs to the projected cash flow calculation are adjusted to best predict cash flows for the future period analysed.

At the same time, Sirower [5] describes the cornerstones of synergy as the four elements of the strategy, in merging companies that need to be put in place to achieve a synergistic effect. The cornerstones of synergy include:

- Strategic vision,
- Operational strategy,
- Operational strategy,
- Power and culture.

The buyer usually allows the competitive advantages of the combined enterprise, including increased revenue, reduced costs or technological improvements over the performance of individual entities. So, the synergistic value is usually created by the buyer.

However, there are exceptions. If the objective has a technology or ownership that the buyer can customize and exploit on its larger volume basis, this "backscatter" synergy is mainly generated by retailers. There are two types of synergy: operational synergy and financial synergy. Operational synergy also takes two forms: increasing revenue and reducing cost.

These revenues and effective profits are derived from vertical or horizontal mergers. Financial synergy refers to the possibility that capital costs may be reduced by a combination of one or more companies.

Operational synergy in the form of returns is very difficult to achieve. Analyses [6] indicate that 70% of mergers will not achieve operational synergies in the form of revenue increases. Increasing revenue can take several forms:

- price power,
- combination of functional forces,
- growth from fast-growing or emerging markets.

Financial synergy reflects the impact of mergers and acquisitions on the cost of the acquiring company or partners in mergers and acquisitions. Depending on the extent to which financial synergy exists in various combinations in companies, then capital costs may be lower. A merger of two companies may reduce the risk if the company's cash flow is not correlated perfectly.

2.1 Sources of synergy

Synergistic benefits [3] generally arise from five possible sources:

- Increase revenue,
- Cost reduction,
- Improvement of processes in society,
- Financial efficiency,
- Risk reduction.

Revenue gains may be the result of higher unit sales, which are usually achieved by a combined entity operating in a wider market or offering an extended product line or both. An increase in the price of products can be achieved, especially if the combined unit creates strategic advantages, for example, that it is the only supplier of technology or product. The projected increase in revenue should be considered with caution as it depends on many external variables, in particular the response of customers and competitors. Both can be difficult to predict and are largely outside the control of the combined unit. For example, customers may have a policy that prevents excessive reliance on any source of supply. A competitor's response should also be envisaged, including new product offers and price discounts. An increase in revenue can be achieved when the combined company offers a wider range of products or services, often by using the new entity's distribution system.

Estimates of the second source of synergy, cost reductions, tend to be more predictable and more reliable than revenue increases. The consolidation of functions eliminates positions and related fixed assets and overheads. The scope of this advantage tends to be greater if the objective is similar to the acquirer in the operations and markets served. Particular attention needs to be paid in advance in order to succeed with cost savings. In particular, wages and salaries require vigilance because, while jobs may be reduced, individuals who held them sometimes survive in a new entity in another department or job.

Process improvements occur when the combined entity adopts more efficient procedures used by the target acquirer. These improvements often result from technological or process improvements that can be utilized on the wider base of the combined unit. Improvements can lead to revenue gains or cost reductions, as well as more efficient operations or more efficient marketing and distribution.

The fourth source of synergy, financial efficiency, is often misunderstood. The value of the target cannot be increased by allocating it lower capital costs through the use of debt financing. Since any acquirer could achieve this advantage, such financial manipulations rarely have any real value creation potential. However, the combination may reduce the financial cost of the combined entity and may allow efficiency in terms of lease, cash management and working capital management. The combined entity may also generate certain tax benefits. Acquirers should be aware that financial performance is not capable of substantially improving the company's strategic position and is rarely the driving force behind transactions to bring companies together.

The target cost of capital can be reduced through the acquisition of a larger company that eliminates many of the risks that exist when the company operates independently. A taller company can eliminate several of the risks that typically occur in smaller companies - such as key personality, customer concentration, pace of technological change, and position compared to its competitors - by becoming a segment of a broad-based company markets.

2.2 Key variables in assessing synergies

In assessing the potential savings from each of these synergistic sources, the merger management should concentrate mainly on three variables, which can significantly affect the accuracy of the estimated synergy and value calculation significantly.

The magnitude of the benefits of synergy. The value of the synergy should be quantified in the net cash flow forecast, which includes estimates of income, costs, financing and taxes and investments in working capital and fixed assets. Every component of the forecast, in particular any estimated improvements, must be questioned consistently. The management of the acquisition team [6] must resist the natural effort to make emotional purchases, which often leads to overly optimistic revenue and expenditure estimates. Each element in the forecast must be accurately estimated.

The probability of achieving synergy. The merger will bring various benefits, some of which have a very high probability of success, while others may have negative consequences [7,8,9]. For example, the probability that the administrative costs associated with the target group board can be excluded is approximately 100%. Conversely, the achievement of certain sales targets against fierce competition is likely to be much less clear-cut. These differences should be recorded and taken into account in the forecast. Calculation of the probability of different results, for example optimistic, expected and pessimistic, can be done using sensitivity analysis, Monte Carlo simulation, helping to quantify the range of possible

Timing the benefits of synergy. Merger and acquisition buyer management needs to recognize that, while an acquisition is usually a single transaction, its benefits accrue over an estimated period of time, which may span many years. The value of the acquisition and its success are critically linked to achieving improved cash flows according to the expected timing. Any delays move cash flows further into the future and reduce their present value. Attempts to accelerate the timing of revenue improvements or cost savings should be avoided, and the timing of each assumption must be questioned as sums. The history of mergers and acquisitions is full of stories about how unrealistic acceleration of improvements to increase the attractiveness of an acquisition has led to an overestimation of the synergy value.

The clear point is to emphasize the importance of objectivity and diligent due diligence in examining the foreseen synergies. Investors expect to improve the performance of the acquirer as well as the target value they set for each company as separate entities. Acquisition-related synergies must reflect improvements beyond those already foreseen. The value of these synergies must exceed the premium above the fair market value of the acquirer in order to create value. Therefore, any envisaged synergy must be thoroughly analysed in terms of the estimated amount, likelihood of achievement and when this benefit will occur. Companies that overlook this process may experience unpleasant surprises and disappointments in the future.

3 Results and Discussions

Companies that want to survive in a competitive environment must choose a strategy to achieve their goal. One such strategy is mergers and acquisitions. Many companies neglect to prepare the merger process. The main reasons are its complexity, slowness and time preferences of the managers' subjective opinions on the percentage of official controls transactions that refuted the misconception about the success of the executed transaction. Merger and acquisition problems are underestimated by the lack of preparation and underestimation of some important aspects of the process. The result may be quite the opposite of that expected from such an operation. Instead of improving the company's position, efficiency and profitability, there may be a loss of efficiency, inflexibility and loss of competitiveness. In spite of all the risks that Mergers and Acquisitions face, there are thousands of business connections worth billions of euros worldwide each year.

The main motivation for merging and acquiring companies is to achieve a synergistic effect. This category was the subject of our research. Mergers and acquisitions companies must take into account the fact that each merger is something special and must be given particular attention. Achieving synergy begins with a clear understanding of what it is. Defining synergies as "a combination of enterprises of which two plus two equals five" or "the amazing benefits of integration from combined strategies and economies of scale" is inaccurate and misleading.

Thus, the buyer usually approaches the assessment of the potential M&A target in a similar way to the choice of investment or investment project. The company usually analyses the potential added value of the target company. In order to properly evaluate the choice of investment, it is necessary not only to address the issue of financial synergy (cost of capital), but also operational synergies related to increasing revenues / reducing costs. Assessing synergies in merging companies is essential to analyse this process through key factors that significantly affect the accuracy and accuracy of the estimated synergies.

This work was supported through the Slovak Science Foundation (VEGA) under projects VEGA 1/0007/19 "Asset allocation in a low interest rate environment in the financial and non-financial companies in the Slovak Republic".

References

1. A.P. Gaughan. Mergers, Acquisitions, and Corporate Restructurings. 5th ed Willey, 2010. s.230-236. ISBN 10:0470561963 (2010)
2. N. G.Greg, I. Renneboog. International Mergers and Acquisitions Activity Since 1990: Recent Research and Quantitative analysis. Academic press, Elsevier, MA, USA ISBN 978-0-7506-8289-3 (2007)
3. CH.M. Mellen, F.C. Evans. Building Value in Private Companies. John Wiley and Sons, Inc. ISBN 978-0-470-60441-0 (2010)

4. D. Schweiger, N.E. Csiszar, N.K.Napier. A strategic approach to implementing mergers and aquisition, The Management of Corporate acquistions, N.Y. (1994)
5. M.L. Sirower. The Synergy trap: How Companies Lose the Acquistion Game (New York: The Free press, 1997,2000), p. 20-29 (2000).
6. S.Sudarsanam. Creating value from mergers and Acquisitions. Harlow, England: Pearson Education limited, (2010).
7. R.Thomson, S. Dettmar, M.Garay, (2020, 19 March). *The state of the deal: MERGER AND ACQUISITION TRENDS 2019.* Retrieved from https://www2.deloitte.com/us/en/pages/mergers-and-acquisitions/articles/ma-trends-report.html [cit.20.12.2018].
8. Cambridge Dictionary, (2020, 21 March). *Synergy*. Retrieved from https://dictionary.cambridge.org/dictionary/english/synergy?fbclid=IwAR2MaabCldBZnTwzxY7pzPsvthbaXrLbu1hjHYjPt13GS-zi3ep2HlXX0So
9. Corporate Finance Institute, (2020, 25 March). *Types of Synergies: The different sources of synergies*. Retrieved from https://corporatefinanceinstitute.com/resources/knowledge/valuation/types-of-synergies/?fbclid=IwAR2iQvsAFxOb8K4FuWeMnkCBkRls-M0u8JhovwOcYMt8kxVaI1dcScYib5A

Industrial revolution 4.0 in industrial laundries and their information technologies landscape

Martin Polívka[1,*], and *Lilia* Dvořáková[1]

[1] University of West Bohemia in Pilsen, Department of Finance and Accounting, Univerzitní 8, 30100, Pilsen, Czech Republic

Abstract. This paper presents the research results in the field of implementation of fourth industrial revolution technologies into the service sector, with a specific focus on the integration of Industry 4.0 technologies into corporate processes of industrial laundries. The theme covered was found topical following the analysis of practical needs and based on the specific nature of industrial laundry processes and only marginal interest paid in professional theory and discussions to the possibility and efficiency of the Industry 4.0 technologies implementation into non-production companies. Based on the analysis and evaluation of secondary data, advanced Industry 4.0 technologies have been identified which might lead to the increase of industrial laundries efficiency. Further, the methods of monitoring, structured interview and case study have been combined to design the integration of Industry 4.0 technologies into the system of existing information and communication technologies and systems. Based on a quality comparison, the case study and structured interview results have proven the benefits of implementation of automated identification and data collection technologies including the implementation of vertical integration of corporate information system and horizontal integration leading to the interconnection with customers in industrial laundries corporate processes.

1 Introduction

The current professional literature mentions frequently that this is the era of the fourth industrial revolution, sometimes also called the "Revolution 4.0" (see more e.g. in [1] , [2], [3]). The first industrial revolution dates back to the late 18th and early 19th centuries and it was linked to the invention of a steam engine, the second industrial revolution dates back to the early 20th century and it was linked to the development of production lines and power engineering and the third industrial revolution started in the 1970's due to development of information technologies and computer-controlled manufacturing; the current extremely rapid development of information and communication technologies, Internet of Things and other modern technologies, such as additive production, system integration, artificial intelligence, virtual reality etc., are considered to be the makers of the fourth industrial revolution. Although many of these technologies are not really new (for example, EDI technology which still plays an important role in system integration dates back to the 1970's), the mass spreading of these technologies enabled by an ever-increasing performance of standardly available IT equipment and their mutual integration clearly brings the potential of large changes in many industries. Therefore, in 2011 at the Hannover fair, the term Industry 4.0 [4] was presented as a designation of an industrial sector of the

* Corresponding author: polivkam@kfu.zcu.cz

future in which these modern technologies will be integrated. This term promptly became largely discussed both in the academic circles and in practice (see [5], [6]).
Soon, efforts to implement the fourth industrial revolution technologies in other economic areas emerged, gradually giving rise to the concepts like "Services 4.0" [7], "Agriculture 4.0" [8], "Healthcare 4.0" [9], "Logistics 4.0" [10], or, respectively at the general level "Society 4.0", or also "Society 5.0" [11]. However, the fourth industrial revolution is still generally seen as a phenomenon which will affect mostly production companies; you can find the proof in a number of professional studies covering these issues. For example, as at 22 February 2020, in the quotation database Scopus, total 1 577 entries were found containing the combination of key words "Industry 4.0" and „Manufacturing," while there were only 32 entries containing the key word "Logistics 4.0", 19 entries containing "Agriculture 4.0," 15 entries for "Healthcare 4.0" and 5 for "Services 4.0". A similar disproportion among studies covering the fourth industrial revolution in production and other industries applies also in full-text scientific databases such as Science Direct or SpringerLink. So far, only a marginal interest has been paid to the possible implementation of fourth industrial revolution technologies into non-production companies in professional discussions. The objective of this paper is therefore to fill in this gap in the present state of knowledge.

2 Research scope and methods

This paper presents the result of research of potential and real implementation of the fourth industrial revolution into industrial laundry environment, or rather into its information system. The research objective was firstly to assess which technologies considered to be a part of the fourth industrial revolution might make the industrial laundry operation more efficient. The second research objective was to design how to implement these fitting technologies into the laundry's physical operation and how to integrate them into its IT landscape, i.e. into the set of information and communication technologies and systems (see more in [12]). Laundry operation was chosen for the analysis mainly because one of the authors of this paper has been practically involved with the industry technologies for a long time and has current knowledge of practical needs of companies in the industry. Also, it is an interesting industry which, although a part of the service sector, has many elements of industrial companies owing to its character (the size of operations and quantity of machine equipment, high automation level, demanding for materials and powers).
A combination of desk research, monitoring, structure interviews and case study methods was chosen as a research method. Based on studying the literature and other resources, individual Industry 4.0 technologies were first identified and assessed as to their appropriateness for increasing the laundry operation efficiency. Then, a case study was prepared, showing the existing solutions for industrial laundries using any of these technologies, and its impacts were assessed. The subsequent discussion then covers potential extension of the existing solution by other fourth industrial revolution technologies.

3 Results

3.1 Technologies of Industry 4.0

Although considerable attention is currently being paid to the issues of the fourth industrial revolution and Industry 4.0, it has not been clearly specified what exactly these terms mean

– as stated in [13], the term "Industry 4.0" is currently subject to inflation and it is present almost everywhere, without its content having been clearly described and structured. Trying to capture the term, professional literature usually uses either the Reference Architecture Model Industrie 4.0 (abbreviated as RAMI 4.0), or an exhaustive list of technologies which should be subject to it – see e.g. [2, page 26 – 27], [4] or [14].
In the RAMI 4.0 model, the conception of Industry 4.0 is described rather generally and its objective is to create a structured aid usable for the purpose of considerations of its introduction in a company [15]. For the purposes of a more practically oriented analysis of the elements of Industry 4.0 which are capable of being implemented into the laundry operation, it is therefore more suitable to use its definition based on the list of particular technologies. The following nine basic technologies covered by the conception of Industry 4.0, frequently referred to also as its "9 pillars" [16] are considered to be:

1. **Big data**
2. **Autonomous**, mainly **collaborative robots**
3. **Simulations**
4. **System integration**, in the form of Horizontal, Vertical or End-to-end integrations
5. **Internet of Things** (referred to also as **IoT**)
6. **Cyber-Physical Systems** (referred to also as **CPS**)
7. **Cloud technologies**
8. **Additive production**
9. **Extended reality**

Based on an analysis made in the publication [6], we believe it is appropriate adding a tenth technology, currently playing an irreplaceable part in the current corporate practice, to these "9 pillars of Industry 4.0", the contents of which are described in more detail in the referenced literature:

10. **Automatic identification and data capture** (referred to also as **AIDC**).

This term contains a set of technologies which allow for object identification, followed by automated capture and storage of data on the objects. For these purposes, a wide spectrum of methods is currently being used, most of which are based on an initial labeling of a monitored object by a machine-readable identifier, followed by reading of the identifier – including but not limited to labeling with bar codes or QR codes and also radio frequency identification (usually abbreviated as RFID) which is based on the principle of information transmission using radio waves. Some methods such as machine recognition of audio or video do not use any identifiers.

3.2 Process improvement in the industrial laundry

In this section, the potential of individual Industry 4.0 technologies to make the industrial laundry operation more effective will be assessed. Industrial laundry activities involve washing laundry for corporate clients such as industrial companies, hospitals or hotels. For these activities, several specifics are characteristic:

- It is a large-volume operation and hundreds of pieces of laundry may pass through an industrial laundry, depending on its size. Industrial laundry operation has also high demands for fixed capital in the form of machine equipment which capacity should be therefore sufficiently utilized.
- Pieces of laundry keep circulating between the customer and the industrial laundry. A certain piece of laundry does not necessarily need to be reserved for a particular customer only – in some cases (unless the piece of laundry is personified, e.g. using a sewn logo), a specific piece of laundry may be delivered to a different customer after each washing. This applies mainly for such laundry where size is not the issue, such as

pillow cases, sheets etc. (the term "flat laundry" is often used to describe these types of laundry).

- Individual pieces of laundry of the same type may not be visually told apart from one another.

These specifics make it practically impossible to keep manual records of laundry containing a sufficient extent of information about each piece of laundry and being sufficiently reliable at the same time. This leads to a situation that the laundry operation works as a "black box" into which laundry accepted from customers enter on one end, being characterized only by the number of pieces of individual types of laundry, and similarly, dispatches of laundry to the customers exit on the other end. Particular pieces of laundry are not identified or recorded in such a system which results in the following problems:

- It is not possible to trace the location of a particular piece of laundry at a given time, whether it is at the customer's or in the industrial laundry. It means that if pieces of laundry are lost, it is not possible to tell where they were lost and whether it is the customer or the laundry that is responsible for the loss.
- No information as to how long a particular piece of laundry has been in the laundry or at the customer is available. Also, the number of washing cycles to which any particular piece of laundry has been subjected is not recorded in any way. This might present a problem in terms of the operation efficiency and also in case that the age or number of washing cycles should be controlled from legislative reasons (such as for protective working aids).
- It is even not possible to detect reliably when a piece of laundry gets completely out of circulation, either permanently (e.g. as a result of theft) or temporarily (when a piece of laundry gets temporarily lost at the customer) because individual pieces are not identifiable.

The severity of these problems is even increased by the fact that the industrial laundry often does not only provide the washing service to its customers but it rents the linen or clothes which are in such case owned by the industrial laundry that is bound to deliver the determined quantities of clean and undamaged laundry to the customer on the agreed dates. The industrial laundry's capital is thus allocated in assets over which it has almost no control because of the above-described specifics. Incorrect records and losses of pieces of laundry make it necessary for the industrial laundry to keep more pieces of laundry (and capital bound in it) for each customer than it would be otherwise necessary.
In regard to the above-described specifics and problems, making the laundry operating processing more effective means in particular:

1. **Making laundry tracing records more extensive and accurate**. This change should result in the laundry workers being able to trace the relevant data for each piece of laundry at any moment (such as age, number of washing cycles, to whom it is allocated, what is its current location etc.). Ultimately, it should result in the drop of losses of laundry and reduction of the associated costs and in the improvement of communication with customers.
2. **Expediting laundry passage through the industrial laundry.** This change brings about more expedient order processing, more efficient use of the industrial laundry's fix capital, reduction of the number of operation workers and the associated reduction of salary costs, cutting down of the cash-to-cash cycle and other positive effects.

3.3 Applicability of "Industry 4.0" technologies in industrial laundry

Based on an analysis of the substance of individual Industry 4.0 technologies, we have identified those which might be used for reaching these objectives:

- **Autonomous collaborative robots**. Even today, transporters moving laundry in bags between individual stages of the washing process are being routinely used in industrial laundries. Individual stages are automated, too, using equipment such as tunnel washing machines or finishers. Some operations, such as receipt or dispatch of laundry or moving of laundry from the transporter into particular equipment and back still depend on human work, although these are often physically demanding tasks. Modern autonomous robots able to work in a changing environment could replace human workers in these activities, or operate together with them, and this should lead to expediting the passage of laundry through the industrial laundry.
- **System integration** in all forms as mentioned above. Vertical integration could be used to interconnect the laundry operation control IT system with other systems used in the laundry, such as its ERP system. Horizontal integration might serve for interconnection with the customers' IT systems. Automated data transfer without the need for their repeated manual entries should lead to office labor savings and to expediting the whole process from the order receipt to the issue and accounting for the resulting invoice. End-to-end integration might be used for recording information about a particular piece of laundry during its whole lifecycle; a necessary precondition for such use is that it would be possible to clearly identify each piece.
- **Internet of Things and Cyber-Physical Systems.** These Industry 4.0 technologies are mentioned together because their potential uses are closely linked together. Connection of all equipment in the laundry into a network would allow for their mutual automated coordination. Communication of all pieces of equipment with the laundry operation control IT system would allow for the creation of a cyber-physical system in which all physical equipment would have their virtual representations and might be controlled centrally, which should minimize downtime and inefficiencies and expedite the passage of laundry through the industrial laundry. A more accurate representation of reality in a virtual environment might be obtained if individual pieces of laundry would be labeled with machine-readable identifiers, thus being incorporated into the cyber-physical system as well.
- **Automatic identification and data capture**. As already explained in the previous text, precise manual records of particular pieces of laundry are practically unfeasible. Automated identification of pieces of laundry is a precondition for keeping these records. Automated data capture should also lead to significantly expediting the receipt and dispatch of laundry because it would eliminate the need to manually count the received or dispatched pieces.

These five technologies were thus assessed as appropriate for making the industrial laundry operation more effective. As far as the remaining Industry 4.0 technologies are concerned, their use in the industrial laundry is certainly possible, perhaps except for additive production (for example, some IT systems might be placed in a cloud instead in local devices, big data analysis or use of simulations for ergonomic optimization of the laundry), however, it would not be directly related to operating issues.

3.4 Case study

To prepare a case study, an industrial laundry with more than 200 employees was selected, being one of the largest companies of this type in the Czech Republic (based on EU classification, however, it is a medium-size enterprise). The enterprise provides services to different types of companies such as industrial enterprises, hospitals or hotels (i.e. B2B segment) mainly by means of renting the linen and clothes, including both flat linen and personal clothes.

For the sake of expediting the passage of laundry through the industrial laundry and precise records and traceability of individual pieces of laundry across its whole lifecycle, the enterprise implemented a project involving introduction of several Industry 4.0 technologies. The project was based on combining the following technologies:

- IT system for laundry operation control.
- Labeling each piece of laundry using RFID UHF chip which allows for automated data capture and processing.
- System integration of laundry IT system with other software.

In the laundry operation control IT system, operation data such as receipts, dispatches or repairs of laundry are recorded and individual orders are created and monitored in the system. It is in fact a specific version of an MES (Manufacturing Execution System). The IT system has a two-layer architecture where the database common for both establishments is located on a server and client applications on individual working stations. A unique identifier has been allocated to each piece of laundry, which allows the system to keep accurate records of its current status – age, number of washing cycles, current location (in the laundry or with a particular customer), and in case of personal clothes, the person to whom they are currently allocated.

Data capture is automated – individual pieces of laundry are not counted manually at their receipt or dispatch but they are read using antennas and UHF readers. These devices are able to read hundreds of chips per second and the data read are transmitted via serial port interface into the IT system where they are further processed. As reading dirty laundry at the receipt is less reliable because of higher humidity, two-level reading is in place at the receipt – the laundry first goes through the reading portal at the entry into the industrial laundry and then, it is read again by devices placed over each washing machine. This measure allows for reliability of even the entry reading of more than 99%; the exit reading has similar reliability with single-layer reading only as the laundry is dry and folded.

Laundry control IT system is connected to other system, using system integration. The first of these systems is ERP into which supporting materials for invoicing are exported in the form allowing for an immediate issue and accounting for an invoice.

Laundry control system is also connected to a web application available on the Internet, which is used by corporate customers for ordering linen and clothes. In this application, customers may enter the types and quantities of clean laundry which they request at the next delivery and at the same time, they are able to see how many pieces of laundry of each type they currently have from the laundry and how many they can still order, based on their current contract.

The last software with which the laundry control system is integrated is a mobile application operated on industrial portable terminals equipped with UHF chip readers. Mobile readers are used both by the laundry operation employees and by selected customers to whom the laundry lends the readers. The laundry employees use the readers for laundry deliveries to customers to verify that the delivered quantity corresponds with the quantity declared on delivery notes, and also for making periodic inventory-takings of linen and clothes with individual customers. Customers' employees may obtain particular information about each piece of laundry using the readers – e.g. how old is the piece, how many washing cycles it has been in or (in case of personal clothes) the person to whom it has been allocated.

As far as the case study assessment is concerned, the studied industrial laundry does not have any statistics available which would allow for quantifying the comparison of the state before and after the implementation of the discussed technologies. Based on the results of fifteen structured interviews with the company management and customers and quality comparison with the previous state, the following benefits of implemented technologies have been confirmed:

- AIDC technology brought a significant expediting of laundry handling because it is no longer necessary to count each piece manually – the laundry is read automatically in bulk within seconds.
- AIDC technology also allowed for creating complete and accurate records of individual pieces of laundry and for their tracing.
- Vertical system integration between the laundry operation control system and ERP resulted in savings of billers' and accountants' work, expediting the invoice issuing and removing any human factor errors resulting from manual data retyping among individual systems.
- Horizontal system integration to the laundry ordering web application simplified and expedited the communication with customers, just like lending mobile terminals to the customers did.
- The use of mobile terminals also lead to making the laundry records and traceability even more precise, it allowed for establishing a double control at laundry deliveries where the pieces of laundry are read both by a stationary antenna and reader in the laundry and at delivery at the customer's by a mobile terminal. Comparison of both results makes sure that laundry does not get lost on the way between the industrial laundry and the customer.

In general, the implementation of laundry control IT system based on AIDC technology, horizontally and vertically integrated to other applications led to expediting the passage of laundry through the industrial laundry and to expediting and simplifying the administrative processes. It also allowed for complete and accurate records keeping concerning individual pieces of laundry, which in turn led to reduction of losses and, ultimately, of the quantity of laundry which must be kept for each customer.

4 DISCUSSION AND CONCLUSION

In the previous sections in this paper, specifics of industrial laundry operation and issues faced by enterprises in this industry have been described. At the next step, technologies of Industry 4.0 have been identified which might be used for solving these issues. The case study presented in the previous section verified that at least two of these technologies – automated identification and data capture and system integration – might be successfully used to make the industrial laundry processes more efficient.
As far as other Industry 4.0 technologies identified as appropriate for launch in a laundry operation are concerned, they have not been included in the presented case study. The authors of this paper believe, however, that they might be efficiently interconnected with the technologies already in use. The current machine equipment used in industrial laundries standardly provides network communication interface even now, so it would be possible to connect the machines into the Internet of Things. As each piece of laundry in the described operation is identified with a unique, machine-readable identifier, a complex cyber-physical system could be created in which each machine and each piece of laundry would be represented. Autonomous collaborative robots might be connected in such a system, assisting human workers with handling the laundry, or replacing them completely in some operations. This could result in a fully automated operation controlled through its virtual model in the IT system. "Smart Factory" which represents a central conception of Industry 4.0 [15], can exist in industrial laundries as well.

The paper was created under the ÉTA Program of the Technological Agency of the Czech Republic, project no. TL02000136 Knowledge-intensive services sector adaptation to the conditions of Society 4.0.

References

1. J. Vacek, L. Dvořáková, M. Černá, J. Horák, Z. Caha, V. Machová, *Identifikace, analýza a hodnocení principů, postupů, metod a nástrojů pro adaptaci sektoru služeb na technické, ekonomické, sociální a environmentální podmínky Společnosti 4.0.* (Plzeň: NAVA, 2019)
2. V. Mařík et al., *Průmysl 4.0: Výzva pro Českou republiku.* (Praha: Management Press, 2016)
3. E. Hofmann, M. Rüsch, Industry 4.0 and the current status as well as future prospects on logistics. *Computers in Industry*, **89**, 23-34 (2017)
4. A. Rojko, Industry 4.0 Concept: Background and Overview. *International Journal of Interactive Mobile Technologies (iJIM)* **11**, 77-90 (2017)
5. Y. Lu, Industry 4.0: A survey on technologies, applications and open research issues. *Journal of Industrial Information Integration*, **6**, 1-10 (2017)
6. M. Polívka, L. Dvořáková, Využití automatické identifikace a sběru dat prostřednictvím radiofrekvenčních technologií v prostředí Průmyslu 4.0. *Trendy v podnikání*, **9**, 53-61 (2019)
7. T. Paschou, F. Adrodegari, M. Rapaccini, N. Saccani, M. Perona, Towards Service 4.0: a new framework and research priorities. *Procedia CIRP*, **73**, 148-154 (2018)
8. L. Klerkx, D. Rose, Dealing with the game-changing technologies of Agriculture 4.0: how do we manage diversity and responsibility in food system transition pathways? *Global Food Security*, **24**, 1-7 (2020)
9. G. Aceto, V. Persico, A. Pescapé, Industry 4.0 and Health: Internet of Things, Big Data, and Cloud Computing for Healthcare 4.0. *Journal of Industrial Information Integration*, **18**, 1-13 (2020)
10. L. Barreto, A. Amaral, T. Pereira, Industry 4.0 implications in logistics: an overview. *Proceedia Manufacturing*, **13**, 1245-1252 (2017)
11. K. Fukuda, Science, technology and innovation ecosystem transformation toward society 5.0. *International Journal of Production Economics*, **220**, 1-14 (2020)
12. Ch. Leyh, T. Schäffer, K. Bley, S. Forstenhäusler, Assessing the IT and Software Landscapes of Industry 4.0-Enterprises: The Maturity Model SIMMI 4.0. In Ziemba, E. (Ed.) *Information technology for management: new ideas and real solutions.* (Cham: Springer International Publishing AG, 2017, pp. 103-119)
13. D. Inkerman, D. Schneider, N. L. Martin, H. Lembeck, J. Zhang, S. Thiede, A framework to classify Industry 4.0 technologies across production and product development. *Procedia CIRP*, **84**, 973-978 (2019)
14. M. Herman, T. Pentek, B. Otto, Design Principles for Industrie 4.0 Scenarios. In *Proceedings of the 2016 49th Hawaii International Conference on System Sciences (HICSS).* Washington DC: IEEE Computer Society, 3928-3937 (2016)
15. V. Alcácer, V. Cruz-Machado, Scanning the Industry 4.0: A Literature Review on Technologies for Manufacturing Systems. *Engineering Science and Technology, an International Journal*, **22**, 899-919 (2019)
16. S. Vaidya, P. Ambad, S. Bhosle, Industry 4.0 – A Glimpse. *Procedia Manufacturing*, **20**, 233-238 (2018)

Changes in the start-ups business models

Katarína Procházková[1,*]

[1]UEBA, Department of management, Dolnozemska cesta 1, 852 35 Bratislava, Slovakia

Abstract. Research paper is summarizing results of project focus on business models and strategies of startups. The research was focused on identification of business models and strategies of startups. Research sample was creating from more than 70 startups at the beginning of the research and we finished research 3 years later with 53 startups. This research paper is partial output that analyze startups in global business environment in terms of doing business abroad and making strategic steps. The aim of paper is identify basic strategic steps and decision making of startups through selected indicators such as market presence and entry modes focusing on the last stage of research. Startups business model were also analyze and part of the output is mapping the shifts in business model Canvas of startups.

1 Introduction

Startups should be able to serve exponentially growing demand without a proportional increase in costs. The exponential growth possibility is also a typical feature of startups. Dave McClure, 2014 [1] claims that startup is a company that does not know exactly what its product is, who its customer is and how to make money. As soon as startup finds the exact answers to these three questions, it ceases to be a startup and becomes a real business Osterwalder and Pigneur [2] define a business model as a fundamental principle of creating, capturing, and offering the value to customers. The business model, the characteristics of business activities, and the specific strategy that should be global or at least regional, especially in the case of a startup from small local economies, are crucial in term of building the successful startup. The success of a startup is a scalable and saleable product that should be associated with effective market targeting and also branding and positioning. Building brand value today appears to be one of the basic strategic tools for the competitiveness of any business. It is also important to find the right investor for startup, and if you are getting financial resources, do not just rely on one source, but also search for other options like government programs, investors or venture capital.

* Corresponding author: katarina.prochazkova@euba.sk

2 The international strategies and business models of startup

The development of new technologies and the rapid development of the Internet have brought the trend of global businesses that have managed to sell their products and services throughout the world over a short period. These businesses are called startups. Author Steve Blank [3, 4] define a startup as a temporary institution created to find a repeatable and scalable business model

One of the strategies how to build a relationship with customers and create customer experience is through unique product or service and strong brand. Beyond building minimum valuable product is actually has to be useful and have people want to use it. Viability is not enough for small business software, services, hardware, or retail and customer products. Delightful products can adopted faster, encourage better word of mouth, and trigger better satisfaction. Users fall in love with delightful products. There are three area must come together to build this product: whole product approach, elegant design and focused quality. A great way to create Minimum Delightful Product is through an Experience Design Prototype, Fischer – Duane, 2016 [5]. A prototype is an early sample or model, which is launch to test a concept or process. Prototyping brings many advantages:

- Reduced time and costs: Prototyping can improve the quality of the specifications and requirements provided to customers. With prototyping, customers can anticipate higher costs, needed changes and potential project hurdles, and most importantly, potential end result disasters. Strong prototyping can ensure product quality and savings for years to come.
- Improved and increased user involvement: Most customers wants to feel like they are involved with the intricate details of their project. Prototyping requires user involvement and enables them to see and interact with a working model of their project. With prototypes, customers can give their immediate feedback, request project changes and alter model specifications. Prototyping most importantly helps eliminate misunderstandings and miscommunications during the development process.

On the other hand, a good product in good quality is always connected with effective branding. In the case of startups, marketing strategies and branding may not be a strength. A key startup recommendation is to focus on creating Minimum Viable Brand (MVB). A minimum viable brand is comprised of the core elements of a brand that are necessary to ensure internal focus and alignment as well as external relevance and differentiation. A framework for defining and developing MVB is the "6 what's" published by Yohn, 2014 [6]. These 6 "what´s" are: what we stand for – our brand essence, what we believe in – our defining values, what people we seek to engage – our target audience(s), what distinguishes us – our key differentiators, what we offer – our overarching experience, what we say and show – our logo, look, and lines (messaging). According to an article by published Garber, Fund [7] minimum viable brand for startups is based on these key components: purpose, people, proposition, product and personality

All these strategic steps and decisions about product development, marketing, building brand, get a new users, keep and manage the existing customer, managing the customer segments, decision making about entry modes and how to scale are often gradual strategic steps that shape strategy but also trigger changes in business models.

According to Allan Afuah, 2003 [8], the business model is influenced by the activities the company performs and the resources it uses. The Allan Afuah's business model concept divides the business model into four key components: industry factors, resources, position, costs and activities.

According to W.M. Johnson, C.M. Christensen a H. Kagermann, 2008 [9], a business model consists of 4 elements that are interconnected and create value in the enterprises: benefit of customer value, profit formula, key resources and key processes. One of the most popular is business model Canvas designed by Osterwalder and Pigneur, 2010 [10]. This business model analyze business through nine blocks:

1. Key partners (Who are our key partners? Who are our key suppliers? Which key sources do we get from partners? Which key activities do partners hold?)
2. Key activities (What key activities are creating value? Our distribution channels? Customer relations? Revenue streams?)
3. Key resources (What key sources require our values? Our distribution channels? Customer relations? Revenue streams?)
4. Value proposition (What value do we bring to the customer? Which of the customer's issues do we help solve? Which customer needs do we satisfy? What packages of products and services do we offer to each customer segment?)
5. Channels (Through which channels do we reach our customers? How do we achieve them now? How are our channels integrated? Which of them work best? Which are the most cost-effective? How do we integrate them into our customer routine?
6. Customer relationship (What type of relationship do we expect from each customer segment to create and maintain with them? Which of them have we already created? How expensive? How are they integrated with the rest of the business model?)
7. Customer segments (For whom do we create value? Who are our most important customers?)
8. Costs structure (How much are you planning to spend on the product development and marketing? What is your current costs structure? What are key costs? What is your type of cost structure?)
9. Revenue stream (What value do customers want to pay for? What do they actually pay for? How do they pay? How would they like to pay? How much does each stream contribute to total revenue?

In this paper we focused on mapping of changes and shifts in business model Canvas of startups in time (synthesis aggregates results during 3 years period, research started with more than 70 startups and was finished with 53 startups).

3 Research sample and research design

During 2015 - 2018, 76 startups in Slovakia were analyzed in three stages, and their business models were questioned. The last stage of research was finished with 53 startups; the basic structure of samples is shown in Table 1.

Table 1. Structure of research sample – the last stage of research

	Number (N)	**%**
Number of startups	53	100
Number of team members		
1-3	9	17
4-6	19	37
7-9	14	27
10 and more	10	19
Phase of business idea development		
Idea	0	0
Developmental stage	0	0

Prototype	9	17
First revenue	16	30
Growing revenue	28	53

Source: research results

For purpose of this research paper is used a general description of the data that we acquired in our research. In next chapter based on descriptive statistics are data aggregated for the last stage of research, general indicator as market presence and entry modes were used. In the last stage of research, the part of questionnaire was focused on startup position in external environment (table 4); scale from 1 to 5 was used for these indicators as follows:

- Action radius (from 1 – local to 5 – global)
- Segmentation level (from 1 – no segmentation to 5 – customization)
- Position in the external environment (from 1 – weak/on the edge to 5 – market leader)

A key part of questionnaire was focus on mapping changes and shifts in business models. Structure of nine blocks designed by Canvas business model was used. At the end of next chapter are stated the changes of startups business models.

4 Research synthesis and discussion

The results of the research show that startups are largely present on the market, in terms of first sales and they are expended transnationally (75,5 % operate on international markets).
The foreign markets, into which, the Slovak startups are operating, are the European Union countries (Czech Republic, Austria, Germany, Spain, Great Britain, France), the North American markets (USA, Canada), Russia, China, Australia and Brazil (a survey of these countries is the result of a qualitative analysis of the questionnaires throughout the research period). More than 70% of startups operate on foreign markets, at least within the Central European market. Creating global business from a local business is not an easy or common matter. The process of business internationalization (regardless of whether it is a classic business or a startup) requires not only geographic expansion (like physically occupying new markets) but also expansion associated with the using of information technology. In Table 2 can be observed than more than 80 % of startups after 3 years of existing are in the first sales stage. These startups were able to transform the business idea into a real business.

Table 2. Indicator - Market presence (the last stage of research)

Structure of indicator	**number (N)**	**%**	**cumulated %**
Preparing to enter the market	**2**	3,77	3,77
Preparation completed	**7**	13,2	16,98
Market entering	1	1,89	18,87
First sales	43	81,1	100

Source: research results

The effective and suitable entry mode depends largely on the financial possibilities of the startup, the nature of the product or the service, the distance of the market which startup wants to enter. The selection of the suitable entry mode is influenced by the time possibilities and the growth strategy itself, and preferences of startup founders. Table 3 provides an overview of the structure of basic entry forms that are used by startups. The majority of startups (almost 84%) are entering foreign markets through cross-border sales, which are largely coordinating in their own direction (more than 67% of startups use direct sales across borders).

Table 3. Indicator - Entry modes (the last stage of the research)

Structure of indicator	number (N)	%	cumulated %
indirect exporting	8	16,3	16,33
direct exporting	33	67,4	83,67
franchising	1	2,04	85,71
licensing	6	12,2	97,96
sale of startup	1	2,04	100

Source: research results

Running a business in the global economy is now a current phenomenon. Through a detailed analysis of the selected indicators, we tried to clarify the situation of Slovak startups in the international business space. The results identified the fact that even the Slovak startups have the potential to become globally successful, based on the results of the last phase of the research, 76% of research sample startups are already operating outside the Slovak Republic and about 30% of the startups are at the stage of making a profit.

The global expansion represents a huge revenue growth opportunity for companies, and that is the main reason why startups should take the chance. It also allows them to diversify revenue sources, and therefore, the risks. Finally, companies that go global quickly can keep their strategic position as a technical leader and sustain their competitive advantage in an increasingly aggressive market. According the data and output of our research, Slovak startups are able to go international very quickly. Due to indicator "action radius", almost 76 % of sample run international business. The existence of startups in the global economy in very common now. In a detailed analysis of selected indicators, we tried to clarify the situation of Slovak startups in the international business environment. The results show that even Slovak startups have the potential to become globally successful. Many startups in Slovakia are looking for clients and customers abroad directly (through direct sales), the question is how far and in what quality they can analyze the foreign market and capture all the potential of global market opportunities. Slovak startups′ most prefer entry form is an export form (mainly indirect sales – 70 % of startups and direct sales of 9 % startups), the licensing is using 11 % startups. Time factor, financial possibilities and the growth strategy of the company, respectively, ideas and preferences of founders and investors usually affect the choice of the suitable entry mode mostly.

According to a recent report from Crane Venture Partners on the internationalization of high-growth startups (more than 88% being funded or headquartered in the US), almost 90% of them have an international presence. Even if takes 5.5 years on average for those companies to launch internationally, we have noticed that the more recent the startup is, the sooner it accelerates its global presence. It took more than 10 years for a company

established before 2000 to go global, while businesses funded between 2006 and 2011 only waited for 3 to 4 years. According to Chotard, 2017 [11] very interesting is how US and non-US companies approach internationalization very differently: "Non-US companies take between 1 and 2 years on average to expand globally. This can be explained by both the smaller size of their home markets and a typically longer technology adoption curve. It's no surprise that the United States still leads the market for tech startups."
The global expansion represents a huge revenue growth opportunity for companies, and that is the main reason why startups take the chance. It also allows them to diversify revenue sources, and therefore, the risks. Finally, companies that go global quickly can keep their strategic position as a technical leader and sustain their competitive advantage in an increasingly aggressive market. We also further analyzed the startup position in the external environment. The following Table 4 gives the general statistical results of three indicators through which was analyze the external environment of startups.

Table 4. Startups position in the external environment

Indicator	Number	Average	Standard deviation	Min.	Max.
The last stage of research					
Action radius	53	3,57	1,35	1	5
Segmentation level	53	2,94	0,89	1	5
Position in business environment	53	3,09	1,43	1	5

The startup action radius largely duplicates the geographical spread of the startup action. Exactly 71.70% of startups have marked their action radius for Central European, European or Global. Startups are trying to segment their customers largely, almost 80 % of startups segment to several or more segments, 7.55% startups even use the customization (adaptation of the product to individual customer requirements). The current situation about the position of startup in the business environment is described by the following results: a third of the startups ranked their position as average, one third is considered as a leader and one-fifth of the startup marked their market position as weak. [12]

Table 5 Development of startup business model

Key Partners	**Key Activities**	**Value Propositions**	**Customer Relationships**	**Customer Segments**
Stage 1: 3,83 Stage 3: 4,51 **increase 0.70 point**	Stage 1: 3,83 Stage 3: 4,51 **increase 0.68 point**	Stage 1: 4,03 Stage 3: 4,49 **increase 0.46 point**	Stage 1: 3,92 Stage 3: 4,40 **increase 0.48 point**	Stage 1: 4,15 Stage 3: 4,45 **increase 0.30 point**
	Key Resources		**Channels**	

<table>
<tr><td></td><td>Stage 1: 3,98
Stage 3: 4,25

increase
0.27 point</td><td></td><td>Stage 1: 3,58
Stage 3: 4,15

increase
0.57 point</td><td></td></tr>
<tr><td colspan="2">Costs Structure
Stage 1: 4,05
Stage 3: 4,44

increase 0.39 point</td><td colspan="3">Revenue Streams
Stage 1: price 3,81 quantity 3,46
Stage 3: price 4,40 quantity 4,06

increase of price 0.59 point
increase of quantity 0.34 point</td></tr>
</table>

Table 5 summarizes the overall changes in the Canvas business model by comparing the first and final stages with the fact that in each block have been stated a positive development (the calculated change was always a positive increase). The most significant changes have been made in the identification of key partners (0.70 points), key activities (0.68 points) and revenue streams especially in the identification of product/service price. The lowest positive changes were identified in the block of key resources (0.27 points) and costs structure (0.29 points).

Startups that operate globally also have very high dynamics and speed of business action. Results also confirmed that globally operating startups sell their products and services at high prices (or at least higher or comparable to the competition they know about). B. Zagorsek, 2016 [13] confirmed this result, in his research output, stated: "the most interesting result was the identification of a higher price strategy than the competition, which positively influenced the higher number of paying customers." From the Canvas business model, the following fact has been identified: the presence of resources and the presence of processes does not have a significant impact on the startup radius of action. Running a business in the global economy is now a common phenomenon. Through a detailed analysis of the selected indicators, we tried to clarify the situation of Slovak startups in the international business space.

According to the Cambridge Associates investment firm, 40% of the startup business has a chance for business success [14]. According to the European Association of Business Angels (2017) there are launched about 50 million new projects every year (137,000 per day), but 90% of them fail.

The results identified the fact that even the Slovak startups have the potential to become globally successful, based on the results of the last stage of the research, 76% of research sample startups are already operating outside the Slovak Republic and about 30% of the startups are at the stage of making a profit.

Acknowledgement

This research paper is partial output of project VEGA MŠ SR č. 1/0631/19 "Metamorfózy startupov na ceste za podnikateľským úspechom. Metamorphoses of start-ups on the way for the business success", 2019 – 2021 in 100% range.

References

1. D. McClure, Startupquote. [accessed 08.06.2019] startupquote.com/post/51277392517
2. A. Osterwalder, Y. Pigneur. Business Model Generation. Handbook for visionaries, game changers, and challengers. John Wiley and Sons; 1st edition. ISBN 978-0470876411 (2010)
3. S. Blank. What's A Startup? First Principles. [accessed 08.06.2019] https://steveblank.com/2010/01/25/whats-a-startup-first-principles. (2010)
4. Blank, S. Why the Lean Start-up Changes Everything? In *Harvard Business Review*. Boston: Harvard Business Publishing. ISSN 00178012 (2013)
5. S. Fisher, J. Duane. The Startup Equation. A visual guidebook for building your startup. USA: McGraw-Hill Education, 2016. ISBN 978-0-07-183236-6 (2016)
6. D. L. Yohn, Start-ups needs a minimum valuable brand. In *Harvard Business review* (2014). [accessed 15.3.2020] https://hbr.org/2014/06/start-ups-need-a-minimum-viable-brand (2014)
7. M. Garber, J. Fund. Minimum viable brand for startups. [accessed 12.3.2020] https://medium.com/rga-ventures/building-minimum-viable-brand-for-startups-4e834d0608fc
8. A. Afuah, A. Business Models: A Strategic Management Approach. New York: McGraw-Hill/Irwin. ISBN 978-0-07-288364-2 (2003)
9. M. W. Johnson, C. M. Christensen, H. Kagermann. "Reinventing Your Business Model." *Harvard Business Review* 86, no. 12 (December 2008).
10. A. Osterwalder, Y. Pigneur. Business Model Generation. Handbook for visionaries, game changers, and challengers. John Wiley and Sons; 1st edition. ISBN 978-0470876411. (2010)
11. Chotard, L. International Success of Startups Worldwide. [accessed 15.3.2020] https://www.textmaster.com/blog/internationalization-startups-worldwide/ (2017)
12. R. Bednár, K. Procházková, B. Zagoršek. Strategies and Business Models of Startups. Reviewers: Michal Varmus, Katarína Chomová. 1st Edition. Brno: Tribun EU (2018). 122 p. ISBN 978-80-263-1508-7. (2018)
13. B. Zagoršek. Využitie stratégie na dosiahnutie úspechu. In Podnikateľské modely a podnikateľské stratégie startupov II. *Recenzovaný zborník vedeckých prác. Bratislava*: Ekonóm, 2016. s 162 - 169. ISBN 987-80-225-4328-6 (201).
14. E. Griffith. Conventional Wisdom Says 90% of Startups Fail. Data Says Otherwise. [accessed 12.3.2020] http://fortune.com/2017/06/27/startup-advice-data-failure

Country branding comparison between Slovak Republic and United Kingdom in Future Brand Country Index

Iveta Kufelová[1,*] and *Monika* Raková[2,]

[1] University of Economics in Bratislava, Faculty of Business Management, Department of Business Economy, Dolnozemská cesta1, 852 35 Bratislava, Slovak republic
[2] University of Economics in Bratislava, Faculty of Business Management, Department of Business Economy, Dolnozemská cesta1, 852 35 Bratislava, Slovak republic

Abstract. Nowadays are known many indexes, which are rating specific countries by their performance in chosen areas. This work will be concentrated on the brands of countries valuated by brand indexes. For purposes of job we have chosen the Future Brand Country Index in order to achieve the most objective result of comparison between Slovak Republic and Great Britain. Although they are known certain number of indexes, factors used in Future Brand Country Index are considered by us as the best suitable for this work. This index valuates strength of perception of the country brand across certain dimensions, which are Value System, Quality of Life, Business Potential, Heritage and Culture, Tourism and Made in. These dimensions consist of numerous associations with specific country which will be specified later. Purpose of our work is to analyse results of Slovak republic and Great Britain, their position in Future Brand Index in 2014 and in 2019 and find reasons of their improvement or deterioration. Since Future Brand Index does not publish data and analyses of these countries, our work will be concentrated on analysing key dimensions of index which could affect position of chosen countries in Future Brand index.

1 Introduction

Nation branding should be interpreted as the practice of building and communicating the country image to the rest of the world, through diplomacy, trading, export and tourism. First mention of this term was in 1996 by the British brand practitioner Simon Anholt [15]. In accordance Miazhevich, nation branding means presenting a country as a brand or a product to an external audience through forms of strategic communication [17]. By the author, Branding includes the visual, verbal, textual and behavioralpresentations and expressions of an organization's identity as well as decisions and strategies underpinning such expressions [19]. Many researchers explore a region where nation branding has rapidly proliferated: Central and East European countries, faced with the need to (re)formulate themselvesand

* Corresponding author: iveta.kufelova@euba.sk

gain global visibility as nations after the collapse of communist regimes, while alsoundergoing economic and political transition and coming to terms with a traumaticpast. These factors caused'a state of inadequacy'affecting both individual psychologiesand political decisions [16]. The two predominant approaches in place branding research are tourism based that focus on branding the country as a tourism destination and product based, that focus on associating countries with products [18]. The perception of visitors is a key element in building the country's image, and should become a new perspective in the study of nation branding states that the image of a country directly affects the behavior of foreign clients: the final consumers and investors. A country image is a set of connotations that take shape in the minds of people based on what they hear, say and remember about a specific destination. [20]

2 Methodology and data

As we have already mentioned above, this index consists of 6 dimensions which include numerous associations stated in the Table 1.

Table 1. Own elaboration of Future Brand Index dimensions with connected associations

DIMENSION	ASSOCIATION
VALUE SYSTEM	1. Political Freedom 2. Environmental standards 3. Tolerance
QUALITY OF LIFE	1. Health and Education 2. Standard of Living 3. Safety and Security 4. Would like to live/study there
BUSINESS POTENTIAL	1. Good for Business 2. Advanced Technology 3. Good Infrastructure
HERITAGE AND CULTURE	1. Historical Points of Interest 2. Heritage, Art and Culture 3. Natural Beauty
TOURISM	1. Value for money 2. Range of Attractions 3. Resort and lodging options 4. Would like to visit for vacation 5. Food
MADE IN	1. Make products that are authentic 2. Make products of high quality 3. They create unique products 4. Would like to buy products made in that country

Table 1. Own elaboration of Future Brand Index dimensions with connected associations base on https://www.futurebrand.com/uploads/Country-Brand-Index-2015-15.pdf

With respect to the limited space, we have chosen just specific associations considered as the most suitable for index results of chosen countries. Both versions of indexes, also from 2014 and from 2019, use the same methodology and valuation of the countries. Difference of versions are their objectives for specific years, which are not considered in our work due to respect to no relevance for overall ranking and limited space. In 2014 was Slovak republic on 44th place from 118 countries but in 2019 was position of Slovak republic better and end up on 35th place. Our second country of interest is Great Britain, which was in 2014 on 12th place but in 2019 fell down on 19th place. [1]

2.1 Value System

Now we will analyse reasons why their position dropped or rised and we will start with the first Value System dimension, where we chose environmental standards. For this association we used data of ecological footprint for stated countries and their change through years 2013-2016. This footprint is measured as a biocapacity reserve or deficit. According to Global Footprint Network, it is approximately 12,2 billion hectares of the land, which can be productively used and that means, it is 1,6 hectare of productive land per person. If country will exceed this land per person, country has ecological deficit and that means, that country uses more land than can planet effectively recover and if uses less, it has ecological reserve. In 2014 had a Great Britain biocapacity deficit -3,5 global hectares. Main reason of this deficit was the production of CO2 by Great Britain, which was in 2014 195 160 525.31 metric tons. In 2016 Great Britain showed small improvement and biocapacity deficit dropped on -3,3 global hectares. This drop could be affected by decrease of producing CO2 by more than 10 million tons. This deficit is considered as one of the highest in the world also with USA, Saudi Arabia, Italy and et cetera. As we can see on the Figure 1 Slovakia got better in these years but still had a biocapacity deficit. In 2014 the deficit was -1,4 global hectares and we can assume, that it was also caused by production of the CO2 in the amount of 13 951 632.83 metric tons.

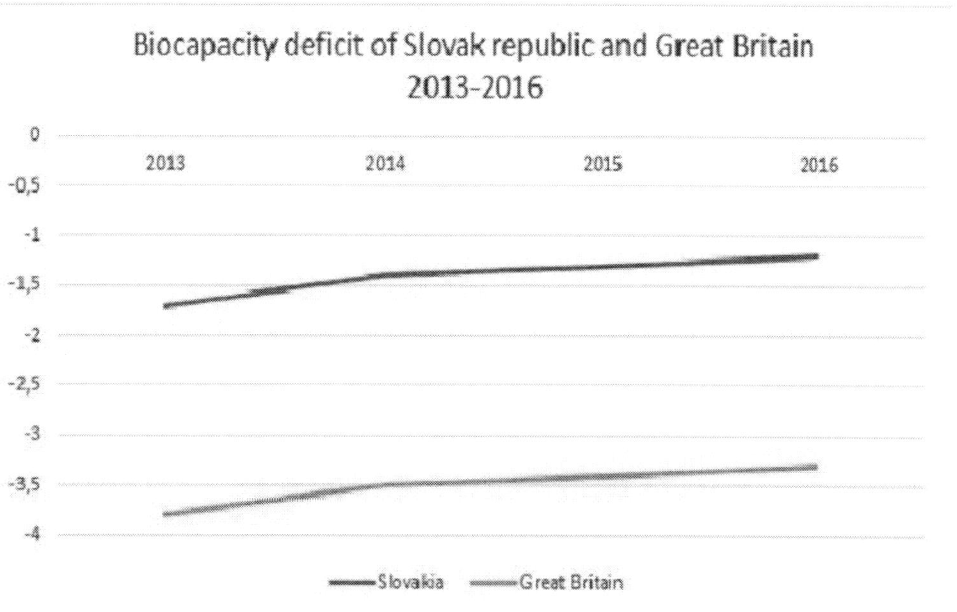

Fig. 1 Own elaboration of biocapacity deficit of Slovak republic and Great Britain

In 2016 Slovak republic also indicated small improvement, when deficit fell on -1,2 global hectares. This result could be affected by decrease of the production of CO_2 and decrease of cropland use. Reason, why were these countries decreasing their CO_2 emissions was their participation in Kyoto protocol. [2][3]

2.2 Quality of Life

Second dimension is Quality of Life. The concept of social quality of life is used to analyze the degree of people's diverse needs satisfaction; they are spiritual, material and social needs. This category, being an interdisciplinary one, is an object of study of not only economy, but also philosophy, and sociology. [24] In the paper we specifically concentrate on safety and security. For purposes of this association we have used data from Global Peace Index. This index rates countries in order to their participation in conflicts, stability inside their country thus for example level of violent crime or violent demonstrations, but also their participation to world peace as a contribution to UN peacekeeping missions. This is one of the associations where Slovak republic had a better position than Great Britain. As we can see on the Figure 2, Slovak republic and Great Britain keep stable position in Global Peace Index with a difference, that Slovak republic has since 2014 till 2019 better results than Great Britain.

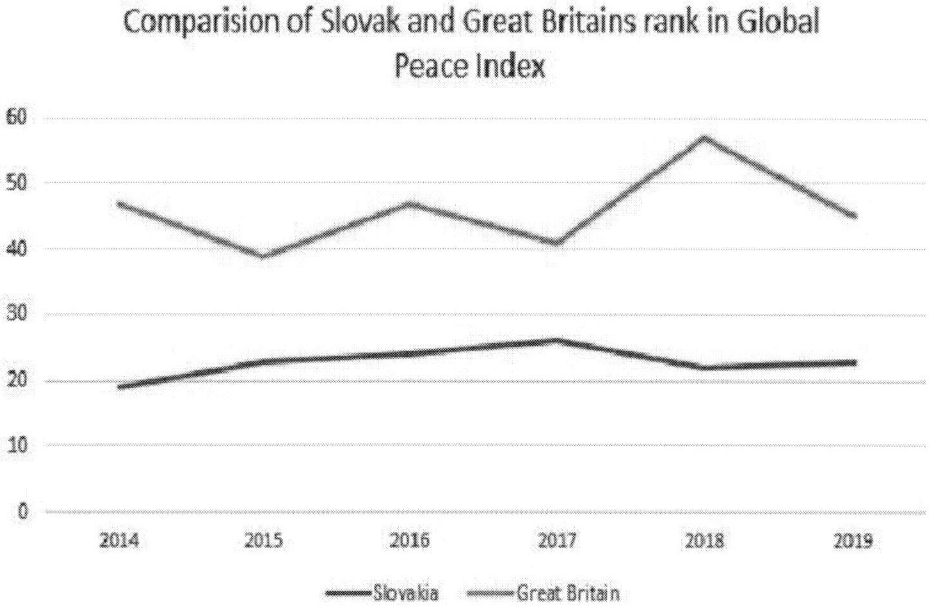

Fig. 2. Own elaboration of comparision of Slovak and Great Britain rank in Global Peace Index

The worst rank of the United Kingdom was reached in 2018, when from 163 countries United Kingdom end up on 57th place. According to the Global Peace Index 2018 report, it was affected by„*relation to rising nationalism*" .Slovak republic has the worst score in 2017 what could be affected by murder of the journalist and small percentage of change in acceptance of the rights of others. [4][5][6]

2.3 Business Potential

Next dimension is business potential, where we analyse association Good for Business. For this association we chose Doing Business Index for the rank chosen countries through the years 2014-2019. The Doing Business Index (DBI) is a multidimensional measurement of the aspects of business regulation, which affect domestic small and medium-size firms. It is developed by the World Bank who acknowledged the importance of ease of doing business for the economic development of countries. [23]

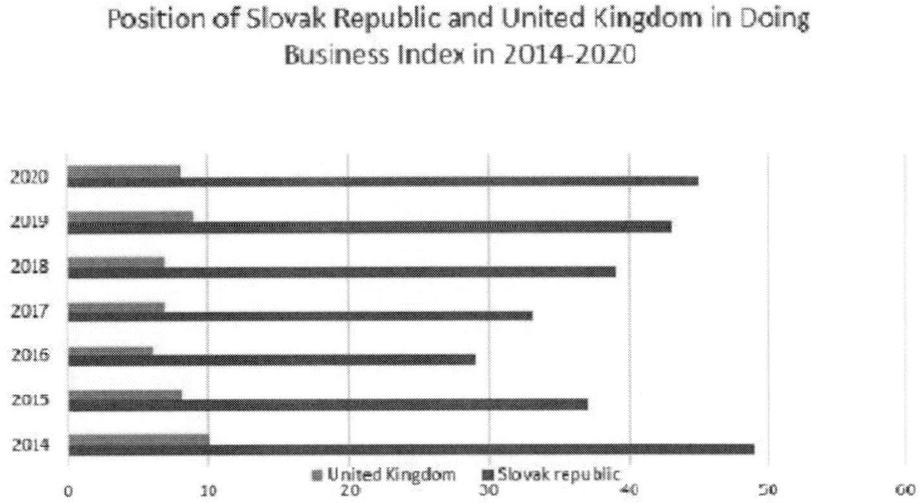

Fig. 3. Own elaboration of comparison of Slovak republic and United Kingdom in 2014-2020

As we can see on the Figure 3, United Kingdom is since 2014 in the first ten countries where it is easiest doing business in. United Kingdom has achieved these results mainly thanks to the good score in topics of Starting a Business, Getting Electricity and Trading across Borders. The weakest parts of the score are in topics of Enforcing Contacts and Getting Credit. If we look on the case of Slovak republic, we can easily understand why United Kingdom has a better rank. If we want to compare business environment in United Britain and Slovak republic, we can compare for example Starting a Business. In Slovak republic it takes 21,5 day to get through bureaucracy to start a new business, whereas in United Kingdom 4,5 days. Getting electricity to company building can take in United Kingdom 46 days, whereas in Slovak republic almost twice more. Only one topic, where was Slovak republic better than United Kingdom was Trading across Borders, where Slovak republic has first place and full score. [7]

2.4 Heritage and Culture

Fourth dimension is Heritage and Culture. The working definition of cultural heritage is the definition of a set of resources that are protected, preserved or restored because they are worthy of transfer to new generations. These include monuments, including archaeological works, works of art, films, books or documents held by museums, libraries and archives, as well as intangible items such as skills, rituals and ceremonial events. [21] We analyse whole three associations, thus Historical Points of Interest, Heritage, Art and Culture, Natural Beauty with number of properties inscribed on the World Heritage List. We analyse these association this way because it is no index or a better way how to measure Culture and Heritage of the country. Slovak republic has 7 properties which are inscribed on the World Heritage List as Bardejov Town Conservation Reserve or Wooden Churches, whereas Great Britain has 32 properties as a Stonehenge, Town or London or Henderson Island. It is need to say, that it can be more reasons of these differences, for example diversion of Slovak and the Czech Republic in 1992, or the fact that Great Britain has almost 5 times bigger area than Slovak republic. [7] [8]

2.5 Tourism

Fifth dimension is tourism where we will analyse multiple associations. There is agreement in the scholarly literature on the tourism sector that it is difficult to define and specify the concept of tourism competitiveness due to the influence of multiple factors or dimensions that influence a destination's success. [22] For better understanding we compared number of tourists of both countries per year in Figure 4 below.

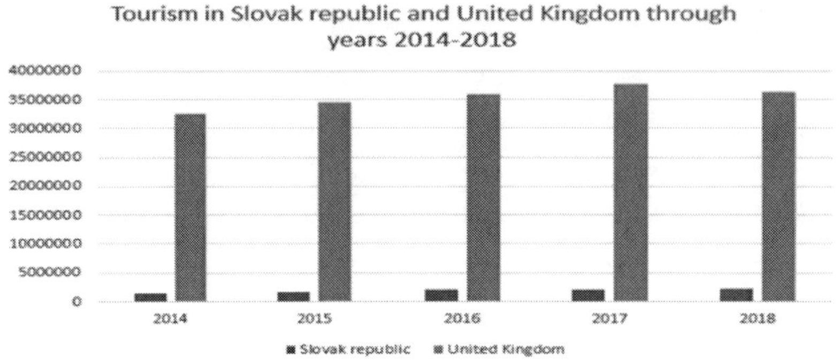

Fig. 4. Own elaboration of comparation of tourism in Slovak republic and United Kingdom

According to World Bank, in 2018 was Great Britain visited approximately by16 times more tourists than Slovakia. This fact is highly affected by smaller size of Slovak republic area, power of the economy and power of the country brand in the world. Tourism in Slovak republic is rising every year also thanks to project Good Idea Slovakia , which is represented in the homeland but mainly in abroad at political events but also on the business missions by SARIO agency. According to Travel in Slovakia, the most interesting regions to visit are Bratislava, Košice, Horehronie and Tatras and Northern Spiš. Slovak republics range of attractions thus include mostly visiting nature and countryside, cultural monuments and tourist life in bigger cities. Tourism in Slovak republic also includes beer tourism, which is type of tourism motivated by experiencing new tastes of beer flavours, types of technology of beer manufacture and getting familiar with history of beer and beer region. Second important type of tourism in Slovak republic is food tourism. Nowadays, it is more and more important to get countryside and its food tradition to attention of the tourist. This process can help to restore food tourism in Slovak republic. Slovak republic context of food tourism contains mainly from sheep's milk diary production, which is considered as a part of Slovak identity. Culinary tourism in this area should be supported also because of its long tradition in manufacturing these products, whereas first signs of coagulated milk leads to the prehistoric times 2000 years BC. If we will analyse case of the Great Britain, first we notice that as was mentioned, Great Britain is visited by tourist 16 times more than Slovakia. According to Statista, Great Britain is mostly visited to experience British big city life. This fact indicates number of overseas visits to the Great Britain in 2018, when 19 090 000 people visited London and Lothian, which is more known for wonderful nature, was visited 4 times less by 2 433 000 tourists. According to Visit Britain, the country has big range of attractions including seeing traditional British cultural monuments, for example Tower of London, Stonehenge or Pontcysyllte aqueduct and Canal which are also inscribed on the UNESCO World Heritage List. Tourist have opportunity to visit also royal part of the Great Britain by visiting official royal residences, royal events or parks. Food and drink tourism in the country can be for tourist interesting because of tradition of afternoon tea, which is in Great Britain famous for centuries. Tourists use to visit also British pubs and cocktail bars. Food tourism in Great Britain is highly affected by globalisation which brought food of whole world in there, but it is easy to experience also the traditional food tourism. United Kingdom has a long tradition to serve cornish pasties or famous fish and chips. [9] [10] [11]

2.6 Made in

In the last dimension, made in, we concentrate on the products from Slovakia and United Kingdom that are unique, high quality and authentic. We start with Slovak products, specifically with the mead. Mead is an alcoholic beverage which is manufactured by fermenting honey mixed with water and fruit. This beverage is for Slovak republic significant and has a manufacturing tradition, which leads to its high quality. Quality of the Slovak mead was approved by prestigious Mazer cup several times, where it was awarded as the best mead in world. Next unique product is Slovak Lyra Chocolate, which was awarded in multiple competitions as Great Taste, International Chocolate awards but also by World Packaging Organisation. It is difficult to compare Slovak republic and United Kingdom in the area of a manufacture. Whereas Slovak republic is producing lot of unique products mainly in the food industry, Great Britain has a bigger range of production. Of course, we can also mention British gin manufacturing tradition but we can see this association as in the wider range, since Great Britain is a country, which is the homeland for the famous, significant and luxury cars as a Rolls Royce or Jaguar. This fact has given to United Kingdom significant know-how in the world, which is that cars from the Great Britain are high-quality and unique and that allows these companies push the prices high. [12][13][14]

Conclusion

We have compared in this article two countries, which are different in numerous aspects. It is power of the economy, history, mentality and culture, approach to the environment and tradition in manufacture. Goal of our index was to find out why has Great Britain in long term better rank than Slovak republic and what can Slovak republic do to change its rank. We have analysed data from all six dimensions and our conclusion is, that although Slovak republic is more peaceful country and can care less about our planet, next dimensions are putting it down. Some dimensions can be affected by country just really hard. Inscription in the UNESCO Heritage List can't be affected by higher investments or by promoting in the abroad, but tourism can be. Good Idea Slovakia is the tool how can Slovak republic become more interesting for tourists, but it can't be the only tool. Simple tool how to reach better score in Doing Business index can be in Slovak republic decreasing the bureaucracy, which is long term complication in the country, but it can't be the only tool. It needs also improve the infrastructure to get even more interesting for investors, although Slovak republic made significant changes that pulled forward direct investments in the country. And as the last one, Slovak republic needs to promote their production in abroad more. It is not enough for rest of the world to know, that in Slovak republic Volkswagen cars are made. If Slovak republic will start to promote their products, which are on the high-quality level, it can lift up the motivation for Slovak citizens to be creative too and eventually lift up level of living in the country. If it will succeed, in one dimension, it can pull up also other ones and with right directed investments for the support of these dimensions, it can change the position and name of the Slovakia in the Future Brand Index in the future.

This contribution is the result of the project VEGA no. 1/0368/18 "Cenové stratégie v prostredí pôsobenia efektívnych regulačných mechanizmov na nadnárodných trhoch sieťových odvetví slovenskej ekonomiky" in extent 50 % and the project VEGA no. 1/0708/20 "Sociálno-ekonomické determinant trvalo udržateľnej spotreby a výroby z hľadiska vplyvu na výkonnosť a konkurencieschopnosť podnikov" in extent 50 %.

References

[1] FutureBrand, online at https://www.futurebrand.com/uploads/Country-Brand-Index-20 14-15.pdf (2015-2019)

[2] Global Footprint Network, online at http://data.footprintnetwork.org/#/? (2019)

[3] European Commission, online at https://ec.europa.eu/commission/presscorner /detail/ en/IP_13_1035 (2013)

[4] Vision of Humanity, online at http://visionofhumanity.org/indexes/global-peace-index/ (2019)

[5] The Institue for Econoics and Peace, Global Peace Index. online article at http://econ omicsandpeace.org/wp-content/uploads/2016/06/GPI-2016-Report_2.pdf (2016)

[6] D. Boffey, Slovakian journalist investigating claims of tax fraud linked to ruling party shot dead. online article at https://www.theguardian.com/world/2018/feb/26/slovakian-journalist-investigating-claims-of-tax-linked-to-ruling-party-shot-dead (2018)

[7] The world bank, online at https://www.doingbusiness.org/ (2020)

[8] Unesco. online at https://whc.unesco.org/en/about/ (2020)

[9] J. Jablonská, T. Pôbiš, G. M. Timčák, Beer tourism in Slovakia. Institute of Geotourism. FBERG, Technical University of Kosice (2018)

[10] K. Matlovicova, The Culinary Tourism in Slovak case study of the traditional local milk products in the regions of Orava and Liptov. GeoJournal of Tourism and Geosites, Oradea – Gdansk, **12**, 3 (2013)

[11] Statista, online at https://www.statista.com/statistics/515696/inbound-tourism-most-visited-uk-regions-united-kingdom/ (2018)

[12] Visit Britain. online at https://www.visitbritain.com/gb/en (2020)

[13] Apimed, online at https://www.apimed.sk/o-firme/ocenenia?lang=en (2020)

[14] Lyra, online at https://www.lyrachocolate.com/en (2020)

[15] Cheregi, Bianca-Florentina, Nation Branding in Transition Countries: A Multimodal Analysis of Romania and Moldova Tourism Campaigns. *Journal of Entrepreneurship, Management and Innovation* **14**, 4, 81-106 (2018)

[16] A. S. Zemanek, Nation Branding in Contemporary Taiwan. A Grasroots Perspective. Culture, Theory and Critique **59**, 2, 119138 (2018)

[17] G. Miazhevich, Nation branding in the post-broadcast era: The case of RT. *European Journal of Cultural Studies* **21**, 5, 575-593 (2018)

[18] Y. Dubinsky, Analyzing the Roles of Country Image, Nation Branding, and Public Diplomacy through the Evolution of the Modern Olympic Movement. *Physical Culture and Sport. Studies and Research* **84**.1: 27-40 (2019)

[19] H. L. Sataøen, Sub-sector branding and nation branding: the case of higher education. *Corporate Communications: An International Journal* (2019)

[20] L. Echeverri, E. ter Horst, G. Molina, Z. Mohamad, Nation Branding: Unveiling Factors that Affect the Image of Colombia from a Foreign Perspective, Tourism Planning & Development (2017)

[21] E. Vodopianova, L. Korobeynikova, Vestnik Tomskogo gosudarstvennogo universiteta. Kul'turologiya i iskusstvovedeniye. *Tomsk State University Journal of Cultural Studies and Art History* **35**, 215–221 (2019)
[22] S. Fernández, J. Antonio, P. Azevedo, M. Jose, Determinants of tourism destination competitiveness in the countries most visited by international tourists: Proposal of a synthetic index. Tourism Management Perspectives (2019)
[23] M. Maričić, M.Bulajić, Z. Radojicic, V. Jeremic, Shedding Light on the Doing Business Index: a Machine Learning Approach. *Business Systems Research Journal*, **10**, 73-84 (2019)
[24] L. Kilimova, Quality of life as a human development determinant in the context of economic instability. Economic Annals-XXI, **157**, 59-61 (2016)

The Impact of digitization and automation of production on the role of the workforce in companies in Slovakia

Julia Rakovskál,[1]

1 University of Economics in Bratislava, Faculty of Business Management, 85235 Bratislava, Slovakia

Abstract. The aim of this article was to evaluate the current view of domestic and foreign authors on problems and trends of production logistics in the conditions of globalization and to explore possibilities of development of the concept of Industry 4.0 in Slovakia. We also focused on the digitalization of processes, which are one of the basic prerequisites for the company's operation in the Industry 4.0 concept. The current conditions in the Slovak Republic, resulting from the functioning of the market economy, create new opportunities and the need to apply business approaches in production management. It is a prerequisite for the success not only of manufacturing companies to find balance between modern trends and technologies and people with necessary knowledge and skills.

1 Introduction

Globalization affects every social area, including manufacturing. Traditional production processes are squeezed out by new, more innovative solutions, which customers expect to be more efficient. Along with intense and rapid changes, not only a technological, but mainly a social question resonates in the society: "How will a person's position change in the manufacturing process of the future?" The effects of robotization and digitization on jobs, which are characterized by monotony, low educational requirements and high frequency or injury risk, are evident already.

The advent of digitization has begun to show fundamental change in all areas of business. New global markets, business models, ICT advances and innovation in every business area are emerging.

The goal of the paper is to identify activities and aspects of the corporate environment, as well as the external environment of companies in Slovakia, which needs to be developed in order to implement and apply the principles of digitization and automation of production. The concept of Industry 4.0, its substance and the possibilities that exist for its implementation into industrial companies with a potential impact on employees, can also be considered as an object of research. Among several research methods, mainly theoretical methods and out of these, general (logical) methods, a method of abstraction, method of analysis and synthesis, method of generalization, comparative method and the method of induction and deduction have been applied.

2 Current state of the area under examination

First of all, it is necessary to define the concept of the manufacturing process so that we can address its automation and robotization in the next sub-chapters.

The core activity of the company is production. In the broadest concept, 'production' means the combination of production factors, such as labor, soil, capital, in order to obtain certain products. Production means a business process that ensures the conversion of inputs into the expected outputs. Therefore, the production process could be defined as a transformation process. In other words, it is the process of making products or providing services. The production process is not intended for any products, but only those that are viable on the market. The conversion of inputs into outputs must take place as efficiently as possible, i.e. with optimal consumption of production factors.

Several elements and components are involved in the production of specific products of a material nature and thus, in the working process. The basic components of the production process, which affect the level and, consequently, the results achieved, are considered to be labor, labor force and work equipment. The work items include raw materials, basic and auxiliary materials, energy, fuel, etc. Work equipment is machinery and

[1] Corresponding author: julia.rakovska@euba.sk

appliances, the role of which in production processes is increasing with the development of mechanization and automation. The workforce is a social element. It can be said that it is a crucial element in production. The workforce affects all other elements of the production process. The three basic elements of production consist only of a potential production process, which is realistically only formed by interaction between all three key components.

2.1 Automation of production

The concept of automation began to be used from about 1947 in the automotive industry to describe the increased number of automatically functioning devices and controls in mechanized production. The origin of the word is attributed to D.S. Harder, who was an engineering manager at Ford Motors Company.

Automation, like many other concepts, does not have a completely precise definition, but the basis is the same. Automation, which is the successor to mechanization from the point of view of industrialization, means that people's control activities with technical equipment or vending machines are replaced. An automatic machine is a machine that itself performs predetermined tasks with minimal assistance of people, mostly with using artificial intelligence.

The Encyclopedia Britannica explains automation as using machines in tasks performed by humans, or more often, in tasks that would not be possible without them. Although the term mechanization is more often used to express the replacement of humans by machines, automation is generally used to express the integration of machines into autonomous systems. It also says that automation can be defined as technology for performing processes using programmed commands combined with automatic feedback management to ensure that instructions are executed correctly. The resulting system is able to operate without human intervention. Development of this technology is increasingly dependent on the use of computers and computer technologies.

The World Bank's report 2019 on the world development shows that economic effects of workers replaced by automation are outnumbered by new sectors and jobs in the technology sector.

The first references come from ancient Egypt, where, for example, automatic grain milling systems and field irrigation were adopted. It was also the automatic closing of the bronze gate in Alexandria, which was described by Herón in his work *Pneumatica*. The most famous vending machines from the Middle Ages are the clocks, wind chimes and others.

In 1905, an automatic machine for blowing glass bottles was introduced. The machine, operated by two workers, could produce 17,280 bottles in 24 hours, compared with 2,880 bottles produced by six men. The costs for making bottles by the machine was 10 to 12 cents, compared to $1.80 for man-made hand operations.

In December 1913, Ford Motor Company introduced an assembly line for car manufacturing, which is considered to be one of the pioneering types of automation in the manufacturing industry. The production time of one car dropped from more than 12 hours to just 93 minutes, the working time was reduced by one hour per day to make the three-change operation easier to implement.

In the twentieth century, significant developments in several areas occurred. Development of an electronic digital computer has made the control function more sophisticated and faster in automation. The current trend in automation is to connect individual control computers to local computer networks, in which they can communicate and coordinate with each other in order to streamline the entire production process and not just the individual parts of it.

2.2 Robotics and robotization

Automation technology has matured to the point in which it has developed a number of other technologies that have achieved their own recognition and status. Robotics is one of those technologies. It is a specialized automation industry. That is why the concept of robotics needs to be clarified in order to understand the difference between robotics and robotization.

Robotics is an automation technology that has been developed since about 1960. It is a science dealing with robots. In robotization, it is nothing more than the introduction of robots into industrial tasks or production processes.

In both cases, however, these are the robots that form the essence of the entire robotization. The word *robot* comes from the Czech word *robot*. The first reference of the word robot was used in 1920 in a game by Karl Capek R.U.R. (Rossum's Universal Robots). The robot is known either as a machine, a device performing complex operations according to a specified program, or as an artificial human, a human-like machine.

The most typical human characteristics of a modern industrial robot is powered by a mechanical arm. The robot arm can be programmed to move a sequence of movements to perform necessary tasks. Industrial robots are usually used as a substitute for human workers in factories.

According to the Oxford Dictionary, a robot is a machine that is programmable by a computer, capable to automatically perform complicated series of operations. A more complicated but more precise definition is provided by the ISO standard 8373:2012, according to which the robot-controlled mechanism is programmable on two or more axes, with a certain degree of autonomy, moving in a space to perform specified tasks.

Thus, robotization means using robots to perform tasks for which they have been pre-programmed. In this paper, we also consider as necessary to compare the robot and the human, so that we could have a better idea of how robots can influence the manufacturing process and the position of the workforce in it.

The advantage of robots is that they can perform repetitive tasks evenly, which are often monotonous and difficult for humans to stay focused on. Thus, we are getting to their next advantage, which is a constant full concentration. The big advantage is also that robots, as they are not living organisms, can work in a dangerous, hazardous environment, and also, their strength and load capacity can be much higher than in humans. Last but not least, robots have essentially unlimited capacity to store the amount of information which they can recover at any time. Unlike robots, a human have empathy, they can solve abstract problems, they can better cope with changes. As opposed to that, robot would have to be re-programmed. In addition, humans are able to deal with unexpected situations and they are flexible.

Cost-effectiveness means reducing production costs by eliminating, for example, the cost of workers' wages. It is also an advantage that robots can operate in a repetitive cycle, ensuring increased productivity. In addition, thanks to this ability, the quality of products is also improved. This means that workers who produce a given product by repetitive activities often make production errors, which also triggers increasing costs. Among other things, we know that there are also injuries in workplaces, especially in dangerous environment, which is not the case with robots.

In regard to disadvantages, it is important to note in particular how robotization affects or could adversely affect the society.

Usually, the biggest hurdle that decides whether or not a company will invest in robotic automation is the high initial investment. In some countries, longer return of investment may also be discouraging. Then we see the downside of increasing skilled staff, which is an advantage on the one hand, as this can provide new jobs. On the other hand, it is difficult to find a skilled workforce which entails the cost of additional training. One of the biggest problems in implementing robotic automation is the impact of jobs on workers. If a robot can work faster and more consistently, then there is a risk that humans may not be needed at all. While these concerns are understandable, they are not really accurate. During the early years of the revolution, people were also worried about job losses, but as history has shown us, these fears did not come true and so people continue to play a critical role in manufacturing.

2.3 Impact of automation and robotization on society

Technological change always raises fears of imminent mass unemployment, when innovation will be able to replace human labor at an ever-increasing pace. In addition to traditional industrial automation and advanced robots, new generations of more capable autonomous systems are emerging, from autonomous vehicles on the road to automated controls in grocery stores (self-service cash registers).

In this section, we will focus in particular on how automation and robotization affect society today and how it can affect it in the future, as it is still in the process of development. On the one hand, Slovakia is one of the world leaders in terms of automation and robotization. In terms of main barriers that make this situation difficult, the following aspects have been included:

- inefficient lifelong learning system which involves less than 3% of adults in Slovakia, while the EU average is 10.8%,
- imbalance between acquired knowledge during formal study and labor market needs.
- lack of knowledge of the labor market needs by the education system and the unpreparedness of universities and secondary schools, but also of primary schools for changes in labor market requirements,
- lack of skilled labor in key sectors of the national economy,
- inflexible labor market, lack of support for flexible forms of work.

In Slovakia, a third of people work in the industry, which ranks the country second in the most industrialized countries in the European Union. It is said that it is the industry that is one of the most automated sectors, which gives us a negative position. The survey shows that in Slovakia, up to 70% of jobs can succumb to automation and robotization, and a further 32% will face a significant change in the way the work will be done. The probability that one job will be automated and cancelled is at 62%, which is in 14 percentage points higher than the OECD average.

Automation is concentrated mainly in the automotive industry in Slovakia. 4 car makers are based in Slovakia, namely Volkswagen Slovakia in Bratislava, PSA Groupe in Zavar, Kia Motors Slovakia in Teplička nad Váhom and Jaguar Land Rover in Nitra. In total, they produced more than 1.25-million vehicles last year, in 2019. The automotive industry employs directly and indirectly approximately 275 thousand people. It is the structure of the Slovak industry, where assembly activity prevails over research and development, marketing, and services, that is one of the reasons why our country is at the forefront.

One of the benefits of automation is also reducing the costs. We demonstrate this by an example. In Fontan in Dolný Kubín, they have a melting furnace that cannot be shut down. Therefore, there is a non-stop operation running. In May last year, higher wages for weekends, nights and public holidays were approved. Personal costs would increase by 60 thousand euros, so they decided to robotize their production more intensively. This decision has helped them increase productivity without having to look for new employees. They needed six employees at milling, but now they only need three. However, this does not automatically mean a reduction in the number of employees. An example is the car factory Volkswagen, which has about 1,500 robots, but the number of employees is growing year-on-year.

Robotization in Slovakia is more visible from year to year. In terms of the number of robots, there were 165 industrial robots per 10,000 employees in 2018. The world average is 99 robots per 10,000 employees, which ranks Slovakia at the 16th place. In the following chart (Fig. 1.) the evolution of the number of robots in Slovakia over 4 years can be observed.

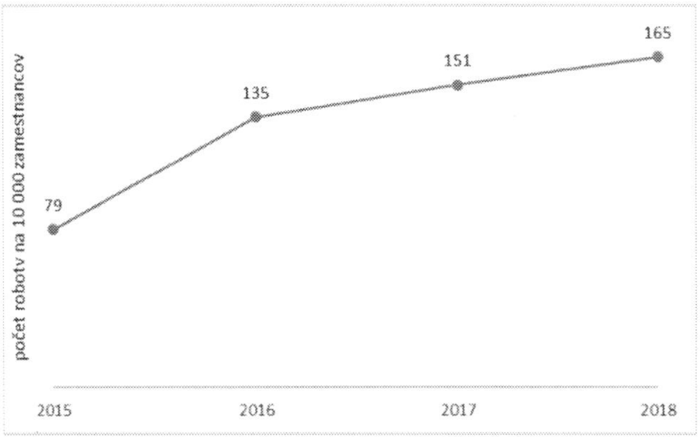

Chart 1. Development of the number of robots in Slovakia

2.4 Possible threats to the labor force due to automated production

Although automation and robotization are not the latest phenomena, there are still a few questions and, in particular, concerns about them. The most worrying impact is unemployment, i.e. the loss of a job due to automation or replacement of human labor by a robot.

Theoretically, around 50% of current jobs around the world could be automated. In case of around 60 % of professions, at least 30 % of the core activities could be automated. More than 5 million jobs in 15 developed and emerging economies could be lost in 2020. Some 45-60% of Europe's employees are at risk of being relocated to another job due to automation.

In their study 2013, Frey and Osbourne hypothesized that automation tends to focus on automating certain tasks rather than on the entire professions. Professions usually consist of performing a set of tasks, some of which are not easy to be automated. Therefore, the potential for automation of coherent occupations and workplaces could be much lower than indicated. Furthermore, it is important to understand that the technological possibility of using machines to provide certain tasks does not always mean losing a job. Replacing people with a machine may not happen at all, or people will still be needed to operate it. The paradox of automation is that the more efficient the automated system, the more important is the human contribution of operators. People are involved less, but their involvement is becoming more critical.

Most of the changes during automation occurred mainly in positions that employed young or uneducated workers. They were the ones who often performed simple, monotonous, and physically or mentally demanding work that is handled by a machine or a robot much better. Then, people can focus more on other work that requires more concentration. Therefore, it is likely that low-skilled workers will be burdened with adaptation costs since the automation of their work is higher than those of highly qualified workers. The most probable challenge for the future is to manage the growing inequality in education and ensure sufficient re-training, especially for low-skilled workers.

However, re-trainings and trainings mean spending money for employers. The best solution is to start developing scientific and technical literacy in schools. It is also important to create a flexible social system to help those whose jobs disappear and thus, they will have nothing to do. This system should help them prepare for another position or to work in a completely different segment.

2.5 Expected implications in production

In its study, the European Parliament's Research Service published the following expected implications in production:

- a higher flexibility through automation, data management, integration of multifunctional robots enables the production of a wider portfolio of products in the existing production hall,
- mass customization (customization of the product exactly according to the requirements of the customer) will ensure economically efficient production of small numbers of products, even individual pieces, owing to a rapid reconfiguration of machines and an additive production. The company's director of Siemens Industry Automation explains the mass customization by the fact that customers' demands for individual customization are increasing (vehicle production is an example), which makes an effort to create factories that do not produce only in mass, but at the same time on a customized level,
- the speed of production will increase because of digital models and simulations, data-driven supply chains will speed up the production process by 120 % by shortening the time from the purchase order to the start of production,
- a high potential has been identified in the increase of quality by reducing the error-rate due to integrated sensors and machines that acquire intelligence through access to large data. The Siemens Amberg plant, which achieves 99.99885 % faultlessness, can be mentioned here as an example. The quality plays a significant role in companies' costs, with the EU estimating a total of €160 billion in 100 major European producers.
- Productivity will increase due to a number of Industry 4.0 elements, for example, the productivity increase, after predictive maintenance are incorporated is defined at 20% as a consequence of 50% reduction in downtime. Another type of productivity increase is staff number optimization and their utilization,
- the concept also provides a higher level of customer involvement in the product development process, because the process of returning European production from low-income countries (if Europe is the primary outlet) will bring the customer closer,
- a change in business models intended to move away from cost-based competition purely to a competitiveness scheme that includes, in addition to costs, innovation speed, flexibility in the degree of customization.

"Strategic Policy Forum on Digital Entrepreneurship" (from the English phrase, refers to an institution, company, or research group, typically in the field of politics or economics) has been established by the European Commission. It predicts that the "internet economy" will create 400,000 - 1.5 million of new jobs. It also estimates the growth potential of companies that can use modern technologies. Such businesses achieve ten times better results than their competitors.

The Forum is the center of representatives from entrepreneurs, universities, organizations, and politics. The report looks at new opportunities emerging from digital technologies. It identifies key European challenges and submits thirteen recommendations to governments and entrepreneurs that should ensure that Europe is better in the availability of labor and the subsequent growth.

3 Research results

An analysis of the current state, problems and challenges associated with the implementation of Industry 4.0 rules and the digitization of the processes constituting the added value, among other things, has highlighted a number of important facts:

- almost three quarter of businesses do not expect digitization to have an impact on the number of staff, but there is a noticeable difference between small and medium-sized enterprises (SMEs) and large enterprises.
- 15% of businesses see a risk in the qualification of employees, in data security and the level of investment which is needed for the greatest risks and obstacles.

3.1 The way of processing and collecting the input data

In the following section of the paper we provide selected facts from a survey that we have elaborated on the basis of theoretical knowledge of the Industry 4.0 initiative and on the basis of consultation with an adviser in IPA Slovakia.

The questionnaire was sent to 150 industrial companies in Slovakia electronically. Business contacts have been accumulated in several ways. Electronically, via the Internet and electronic communications, individually thanks to the contacts of the author and, last but not least, thanks to consultants and author´s fellow students who helped find contacts in industrial companies willing to answer the survey on Industry 4.0.

We have obtained 92 duly completed questionnaire forms. In spite of the fact that there are only slightly more than a half of respondents (exactly 61.33%), the data range could be considered to be a success, since previous experience with the response to electronic questionnaires was below 25% of respondents. Then, we analyzed the 92 duly completed responses.

In this text, only selected issues related to the expected impact of automation and digitization on the workforce position in companies will be addressed.

Among the companies concerned that sent a duly completed questionnaire form were 32 micro-companies (1-9 employees), 28 small enterprises (10-49 employees), 19 medium-sized enterprises (50-249 employees) and 13 large enterprises (250 or more employees). Percentage values have been rounded to integers.

Industry 4.0 is a "future trend" in Slovak conditions – for many businesses it will represent a substantial part of their strategic agenda in the near future. Already today, businesses trying to overtake competition in Slovakia combine advanced connectivity capabilities, advanced automation, cloud computing, sensors and 3D printing, computer-controlled processes, intelligent algorithms, and the Internet of things to transform their business. More than 50% of enterprises have said that digitization and integration processes are at least sufficiently developed and, given the trend in digitizing processes in the surrounding world (and which at the same time indirectly affects Slovak businesses), it can be said that this number will continue to grow.

3.2 Expected impact of digitization on the number of employees in companies

As many as 71% of the companies surveyed do not expect a change in the number of employees related to the implementation or development of digitization. It should also be added that this percentage is so high, in particular for the number of small and medium-sized enterprises (SMEs) that have answered this question. If we look at the response structure, we can see a difference in the perception of the effect of digitization on employment between SMEs and large enterprises.

Large enterprises are clearly more aware that digitization should have an impact on employment. Their expectation is that the number of employees could fall. In this case, it is about working in administration or working with data from the production. Digitizing processes working with data is extremely beneficial for large companies because computers can collect, analyze, and model different types of data incomparably faster than a person would do by means of manual computing. The amount of data in large companies that are processed by computers is a more disjointed factor because a computer can process data for 100 or more workers in less time. The larger the database that a business needs to work with, the more convenient it is for it to digitize the value-added creation processes.

In case of SMEs, management trends are not expected to change in more than three-quarters of responses. This can be explained in particular by the fact that the database used by SMEs is significantly smaller than with large companies, and therefore, even if a certain number of employees are replaced by computers, roughly the same number of new employees will have to be recruited for the management, maintenance and use of information technology (ICT).

Chart 2. Impact of digitization on the number of employees

	Large enterprises	Small and medium-sized enterprises	Total enterprises
Not expected to change	61 %	78 %	71 %
The number of employees rises	8 %	14 %	13 %
The number of employees drops	31 %	8 %	16 %

With the consistent application of Industry 4.0 principles, certain jobs will not be sought by employers and thus, they will cease to exist. However, this cannot be seen unilaterally and predict a rapid decline in employment. After all, there are dozens, if not hundreds of jobs that have already disappeared over the historical development. On the contrary, Industry 4.0 is expected to increase the overall need for the workforce. This idea is also shared at the European Union level, which has identified a huge potential in the internet economy and expects to create 2.6 new jobs for every offline vacancy cancelled.

In the context of the positions at risk, these are mainly jobs which are not of great interest in the Slovak Republic, so foreign workers are recruited for these positions. Therefore, the will of the Ministry of Economy should be the introduction of elements of Industry 4.0 which eliminate professions with a lower level of qualification.

4 Conclusion

Slovakia has great potential to take advantage of the fourth industrial revolution because of a highly developed automotive and electrotechnical industry. There is a very strong structure of production companies and their supply chains, which is suitable for experimenting with new practices and techniques brought about by the digitization of industrial processes.

The fourth industrial revolution has also the potential to fulfill the ambition to increase global production to meet growing human needs without harming the environment. In other words, it helps to increase environmental sustainability by continuously improving production and non-production processes.

However, the digitization of companies will change the nature of work and increase the demands and requirements for skilled workers. Demand for workforce will continue to increase as businesses will develop new professions with demands to reflect the digitization of production and logistics.

Note

This paper is the partial output of the Vega project solution MŠ SR no. 1/0375/20 "New dimension of development of production and logistics management under the influence of Industry 4.0 in enterprises in Slovakia." in the range of 100 %.

References

1. A. Dupaľ, P. Richnák, Ľ. Szabo, M.M. K. Porubanová, Modern Trends in Logistics of Agricultural Enterprises. Agricultural Economics. *Prague:* Czech Academy of Agricultural Sciences, **65 (8)**, 359-365 (2004)
2. C. Fischer, W.H. Parry, W.A. Pizer, Instrument choice for environmental protection when technological innovation is endogenous. *Journal of Environmental Economics and Management* **45**, 523 (2003)
3. Energy, transport and environment statistics 2019 edition, *The Statistical books, Eurostat, Printed by Imprimeries Bietlot Freres in Belgium*, European Union, (2019)
4. European Commission, Stimulating growth and employment: an action plan for doubling the volume of e-commerce in Europe by 2015 [online]. 11.1.2012 Available at: http://europa.eu/rapid/press-release_IP-12-10_en.htm?locale=en
5. Ironies of automation. *Automatica.* **19 (6)**, 775-779 https://www.sciencedirect.com/science/article/pii/0005109883900468?via%3Dihub
6. J. Dobránsky, M. Kocisko, P. Baron, V. Simkulet, L. Běhálek, E. Vojnová & L. Nováková-Marcinčinová, Evaluation of the Impact Energy of the Samples Produced by the Additive Manufactoring Technology. *Metalurgija* **55 (3)**, 477-480 (2016)
7. J. Pagáč, Problems and trends of the development of production logistics in conditions of globalization. Thesis. (Tutor: J. Rakovská), Bratislava: University of Economics in Bratislava, Faculty of Corporate Management. (2017)
8. M.A. Marinov, S.T. Marinova, J.A. Larimo, T. Lepovsky, Characteristics and Future Trends. *Internetional Business and Emerging Economy Firms* **1**, 1 (2020)
9. P. Richnák, K. Porubanová, Innovations in Logistics: Digitalization Challenges in Logistics. *Prague: Wolters Kluwer.* 144 (2019)

How a pricing decisions are decided and executed

Katarína Remeňová[1,*], *Jakub* Kintler[2]

[1]University of Economics in Bratislava, Faculty of Business Management, Department of Management, Dolnozemská 1/B, Bratislava, Slovakia
[2]University of Economics in Bratislava, Faculty of Business Management, Department of Business and Administration, Dolnozemská 1/B, Bratislava, Slovakia

Abstract. Marketing and management studies are rich in the theoretical backround related to price strategy and pricing. However, consistent with the lack of interest in complexity of decision making process in price management. The purpose of the research study is to identify all stages in decision – making process in price management in Slovak enterprises and to measure their mutual association and their dependence on elements of organizational parameters. Based on the results of descriptive statistics on decision making process in price management, we can say that companies go through an average of four steps in the decision-making process (M = 4.72), with higher variability (SD=3.02), compared to the mean.

1 Introduction

The need to define a decision-making process in price management has its roots in the firm's behavioral theory, which is based on decision-making processes in terms of the enterprise's production economy, focusing on how organizations make decisions about what process to follow in managing production and pricing. In the development of behavioral theory, pricing theory is emphasized as an equally important theory within the theories of the company. The decision-making process in price management (determination of price outputs and their changes) is placed within the nine-step decision-making process [1]. At the heart of behavioral theory by Cyert and March is the economic decision-making in terms of their process [2]. A special place in their work took the price setting topic as a decision – making process. The authors' research was focused on the development of process - oriented company models. Company decisions are viewed as well-defined behavioral steps [3].

Currently, only few scientific research studies about pricing is devoted to the decision-making process in price management. This issue has become the domain of businesses that must base price management on a formal decision-making process.

Decision making is concerned with evaluating and ranking possible alternatives of action [4]. Managerial decision making in pricing deals with decisions relating to setting or

* Corresponding author: katarina.remenova@euba.sk

changing prices. Which includes price positioning and product decisions introducing new pricing points of products or service mix [5, 6].

Price management decision – making problems affect whole organization. Pricing decisions are made in a company based on overall marketing plan [7, 8, 9] because they influence their relationship with customers [10]. This topic is complex by the fact that there are various biases affecting the process of price setting within a company [11, 12, 13]. According to Smith [14] pricing is a continual process, that can't be done in isolation. Moreover, specialists in price management claim, that pricing should be seen as a process of deciding. Where the process must deliver the goal of making pricing decisions repeatedly and reliably, given the information. Some researchers perceive price management only through the optics of the price setting process, where they mention some steps of price management [15, 16, 17, 18].

2 Methodology

The purpose of the research study is to identify all stages in decision – making process in price management in Slovak enterprises and to measure their mutual association and their dependence on elements of organizational parameters. The statistical sample (N = 250 respondents) is comprised of managers operating in tactical and top management. The sample was created in PSPP by random selection. The researchers have ensured the measurement objectivity by using data collection tools in electronic form to prevent influencing the research subject. The participants were instructed in writing by one researcher. The observation survey was conducted using a questionnaire on the decision-making in price management.

Descriptive decision - making theory provides knowledge about decision-making, where the dominant topic is the decision-maker, with its approach to decision-making and the ability to assume the competence and responsibility to decide. In the research sample we observed the role of the respondent in the decision-making process in price management, namely in the implementation of the strategy. Most respondents define company's pricing strategy (46.53%) and taking responsibility for its results. A third of the respondents have an impact on the performance of the strategy at the middle and top management level. Almost twenty-four percent of respondents have no influence on the definition and implementation of the pricing strategy (Table 1).

Table 1. Structure of respondents in terms of their impact on pricing and inclusion in the organizational structure.

Level of impact on pricing	*Frequency*	*%*	*% cum.*
I define the price strategy	114	46.53	46.53
No impact on pricing	58	23.67	70.20
Implementation – middle level management	46	18.78	88.98
Implementation – top level management	27	11.02	100.00
Total	245	100.0	

Source: own processing in PSPP

2.1 Data analysys

The data obtained through the questionnaire method are of a nominal and cardinal variable. Thus the type of the variable also made the selection of statistical methods conditional. Two-dimensional inductive statistics methods and methods of descriptive statistics were used to test the dependence of the nominal variables. The data were analyzed in PSPP statistical software. Hypotheses were tested at a significance level of $p \leq 0.05$; while maintaining the primary rule of the Chi-Square Test of Independece, where the theoretical frequencies did not fall below a value of 5 in 80%, and for other values $X > 1$ applied.

3 Results and Discussion

Researchers have ensured the objectivity of measurement using data collected trough the on-line form of the secondary sources, that represent the financial statements, annual corporate reports and analysis of external entities. All variables represented interval or nominal data. Individual consultations and structured interviews with individual companies and researchers in the price management field led us to focus on selected areas of issue outlined in the final version of the questionnaire (in the pilot phase of the research). Therafter, we had proved the content validity and reliability of the research tool using Lawshe CVR for validity measurement [19]. Content validity was assessed using a panel of seven experts with professional expertise in the pricing. Based on the Likerts scale we had collected and analyzed their viewpoints on the relevance of the 18 questionnaire items to measure the construct defined by these items to ensure the content validity of the instrument [20]. For the assessment, the following scale was used: „1= the item isn´t representative of revenue model“, „2= the item needs major revisions to be representative revenue model“, „3=the item needs minor revisions to be representative of revenue model“, „4=the items representative of revenue model“.

The CVR of each item didn´t decrease below the level of 0.7 and overall CVI reached the level of .95. Thereby we had found this result representative.

The next step in the validation process of the questionnaire was to assess the reliability of our research tool. Overall the internal reliability measured by Cronbach's alpha reached the level of 0.71, which were considered as acceptable.

Decision – making steps in price management

Decision making can be a one-time, mostly routine action or a process of progressive and logical steps, a so called decision making process. It is the most important part of decision making because it provides a normal progress of a company. We can define it as a collection of eight steps, starting with problem recognition and identification of criteria, then allocating weight to criteria, develop alternatives, analysis and selection of alternatives, implementation of the best alternative and finally evaluation of decision effectiveness.

Graph 1. Decision – making steps in price management.

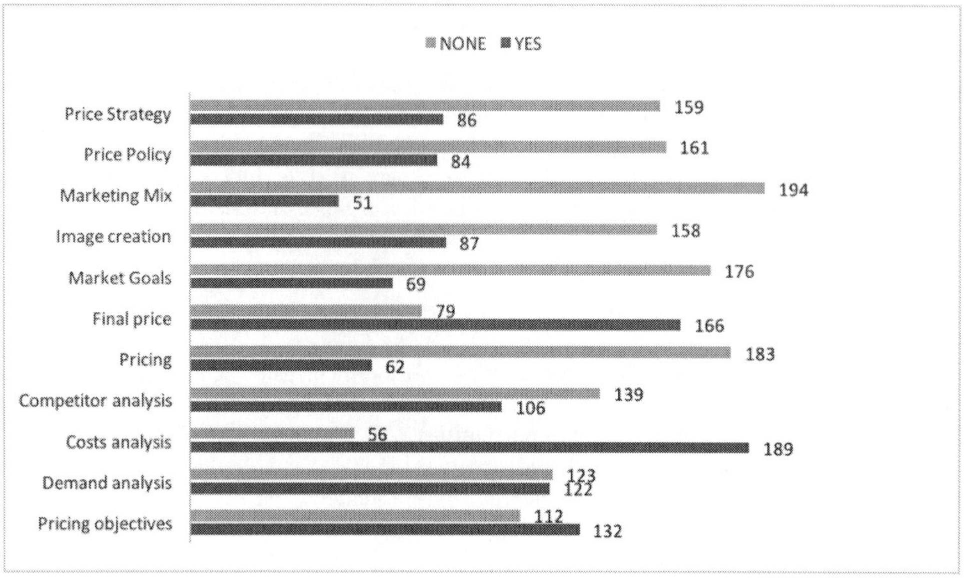

Source: own processing

Based on the results of descriptive statistics on decision making process in price management, we can say that companies go through an average of four steps in the decision-making process (M = 4.72), with higher variability (SD=3.02), compared to the mean. The distribution normality was monitored through kurtosis and skewness. Based on the kurtosis data, which actually describes the shape of a data distribution K = -.33, which corresponds to platykurtic distribution. The skewness also points to the non normal distribution of data in the decision-making process with skeweness = .82, which means, that score on the left arise from low point responses.

Table 2. Descriptive table for Number of steps in decision making process in price management

Variable	*N*	*Mean*	*Std Dev*	*Variance*	*Skewness*	*Kurtosis*	*Range*	*Min*	*Max*
No. of steps	245	4.72	3.02	9.14	.82	-.33	10.00	1.00	11.00

Source: own processing in PSPP

Another fundamental obstacle to an effective decision-making process in price management is its organization. The individual steps of the decision-making process ensure different functional areas of the company without a central coordinating element, often in the absence of a clear concept of price strategy. In the following graph we present a view of individual functional areas of control with the competence to define the price strategy or its implementation at various levels of management. Based on the results of the analysis, in less than 62% of these, the pricing strategy is dealt with by the marketing department, which includes product management and sales. Only 15.5% of respondents who are also CEOs or business owners declare that they are responsible for defining a pricing strategy.

Graph 2. Impact on price management by functional areas of control

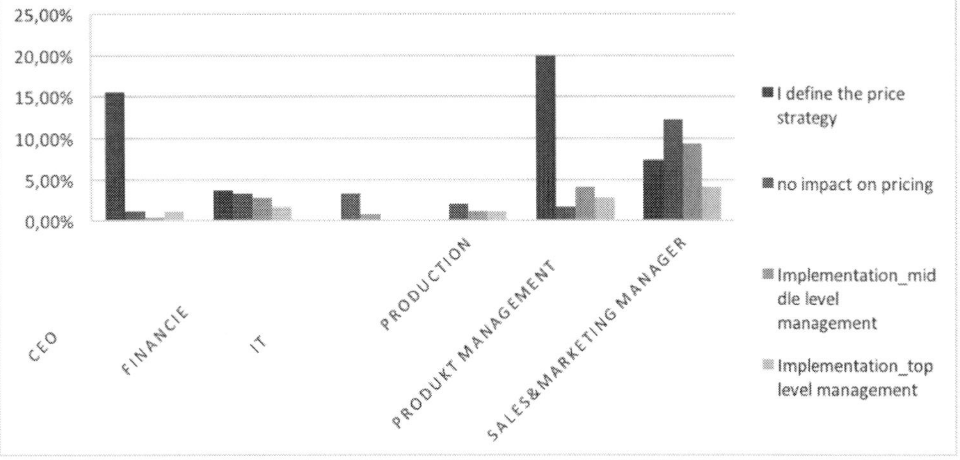

Source: own processing

Conclusion

Decision-making process in price management many researchers and entrepreneurs identify with the process of price setting. However, it is only a thread of the decision-making process in price management. The survey in small, medium and large companies operating on the Slovak market was carried out in order to identify and formalize the decision-making process in price management. The entrepreneurial view of price management is different and the individual steps of the process are limited to product pricing. Two key reasons for this are: lack of knowledge and experience in the price management of Slovak managers and the individual threads are divided in different organizational units.

The decision-making process in price management in Slovak companies is not unambiguously formalized, mainly due to insufficient information of the decision-makers. Its phases are applied in a non-sequential manner and many are omitted in the process. Rather, businesses benefit from a simplified pricing process focused on a cost-oriented approach, without the price optimization. These are small companies that do not have the necessary knowledge or the staffing capacity of specialists.

The most important stages in the decision-making process are for companies "Cost analysis" and "Setting final price".

The authors acknowledge that this paper is the partial result of the project VEGA No.: 1/0017/20 Changes in the application of managerial functions in the context of the fourth industrial revolution and adaptation processes in Slovakia for financial support to carry out this research.

References

1. K. A. Cyert, G.March, *Models in a behavioral theory of the firm*. Pittsburg: Carnegie institute of technology (1959)
2. G. Schanz, *Grundlagen der verhaltenstheoretischen Betriebswirtschaftslehre*. Tůbingen : Deutsches Forschungsgemeintschaft, (1977)
3. R.M. Cyert, G. March, *Behavioral Theory of the Firm*. Milano: Wiley-Blackwell, (1992)
4. B. Malakooti, Decision making process: typology, intelligence, and optimization. *Journal of intelligent manufacturing*, **23**(3), 733-746 (2012)
5. T.J Smith, *Pricing Strategy*. Mason: Cengage Learning, (2012)
6. S. Liouzu, *The Pricing Journey. The Organizational Transformation Toward Pricing Excellence*. Stanford: Stanford Business Books, (2015)
7. G.R. Foxall, The logic of price decision – making. *Management Decision*, **18**(5), 235-245 (2007)
8. Ch.Homburg, O. Jensen, A. Hahn, How to Organize Pricing? Vertical Delegation and Horizontal Dispersion of Pricing Authority. *Journal of Marketing*, **76**(5), 49-69 (2012)
9. L. Sha, T. Fan, Research on Pricing Decision-Making for Fresh Agricultural Products with Strategic Customers and Value Decreasing Rate. 2ND international conference on sustainable energy and environmental engineering, DEStech Transactions on Environment Energy and Earth Sciences, Proceedings Paper, Xiamen, (2016)
10. Y. Polo, F.J. Sese, Strengthening Customer Relationships: What Factors Influence Customers to Migrate to Contracts? *Journal of service research*, **16**(2), 138-154 (2013)
11. A. Hinterhuber, Violations of rational choice principles in pricing decisions. *Industrial marketing management*, **47**, 65-74 (2015)
12. M. Carricano, Pricing myopia: do leading companies capture the full value of their pricing strategies? *Management decision*, **52**(1), 159-178 (2014)
13. W. Georges, R. W. McGee, *Analytical Contribution Accounting: The Interface of Cost Accounting and Pricing Policy*. Greenwood Publishing Group, (1987)
14. T.J. Smith, *Pricing Done Right: The Pricing Framework Proven Successful by the World's Most Profitable Companies*. Hoboken: John Wiley & Sons (2016)
15. T. T. Nagle, J. Hogan, J. Zale, *The Strategy and Tactics of Pricing: New International Edition*. Oxon: Routledge, (2016)
16. S. L. Barber, L. Lorenzoni, P. Ong, *Price Setting and Price Regulation in Health Care*. Paris: OECD Publishing, (2019)
17. J. Ronen, Y. Varda, *Earnings Management: Emerging Insights in Theory, Practice, and Research*. Vienna: Springer Science & Business Media (2008)
18. S. Liozu, A. Hinterhuber, *The ROI of Pricing: Measuring the Impact and Making the Business Case*. Oxon: Routledge, (2014)
19. C.Ayere, A. J. Scally, Critical Values for Lawshe's Content Validity Ratio: Revisiting the Original Methods of Calculation. *Measurement and Evaluation in Counseling and Development*, **47**(1), 79–86 (2014)
20. A.P. Singun, Heuristics as Mental Shortcuts in Evaluating Interactive Systems. *International Journal of Engineering Pedagogy*, **8**(4), 143–152 (2018)

Modern logistics technologies in the conditions of Slovak enterprises

Patrik Richnák[1,*]

[1]University of Economics in Bratislava, Department of Production Management and Logistics, Dolnozemská cesta 1, 852 35 Bratislava, Slovak Republic

Abstract. We are living in a time of technological progress, the development of new technologies and the digital transformation of industry. Customers continue to focus on product quality and enterprises are looking for solutions to reduce production costs, speed up and simplify the manufacturing process. One effective way to do this is to use modern logistics technology. The main aim of the paper was to examine logistics technologies used in Slovak enterprises. The intention of this paper was to analyse the issue under consideration the literary research and the survey, which was conducted in enterprises in Slovakia. To provide a comprehensive view, it was necessary to define terms and concepts in the first part of the paper - technology, logistics technology and types of logistics technologies. Subsequently, in the theoretical part of the paper we focused on the analysis of selected modern logistics technologies. From these we chose Efficient Consumer Response, Radio Frequency Identification and Quick Response. To provide a comprehensive view, we have selected modern logistics technologies under the conditions of Slovakia.

1 Introduction

Today's world is characterized by technical progress, development of new technologies, demanding requirements of customers who emphasize the quality of products. New global markets, business models, advances and innovations are emerging in every business area. This transformation will dramatically influence how organizations will be managed according to the new incentives, and environmental and context configuration [1]. Enterprises are looking for solutions to reduce costs in production and logistics, accelerate and simplify manufacturing and logistics processes. One effective way to achieve this is to use logistics technology.

In today's revolutionary conditions, logistics technologies are a prerequisite for efficient results in business management in the form of economy and efficiency. Their use and continuous improvement must be one of the necessary management activities in the enterprise. Nowadays we are witnessing a different theoretical understanding of logistics technologies and the use of a wide range of technologies by enterprises. In the paper, we first defined which are necessary to understand the issue. Subsequently, in the next part of

* Corresponding author: patrik.richnak@euba.sk

the paper, we analysed selected modern logistics technologies on the basis of a literary research and through the survey we found their application and use in the Slovak enterprises.

2 Literature review

Modern technologies that reduce the need for jobs are part of everyday life and every society. Multinational companies are coming to the market to compete for global market share and to do anything to succeed. This is associated with a change in the economic integration regime throughout the world and the old markets will be transformed into new consumer markets. This implies that if a company wants to survive the current 21st century, it needs to adapt quickly and flexibly to new conditions [2].

Lukoszová et al. characterizes technology as a sum of resources, which are used in production and service processes of the enterprise. It can be understood as a variety of means for a given job related to production, repair or service. Technology is also a collection of documents that contain trade secrets or know-how belonging to an enterprise, which may include plans, procedures, etc. [3].

Cejthamr and Dědina state that the technology contains various activities, machines, equipment, materials and experience used to perform work tasks. Technology can be found in a variety of forms, both in small and large enterprises, in the service sector, government institutions, manufacturing companies, international organizations and local small businesses. New technologies are created in the external environment of the company and can influence it in different ways. Technological development, which is represented by new knowledge and new techniques, can lead to a change in the company's existing production and other activities. It can also significantly affect the nature of the products or services provided by the organization [4].

Řezáč talks about management technologies that focus on the use of methods, approaches and management procedures to achieve corporate objectives. At the same time, they assume internal organization, sequence and harmonization of managerial activities. Managerial technologies can be divided into primary, which are related to material flow and its management processes and secondary, which are related to information flow and information cycle processes, also called digital technologies [5]. The next stage of evolution within these long-standing, leisurely-evolving technologies is here and the new generation of internal logistics is already knocking on the door, waiting to change logistics as we know it today. The revolution will mainly be possible thanks to the noticeably amplifying use of mobile robotic systems; modular, flexible and easily scalable automated logistics equipment; the ubiquitous IoT (Internet of Things) technologies allowing for flexible communication, real-time data transfer and big data analytics; and sophisticated, automated, agent-based control via cloud [6]. Logistics gains a greater vision in terms of mass adoption of smart and connected digital technologies and applications (e.g. mobile, cloud, sensors, data analytics, machine learning, blockchain, IoT) and enhancing vertical and horizontal integration among the supply chain partners [7].

Logistics technologies are understood to be the optimal arrangement of logistics operations into relatively stable processes. The logistics sequences that are created for the non-productive transformation of goods orders into their deliveries to customers do not guarantee optimal flexibility and economy in themselves. Therefore, using appropriate methods, these operations are organized as a whole so as to maximize the performance of the logistics system at a given cost level or to operate the logistics system at the lowest cost possible to achieve the desired performance [8].

Sixta and Žižka of the most important logistics technologies classified: Kanban, Just in Time, Quick Response, Efficient Consumer Response, Hub and Spoke, Cross-Docking,

Warehouse Centralization and Network Concentration, Combined Transport, Automatic Identification, Computer Integrated Technologies in Manufacturing and Circulation and Communication Technologies [9].

In the logistical practice, a constantly growing range of technologies is applied, which are technologies that are shared by multiple disciplines and integrated into different concepts. The choice of appropriate logistics technologies depends on the level of the system or subsystem to which it is to be applied. The techno-technological subsystem includes classical logistics technologies (creation of handling groups, combined transport, Hub and Spoke, concentration of warehouse networks and centralization of warehouses); information and communication subsystems include telematic logistics technologies (automatic identification, radio frequency data communication, electronic data interchange and Internet, technology to support traffic fleet management); management subsystem includes virtual logistics technologies (simulations, graphical visualizations) and the multi-system includes complex logistics technologies (Just in Time, Kanban, Quick Response, Efficient Consumer Response) [10].

There are a couple of new challenges enterprises are facing to, such as changing economic conditions, coupled with decreasing trust in markets, increasing pressures on natural resources and significant changes in customer and employee demographics [11]. On account of that businesses will need to react to meet these new challenges in order to be successful. In our fast-moving world, the winners will be companies that can sense change and respond accordingly [12]. Now the fourth industrial revolution is coming, which can be called the digital revolution, dating back to the middle of the last century. It is characterized by a combination of technologies that eliminate the boundaries between the physical, digital and biological spheres [13]. Based on this fact, the current digital era brings changes and progress in logistics. If an enterprise wants to succeed in the era of digital technology, it must incorporate logistics technologies into the enterprise, which are a significant determinant of all sectors and logistics is no exception. Logistics technologies are rapidly evolving and become part of day-to-day business management by facilitating individual processes, increasing work efficiency and competitiveness of the enterprise. In the next part of the paper, selected modern logistics technologies, which are important and necessary for the proper functioning of business logistics, are discussed in the current digital revolution.

The Efficient Consumer Response (ECR) concept was introduced in 1992 as a result of competition from alternative store formats which highlighted major inefficiencies within the supermarket industry and its supply chain [14].

ECR has been defined more in the sense of a system or philosophy rather than a particular concept and stresses the complete vertical integration of the marketing channels [15].

The basis of formation of the ECR was the rapid development of information technologies, the ever growing competition, the formation and development of international trade organizations, as well as consumers' need for more qualitative, safer, fresher product with better service and broad assortment [16].

Automatic identification of objects has been very prominent in many fields mainly because of ease of use and high efficiency. Auto-ID includes bar codes, smart cards, biometrics and voice recognition, magnetic stripes, RFID, and numerous other devices. However, RFID has many benefits, especially over bar codes; RFIDs are non-line-of-sight and reusable, they can withstand harsh environments, and their high encryption prevents counterfeit [17].

The last decades have witnessed a rapid growth of RFID technology for identification and tracking because of its unique identification (UID). Besides this common usage, an analogue processing of the physical signals related to the reader-tag communication, could permit to achieve much more information about the target without the need for additional electronics or sensors [18].

Radio frequency identification (RFID) technology is based on wireless communication systems and offers easy integration into the Internet cloud system. The potential of RFID tag sensor technologies has been studied in different industrial sectors including healthcare, food safety, environmental pollution, anticounterfeiting of bank-notes and fake medicines, factories, customer shopping behavior, logistics, public transport, and safety [19].

Radio Frequency Identification is contactless communication is linked to barcodes. The carrier medium of information transmission are chips, which are either read or used to write data and read. RFID chips can be divided into two groups on active chips and passive chips. Passive chips are much cheaper and includes a power capacitor, which when capturing the electromagnetic pulses of the transmitter, uses the received energy to charge the capacitor and then send the answer. Passive chips transmit a number or have another memory into which they write or report more information. In addition, active chips have a power supply in addition to passive chips, so this chip can actively send information about itself and can be accurately searched for [20].

Huang et al. has of the opinion that a typical RFID system comprises of components: an active or passive tag which is a tiny programmable electronic circuit - that holds the Electronic Product Code (EPC) and other information to be transmitted; a tag reader which comprises of an antenna and transceiver which communicate with RFID tags and transfer the information to the processing devices [21].

RFID plays an important role in data acquisition systems, tracking of material flow on shop floors and through-out the distribution phase of the product. RFID when implemented in manufacturing firms, enables them achieve complex production system and to perform better life cycle assessment of their products [22].

The QR code is basically a two dimensional which takes matrix form with cell architecture in the form of a square. The QR consists of mainly two parts: function patterns and encoding region. The function pattern is usually not used encoding. The most commonly used encoding algorithm in the QR code is Reed-Solomon (RS) encoding technique [23].

Quick response codes are two-dimensional square patterns that enable the coding or storage of more than 7000 characters that can be accessed via a quick response code scanning application [24].

The quick response (QR) supply chain system has received a great deal of attention in the recent past because of the advances in many new technologies such as RFID systems and mobile computing. Establishment of supply chain practices, such as collaborative planning, forecasting, and replenishment (CPFR), and vendor managed inventory (VMI) has also contributed to the development of QR in industrial practice [25].

Quick response strategy (QRS) has been widely adopted in a supply chain where members collect timely market information for better forecasting, and then respond promptly to the market changes by adjusting initial inventory decision [26].

3 Research design

The main aim of the paper was to examine logistics technologies used in Slovak enterprises. The intention of this paper was to analyse the issue under consideration the literary research and the survey, which was conducted in enterprises in Slovakia.

To provide a comprehensive view, it was necessary to define terms and concepts in the first part of the paper - technology, logistics technology and types of logistics technologies. Subsequently, in the theoretical part of the paper we focused on the analysis of selected modern logistics technologies. From these we chose Efficient Consumer

Response, Radio Frequency Identification and Quick Response. To provide a comprehensive view, we have selected modern logistics technologies under the conditions of Slovakia. The research, which was conducted by questionnaire survey took part were small, medium-sized and large enterprises. The categorization of enterprises by size was based on the EU Commission Regulation no. 651/2014, where we considered as a small enterprise with a number of employees up to 50 and annual turnover up to EUR 10 million. As a medium-sized enterprise we considered the company with a number of employees up to 250 and an annual turnover of up to EUR 50 million. As large enterprise we considered the company with more than 250 employees and annual turnover of more than EUR 50 million.

We used system analysis and comparative analysis to process information and knowledge in the paper on the state and analysis of modern logistics technologies. We also used other classical methods such as: synthesis, induction and deduction. Among the specific methods we used the query, sorting method and visualization method.

4 Research in the Slovak enterprises

The research, which was conducted by questionnaire survey was attended by 150 enterprises. Figure 1 reveals that the Slovak enterprises make the most use of Efficient Consumer Response. The percentage of this technology is 19.3%. The second most used logistics technology is Quick Response with a share of 12%. Radio Frequency Identification is used by the Slovak enterprises with a share of 10.7%. Modern logistics technologies are a big challenge and a financial burden for businesses, therefore we see in Figure 1 high percentages of non-use of logistics technologies - Radio Frequency Identification (58%) and QR (48.7%) by Slovak enterprises.

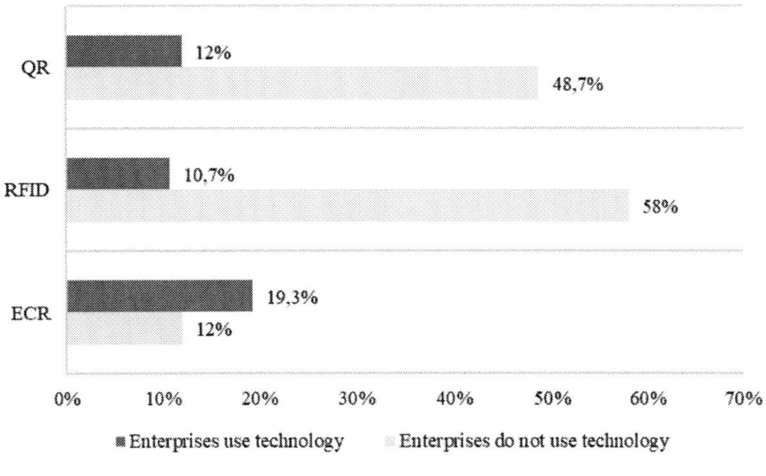

Fig. 1. Usage logistics technologies in Slovak enterprises. Source: own research

Figure 2 shows the use of selected modern logistics technologies according to business size. In large enterprises in Slovakia they use mainly Quick Response (60.7%) and least Efficient Consumer Response (49.6%). Efficient Consumer Response (42.2%) is the most used in medium-sized enterprises and other logistics technologies differ by small percentages. Quick Response is used in medium-sized enterprises with a share of 31.6% and Radio Frequency Identification with a share of 31.2%. Small enterprises are dominated by the use of Radio Frequency Identification (10.1%) and the least are using Quick Response with a share of 7.7%. When comparing selected logistics technologies according to business size, we can see a significant differentiation in the use of selected logistics technologies.

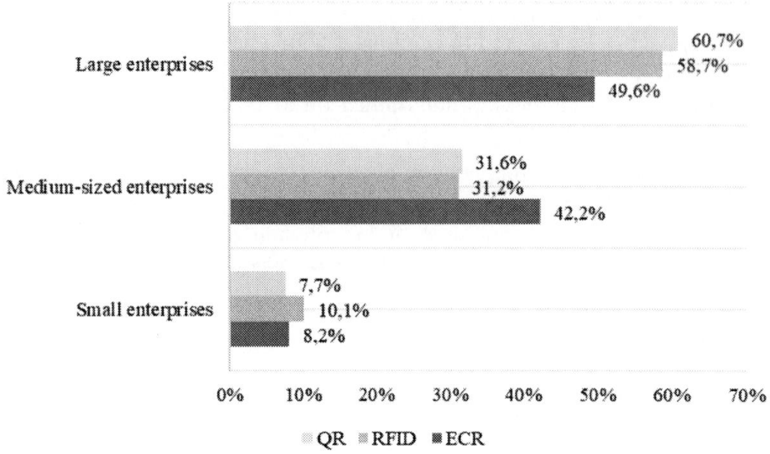

Fig. 2. Usage logistics technologies according to business size. Source: own research

Logistics technology - Efficient Consumer Response is used by enterprises mainly in distribution logistics. This part of business logistics achieved the largest percentage of 49.1%. Enterprises use Efficient Consumer Response in procurement logistics with 38.8% share. The lowest percentage (12.1%) of Efficient Consumer Response usage is in production logistics. Figure 3 shows the percentage.

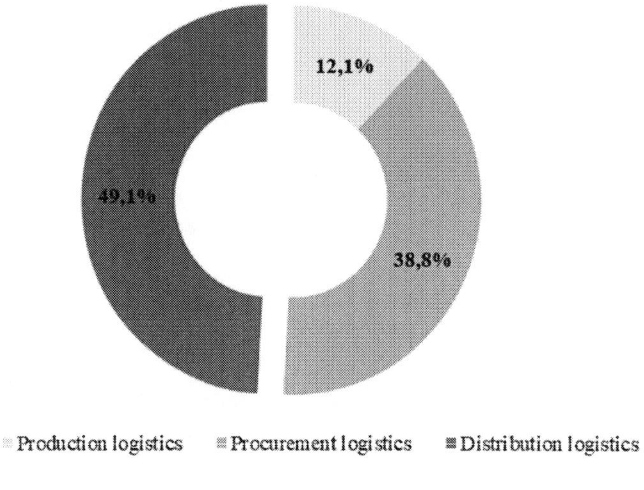

Fig. 3. Usage of ECR in business logistics fields. Source: own research

Figure 4 shows the percentage of usage of Radio Frequency Identification in business logistics of Slovak enterprises. The enterprises usage most Radio Frequency Identification in distribution logistics. The percentage is 53.6%. Radio Frequency Identification is used by enterprises with a 36.6% share in procurement logistics. Radio Frequency Identification is the least used (9.8%) in production logistics.

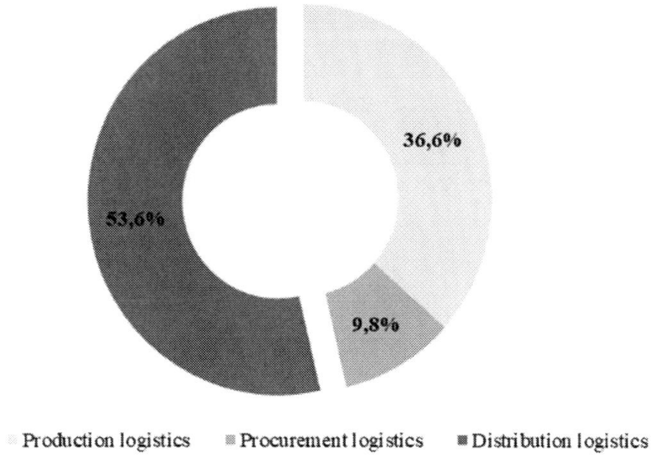

Fig. 4. Usage of RFID in business logistics fields. Source: own research

In Figure 5 we have the last selected logistics technology - Quick Response. The enterprises usage most of Quick Response in production logistics. Percentage achieved value 33.9%. Quick Response is used by enterprises in procurement logistics with a share of 33.3%. Quick Response is used in distribution logistics with a share of 32.8%. The values obtained in this logistics technology are very similar. The percentages do not differ significantly from previous technologies.

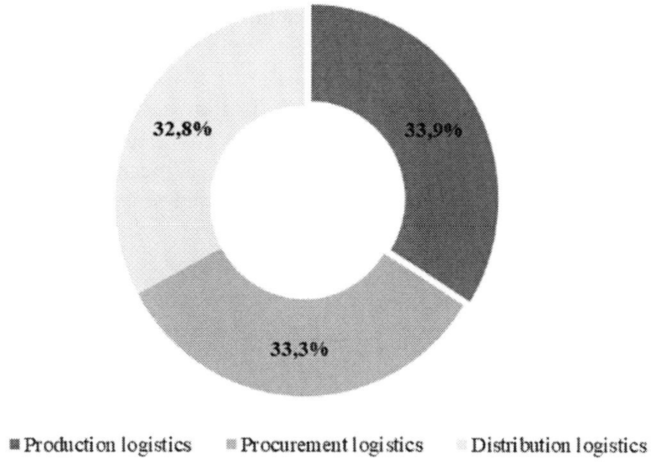

Fig. 5. Usage of QR in business logistics fields. Source: own research

Figure 6 shows the use of Efficient Consumer Response, Radio Frequency Identification and Quick Response in each type of industry. The radar graph was created based on the mean values. Efficient Consumer Response is the most widely used (4.2) in electrical engineering industry. Efficient Consumer Response also recorded a high mean usage in construction industry (3.9) and the automotive industry (3.4). The lowest mean value (1.4) of this technology was recorded in chemical industry. Radio Frequency Identification is the most used in automotive industry (2.0). The use of Radio Frequency Identification is recorded a high mean in electrical engineering industry (1.9). The lowest mean value (0) was recorded in textile industry. Quick Response is the most used in food industry (2.6). The lowest mean value (0) was recorded in glass industry.

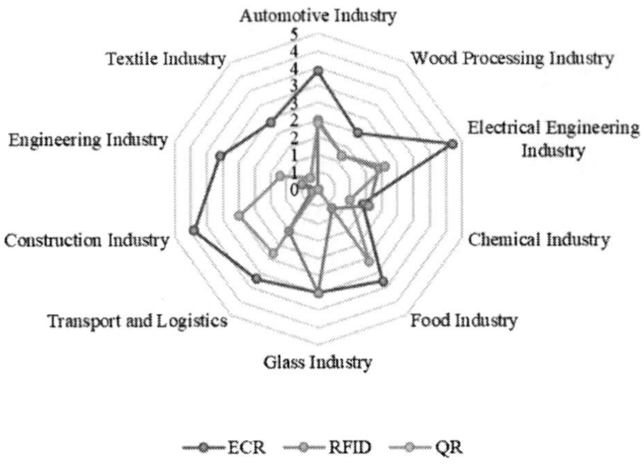

Fig. 6. Usage of logistics technologies in enterprises by types of business industries. Source: own research

Figure 7 shows the use of Efficient Consumer Response, Radio Frequency Identification and Quick Response in enterprises by regions in Slovakia. The radar graph was created based on the mean values. Efficient Consumer Response is mostly used in enterprises in Prešov Region. The mean value achieved 4.0. The enterprises in Bratislava, Trnava and Trenčín Regions obtained a higher mean values (3.8) in the use Efficient Consumer Response. The lowest mean value (3.1) of Efficient Consumer Response was recorded in Žilina Region. Radio Frequency Identification is the most used in enterprises in Bratislava Region. The mean value achieved 2.1. Prešov Region (1.8), Banská Bystrica Region (1.5) and Trnava Region (1.5) recorded balanced mean values. The lowest mean value (0.6) of Radio Frequency Identification was recorded in enterprises of Žilina Region. Quick Response is the most used in enterprises in Bratislava Region. The mean value achieved 3.1. The lowest mean value (0.9) of Quick Response was recorded in enterprises of Žilina Region.

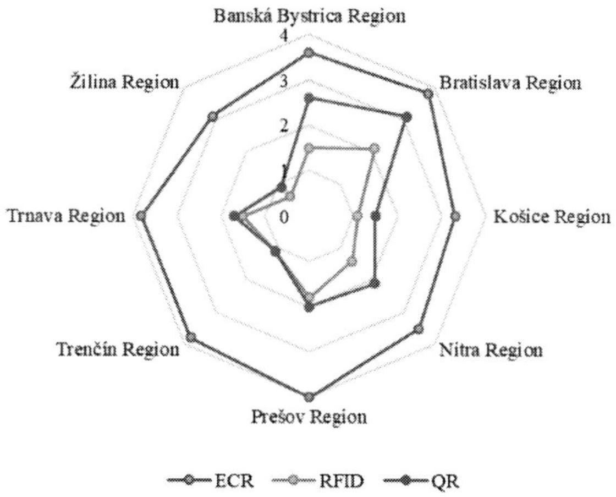

Fig. 7. Usage of logistics technologies in enterprises by regions of Slovakia. Source: own research

5 Conclusion

Nowadays, if companies want to stay at the forefront of the market and meet the needs and wishes of their customers, they need to be flexible, so they can adapt their business processes to demand [27]. This is the reason why the company seeks to create a competitive advantage over other producers and service providers [28].

New technologies fundamentally change the way businesses operate and develop their business, bringing new challenges to IT managers. How successfully the company will adapt to the digital transformation will be directly related to its competitiveness and future market survival [29]. Today, we use new technologies (Internet, self-service systems) to reduce service jobs [30].

The main aim of the paper was to examine logistics technologies used in Slovak enterprises. The intention of this paper was to analyse the issue under consideration the literary research and the survey, which was conducted in enterprises in Slovakia.

The surveyed Slovak enterprises from modern logistics technologies make the most use of Efficient Consumer Response with a share of 19.3%.Of the selected logistics technologies analyzed, companies do not use Radio Frequency Identification with a share of 58%. In large enterprises in Slovakia they use mainly Quick Response (60.7%). Efficient Consumer Response (42.2%) is the most used in medium-sized enterprises. Small enterprises are dominated by the use of Radio Frequency Identification (10.1%). Logistics technology - Efficient Consumer Response is used by enterprises mainly in distribution logistics. The enterprises usage most of Quick Response in production logistics. . Efficient Consumer Response is the most widely used (4.2) in electrical engineering industry. Radio Frequency Identification is the most used in automotive industry (2.0). Quick Response is the most used in food industry (2.6). Efficient Consumer Response is mostly used in enterprises in Prešov Region. Radio Frequency Identification is the most used in enterprises in Bratislava Region. Quick Response is the most used in enterprises in Bratislava Region.

Acknowledgments

The article is a partial output of research project VEGA No. 1/0375/20 „New dimension in the development of production management and logistics under the influence of Industry 4.0 in enterprises in Slovakia".

References

1. L. Barreto, A. Amaral, T. Pereira, Industry 4.0 implications in logistics: an overview. *Procedia Manufacturing*, **13**, 1245–1252 (2017)
2. J. Sixta, V. Mačát, *Logistika : teorie a praxe* (CP Books, 2005)
3. X. Lukoszová, a kol., *Logistické technologie v dodavatelském řetězci.* (Ekopress, 2012)
4. V. Cejthamr, J. Dědina, *Management a organizační chování.* (Grada Publishing, 2010)
5. J. Řezáč, *Moderní management: manažer pro 21. století.* (Computer Press, 2009)
6. T. Gregor, M. Krajčovič, D. Więcek, Smart Connected Logistics. *Procedia Engineering*, **192**, 265–270 (2017)
7. Y. Kayikci, Sustainability impact of digitization in logistics. *Procedia Manufacturing*, **21**, 782–789 (2018)
8. V. Cempírek, R. Kampf, J. Široký, *Logistické a přepravní technologie.* (Institut Jana Pernera, 2009)
9. J. Sixta, M. Žižka, *Logistika: metody používané pro řešení logistických projektů* (CP Books, 2009)
10. P. Pernica, *Logistika pro 21. století : supply chain management.* (RADIX, 2005)
11. A. Čambalíková, Agile way of working and its impact on business performance. *Edamba 2017: International Scientific Conference For Doctoral Students And Post-Doctoral Scholars: University Of Economics In Bratislava, Slovak Republic, 4 - 6 April 2017*, 52-59 (2017)
12. A. Čambalíková, Ľ. Szabo, Modern trends in management and their application in controlling. *Socio-Economic Perspectives In The Age Of Xxi Century Globalization: Proceedings Book: International Conference: Tirana, December 1-2, 2017*, 475-488 (2017)
13. J. Rakovská, Využitie Industry 4.0 vo výrobe a logistike na Slovensku. *Nová dimenzia logistiky v štvrtej priemyselnej revolúcii v podnikoch na Slovensku: Zborník vedeckých statí*, 90-98 (2019)
14. K. S. Bhutta, F. Huq, F. Maubourguet, Efficient Consumer Response — Increasing Efficiency through Cooperation. *Cost Management in Supply Chains*, 195–212 (2002)
15. G. Svensson, Efficient consumer response - its origin and evolution in the history of marketing', Management Decision, **40**, 5, 508–519 (2002)
16. Y. Kato, R. Kiriyama, T. Takenaka, Y. Kurisu, D. Nozaki, K. Yano, F. Sato, T. Iida, Enhanced production of ECR plasma by using pulse mode microwaves on a large bore ECRIS with permanent magnets. Source: 19. international conference on ion implantation technology (2012)
17. E. C. Jones, B. Gray, M. Wijemanne, J. Bolton, *Tracked: What You Should Know about RFID, Big Data, The Internet of Things, and Data Security* (PWD Group Press, 2015)
18. C. Occhiuzzi, S. Caizzone, G. Marrocco, Passive UHF RFID antennas for sensing applications: Principles, methods, and classifcations. *IEEE Antennas Propag.* **55**, 14–34 (2013)
19. R. Singh, E. Singh, H. S. Nalwa, Inkjet printed nanomaterial based flexible radio frequency identification (RFID) tag sensors for the internet of nano things. *RSC Adv.*, **7**, 77, 48597–48630 (2017)

20. W. Macdougall, *Industrie 4.0: Smart Manufacturing for the Future,* (Germany Trade & Invest, 2014)
21. Y. Huang, B. C. Williams, L : Zheng, Reactive, model-based monitoring in RFID-enabled manufacturing. *Computers in Industry,* **62**, 811-819 (2011)
22. A. M. Fet, D. Margrethe Aspen, H. Ellingsen, Systems engineering as a holistic approach to life cycledesigns. *Ocean Engineering*, **62**, 1-9 (2013)
23. S. Nishane, V. M. Umale, Digital image watermarking based on DWT using QR code, *International Journal of Current Engineering and Technology*, **5**, 3, 1530–1532 (2015)
24. J. Zurmehly, K. Adams, Using Quick Response Codes in the Classroom. *CIN: Computers, Informatics, Nursing*, **35**, 10, 505–511 (2017)
25. T.M. Choi, S. Sethi, Innovative Quick Response Programs: A Review. *International Journal of Production Economics,* **127**, 1, 1-12 (2010)
26. H-L. Chan, B. Shen, Y. Cai, Quick response strategy with cleaner technology in a supply chain: coordination and win-win situation analysis. *International Journal of Production Research*, **56**,10, 3397–3408 (2017)
27. K. Porubanová, N. Púčková, Technologické prínosy ekonomickej analýzy výrobného procesu podniku: Technological Benefits of Economic Analysis of the Production Process. *Nová dimenzia logistiky v štvrtej priemyselnej revolúcii v podnikoch na Slovensku: Zborník vedeckých statí*, 67-81 (2019)
28. D. Klátiková, K. Gubová, Kvalita elektronickej služby ako faktor zabezpečenia konkurencieschopnosti podniku. *Manažment podnikania a vecí verejných - Dialógy: vedecko-odborný časopis Slovenskej Akadémie Manažmentu*, **10,** 30, 97-107 (2015)
29. A. Čambalíková, Využitie manažérskych nástrojov Big Data Analytics a Digital Transformation. *Manažment informačnej bezpečnosti v malých a stredných podnikoch: Zborník vedeckých statí*, 35-41 (2015)
30. J. Rakovská, Podnikanie ako poslanie. *Marketing Identity 2013: Zborník z medzinárodnej vedeckej konferencie 4. - 5. November 2013, Smolenice*, 351-358 (2013)

Hospital information systems in the Slovak Republic

Anita Romanová[1] , *Natália* Švejdová[2,*]
[1]University of Economics in Bratislava, Faculty of Business Management, Department of Information Management, 852 35 Bratislava, Slovak Republic
[2]University of Economics in Bratislava, Faculty of Business Management, Department of Information Management, 852 35 Bratislava, Slovak Republic

Abstract. Process-managed organization has much in common with the management of the information system. Management of the company can therefore be compared to the management of the system, while it is necessary to ensure information and communication flows. Effective information processing and knowledge base building are the basis for the functioning of the organization and its strategic intent. The means to achieve this goal is the information system. Informatization is the process of penetrating ICT into all areas of society in order to make it work effectively. An important part of building the information society is eHealth. The Internet, information systems, information technology and healthcare electronisation are important aids to improve and streamline the functioning of the entire health system, as they enable faster and more effective communication between all actors in the health sector (ie pharmacies, patients, healthcare facilities and insurance companies). The aim of this paper is to analyze hospital information systems in hospitals on the territory of the Slovak Republic, to identify the modules they consist of and to define interconnections of these systems with other information systems, not only within the given organization but also from the external environment.

1 Introduction

The planning, implementation and maintenance of information and communication technologies (ICT) in health care is a very demanding process. The health care sector shows certain specifics - lack of funds, substandard demands arising mainly from legislation and requirements of insurance companies and authorities (it is almost impossible to use standard information systems (IS) and therefore they are tailor-made, which is reflected in their financial demands) , the need for safe and reliable ICT (not only from the IT point of view, but also patient-related safety - eg operating room devices and computers must meet stringent requirements and be treated with special preparations to avoid compromising patient health) or the need to process large volumes of data , which must be securely stored but accessible at any time to medical staff. Hospital information systems are complex

* Corresponding author: natalia.svejdova@euba.sk

integrated systems that manage the operation of the hospital and manage all aspects of it, such as health, administrative, financial or legal. In practice, this involves replacing conventional paperwork with a computer system and thus simplifying the administrative work associated with healthcare facilities.

2 Information systems and information technologies in healthcare

Literary sources unambiguously document that the era of computer use in health services emerged in technologically and knowledgeable countries as soon as these devices appeared within the reach of medicine. So we can talk about half a century of history. The past 15 years are particularly significant. ICT development has accelerated, mainly due to the miniaturization of computer technologies and the dramatic decline in their cost. Instead of large mainframes, today we have them on our desktops in the form of personal computers, portable in briefcases (notebooks), and even in the pocket of a work jacket (PDA - a personal digital assistant in conjunction with a mobile phone capable of digitally capturing and transmitting image information).

Computers having high performance are utilized in many industries to increase competitiveness and quality. Informatization has become one of the critical factors determining the success of a company. Information technology has been heavily introduced in many companies, and the highly complex medical industry is no exception. [11] The development of computers and communication technologies in the 1990s can be seen as a breakthrough. In this period, funds were also created that enabled computers to be interconnected (even wirelessly) into information networks, and the potential of using the Internet in the health sector was not fully appreciated. [5] Therefore, it can be argued that it was in this period that a great information and knowledge potential arose - a system of information highways, which creates not only a significant knowledge, but also an organizational phenomenon. In the healthcare industry, information has a significant influence on service efficiency, physician decisionmaking and human health. The information of selfmedication consumer´s behavior is not come from medical diagnosis and is not imposed by physicians. Good information integration are required for providing effective patient services. The healthcare industry should have the best information management because information has a grave impact on physician services. Information flows in the healthcare industry occur for the production of medicine and the manufacturing of medical supplies for hospitals, and they help organizations provide the right patient services. [9] The rapid development of ICT has improved not only the technical, but also the software, enabling the creation of a large database IS accessible to multiple users simultaneously in real time. However, the use of ICT in the health sector was not limited to communication and archiving records. New technologies enable medical devices to be integrated into computers, making them an integral part of healthcare delivery. These include computer and ECG, EEG, EMG, vital signs monitoring, laboratory testing equipment (hematology, microbiology, or biochemistry), or a computer-connected imaging device, such as a computer. X-ray, CT, sonograph and others.

Research of information systems and information technologies used in the field of health care is carried out by a large group of foreign experts. Ehrenfeld and Cannesson point out that the use of ICT in healthcare is essential. For health ICT they use the term health information technology (HIT), which is used to collect, transmit, display or store patient data electronically. HIT is also a concept that involves the use of computerized systems to make healthcare information available to patients, healthcare providers or insurance companies. Using HIT helps reduce doctors' mistakes, costs and paperwork, and increases the efficiency and quality of healthcare. [7]

Sayles states that HIT includes a wide range of products, technologies and services such as remote and mobile medical technology, cloud services, medical devices, telemonitoring tools, assistive and sensory technologies, electronic health records (EHR) and others. These technologies allow users to collect, share and use medical information. [15] Although Health IT (HIT) has tremendous potential for improving quality and reducing cost in healthcare, significant challenges need to be overcome to fully realize this potential. [1]

HIT is the overarching term applied to various information and communication technologies used to collect, transmit, display, or store patient data. HIT, which includes hospital information systems (HIS), electronic health record (EHR), computerized physician order entry (CPOE), electronic prescribing (e-prescribing), clinical decision support systems (CDSS), and bar-coded medication administration (BCMA), has been widely implemented in healthcare organizations to support healthcare services. HIT is expected to increase the quality and safety of healthcare services by improving communication, patient delivery time, and documentation. Despite the benefits, HIT also introduce consequences when it is poorly designed, implemented, and used. Problems related to HIT can disrupt healthcare delivery, harm patients, and even cause death. [14]

In their research (2015), Cresswell and Sheikh confirmed that the use of HIT tools is widespread in healthcare facilities. Given the proven benefits of HIT, the use of these technologies is now essential. However, the balance between the benefits and risks of the use of information technology in healthcare facilities in the coming years is not clear. Therefore, in order to improve the planning and successful implementation of these technologies, they considered the need to use technology forecasting methods. [6]

HIT can effectively improve the quality and efficiency of medical services, distribution of health care resources, safety in health care, and output of scientific research. Therefore, governments of various countries have set up ambitious plans to develop HIT and invested enormous amounts of money in this development, using HIT as an important starting point for the reformation of medical services and medical systems. [10]

Almunawar and Anshari are also involved in health information system (HIS) research. In their research HIS: Concept and Technology (2012), they concluded that the use of ICT in healthcare facilities grew at the same pace as in other areas of industry or services. They state that there is a growing tendency for people to know and participate actively in the prevention and care of their health, and this fact leads to the development of information systems in health care. This trend leads to greater involvement of patients in taking information, decision-making and responsibility for their own health through ICT. This vision is achievable through shared care, which is based on health telematics networks and services, linking hospitals, laboratories, pharmacies, primary health care providers and patients through a "virtual health center", accessible via a single entry point online. [3] However, Hansen concluded that the major problems of HIS and HIT are the exchange of data between users as well as the actual use of data from health systems to provide intelligent decision support. [8]

The development of ICT, as well as artificial intelligence, will eventually manifest itself in medicine. Currently, there are already ITs that remember and effectively apply the knowledge of leading experts. They can therefore be called intelligent or knowledge-based systems, and their use can be found in the process of diagnosis, choice of medical procedures and decision-making processes. Such an intelligent system is e.g. Watson Health from IBM. It is a cognitive system with artificial intelligence, which is used to diagnose the patient's medical records and suggest the recommended treatment. Watson also suggests a treatment that is ineffective and also determines the probability of success / failure of the treatment. After assessing the patient's health status and suggesting a treatment procedure, the attending physician has available scientific articles from around the world for a more professional look at diagnostics from IBM Watson Health. IBM spent more than $ 6 trillion on the development of this knowledge system.

One of the elements of the information system in healthcare facilities is HIS. Moghaddasi et al. in their study (2018) state, that HIS is an integrated information system that not only provides patient information but also supports all hospital activities, including clinical, administrative and financial activities. In addition, HIS plays an important part of the hospital budget. Usually 2 to 5% of the hospital operating budget is devoted to the information system. Studies also show that hospital staff spend most of their time recording information, but only 42% of their time is devoted to clinical activities. [13] Based on a foreign study carried out by a team of experts (Ahmadian, Dorosti, Khajouei, Gohari) in 2015, the human factor can be considered the largest and most important problem of hospital information systems. On the basis of this research, it emerged that HIS is failing because health professionals have a negative attitude towards IS and that there is no incentive to (effectively) use IS. [2]

HIS are often portrayed as precious tools which can enable better quality and safety of healthcare and make the whole process of patient treatment more efficient. HIS are expected to collect, store, analyse, and provide information about the patient to caregivers, thus, enabling better decision-maiking and timely actions with respect to treatments and the provision of healthcare services. Healthcare systems are best described as complex adaptive systems. [12]

The IS of healthcare facilities cannot be understood only on the health level, but these IS must also ensure the normal operation of the entire organizational unit. Last but not least, the IS must interconnect all internal departments and provide feedback, it must be able to adapt sufficiently to changing conditions in the internal and external environment of the company. However, efficiency in terms of cost and achieving the effects of its application is also important. The IS aims to improve the performance of people in organizations and businesses through the use of information technology. For this reason, it is necessary to regard IS in health care as a complex HIS (CHIS) ensuring the operation of the entire organization, including the provision of health care.

The use of ICT in the health sector can not only be viewed from an administrative perspective, but its full potential needs to be fully exploited. The achievements of the 21st century will not only improve health care and reduce healthcare costs, but also ensure more effective treatment or prevention, thus improving and extending people's lives. However, the fact remains that cognitive systems, which largely assist doctors in diagnosing and treating patients (such as Watson Health), are expensive, and few health care facilities (both in Slovakia and abroad) are able to do so. allow to procure.

3 Research design

The main aim of this paper is to analyze HIS, identify the modules it consists of and define the interconnection of HIS with other IS not only within the organization but also from the external environment.

We have analyzed this area by a questionnaire survey. The questionnaires were distributed to 40 respondents (38 state hospitals + two private health care providers).

When compiling the questionnaire, we started from the theoretical knowledge that we gained during the study of this selected issue, and we were also very helpful by experts in the field of hospital information systems, healthcare electronization, network management and personal data protection.

The return rate of the questionnaire was 75%, which represented 30 respondents. We were unable to obtain data from private healthcare providers on the grounds that the questions in the questionnaire were very sensitive in terms of protecting business secrets, IT security or even corporate reputation. We were unable to obtain information from the group of state hospitals from 5 health care providers in order to maintain the information security of the organization or with reference to the optional provision of comprehensive information in the form of a questionnaire. 3 hospitals were eventually excluded from the statistical sample, as they currently only act as a social services home or diagnostic center.

The structure of the analyzed sample of 30 respondents is shown in the following table.

Table 1. Structure of respondents depending on the founder and territorial scope.

Founder	**Region of the Slovak Republic**			**Total sum**
	West	**Middle**	**East**	
Ministry of Health of the Slovak Republic	33.34 %	20.00 %	10.00 %	63.34 %
Higher territorial unit	10.00 %	10.00 %	0.00 %	20.00 %
Other authority	3.33 %	3.33 %	10.00 %	16.66 %
Total sum	**46.67 %**	**33.33 %**	**20.00%**	**100.00 %**

When analyzing the data, it was necessary to differentiate hospital subjects into size categories. Whereas the European Commission Recommendation no. 2003/361 / EC in our case would evaluate 26 subjects as large enterprises and only 4 subjects as medium-sized enterprises, we decided to compile our own categorization of hospitals on the basis of basic characteristics of individual subjects.

The primary criterion for the classification of the subject is the number of employees. Based on our analysis, the closest approximation to this breakdown is the number of outpatient examinations ($r=0.7562$; $r2=57.18\%$; $p<0,001$), then the number of hospitalizations ($r=0.8031$; $r2=64.50\%$; $p<0,001$). Based on these relationships and data analysis using statistical indicators average and standard deviation, Slovak hospitals can be categorized based on the following indicators:

Table 2. Proposed criteria for categorization of Slovak hospitals.

Indicator	**Category of organization**		
	Small	**Medium**	**Large**
1. Number of employees	$X \leq 178$	$178 < X \leq 1\,446$	$1\,446 < X$
2. Number of outpatient examinations	$X \leq 15\,355$	$15\,355 < X \leq 395\,196$	$395\,196 < X$
3. Number of hospitalizations	$X \leq 2\,095$	$2\,095 < X \leq 27\,797$	$27\,797 < X$

One of the above categories falls within a hospital subject if it has the relevant number of employees (the main criterion) and meets at least one secondary criterion (the number of outpatient examinations or the number of hospitalizations). Based on our proposed categorization, the respondents of the research sample can be classified as follows:

Table 3. Categorization of respondents to the research sample.

Founder	**Category of organization**		
	Small	**Medium**	**Large**
Ministry of Health of the Slovak Republic	13.33 %	36.68 %	13.33 %
Higher territorial unit	0.00 %	20.00 %	0.00 %
Other authority	3.33 %	13.33 %	0.00 %
Total sum	**16.66 %**	**70.01 %**	**13.33 %**

In the next sections of this paper we will start with this categorization of hospitals into small, medium and large organizations.

4 Research results

The first part of our questionnaire was focused on the identification of a hospital subject in order to categorize the organization. This was followed by a section dealing with basic information on HIS.

83.33% of respondents stated that they have a HIS that was tailored to the requirements of the organization. The remaining 16.67% use standard HIS, ie. that they were not additionally customized by suppliers. In Slovak hospitals they use HIS, which were implemented on average in 2009 (M=2009; SD=6.61; Min=1995; Max=2019). In practice, unfortunately, we may also encounter systems that are more than 20 years old (two respondents in our research), and these systems are very outdated today and often record their failure rate. 8 respondents of our research have a system not older than 3 years.

Only 3 Slovak hospitals, which are under the founding competence of the Self-Governing Region of Trenčín have a unified HIS. This system was procured for all three hospitals at the same time, consisting of the same modules, extensions and functionalities. These entities perceive the advantages of a unified IS for the possibility of making uniform outputs from the IS for the purpose of management from the Self-Governing Region of Trenčín, the possibility of joint negotiations on contracts with suppliers, quantity discounts, the same system adjustments for all organizations.

Table 4. Number of workstations and users in Slovak hospitals.

Organization	Number of workstations		Number of users	
	M	**SD**	**M**	**SD**
Small	86	12	116	17
Medium	337	218	589	371
Large	792	237	1482	420

From Table 4 we can see that in Slovak hospitals there are two users per workstation. In the case of medium and large organizations we see that a large number of people work with IS, and therefore these systems must constantly and quickly respond to user requirements, process and store a large amount of information (not only about employees, but also patients, assets, economic variables, etc.).

In the next part of the questionnaire we focused on modules, which consist of HIS, IS that make up CHIS, to which specialized devices the HIS is connected and whether the IS of organizations are also connected with external IS.

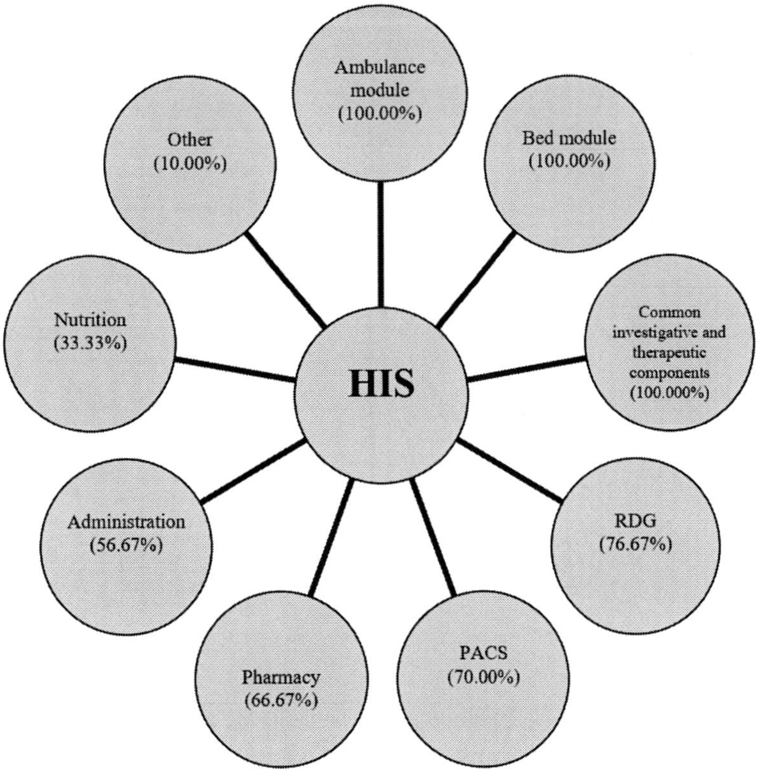

Fig. 1. Modules of hospital information system in Slovak hospitals.

Figure 1 shows which modules consist of HIS in Slovak hospitals. The ambulance, bed and Common investigative and therapeutic components modules are available to HIS at all health care providers. More than 50% of respondents identified administration, pharmacy, PACS and RDG modules.

PACS (picture archiving and communication systems) are clinical information systems used for the acquisition, archival, and post-processing distribution of digital images. An image must either be directly acquired using digital radiography or be digitised from a paper-based format. It can be stored using an electronic, magnetic, or optical storage device.[4]

Nutrition is part of HIS in one third of hospitals and 10% of respondents have other HIS modules (eg transfusion module, transport health service, boiler room and incineration system).

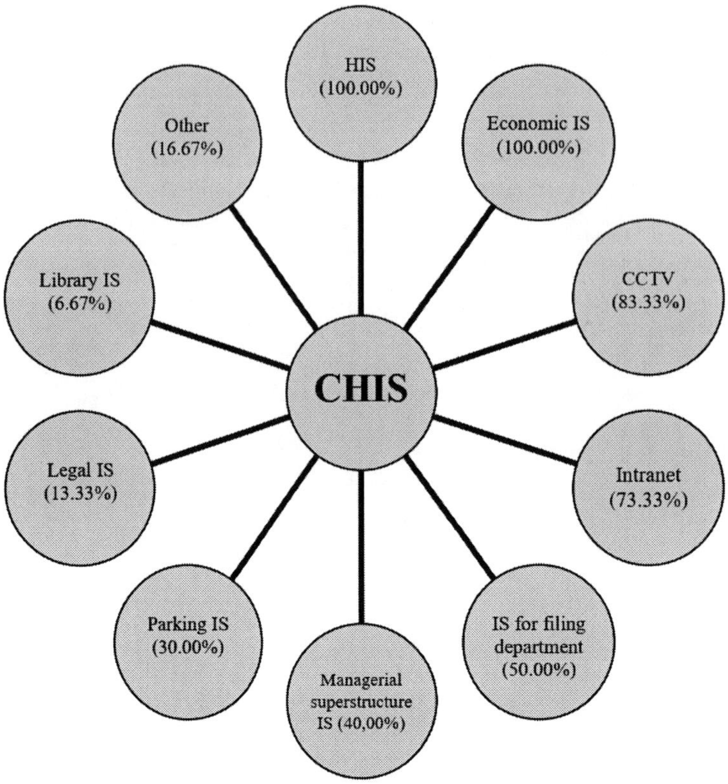

Fig. 2. Modules of complex hospital information system in Slovak hospitals.

As shown in Figure 2 for each respondent of our research, the CHIS module includes the HIS module and the economic IS. More than 70% of respondents have CCTV and intranet. 40% of hospitals have managerial superstructure and one third of organizations have a parking system. The least represented IS are legal IS (13.33%) and library IS (6.67%). As other modules of the CHIS respondents stated eg. attendance system for employees, electronic requests for purchase and repairs, waste records, questionnaire IS or IS for building management.

In the next part of the research we focused on whether HIS are linked to specialized facilities and therefore it is necessary that the IS not only communicates with these facilities, but also processes and evaluates the results of patients. The interconnection of HIS with such specialized devices was observed in 80% of respondents, the most common devices being: digital X-ray (77%), analyzers (70%), sono (53%), CT (47%), MRI (20%) and central drug preparation plant (7%). Two other respondents mentioned other specialized devices to which HIS is linked: sampling weights of hematology-transfusion department, cryo-bifuge, shock freezer, angiographic devices and special urological devices. None of our respondents has cognitive systems (eg IBM Watson).

The last area we analyzed is the interconnection of IS hospitals with IS from external environment. In all of the subjects monitored by us, we noted the link between CHIS and the IS ezdravie (IS of electronic health care in the Slovak Republic), the IS of individual health insurance companies and the IS of the National Center of Health Information. 60% of hospitals have links to the DôveraLab IS (the insurance company Dôvera, which provides access to the results of laboratory examinations) and 47% of the organizations are linked to the Dôvera Návrhy (hospital suggestions for spa treatment of patients). Two respondents reported linking CHIS via online applications to Medirex and Dr.Magnet and linking to external laboratories.

5 Discussion and conclusion

The informatization of health care is currently a basic trend that affects not only health care as a whole, but also individual facilities providing health care. Computerization not only facilitates the work and processing of documents by healthcare professionals, but also plays a major role in communication between doctors, patients, laboratories, hospitals and insurance companies. ICT that is applied in the health sector has the potential to contribute to improving the quality, cost-effectiveness, timeliness and mobility of health services. [12]

By carrying out our research, we managed to define the modules that make up the hospital information systems of individual organizations, as well as the interconnection of these information systems to other systems in hospitals (eg economic IS, parking IS, CCTV, etc.). The fact that CHIS are not closed IS within one organization was confirmed by an analysis of the connection of CHIS to external IS (IS of insurance companies, IS of electronic health care of the Slovak Republic, etc.). We have also been able to demonstrate that CHIS must also ensure the operation and processing of the results of specialized devices, such as: digital X-ray, sono or MRI. Based on these findings, we can confirm the claim of Moghadassi et al., that CHIS is an integrated information system providing not only patient information but providing support for all hospital activities (clinical, administrative, personnel, financial, etc.). Last but not least, these information systems must constantly respond to the changing environment (legislation, conditions of insurance companies and authorities).

A less satisfactory area is the area of interconnection of health care organizations and workplaces. It is e-health (in the territory of the Slovak Republic edzravie) that this problem should be eliminated. The implementation of this project requires that ICTs be installed in appropriate healthcare structures in appropriate structured network connections. The interconnection of IS between hospitals was recorded only in the case of three respondents using a single IS (these are hospitals within the founding competence of the Self-Governing Region of Trenčín).

The use and benefits of ICT in the health sector are certain. As domestic and foreign authors define the positives of the use of ICT resources in medicine, the risks associated with them should not be forgotten. In the health sector, sensitive personal data are processed, which need to be handled not only in terms of acquisition, but especially storage. ICT is a constantly changing field and it is therefore very important that healthcare providers plan their procurement, implementation and maintenance very well. As the ICT field is very dynamic - what is innovative today is already 'out of date' tomorrow - the spending of these modern means must be as efficient and effective as possible.

Notation

The paper was elaborated within VEGA No. 1/0388/20 IT Managemenet in Enterprises in Slovakia: International Standards and Norms Versus Individual Business Processed – proportion 100%.

References

1. R. Agarwal, G. Gao, C. DesRoches, A. Jha, The Digital Transformation of Healthcare: Current Status and the Road Ahead. *Information systems research.* **21**, 796-809 (2020)
2. L. Ahmadian, N. Dorosti, R. Khajouei, S. Gohari, Challenges of using Hospital Information Systems by nurses: comparing academic and non-academic hospitals. *Electronic Physician.* **6**, 4625-4630 (2017)
3. M. Almunawar, M. Ansahri, Health Information Systems (HIS). *Researchgate.* (2012)
4. M. Alohali, F. Carton, Y. O'Connor, Investigating the antecedents of perceived threats and user resistance to health information technology : a case study of a public hospital. *Journal of Decision Systems.* **29**, 27-52 (2020)
5. A. Black, J. Car, C. Pagliari, C. Anandan, K. Cresswell, T. Bokun, B. McKinstry, R. Procter, A. Majeed, A. Sheikh, The Impact of eHealth on the Quality and Safety of Health Care: A Systematic Overview. *Plos Medicine.* **8**, e1000387 (2011)
6. K. Cresswell, A.Sheikh, Health Information Technology in Hospitals: Current Issues and Future Trends. *Future Healthcare Journal.* **1**, 50-56 (2015)
7. J. Ehrenfeld, M. Cannesson, *Monitoring Technologies in Acute Care Environments : A Comprehensive Guide to patient Monitoring Technology.* (2013)
8. D. Hansen, Challenges and Opportunities for Health Information Systems Research. *Health Information Science.* 60-67 (2012)
9. S. Lancharoen, P. Suksawang, T. Naenna, Readiness assessment of information integration in a hospital using an analytic network process method for decision-making in a healthcare network. *International Journal of Engineering Business Management.* **12**, 1847979019899318 (2020)
10. J. Liang, Y. Li, Z. Zhang, D. Shen, J. Xu, G. Yu, S. Dai, F. Ge, J. Lei, Evaluating the Applications of Health Information Technologies in China During the Past 11 Years: Consecutive Survey Data Analysis. *JMIR Medical informatics.* **8**, e17006 (2020)
11. M. Liao, I. Lin, Performance Evaluation of an Information Technology Intervention Regarding Charging for Inpatient Medical Materials at a Regional Teaching Hospital in Taiwan: Empirical Study. *JMIR mHealth and uHealth.* **8**, e16381 (2020)
12. V. Ljubicic, P. Ketikidis, L. Lazuras, Drivers of intentions to use healthcare information systems among health and care professionals. *Health Informatics Journal.* **26**, 56-71 (2020)
13. H. Moghaddasi, et al. Hospital Information Systems : The status and approaches in selected countries of the Middle East. *Electronic Physician.* **5**, 6829-6835 (2018)
14. L. Salahuddin, Z. Ismail, U. Hashim, N. Ismail, R. Ikram, F. Rahim, N. Hassan, Healthcare practitioner behaviours that influence unsafe use of hospital information systems. *Health Informatics Journal.* **26**, 420-434 (2020)
15. N. Sayles, *Health Information Management Technology: An Applied Approach.* (2012)

Assessing progress towards responsible consumption and production

Daniela Rybárová[1,*]

[1]University of Economics in Bratislava, Faculty of Business Management, Dolnozemská cesta 1, Bratislava, Slovak republic

Abstract. The success of building sustainable consumption and production depends on creating a sufficient awareness and above all the involvement of all countries in the process of improving the environmental performance of products throughout their life cycle. Awareness and understanding of the social and economic dimension, influences building of sustainable consumption and production in the widest scale and makes it possible to encourage and motivate to socially responsible behavior. The article focuses on summarizing the basic attributes of sustainable consumption and production, as well as on the comparison of the progress of V4 countries in meeting the goal of "Responsible consumption and production" forming part of the Sustainable Development Agenda until 2030. We focus on indicators with quantitative target, where insufficient progress towards the EU target was achieved or there was a movement away from the target. The indicators are Primary energy consumption and Final energy consumption.

1 Introduction

The challenge today is a quality environment and sustainable development. This can only be achieved by streamlining production and changing consumption patterns [1], enabling optimization of resource use and minimize waste [2]. Problematic of sustainable consumption and production is a broad conception that is integrating application of advanced methods and techniques of business management [3], in all areas, including eco-design and innovation, sustainable marketing, risk management and strategic management with a focus on sustainable development for the environment.

Based on economic theories, the possibilities of satisfying the needs of either the producer or the consumer are limited only by the influence of market forces [4]. Various economists have pointed out the possibility of abusing your own interests in economic processes [5], US member of institutionalism; who has criticized the materialistic tendency of individuals' efforts of appropriation and not the value creation, which is leading to massive waste of creative human work and of limited resources. Some economists [6] in the broader context are considering whether the purpose of human life is merely the accumulation of wealth, which in terms of exhaustibility of limited resources leads to the

* Corresponding author: daniela.rybarova@euba.sk

restriction of resources for the future generations. They point out that there is currently an economic system that is in contradiction with natural laws, expecting someone else to resolve the consequences. In connection with the conditions of sustainable development of human society, the economic theory is faced with questions that need to be answered [7]. These facts lead to the need to develop new concepts of wealth and prosperity, achieved through lifestyle change that will enable a higher standard of living with lower dependency on limited resources.

The term "sustainable development" itself was introduced in 1987 in the report of UN Commission "Our future to all" led by GH Burdtland. The United Nations (UN) has declared fundamental support for sustainable development (SD) at the Conference on Environment and Development in Rio de Janeiro in 1992, at which AGENDA 21 was adopted as a basis for developing SD strategies. Sustainable consumption and production was also part of the SD objectives. In 1994, sustainable consumption and production was defined at the Oslo Symposium, as "use of services and related products that meet basic human needs and provide a better quality of life, minimizing the use of natural resources, hazardous materials as well as releasing pollutants or producing waste throughout the life cycle of services and products into the environment in a way that it will not affect the needs of future generations." In 2015, UN Member States adopted AGENDA 2030, which consists of 17 goals for sustainable development (including sustainable consumption and production) and 169 related milestones that are balancing three aspects of sustainable development - economic, social and environmental [8].

Sustainable development has become the priority of European Union (EU) since the summit of the Council of Europe in Cardiff (June 1998). The European Council summit in Gothenburg (2001) adopted the first Sustainable Development Strategy (SDS) and in 2006 adopted revised Sustainable Development Strategy (SDS) due to the persistency of unsustainable trends in terms of climate change, the use of energy and use of natural resources. The main challenge was to change the currently unsustainable consumption and production models. In 2008, the European Commission issued an Action Plan on Sustainable Consumption and Production (SCP), which included series of proposals on sustainable consumption and production such as eco-design requirements, strengthening of environmental and energy labels, promoting green public procurement, promoting resource efficiency, eco-innovation and promotion of sustainable production and consumption, etc. In 2010 European Council endorsed a strategy for the next period, which was named Europe 2020. In December 2015 European Commission approved EU Action Plan for Circulatory Economy as a tool to achieve goals from AGENDA 2030, particularly Goal 12 "Sustainable Consumption and Production". Subsequently developed strategy Europe 2030 is dealing with long-term visions and focus of sector policies after 2020, where it is assumed that in horizon of 2030, Europeans will need a highly competitive and sustainable social market economy in order to maintain social cohesion and combat climate change. The sustainability of Europe's economic and social model will depend on the ability to restore a dynamic balance between economic, social and environmental dimensions of development. The impact of the nonachievement of the sustainable development goals [9] may expose humanity to forms of global catastrophic risk and existential risk. Some authors [10] intend whether those objectives to support responsible consumption and production.

Sustainable consumption and production have been recognized as an integral part of the Sustainable Development Agenda until 2030. It is identified separately as objective number 12, and at the same time as a link for some other agreed objectives. Goal 12 "Responsible consumption and production" is divided into 11 indicators with a quantitative target.

Eurostat has been entrusted with the task to monitor regularly progress towards the goals of the 2030 Agenda for Sustainable Development in the EU context. To this end, the

Eurostat coordinated the preparation of the EU indicators set and has been tasked with its management. It also works out regular statistical monitoring reports on progress towards sustainable development goals in the EU context. Reports are based on a set encompassing 100 indicators, of which 41 are multipurpose indicators, i.e. they are used for monitoring more than just one goal. The set of indicators is open, and it will be regularly reviewed in line with future political developments. New indicators will also be assessed if data sources and methodology are available.

Over the most recent five years of available data [11], the EU made progress towards almost all the 17 sustainable development goals (SDG). Progress in some goals has been faster than in others, and within goals, movement away from the sustainable development objectives also occurred in specific areas. In the article, we focused on the development of indicators of goal 12 „Responsible consumption and production".

Insufficient progress towards the EU target under Goal 12 was recorded under the indicators: Primary energy consumption, Generation of waste excluding major mineral wastes and movement away from the EU target was recorded under the indicator Final energy consumption. For further analysis, we selected two of these indicators, Primary energy consumption and Final energy consumption. Data from the Eurostat database will be used.

2 Methods

Mathematical methods such as percentage, development index will be used to analyze the indicators, with the base year being 2005. The results will be compared with the targets set by the EU for the given area.

Indicators characteristics:

Primary energy consumption - the indicator measures the total energy needs of a country excluding all non-energy use of energy carriers (e.g. natural gas used not for combustion but for producing chemicals). "Primary Energy Consumption" covers the energy consumption by end users such as industry, transport, households, services and agriculture, plus energy consumption of the energy sector itself for production and transformation of energies, losses occurring during the transformation of energies (e.g. the efficiency of electricity production from combustible fuels) and the transmission and distribution losses of energy) [11]. **Final energy consumption** - the indicator measures the energy end-use in a country excluding all non-energy use of energy carriers (e.g. natural gas used not for combustion but for producing chemicals). "Final energy consumption" only covers the energy consumed by end users, such as industry, transport, households, services and agriculture; it excludes energy consumption of the energy sector itself and losses occurring during transformation and distribution of energy [11]

EU goals for Primary and Final energy consumption indicators:

According to the Directive (EU) 2018/2002 of the European Parliament and of the member States should set their national indicative energy efficiency contributions taking into account that the Union's 2030 energy consumption has to be no more than 1 273 Mtoe of primary energy and/or no more than 956 Mtoe of final energy. This means that primary energy consumption in the Union should be reduced by 26 %, and final energy consumption should be reduced by 20 % compared to the 2005 levels.

3 Results

Insufficient progress towards EU target was recorded for the Primary energy consumption indicator. The development of the indicator in comparison with the base year

2005 in the V4 countries and for comparison also in the EU (Renewable electricity and sustainable development goals in the Renewable electricity and sustainable development goals in the EU [7] is shown in the following figure (Fig. 1).

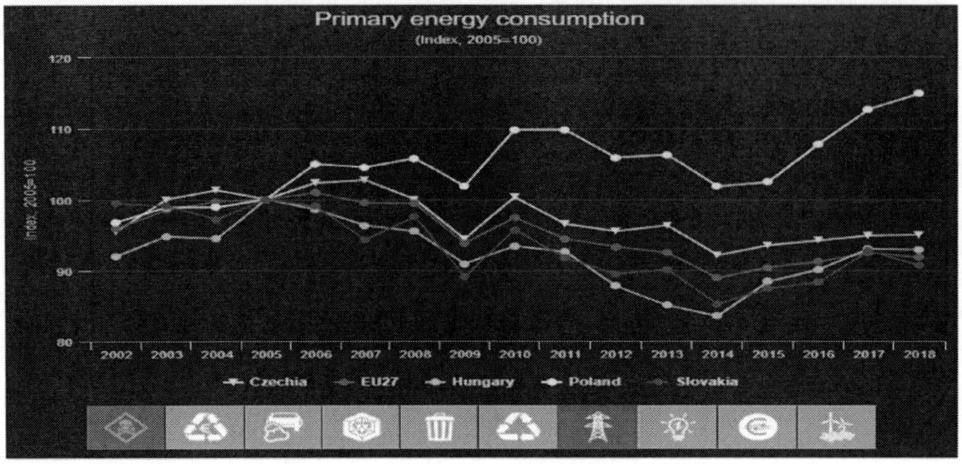

Fig. 1. Primary energy consumption from 2002 to 2018 (index, 2005=100)

Compared to 2005, primary energy consumption was lower each year only in the Slovak Republic and Hungary. Within the EU, there was an increase only in 2006, but in recent years consumption has shown a declining trend. In the Czech Republic, primary energy consumption exceeded the value of 2005 in 2006, 2007 and 2010. In recent years, its amount is lower. In the Republic of Poland, unlike the others, primary energy consumption in each subsequent year is even higher compared to 2005. The positive development of primary energy consumption in the other countries analyzed ended in 2014, when consumption has been on the rise, although not exceeding 2005 levels. These facts are related to the development of national economies after the end of the financial crisis, as well as its second wave in 2012. The percentage change in primary energy consumption is shown in the following table (table 1).

Table 1. Primary energy consumption (percentual change, 2005=100)

geo\time	2005	2006	2007	2008	2009	2010	2011	2012	2013	2014	2015	2016	2017	2018
EU (28 countries)	100	0,6	-1	-1,2	-7	-3,3	-6,8	-7,4	-8,4	-12,1	-10,7	-10,2	-9,2	-9,8
Czechia	100	2,3	2,7	0	-5,6	0,3	-3,5	-4,5	-3,7	-7,9	-6,5	-5,8	-5,1	-5
Hungary	100	-1,4	-3,7	-4,5	-9,1	-6,6	-7,4	-12,2	-15	-16,5	-11,6	-9,9	-7	-7,1
Poland	100	5	4,5	5,8	1,8	9,8	9,8	5,9	6,3	1,8	2,4	7,8	12,7	14,9
Slovakia	100	-1	-5,7	-2,5	-10,9	-4,3	-8,3	-10,5	-9,9	-14,8	-12,6	-11,7	-7,3	-9,3

Since the EU's goal is that primary energy consumption in the Union should be reduced by 26%, compared to the 2005 levels, the decrease (with the exception of the Republic of Poland) by 5 to 10% is insufficient to meet the objective.

Movement away from the EU target shows the indicator Final energy consumption, which, unlike the previous indicator excludes energy consumption of the energy sector itself and losses occurring during the transformation and distribution of energy. Final energy consumption should be reduced by 20% compared to the 2005 levels. We therefore examined its development in the V4 countries and in the EU.

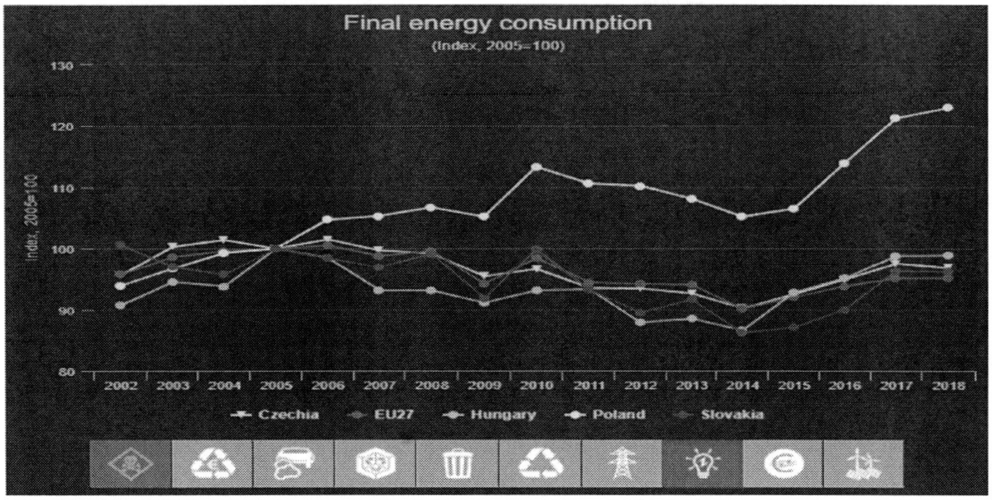

Fig. 2. Final energy consumption from 2002 to 2018 (index, 2005=100)

Final energy consumption of the V4 countries and the EU in 2018 approached the value from 2005. In the Republic of Poland, the consumption has even been on the rise throughout the analyzed period since 2005. The percentage changes in the amount of consumption are shown in the following table.

Table 2. Final energy consumption (percentual change, 2005=100)

geo\time	2005	2006	2007	2008	2009	2010	2011	2012	2013	2014	2015	2016	2017	2018
EU (28 countries)	100	0,3	-1,4	-0,7	-6,3	-2,3	-6,7	-6,5	-6,6	-10,6	-8,7	-7	-5,9	-5,8
Czechia	100	1,5	-0,3	-0,8	-4,6	-3,4	-6,5	-6,6	-7,4	-9,8	-7,5	-5,1	-2,5	-3,2
Hungary	100	-1,5	-6,9	-6,9	-8,9	-6,9	-6,7	-12,1	-11,5	-13,4	-7,2	-4,9	-1,2	-1,1
Poland	100	4,7	5,3	6,7	5,2	13,3	10,6	10,2	8,1	5,2	6,5	13,9	21,3	23
Slovakia	100	-1,6	-3,1	-0,9	-8,1	-0,2	-6,8	-10,5	-8,4	-13,8	-12,9	-10	-3,7	-3,8

As final energy consumption in the Union should be reduced by 20% compared to the 2005 levels, we can speak of a positive development only until 2014. Since that year, consumption has grown and in 2018 approached the consumption from 2005, when except for the Republic of Poland, consumption is only 1% to 6% lower. With this indicator, it can be stated that there has been a movement away from the EU target.

Although the Slovak Republic, together with the Czech Republic, also show insufficient fulfillment of the given indicators, they are closest to the EU average. Based on the given analyzes, the Republic of Poland achieved the worst results as the indicators of primary and final energy consumption show an increasing tendency compared to 2005. We therefore further examined the state of final energy consumption measured in tons of oil equivalent per capita in 2018 compared with other EU countries (picture)

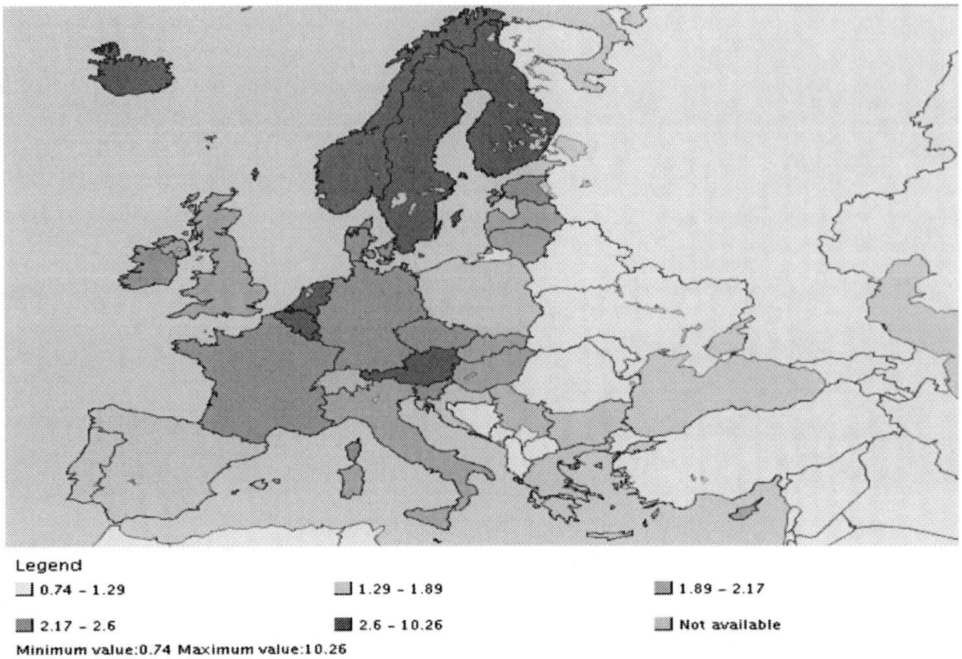

Fig. 3. Final energy consumption, 2018 (Tons of oil equivalent per capita)

The map of final energy consumption for 2018 points to the fact that of the V4 countries, the Republic of Poland has the lowest final energy consumption in tons of oil equivalent per capita, which partly explains the steady growth in consumption compared to 2005. Slovak republic and Hungary are included in the same group and reach the EU average. The Czech Republic has the highest final energy consumption of the V4 countries.

The final and primary energy consumption indicators are clearly linked to the development of the economies of individual countries, we analyzed the development of the economies of the V4 countries using gross domestic product at market prices, where due to the change in statistical reporting we have consistent data since 2008, we chose this year as the base year.

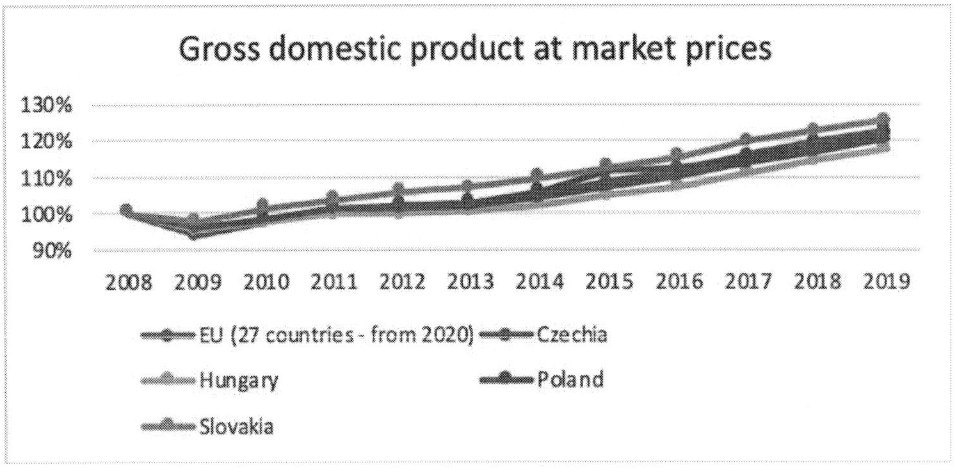

Fig. 3. Gross domestic product at market prices from 2008 to 2019

The development of gross domestic product at market prices was affected by the financial crisis and in 2009 there was a significant decline. Subsequently, in the years 2011 to 2014, it slowed down its growth. After 2014, there is already relatively stable growth. This fact also affected the fulfillment of the analyzed indicators. Among the V4 countries, the lowest growth was again recorded in the Republic of Poland, which means that this growth was also accompanied by an increase in final and primary energy consumption. However, the Slovak Republic, which has the highest GDP growth and showed the best results in meeting the analyzed indicators, was able to reduce final and primary energy consumption the most of the V4 countries. There was a decrease in the energy intensity of the Slovak economy.

According to the Statistical Office of the Slovak Republic for the years 2005 - 2018, the energy intensity of the Slovak economy decreased by 42.9%. This positive trend is the result of GDP growth expressed in p.c.15, which increased by about 57.9% over the same period and a decrease in gross domestic energy consumption, which, on the contrary, decreased by 9.8% over the observed period. Despite the favorable trend, in 2017 the Slovak Republic had the seventh highest energy intensity in the EU 28. There was a 3.6% year-on-year decrease in energy intensity in 2018, which was mainly due to a 4.0% year-on-year increase in GDP. Energy intensity in selected sectors of the Slovak Republic according to final energy consumption had a declining trend in all monitored sectors from 2005 to 2018. The largest decreases were recorded in the agricultural sector (75.9%) and in the industrial sector (55.0%), followed by the household sector (19.7%). The smallest decrease was achieved in the transport sector (9.1%). Energy intensity also decreased in all sectors in the year-on-year comparison.

4 Conclusion

At national level, well-designed national policy frameworks and instruments are needed to enable a fundamental shift towards sustainable consumption and production patterns. The strategic goal of the energy policy of the Slovak Republic is to achieve competitive low-carbon energy ensuring a secure, reliable and efficient supply of all forms of energy at affordable prices, considering consumer protection and sustainable development. Energy policy [12] is complemented by an energy security strategy, the main objectives of which are: a reliable, environmentally friendly and economically efficient energy supply; increasing the use of renewable energy sources, in particular watercourses, biomass, geothermal energy and solar energy; use of domestic primary energy sources for the production of electricity and heat on an economically efficient principle in accordance with the raw materials policy - coal and domestic reserves of uranium ores to reduce dependency on energy imports, as well as reducing energy and raw material intensity of the Slovak economy.

Economists should focus on finding good solutions [13] to managers and decision makers to achieve energy cost savings without sacrificing the productivity which can contribute to sustainable development of the manufacturing industry. All companies within their business models should also contribute to reducing energy intensity [14, 15]. New models of sustainable consumption [16] and production must emphasize the creation of higher added value with fewer inputs, reduced costs and minimized environmental impact. It is beneficial for companies themselves, because they can reduce their operating costs and reduce their dependence on minerals.

Acknowledgment

This paper is an output of the science project *VEGA č.* 1/0708/20 „Socio-economic determinants of sustainable consumption and production in terms of impact on performance and competitiveness of enterprises“ - project share is 100%;

References

1. M. E. Silva, M. D. Figueiredo. Practicing sustainability for responsible business in supply chains. *Journal Of Cleaner Production*, **251**, 119621. (2020)
2. G. K. Amoako. A conceptual framework: Corporate environmental management activities and sustainable competitive. *Management Of Environmental Quality*, **31**(2), 331-347. (2020)
3. J. Elbaz, S. Iddik. Culture and green supply chain management (GSCM): A systematic literature review and a proposal of a model. *Management Of Environmental Quality*, ***31***(2), 483-504. (2020)
4. A. Smith. *Pojednání o podstatě a původu bohatství národů*. (Liberální institut, Praha, 2016.
5. T. Veblen. *Teorie zahálčivé třídy*. (Sociologické nakladatelství, Praha,1999)
6. P.M. Senge, B. Smith, N. Kruschwitz, J. Laur, S. Schley. *The Necessary Revolution: How Individuals and Organizations are Working Together to Create a Sustainable World*. (Nicholas Brealey Publishing, London, 2010)
7. M. Uramová et al. *Hospodárska politika*. (Univerzita Mateja Bela, Ekonomická fakulta, Banská Bystrica, 2003).
8. P. Chowdhury, S. K. Paul. Applications of MCDM methods in research on corporate sustainability: A systematic literature review. *Management Of Environmental Quality*, ***31***(2), (2020)
9. T. Cernev, R. Fenner. The importance of achieving foundational Sustainable Development Goals in reducing global risk. *Research & Management Science*. **115**, 102492, (2020)
10. J. Gunawan, P. Permatasari, C. Tilt. Sustainable development goal disclosures: Do they support responsible consumption and production? *Journal Of Cleaner Production*, **246**, 118989. (2020)
11. European Union. Sustainable development in the European Union – *Monitoring report on progress towards the SDGs in an EU context*. (Publications Office of the European Union, Luxembourg, 2019)
12. Ministerstvo hospodárstva SR. *Integrovaný národný energetický a klimatický plán na roky 2021-2030*. (Bratislava, október 2019)
13. M.H. Ho, F. Hnaien., F. Dugardin. Electricity cost minimisation for optimal makespan solution in flow shop scheduling under time-of-use tariffs. *International journal of production research*. 10.1080/00207543.2020.1715504 (2020)
14. R. B. Swain, A. Karimu.. Renewable electricity and sustainable development goals in the EU. *World Development*. **125**, 104693. (2020)
15. Z. Fu, R. Li. The contributions of socioeconomic indicators to global PM2.5 based on the hybrid method of spatial econometric model and geographical and temporal weighted regression. Science of the total environment. **703**, 135481 (2020)
16. R. Šlosár, Š. Majtán, P. Štetka, N. Grisáková. *Vizuálne podnety a responzívne správanie spotrebiteľov*. (Wolters Kluwer, Praha, 2019)

Human Capital Costs in the Context of Coronavirus Crisis

Slavka Šagátová[1,*]

[1]FPM EU in Bratislava, Department of Business Economy, Dolnozemská cesta,1 852 35 Bratislava, Slovakia

Abstract. Human capital is an important motor of business performance. In this context, the costs of human capital that the company spends in its efforts to ensure the development and achieve growth in its performance play an important role. However, in times of crisis and decline of production, labor costs that cannot be eliminated become unproductive. The sensitive area is especially area of the wage costs that the company has to pay to employees and thus they represent not only the loss for the company but also the cash outflow. The article maps the influence of this factor on individual branches of the Slovak economy. Using data on the volume of wages paid and estimates of scenarios for the possible development of the coronavirus crisis, it identifies the most vulnerable sectors in terms of unproductive labor costs.

1 Introduction

Human capital is an important part of the business economy, and investing in this area of production factors ensures the prosperity of the company. However, with the reduction of corporate production due to the coronavirus crisis, the costs of human capital are becoming an item that reduces corporate profits. A significantly negative feature of these costs is the transformation of a large part of labor costs directly into losses, which further exacerbates the possibility of remediation of damages from the decline in production by enterprises. The high share of labor costs thus becomes a risk factor for the survival of companies. The inability of companies to pay wages can lead to the loss of employees in individual companies. In a broader context, this situation can lead to rising unemployment and, in this context, to further negative effects on the costs economy as a whole.

2 The essence of human capital

The OECD defines human capital as the knowledge, skills, abilities and other characteristics of an individual that are relevant to economic activity. [1] From the company's point of view, therefore, human capital is an essential input into production factors in the form of human resources. Exploring this issue is important as good human

* Corresponding author: slavka.sagatova@euba.sk

resource management practices can contribute to stabilizing and creating good conditions for maintaining a workforce. [2, 3] By inducing interactions to gather and exchange knowledge, human capital contributes to the growth of innovation. [4, 5] To use human resource management as a tool for performance management, it is important to know the basic principles that allow employees to be involved in these management processes. [6]

The human capital costs can be considered as certain investments that are made by companies to improve human activity. It is especially important to match investments into human capital with the needs of the company and the ability to use human potential. [7] It is inefficient if the company invests in developoment of employees and then the acquired skills cannot be used by them. On the other hand, it is very important for the company to invest in necessary education, skills and abilities of employees, which are inevitable for the efficient production activity of the company. The company can make these investments in human capital by obtaining it from the external environment (funds spent on acquiring quality employees) or by investing in the internal environment (providing training to employees, incentive programs for employees - wages, benefits). However, these investments can also be seen in the broader context of building an employer brand, which will enable the acquisition of qualified employees for the company. [8]

The immediate picture of investing in human capital in a company are the personnel costs. The advantage of this parameter is the possibility of its unambiguous quantification. Personnel costs represent the evaluation of the labor force as a basic production factor and other costs related to ensuring its use in the transformation process. For effective cost management, it is first and foremost necessary to know them. In the case of personnel costs, we can talk about direct wage costs and other costs related to production workers, other "overhead" employees as well as personnel department operating costs of the. there are mutual interactions in all these groups when implementing the business activities. It is necessary to be aware of the importance labor costs, which can be an important motivating element in increasing the performance of business.

Although the current performance studies point to the growing importance of social processes and the focus on manager-employee interaction [9], the importance of wages as a motivating factor cannot be overlooked. There are also studies confirming the impact of salary on employee satisfaction and loyalty. [10] However, the degree of sensitivity to remuneration can also be perceived in a broader context, as an indicator of the employee's response to various stimuli from the company, for example in the form of training. [11] It should be noted, however, that the different effect in this area will be influenced by the very perception of wages by the individual in the context of other employment benefits. [12] As a starting point can be considered a combination of internal motivation and managerially well-managed management (job identification, remuneration strategies, etc.). [13]

In case of a loss of revenues during crisis, it is possible to reduce or completely eliminate the expenditure of certain personnel costs, such as costs associated with recruiting or employees training. However, such a solution is not possible at all types of these costs. The main problem may be the wage costs paid to employees, which become unused costs if the company does not produce. As it is difficult for a company to keep the work of employees in stock and consume it later, unused (non – productive) labor costs are a serious problem. It is only possible in the short term not to transmit performance shocks to employees, in line with the notion that firms isolate workers from idiosyncratic shocks consistent with the notion that firms insulate workers from idiosyncratic shocks [14] is possible only for a short time period.

3 Methodology

Nowadays in the context of coronavirus crisis and its impacts on human capital, the currently emerging problem appears to be the labor costs that companies are forced to pay to the employees regardless of the decrease in their profits and a drop in revenues. The aim of the article is, therefore, to point out the possible effects of labor costs in the event of a decline in production, focusing on individual sectors of the Slovak economy.

As a data source, we used the databases of the Statistical Office SR [15] and the government statement on the termination of the economy. Data on the number of employees and the average salary was examined in more detail. The sectoral structure of average salaries calculated from the data for 2019 was used to assess the situation in individual sectors, provided that this structure has not changed in the following period.

The development of production during the crisis willbe estimated based on data publicly presented by government officials and resulting from measures to close selected companies. According to the regulation valid from 16.3.2020 [16], all retail establishments and services were closed except groceries, drugstores, pharmacies, medical device stores, newspapers, post offices, banks, insurance companies, petrol stations, pet shops, and veterinary clinics. Restaurants could have stayed open, but people were not allowed to gather there, but they could have taken takeaway food. International airports were closed, as well as the international train and bus transportation was stopped (except for cargo transports). School facilities, social, cultural facilities, ski resorts, and wellness centers, and aquaparks were also closed. Part of these measures was abolished on 6.5.2020. [17]

4 Results and Discussions

The crisis caused by the coronavirus pandemic has dampened the economy. The overall decrease in production is said to be by 30% (according to the statement of Prime Minister in March 2020) [18]. The amount of labor costs is an important factor playing an important role in terms of the survival of vulnerable sectors, which burden their financial management and are a significant component of cash flow outflow.

In the context of revenues drop, the number of wages that were paid even though they did not contribute to ensuring the company's performance can be considered. Therefore it is possible to talk about unused labor costs. Unused labor costs depend on two parameters: the volume of wages paid and the extent of the company's production downtime. The largest volume of unused labor costs can, therefore, be observed, especially in the sectors with the most massive volume of wages paid. (In order to get a picture of the total amount of costs related to employees' wages, it is necessary to add wage levies to the value of wages paid.)

According to the average data from 2019 (Statistical Office of the Slovak Republic, 2020), the most wages paid were in the industry sector, which employed 27% of all employees, and 27% of all wages were paid in it. In second place are the branches of Public Administration, Defense and Compulsory Social Security, which employed 11% of all employees and paid 13% of all wages. The third is the education sector, employing 11% of all employees out of all, and 10% of all wages were paid. In fourth place is the wholesale and retail sector, which employed 10% of employees, to whom 10% of wages were paid. In fifth place is the sector of healthcare and social assistance sector with 8% of employed persons and 8% of wages paid.

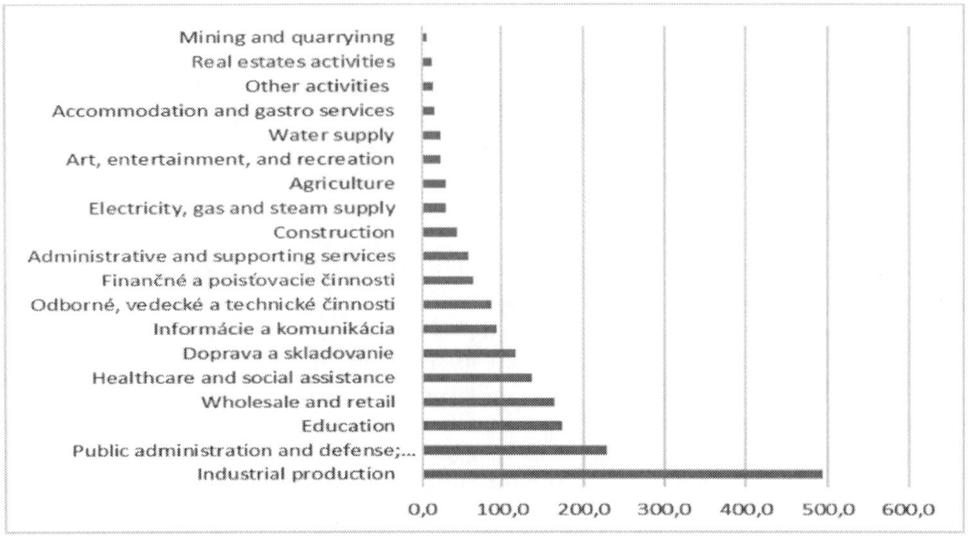

Graph 1 Volume of wages paid in individual branches in 2019 in mil. €
Source: own processing according to the Statistical Office of the Slovak Republic

The second parameter that has already been mentioned, influencing the occurrence of unused labor costs, is production decline. However, not all economic sectors were negatively affected by the crisis. The endangered companies can be divided into primary endangered and secondary endangered. In primarily endangered companies, there is a decline in production due to the closure of facilities by government decision, to ensure the protection of public health. These companies had to cease their activities completely, which meant that no revenues were achieved. In secondary endangered companies, they recorded a decline in revenues due to declining customer demand resulting from their new needs at the time of the coronavirus pandemic. Some companies might also have seen an increase in production volume due to increased customer demand, whether due to the loss of substitute products or purchasing of stocks. The biggest threat of unused labor costs is naturally in companies with the most significant decline in production.

Table 1. Non - productive labor costs in the event of a decline in production

	Non - productive labor costs in % of annual wages	
Production decline by	**Duration of the decline one month**	**Duration of the decline two months**
10%	0,8%	1,7%
20%	1,7%	3,3%
30%	2,5%	5,0%
40%	3,3%	6,7%
50%	4,2%	8,3%
60%	5,0%	10,0%
70%	5,8%	11,7%
80%	6,7%	13,3%
90%	7,5%	15,0%
100%	8,3%	16,7%

Source: own processing

According to an estimate of a 30% slowdown in the economy over two months, the volume of non-productive labor costs can be expected to amount to 5% of annual wages. The highest value would achieve the costs in the industry sector. According to average wages in 2019, it would be more than 24.7 mil. € (after taking into account levies of € 33.4 million). Non-productive labor costs would represent up to 12.5% of annual wages in enterprises in which there has been a complete closure of operations or a production outage for 1.5 months. According to average wages in 2019, it would be, for example, in the arts, entertainment, and recreation sector, almost 3 mil. € (after taking into account levies 4 mil. €) and in the accommodation sector more than 1 mil. € (with levies of almost € 1.4 million)

5 Conclusion

The costs of human capital are an important component of the company's costs. In case of a significant decline in production, there is a non-productive component created, especially in the part of labor costs, which is a loss for the company. In terms of unused labor costs, the most vulnerable sector is the industry. In this sector, a high production outage can be estimated, especially due to the production outage and also the high volume of labor costs. Both of these factors are cumulative, especially in the automotive industry. Businesses that have been closed by government decision, and that have not achieved any revenues for a certain period are also at risk. Compensation provided by the state as compensation for employees' wages can be assessed positively in this situation. This measure will reduce the loss caused by unproductive labor costs of companies and will also improve the cash flow balance.

In terms of human capital costs, not only losses from unproductive labor costs can be perceived negatively, but also costs associated with possible redundancies. This factor would play a greater role in companies with higher wages and higher numbers of employees. Last but not least, a higher number of employees at risk of losing their jobs or a reduction in their income would also have an impact on the macroeconomic parameters of this situation, such as the unemployment rate and purchasing power.

Note

This contribution is a partial output of the VEGA project no. 1/0569/18 "Stabilization on Human Capital in Slovak Enterprises as an Immanent Determinant of their Success, Competitiveness and Sustainable Development " in the range of 100%.

References

1. OECD, Centre for Educational Research and Innovation (CERI) Paris, 113 (1998).
2. L. Onnis, Human resource management policy choices, management practices and health workforce sustainability: remote Australian perspectives, *Asia Pacific Journal of Human Resources* 57, 3 – 23, (2019)
3. M. Amstrong, *Řízení lidských zdrojů: nejnovější trendy a postupy.* 10 vydanie. Praha: GRADA PUBLISHING, 2007, 789 s. (2007)
4. Y. Li, M. Wang, D. D. van Jaarsveld, G. K. Lee, D. G. Ma, From Employee-experienced High-involvement Work System to Innovation: An Emergence-based Human Resource Management Framework. *Academy of Management Journal* **61**, No 5 (2018)

5. S. Kozhakhmet, S. Jayasingam, N. Majeed, S. Jamshed, When investment in employee development promotes knowledge sharing behavior in an uncertain post-Soviet context. *Personnel Review* **49** (2020)
6. S. Kakkar, S. Dash, N. Vohra, S. Saha, Engaging employees through effective performance management: an empirical examination,, *Benchmarking: An International Journal* 27, (2020)
7. D. Rybárová, Selekt problem of human capital management, In : G Dubcova a kol. *Stabilization on Human Capital in Slovak Enterprises as an Immanent Determinant of their Success, Competitiveness and Sustainable Development* Masaryk University, Brno, 113 - 118, (2019)
8. A.Tumasjan, F. Kunze, H. Bruch, I. M. Welpe, Linking employer branding orientation and firm performance: Testing a dual mediation route of recruitment efficiency and positive affective climate, *Human Resource Management* **59**, 83-99 (2020)
9. S. T. Tseng, P. E. Levy, A multilevel leadership process framework of performance management. *Human Resource Management Review* **29**, 100668 (2019)
10. D. Johnson, Ch. J. Lake, Contingent worker monetary influence, work attitudes and behavior, *Personnel Review* 48, (2019)
11. J. C. Gawke, M. J. Gorgievski, A. B. Bakker, Personal costs and benefits of employee intrapreneurship: Disentangling the employee intrapreneurship, well-being, and job performance relationship. *Journal of Occupational Health Psychology 23* (4), 508–519, (2018)
12. D. C. Jones, P. Kalmi, T. Kato, M. Mäkinen, The differing effects of individual and group incentive pay on worker separation: evidence using Finnish panel data. *The International Journal of Human Resource Management* **30**, (2019)
13. C. S. Rigby, R. M. Ryan, Theory in Human Resource Development: New Directions and Practical Considerations, *Human Resource Development International* **20**, (2018)
14. Ch. Juhn, K. McCue, H. Monti, B. Pierce, Firm Performance and the Volatility of Worker Earnings. *Jurnal of Labor Economics* **36**, (2018)
15. R. Pečíková, *Zamestnanci a priemerné mesačné mzdy v SR v roku 2019*. Ústredie Štatistický úrad SR (2020) Retrieved from https://slovak.statistics.sk/
16. Urad verejného zdravotníctva SR, Opatrenie Úradu verejného zdravotníctva Slovenskej republiky pri ohrození verejného zdravia Číslo: OLP12595/2020 Bratislava 15.03.2020 Retrieved from http://www.uvzsr.sk/docs/info/covid19/Opatrenie_UVZ_SR_o_zakaze_a_obmedzeni_prevadzok_15032020.pdf
17. Úrad verejného zdravotníctva SR, Opatrenie Úradu verejného zdravotníctva Slovenskej republiky pri ohrození verejného zdravia Číslo: OLP/3795/2020 Bratislava 05.05.2020 Retrieved from http://www.uvzsr.sk/docs/info/covid19/Opatrenie_UVZ_SR_o_zakaze_a_obmedzeni_prevadzok_15032020.pdf
18. E. Frantová, B. Toma, Premiér neprekvapil. Obchody zostávajú zatvorené, Pravda 14.4.2020 Retrieved from https://ekonomika.pravda.sk/

Is marketing communication really a challenge for companies on Instagram?

Kamilla Sánta[1,*], *Patrik* Baša[2] , and *Renáta* Machová[3]

[1]student, J. Selye University, Faculty of Economics and Informatics, Bratislavská cesta 3322, 945 01 Komárno, Slovakia
[2]student, J. Selye University, Faculty of Economics and Informatics, Bratislavská cesta 3322, 945 01 Komárno, Slovakia
[3]Management Department, J. Selye University, Faculty of Economics and Informatics, Bratislavská cesta 3322, 945 01 Komárno, Slovakia

Abstract. Meeting the challenges of social media is becoming more and more important for companies in the 21st century. If companies want to be successful, they need to be present not only in the real world but also in the virtual world. With this paper, we would like to help companies in their marketing communication on Instagram. We carried out an international research on the most important factors of Instagram business profile and examined their relationship by correlation analysis. Our results are summarized in a model, which includes several factors that companies must simultaneously consider when following a particular social media strategy. Since we have not discovered big differences among the countries we examined, our model could have been used internationally (primarily by Central European companies).

1 Introduction

As we move forward in the 21st century, it is becoming increasingly important for companies to meet the challenges of social media, as it is one of the simplest and perhaps the most effective channels for building relationships with consumers. Companies need to be present and build their brand in the real world, but also in the virtual world. In the first chapter of the paper, we deal with social media and Instagram on a theoretical level. After that we write down our results and demonstrate our model.

1.1 Social media

According to Kaplan and Haenlein the social media "is a group of Internet-based applications that build on the ideological and technological foundations of Web 2.0, and allow the creation and exchange of user-generated content" [1]. Van der Heide and Lim described social media as an environment where anyone can say anything about anything to everyone. [2] An important unique characteristic of social media is networking capability, which is also one of

* Corresponding author: 122111@student.ujs.sk

the most capitalized benefits of it. Consumers are partly in control of how information is created and shared. [3] Social media also differ from other media in the way in which they are experienced, such as satisfying the need to find useful information, filling empty moments, or creating or sharing content with others. [4] Social media are unique with regard to the massive amounts of data they provide. The data provided by social media platforms have great potential to examine and explain consumers' interactions and responses to brand communication in social media in a natural setting. [5]

Social media and television are major influencers when it comes to the purchasing decisions of consumers. The products of tens of thousands of companies entice consumers through various channels, which creates the artificial generation of consumer needs. There have also been changes in the human perception: people measure everything in money, they always want to dress in the latest fashion, call their friends on the newest mobile phones and post about these things on social media. These facts represent a huge opportunity for marketers and offer a virtual marketplace for companies to sell their products. [6]

The forms of social media are the following: social networking sites, creativity works sharing sites (video, photo, music, content, general intellectual property), user-sponsored blogs, company-sponsored websites/blogs, company-sponsored cause/help sites, invitation-only social networks, business networking sites, collaborative websites, virtual worlds, commerce communities, podcasts, news delivery sites, educational materials sharing, open source software communities, social bookmarking sites allowing users to recommend online news stories, music, videos, etc. These media have become a major factor in influencing various aspects of consumer behaviour including awareness, information acquisition, opinions, attitudes, purchase behaviour, and post-purchase communication and evaluation. [7]

Social networking sites are probably the most prominent type of social media. [8] The number of these platforms grows rapidly, as new applications spring up almost every day. [9]

1.2 Instagram and business profile

The strength of social ties plays an important role in consumer responses to brand communication via social media. [10] Brands often use native advertising formats on social media. [11] The message on social media does not necessarily refer to what is being said by the brand, it also refers to User Generated Content. Professionals often use the term content rather than message. [5]

Instagram, a mobile photo and video capturing and sharing application, has quickly emerged as a new medium in spotlight in the recent years. It provides users an instantaneous way to capture and share their life moments with friends through a series of pictures and videos. [12] In the past few years Instagram is getting popular among companies. Nowadays users and companies can switch their personal profile to business profile and take advantage of it on the field of marketing.

Instagram's business profile promises a number of new opportunities for companies. They can view analyzes, provide contact information and specify the price of the products in the photos. Later & HubSpot's e-book details what companies need to pay attention when creating the profile:

- name - user name on Instagram (mostly the name of the company)
- description - a concise but comprehensive description of the company's activity
- keywords - the values represented by the company
- link to website - placing the link to the website of the company
- own hashtags - using the company name and product names in the form of a hashtag
- call to action - encouraging followers to take action on the companie's profile [13]

As more and more companies appeared on Instagram, traces of their marketing activities became more and more noticeable. The three most common marketing tools for companies on Instagram are:

- Giveaways - the most common and simplest form of giveaways is based on the "like and comment" principle. As the name implies, the user has to like a photo or video posted by the company and write a comment. Recently, it has become more and more common to tag some friends in the comment, so they can also participate. The prize of the giveaway is usually a product of the company.
- Influencer-marketing - In every social group, there are people who are better informed than others or express their opinions more often than other members of the group. Because most people are unable, or unwilling, to provide relevant information on a topic, they accept the information shared by those individuals and believe it is trustworthy. [14] Influencers are now the basis of marketing activities in social media.
- Paid Advertising - Within the classic Instagram ads, we know four basic categories that differ in the way they are displayed. This includes advertising by image, story, video, or carousel.

Instagram is a valuable tool for companies who want to get their brand known. [15]

2 Methodology

The research was realised between February 2019 and February 2020. Our questionnaire was prepared online and sent to Slovak Instagram users via social media. People, who filled in the questionnaire were asked to send it to another five Instagram users. In order to obtain answers from another countries, we contacted the partner institutions of the J. Selye University. Thanks to it, we got answers from Hungary and Romania. A total of 754 evaluable responses were received to the questionnaire prepared in Hungarian and English language. Responses were coded in Microsoft Excel and analysed in SPSS statistical software. To test our four hypotheses, we performed correlation analysis. After the analysis, we created a model that includes the relationship of factors belonging to the same group.

3 Results

As mentioned above, a total of 754 evaluable responses were received to our online questionnaire – 242 from Slovakia, 332 from Hungary and 180 from Romania. In terms of gender, 66,31% of the respondents were women and 33,69% are men.

Our questionnaire was filled in by the representatives of the 18-24 age group in the highest proportion (75,73%), followed by the 25-34 age group (15,65%), by the 35-44 age group (3,85%), by the 45-54 age group (1,99%), by those over 55 (1,46%), and finally those under 18 (1,33%).

Our research was based on a series of questions measured on a Likert scale containing 13 factors. Respondents had to decide how important (1-not important at all, 5-very important) a factor was for them on a company's business profile.

Table 1. Means of factors of Instagram business profile

	Slovakia	**Hungary**	**Romania**
High number of followers	2,281	2,130	2,106
Good profile picture	2,508	2,539	2,517
Comprehensive description of the company	2,901	2,883	2,750
Regular posting	2,938	3,123	2,694

Sharing costumer photos	2,707	2,726	2,433
Consistency of images	2,583	2,873	2,572
Creative descriptions of pictures	2,781	2,946	2,722
Placing link to the website	2,967	3,120	2,789
Ordering option in message	2,517	2,413	2,444
Indication of the price	2,955	3,060	2,900
Regular Story	2,227	2,202	2,061
Giveaways	2,380	2,419	2,250
Collaborate with famous people	2,475	2,304	2,417

There are no major differences between the three countries, but minor differences can be noticed. Based on the means of the individual factors, Instagram users in Slovakia consider the placement of a link to the corporate website, the indication of the price of the product, the regular posting and a comprehensive description of the corporate activity to be the most important. For the Hungarian respondents, the most important factor of the business profile is the regular posting, followed by the placement of a link to the company's website, the indication of the price of the product and the creative description of the uploaded images. In the case of Romania, the most important factors are the indication of the price of the product, the placement of a link to the company's website, a comprehensive description of the company's activities and a creative description of the uploaded images. Instagram users in all three countries consider regular Story sharing and a high number of followers to be the least important factors.

In our research four hypotheses were formulated. When examining the hypotheses, we performed correlation analysis, as each intervale-scale variable was measured on a metric scale. In the course of correlation analysis, we can get an answer to the question of whether there is a relationship between the examined quantitative variables and, if so, how close it is. The closer the absolute value of Pearson's correlation coefficient (r) is to 1, the closer the relationship. According to Sajtos-Mitev's [16] SPSS data analysis manual, if $|0,2| < r < |0,7|$, the relationship is moderately strong, and if $|0.7| < r < |1|$, the relationship is strong.

In the analysis of the first hypothesis, we examined the relationship between "creative factors" related to the business profile.

Table 2. Correlation analysis – Creative factors

		Good profile picture	Comprehensive description of the company	Consistency of images	Creative descriptions for the pictures
Good profile picture	Pearson Corr.	1	**,518****	**,268****	**,280****
	Sig. (2-tailed)		,000	,000	,000
	N	754	754	754	754
A comprehensive description of the company	Pearson Corr.	,518**	1	**,306****	**,430****
	Sig. (2-tailed)	,000		,000	,000
	N	754	754	754	754
Consistency of images	Pearson Corr.	,268**	,306**	1	**,526****
	Sig. (2-tailed)	,000	,000		,000
	N	754	754	754	754
Creative descriptions for the pictures	Pearson Corr.	,280**	,430**	,526**	1
	Sig. (2-tailed)	,000	,000	,000	
	N	754	754	754	754

We identified as "contact factors" the factors that target keeping in touch with followers, customers, and influencers.

Table 3. Correlation analysis – Contact factors

		High number of followers	Share customer photos	Giveaways	Collaborate with famous people
High number of followers	Pearson Corr.	1	**,235****	**,229****	**,371****
	Sig. (2-tailed)		,000	,000	,000
	N	754	754	754	754
Share customer photos	Pearson Corr.	,235**	1	**,385****	**,311****
	Sig. (2-tailed)	,000		,000	,000
	N	754	754	754	754
Launch giveaways	Pearson Corr.	,229**	,385**	1	**,467****
	Sig. (2-tailed)	,000	,000		,000
	N	754	754	754	754
Collaborate with famous people	Pearson Corr.	,371**	,311**	,467**	1
	Sig. (2-tailed)	,000	,000	,000	
	N	754	754	754	754

"Business factors" included factors that were closely related to the purchase of products and the use of services.

Table 4. Correlation analysis – Business factors

		Placing a link to the website	Indication of the price of the product	Order option in message
Placing a link to the corporation´s website	Pearson Corr.	1	**,449****	**,449****
	Sig. (2-tailed)		,000	,000
	N	754	754	754
Indication of the price of the product	Pearson Corr.	,449**	1	**,416****
	Sig. (2-tailed)	,000		,000
	N	754	754	754
Order option in message	Pearson Corr.	,449**	,416**	1
	Sig. (2-tailed)	,000	,000	
	N	754	754	754

"General factors" included regular posting and regular Story sharing.

Table 5. Correlation analysis – General factors

		Regular posting	Regular Story Sharing
Regular posting	Pearson Corr.	1	**,367****
	Sig. (2-tailed)		,000
	N	754	754
Regular Story Sharing	Pearson Corr.	,367**	1
	Sig. (2-tailed)	,000	
	N	754	754

A significant result was obtained when analyzing the factors, as r> 0,2 in all cases. The relationship is positive in all cases and moderately strong. Hypotheses H1, H2, H3, and H4 were accepted and null hypotheses were discarded.

Since all hypotheses were accepted, we could have created the relationship model of the factors belonging to the four groups:

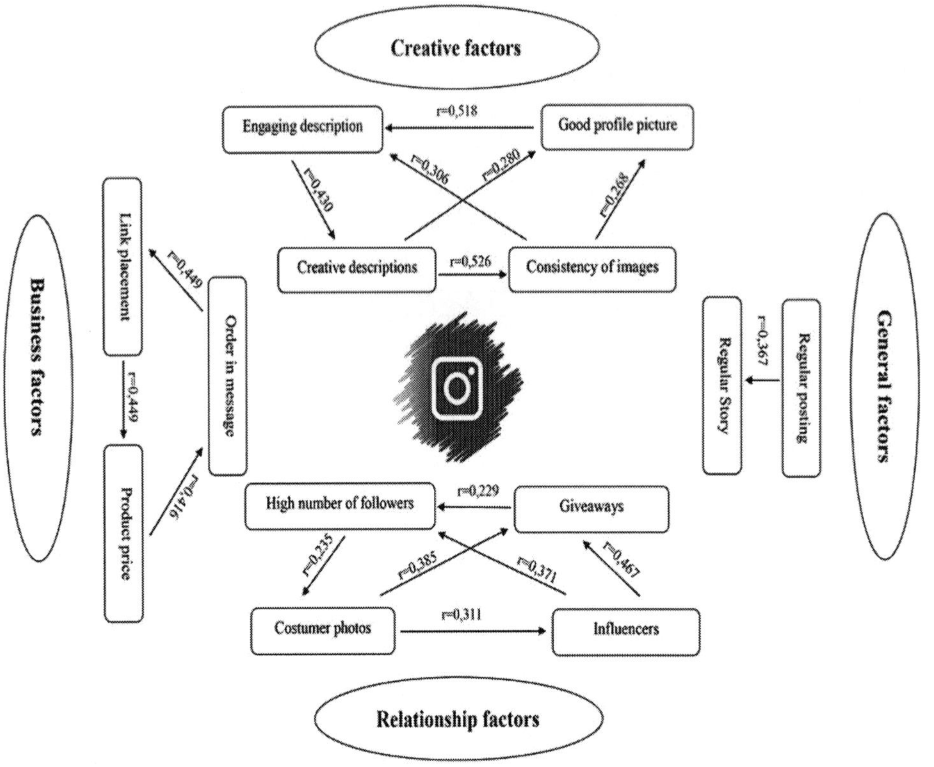

Fig. 1. The model of the most important factors on Instagram business profile

If a particular factor is considered important by the followers of the company, with our model the companies can already know which other factors are also important to these consumers. Based on the model, companies can develop their Instagram marketing strategy and focus on the factors of that strategy (creative, relationship, business and general).

We also examined whether there is a difference between the means of responses in Slovakia, Hungary and Romania, but since the difference is minimal, the model can be applied internationally by companies operating (primarily) in Central Europe.

4 Discussions

Marketing communication on social media is no longer such a challenge for companies. The model we have created allows the companies to choose a type of strategy they want to focus on in their marketing communication. It is the responsibility of companies that focus on creative factors to focus on the visual elements of the business profile: have a good profile picture, have visually consistent photos, have compelling description of the company's activities and write creative descriptions under the uploaded photos. In a relationship-based strategy, the company needs to build the highest possible follower base, share customer images on a regular basis, collaborate with well-known people (influencer marketing) and organize giveaways. The essence of the strategy focusing on business factors is to place a link to the company website, to ensure the possibility of ordering via private message and to indicate the price of the products on pictures. Companies that use the strategy of general factors related should pay attention to regular posting and regular story sharing. The four groups (strategies) do not mean that a company can use only one of them, but it is recommended that the company select just one, and chooses another one once its activities related to those factors are working effectively. As we examined the responses of three countries and did not find big differences between the responses (means) in Slovakia, Hungary and Romania, the model can be applied internationally (primarily) in Central Europe.

This work was supported by the by the Collegium Talentum 2019 Programme of Hungary.

References

1. A. M. Kaplan, M. Haenlein, Users of the world, unite! The challenges and opportunities of social media. *Business Horizons* **53** 59-68 (2010) DOI: 10.1016/j.bushor.2019.09.003
2. B. Van Der Heide, L. Young-Shin, On the conditional cueing of credibility heuristics: The case of online influence. *Communication Research* **43**, 672 (2016) DOI: 10.1177/0093650214565915
3. S. Okazaki, C.R. Taylor, Social media and international advertising: Theoretical challenges and future directions. *International Marketing Review* **30**, 56-71 (2013) DOI: 10.1108/02651331311298573
4. H.A.M. Voorveld, G. Van Noort, D.G. Muntinga, F. Bronner, Engagement with social media and social media advertising: The differentiating role of platform type. *Journal of Advertising* **47** 38-54. (2018) DOI: 10.1080/00913367.2017.1405754
5. H.A.M Voorveld, Brand communication in social media: A research agenda. *Journal of Advertising* **48**, 14-26 (2019) DOI: 10.1080/00913367.2019.1588808
6. L. Mura, D. Halasi, The marketing is the key factor? – Modern marketing elements are the keys of the corporate success. *Selye E-studies* **7** 1-11 (2016)
7. W.G. Mangold, D.J. Faulds, Social media: The new hybrid element of the promotion mix. *Business Horizons* **52**, 357-365 (2009) DOI: 10.1016/j.bushor.2009.03.002
8. D.H. Boyd, N.B. Ellison, Social network sites: Definition, history, and scholarship. *Journal of Computer-Mediated Communication* **13**, 210-230 (2007) DOI: 10.1111/j.1083-6101.2007.00393.x
9. B.J. Phillips, J. Miller, E.F. McQuarrie, Dreaming out loud on Pinterest: New forms and indirect persuasion. *International Journal of Advertising* **33**, 633-655 (2014) DOI: 10.2501/IJA-33-4-633-655
10. J.L. Hayes, Y. Shan, K. Whitehill King, The interconnected role of strength of brand and interpersonal relationships and user comment valence on brand video sharing behavior. *International Journal of Advertising* **37**, 142-164 (2018) DOI: 10.1080/02650487.2017.1360576
11. S.C. Boerman, L.M. Willemsen, E.P. Van Der Aa, This post is sponsored: Effects of sponsorship disclosure on persuasion knowledge and electronic word of mouth in the context of Facebook. *Journal of Interactive Marketing* **38**, 82-92 (2017) DOI: 10.1016/j.intmar.2016.12.002
12. Y. Hu, L. Manikonda, S. Kambhampati, What we Instagram: A first analysis of Instagram photo content and user types. *Proceedings of the Eight International AAAI Conference on Weblogs and Social Media* 595-598 (2014)
13. Later&Hubspot, Instagram for business in 2018: Using Instagram stories, carousels, ads, influencers, and more. 4-22 (2018)

14. R. Bačuvčík, L. Harántová, *Sociální marketing.* (Vědecká redakce nakladatelství, Prague, 2016)
15. J.G. Miles, *Instagram power: Build your brand and reach more customers with the power of pictures*. (McGraw-Hill Education, New York, 2014)
16. L. Sajtos, A. Mitev, *SPSS Kutatási és adatelemzési kézikönyv* (Alinea Kiadó, Budapest, 2007)

The effect of formalization in the enterprise

Zuzana Skorková[1,*]

[1]University of Economics in Bratislava, Faculty of Business Management, Department of Management, Dolnozemská cesta 1, Bratislava, Slovakia

Abstract. Formalization is one of the pillars of a bureaucratic company. It is manifested by a high number of formally given rules, procedures and written guidelines. These documents guide the way of employees' conducting and their activities. Precisely defined rules determine the way of work and based on rules the activities are divided among the employees of the company. They feel that the rules are readable, clear and valid for everyone. Research shows a positive correlation between the effectiveness of business activities and the degree of formalization. On the other hand too detailed rules prevent creativity, initiative and self-development. Employees require rules and written materials from the sovereign' supervisor to solve every problem, often one that the employees themselves can easily solve. Our contribution is dedicated to formalization and its impact on employee initiative to make changes in the organization. In a sample of 190 companies, we confirmed that with the increasing formalization, the initiative of employees to change has been decreasing.

1 Introduction

Formalization is represented by the rules, regulations, procedures and other written documents that organizations create to support and coordinate different and challenging tasks through behavioural regulation. The more tools to streamline the activity and the more definitions, descriptions and internal procedures the organization uses, the higher the level of formalization. A high level of formalization is typical for a bureaucratic enterprise. The opposite of this is that management trusts employees - their knowledge, skills and judgment - and does not insist on such a high degree of formalization. [1]

Different theorists characterize formalization in different ways. Weber [2] himself defined it as a way of regulating activities in an organization through formal rules. In Weber's concept of bureaucracy, there is real authority in regulations. The power of superiors is precisely defined by regulations. But Organ and Green [3] say that formalization is a way of controlling individual activities through rules and procedures.

* Corresponding author: zuzana.skorkova@euba.sk

Mintzberg [4] claims that organizations formalize employee behaviour in order to limit their variability and thus better anticipate and control it. According to him, formalization serves for easier coordination of activities. However, it also gives other reasons for formalization. These include the need for keeping the things in order. According to Mintzberg [4] highly formalized structures are above all neat and "warm the hearts of people who like to see things organized". The definition by Armandi - Mills [5] is very different, as they say that formalization is a degree of standardization of individual tasks in an enterprise.

Given the formal and informal part of the organization, we can assume two absolutely extreme cases. There are companies with a high degree of formalization that are very austere and inflexible. In highly formalized companies, there is a well-defined written procedure for almost every work activity that determines what and how to be implemented and who is responsible for what and to what extent. Especially bureaucratic companies are characterized by a high degree of formalization. However, the organizations without any formalization are also known as their counterpart. We define these as informal and very volatile, where there are no officially established rules or procedures. Formalization leads to system stability, simplifies coordination and control. Employees know exactly what to do and how to do it. From a management perspective, it is certainly easier to manage a highly formalized system than an organization without any official rules.

But organizations are based on natural human communication, which is not always formal. Formalization within an organization creates a framework and boundaries that determine the system's ability to maintain its internal order. However, too high a degree of formalization can hinder the development of employee initiative and critical thinking, which is often criticized for bureaucracy. [6-8] Formalization is measured simply by counting the pages of the documentation governing the relationships between employees or working procedures, instructions, rules within the enterprise. It can be stated that the larger enterprises are the more formalized they are. Large enterprises tend to be highly formalized because they have dozens of written rules determining the required behavior and actions of employees. On the other hand, a small, family-run business may have almost no written rules and will be considered informal, yet it can work without problems and very successfully. Nowadays, it is a great challenge for managers to determine the optimum degree of formalization that would ensure order and stability while avoiding cumbersome. As Sedlák [9] argues, non-essential, special or individual things need to be left to the initiative of its members, as a large number of regulations restrict human beings, reduce or eliminate the initiative of lower organizational units and hinder the reasonable use of all human abilities. The author also adds that a pre-formalized organization is becoming cumbersome. Therefore, the formal basis of the organization should be limited to an appropriate level, which means knowingly leaving the possibility of informal relationships and activities to emerge.

The degree of organization is closely related to formalization. The degree of organization expresses the relationship between what is already organized in the form of rules and what is dealt with spontaneously in the immediate context of the implementation of a specific activity. If the organization has a degree of organization too low, the manager often has to deal with tasks of an operational nature. Of course, this takes him time to devote to conceptual issues. The opposite of low organization is too high organization, where there are many rules in the company that are extremely detailed - which is typical of bureaucracy. It can be very time consuming to create such a rule structure. In addition, the organization is gradually becoming an overly rigid apparatus [10], which is only concerned with compliance and is unable to respond flexibly to changing conditions. This is also evidenced by research by Hofstede - Hofstede (2006) [11], where a strong negative correlation between business formalization and results orientation has been demonstrated.

Therefore, in their study they conclude that high formalization leads to a more process-oriented than result-oriented approach.

The main purpose of in-house standards is to determine a uniform and binding structure of rules of different nature according to the needs of the business. As Sedlák [9] says, organizational standards used in one organization form their system, which is hierarchically arranged. It is a purposefully arranged set of groups of organizational standards that consist of individual organizational standards. The system of organizational standards of organizations can be divided mainly from the hierarchical point of view, according to organizational levels (company, plant, etc.) and from the generic point of view.

Standards in organizations can form a comprehensive code, respectively. system of organizational standards. These organizational standards vary in scope, commitment (for all or for a specific group), validity, scope (affects all or part of a company) and diction (binding or informative. The formulation of sentences is also important, which should reflect the word order of the type: "Who - where - when - with whom - why and why." [12]

Formalization is one of the basic manifestations of bureaucracy in a company. Its advantages and disadvantages are summarized in the following table.

Table 1. Advantages and disadvantages of formalization.

Advantages of formalization	Disadvantages of formalization
Precisely defined rules determine the way work and activity are divided among the employees of the company. They feel that the rules are readable, clear and valid for everyone.	Too detailed rules prevent creativity, initiative and self-development. Employees require rules and written documentation from their supervisor to solve any problem (even one that they can solve themselves).
The scope of work is predetermined by formal rules, which is understandable for employees.	Clinging to old, established and often inefficient ways of doing business, since only these are accurately captured in procedures.
The formal rules cover all areas of life - from wage conditions to disciplinary rules. They lead to order, harmony, readability.	The process itself is more important to many employees as a result of the activity. [11] Many employees do not know the business strategy or added value that their product / service represents to the client. They are only fixed to the correct execution of the process and under all circumstances. Deviation from the result.
Research shows a positive correlation between the effectiveness of business activities and the degree of formalization. [13]	Employees do not want to take responsibility [8] and do not feel autonomous. Juillerat [14] confirmed the negative correlation between the degree of formalization and the autonomy of employees.
Standards promote a culture of continuous improvement throughout the value chain. The use of standards helps to identify and minimize risks, simplifying control.	Standards require regular revision. Practice often proves the existence of conflicting regulations.
Rules are the standards by which it is possible to continuously measure and improve the quality of services. Formal performance appraisal and a formal R&D function contributed positively for the performance of only established firms. [15] Formalization of performance appraisal has a positive effect on change in trust in leadership. [16]	Too many formal rules make the system unclear. It gives power to those who know the rules and often abuse their knowledge against those who do not master the formal rules.

Source: Own processing.

2 Methods

Our contribution is dedicated to formalization and its impact on employee initiative to make changes in the organization. In a sample of 190 companies, we confirmed that with the increasing formalization, the initiative of employees to change has been decreasing. The statistical sample (N = 190 respondents) is comprised of managers from enterprises running their business in the Slovak Republic. We conducted the research through a questionnaire data collection during the last 2 months of 2019. The data obtained through the questionnaire method present a nominal and ordinal variables. The data were analysed in PSPP statistical software. Hypotheses were tested at a significance level of $p \leq 0.05$.

3 Results

In our study, we were focused on the manifestations of formalization in the company. We were especially interested in the relationship between formalization and employees' initiative to make changes in the company. Our companies were from following business area: Transport 3,7%, Energetics 1,1%, Finance 7,4 %, Other 36,3%, Commerce 19,5%, Agriculture 2,1%, Service 21,6%, Education 2,1%, Telecommunication 2,1%, Tourism 1,1%, Health service 3,2%. Following table describes the companies by the number of employees, total number of managers and the span of control.

Table 2. Companies by the number of employees.

		Total number of employees	Total number of managers	Span of control
Mean		7302	531	13,9
Median		72	8	8,6
Std. Deviation		39746	2937	14,8
Minimum		2	1	2,0
Maximum		380300	30000	90,0
Percentiles	25	20	3	5,0
	50	72	8	8,6
	75	728	50	15,1

Source: Own processing.

Relating to the formalization we asked the managers about the degree of formalization in their organization. The results are seen in following table n. 3.

Table 3. Degree of formalization in the company

What is the degree of formalization in your company?	Frequency	Percent
There are no formal rules, procedures, internal regulations.	3	1,6
There are very few procedures, rules and internal regulations.	33	17,4
Internal procedures, rules and regulations are present in each area.	86	45,3
We have a large number of internal rules, regulations and procedures.	30	15,8
Internal rules, procedures and regulations exist for almost every activity in the company.	38	20
Total	190	100%

Source: Own processing.

Relating the changes we asked the managers if their subordinates are encouraged to submit improvement proposals, changes or innovations. The results are included in following table.

Table 4. Support employees to submit proposals for change.

Are your employees encouraged to submit improvement proposals / changes / innovations?	Frequency	Percent
Proposals for change are not desirable.	6	3,2
It is not customary for us to submit proposals for change.	51	26,8
Our superiors invite us to propose changes.	89	46,8
We are strongly motivated to propose the change and innovation.	44	23,2
Total	190	100%

Source: Own processing.

Next question was aimed to the frequency the managers submit proposal for change or improvement. Our results are illustrated in table n. 5.

Table 5. Frequency of proposal to changer/improvement at work.

When did you made a proposal to change or improve at work for the last time?	Frequency	Percent
Lat week.	44	23,2
Last month.	68	35,8
Last six months.	46	24,2
Last year.	21	11,1
I haven't given any suggestion for over a year.	11	5,8
Total	190	100%

Source: Own processing.

Last questions relating the innovation and change was aimed to the time spent for innovation or change proposal creation. Our results are included in table 6.

Table 6. Time for innovation or change proposal creation.

How much of your working time do you spend on innovation, change, implementation of news?	Frequency	Percent
I have no time for innovation.	7	3,7
I have very little time for innovation and change.	59	31,1
I have adequate scope for innovation and change.	87	45,8
I spend a lot of my time working on changes.	33	17,4
I spend most of my time on changes and innovative projects.	4	2,1
Total	190	100%

Source: Own processing

Based on these results we created following hypothesis to be tested :

H1: As formalization increases, the initiative to change decreases.
H0: There is no statistically significant association between the formalization and the initiative to change.

We used regression of summary formalization and innovation variables. The significance level was 5%. Our results are presented in following tables.

Table 7. Parameter Estimates – Testing hypothese

Parameter Estimates					
Dependent Variable: Innovations					
Parameter	B	Std. Error	t	p	Partial Eta Squared
Intercept	,644	,026	24,595	,000	,763
Formal	-,166	,058	-2,855	,005	,042

Source: Own processing.

Table 8. Hypotheses testing – Tests of Between-Subjects Effects

Tests of Between-Subjects Effects					
Dependent Variable: Innovations					
Source	Type III Sum of Squares	df	Mean Square	F	p
Corrected Model	,279[a]	1	,279	8,151	,005
Intercept	20,722	1	20,722	604,933	,000
Formal	,279	1	,279	8,151	,005
Error	6,440	188	,034		
Total	70,629	190			
Corrected Total	6,719	189			
R Squared = ,042 (Adjusted R Squared = ,036)					

Source: Own processing.

The dependence of the degree of innovativeness on the degree of formalization is significant in the negative direction (p = 0.005). Thus, the hypothesis H1 was confirmed.

4 Discussion

Hamel-Zanini [8] point out that bureaucratic structures are resistant to change. Bureaucracy destroys the value system of businesses in a myriad of ways, including slowing down problem resolution, discouraging innovation, and redirecting a great deal of time to internal policy making and "system functioning" solutions. The authors (Sherf - Tangirala - Venkataramani, [7] prove that managers in bureaucratic enterprises themselves do not support innovation by their subordinates because they have to follow a system of formal rules and communicate every change with senior management. But if managers had real autonomy in decision making, they were up to 30 percent more likely to support innovation, ideas, and demand feedback from their subordinates. In his study, Cochran [17] also deals with bureaucracy and innovation. It argues that in promoting innovation, it is imperative to allow people to make mistakes, but this does not allow or eliminate bureaucracy through the introduction of a high level of formalization. Hamel [18] proved that the compromise between control and freedom is possible. In our research, we looked at the relationship of formalization and employee initiative to make changes. We have found that almost 36% of businesses report a high level of formalization, while reporting large or high levels of formalization. full coverage of its activities by internal procedures and formal documents. Almost 35% of managers confirmed that they have no time, respectively. very little time to propose amendments. In addition, up to 30 percent of all businesses reported that proposals for change are not desirable and it is not customary for us to submit proposals for change. Through regression analysis, we have shown that, as the level of formalization increases, the rate of innovation and change in the enterprise is decreasing. Therefore, we agree that excessive formalization as one of the pillars of bureaucracy is an obstruction in change and development in the enterprise.

Conclusion

Our contribution is dedicated to formalization and its impact on employee initiative to make changes in the organization. In a sample of 190 companies, we confirmed that with the increasing formalization, the initiative of employees to change has been decreasing. We have found that almost 36% of businesses report a high level of formalization. Almost 36% of companies confirmed full coverage of its activities by internal procedures and formal documents. Almost 35% of managers confirmed that they have no time, respectively very little time to propose any change or improvement. In addition, up to 30 percent of all businesses reported that proposals for change are not desirable and it is not customary for us to submit proposals for change. We can state that excessive formalization as one of the pillars of bureaucracy is an obstruction in change and development in the enterprise.

The research results published in the article are the results of the project of the grant scheme VEGA 1/0412/19 Human Resource Management Systems in the 4. Industry Era.

References

1. J. Váchal, M. Vochozka, *Podnikové řízení (*Praha: Grada, 2013)
2. M. Weber, M. *Autorita, etika a společnost: pohled sociologa do dějin (*Praha: Mladá fronta, 1997)
3. D. Organ, Ch. Greene, The Effects of Formalization on Professional Involvement: A Compensatory Process Approach. *Administrative Science Quarterly* **26,** 2 (1981)
4. H. Mintzberg, *Structure in Five: Designing Effective Organizations.* (Englewood Cliffs NJ: Prentice Hall, 1983)
5. R. B. Armandi, W. Mills, W. Edgar jr., Bureaucratic and Personalized Strategies for Efficiency and Organization: An Investigation of Structures and Efficiency in a Set of 104 Profit-Seeking Firms, *The American Journal of Economics and Sociology* **44**, 3 (1985)
6. R. Gulati, The Soul of a Start-Up. [online]. *Harvard Business Review,* July-August, 2019. [accessed: 2020-02-02]. Available at: https://hbr.org/2019/07/the-soul-of-a-start-up
7. N. Sherf, E. Tangirala, S. Venkataramani. Research: Why Managers Ignore Employees‘ Ideas. *Harvard Business Review,* April, 2019. [accessed: 2020-02-02]. Available at: https://hbr.org/2019/04/research-why-managers-ignore-employees-ideas
8. G. Hamel, M. Zanini, The End of Bureaucracy. *Harvard Business Review,* November-December 2018. [accessed: 2020-02-02]. Available at: https://hbr.org/2018/11/the-end-of-bureaucracy
9. M. Sedlák, *Základy manažmentu. 2. preprac. a dopln. vyd.* (Bratislava : Iura Edition, 2012)
10. P. G. Aquinas, *Organization structure and design, applications&challenges*, (New Delhi, Excel Books, 2008)

11. G. Hofstede, J. Hofstede, *Kultury a organizace : software lidské mysli : spolupráce mezi kulturami a její důležitost pro přežití.* (Praha : LINDE nakladatelství s.r.o., 2007)
12. L. Neščáková, L. Marelová *Vnitřní závazné předpisy zaměstnavatele a jiné pracovněprávní úkony*, (GRADA Publishing, a.s., 2013)
13. E. Basol, O. Dogerlioglu, Structural Determinants of Organizational Effectiveness *Journal of Organizational Management Studies*, (2014)
14. T. Juillerat, Friends, not foes?: Work Design and Formalization in the Modern Work Context. *Journal of Organizational Behavior* **31**, (2010)
15. P. Neves, A. Lopez, M. Cunha, A High-Growth Firm Contingency Test of the Formalization-Performance Relationship. *Journal of Small Business Management* **57**, (2019).
16. C.A. Belsito, C.R. Reutzel, SME employee performance appraisal formalization and trust in leadership change, *International Journal Of Organizational Analysis* **28,** (2019)
17. T. Cochran, How I Led Change in the U.S. State Department Bureaucracy. *Harvard Business Review*. January 04, 2017. [accessed: 2020-02-02]. Available at: https://hbr.org/2017/01/how-i-led-change-in-the-u-s-state-department-bureaucracy
18. G. Hamel, General Management, First, Let´s Fire all the Managers, *Harvard Business Review, December 2011* [accessed: 2020-02-02]. Available at: https://hbr.org/2011/12/first-lets-fire-all-the-managers

Business ideas in start-ups

Štefan Slávik[1*]

[1]University of Economics in Bratislava, Department of Management, Dolnozemská cesta 1/b, 852 35 Bratislava, Slovakia

Abstract. Start-up is a modern entrepreneurship form designed to realize original business ideas, mostly based on new technologies and the Internet. It evolves in the development cycle, which is determined by the business idea development cycle and the financing cycle. The purpose of the paper is to describe and analyse the business idea. The business idea is characterized by its content, circumstances of its origin, degree of originality and evidence of this originality. Start-ups are dominated by business ideas based on the application of information and communication technologies, the business idea is most often created by combining professional and business experience, but its originality is from the international point of view only average and the level of legal protection is quite rare.

Introduction

The transformation of a business idea into a satisfying need can best be studied on the basis of a start-up which is a small, young, beginning enterprise with a growth perspective [1, pp. 63-72], [2, p. XVII], [14, pp. 27], [17, pp. 10-11]. Its relatively simple structure, frequent metamorphoses and rapid feedback provide real-time living laboratory conditions. Knowing this new business phenomenon should increase the likelihood of business success. The very creation of an idea, its transformation into a product or service, its acceptance by customers is a complex process, it has many dead ends, inefficient turns, returns and reversals, failures topped by destroyed resources, frustrated entrepreneurs, disappointed investors, embittered customers, sometime by moderate success and rarely the business feat of the decade [6, pp.6-35].

Creation of a business idea

Starting a business is not easy and it is even harder to run [13, pp. 1-17]. Although a business person has a good idea, he is discouraged by situational and psychological

* Corresponding author: stefan.slavik@euba.sk

reasons [5, p. 10]. Situational causes are the need for regular income to secure the family, fear of losing capital or lack of capital. Psychological background has risk aversion and doubts about one's own abilities. Despite these hurdles, many people decide to do business because they are driven by incentives that push them situationally into self-employment. The most common are job losses, different opinions on the content and performance of work in the current employment relationship, personal disagreements in the workplace or unemployment, e.g. due to loss of qualification [4]. Other people are drawn into a business for more positive reasons. They desire for independence, recognition, personal development and wealth too. These factors have a psychological origin because they are derived from character traits.

Finding a good business idea is not just a matter of chance and fortune. There are also systematic ways to work out the idea. The entrepreneur must have the skills, abilities and knowledge necessary for the business. Experience with an industry is very valuable [8, pp.1-17].

Good business ideas are based on commercial opportunities created by market needs. *Opportunities can be created* through radical or incremental product or market innovations, with little or no evidence of market need. *Opportunities can also be spotted* in finding unsatisfied customer needs. Both approaches are only successful if they are connected with the needs of customers (Fig. 1). The process of creating ideas has three stages:

1. An entrepreneur is exposed to as many diverse and different ideas, influences and people as possible and aware of them.
2. An entrepreneur recognizes market opportunities at observing the daily lives of customers and asks them whether their needs can be met better or otherwise. He doubts and experiments. He questions the status quo and asks if things can be done differently.
3. He formulates and reformulates the business idea to make it a viable business model. Not all products work since their inception, and not all ideas are immediately commercially viable, and therefore an entrepreneur is experimenting to improve a product or service and business model [7, pp. 200-227].

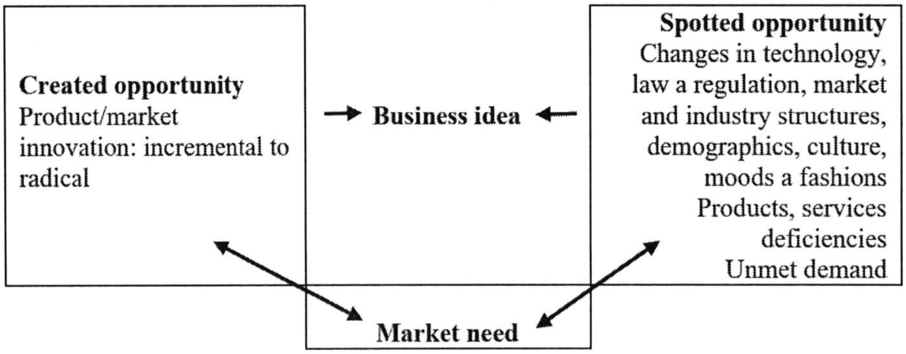

Fig. 1. Generating a business idea [5, p. 52]

Creation of idea and creativity is encouraged by the discovery skills, they are associating, questioning, observing, experimenting and networking. The whole process takes some time. It is a time of thinking, deepening, incubating ideas. Generating ideas is like a number game. The more ideas arise, the more likely one of them will be viable. It is appropriate to distinguish between techniques that offer quantity and those that increase quality. Some techniques are more suited to spotting opportunities, while others create opportunities. The most common and not too demanding techniques are brainstorming, SLEPT analysis, thinking about the future, mind maps, analogy, attribute analysis and gap analysis.

Best business ideas generate high profits and are burdened with low risk [11]. The occurrence of such opportunities is rare, but they can be assessed according to the following criteria, which are derived from entrepreneur aspirations and capabilities and the commercial viability of the idea [5, p.66)]:

1. Identified market need/gap.
2. Identified customers and marketing strategy.
3. No or few existing competitors.
4. Not easily copied.
5. Growing market.
6. High margins.
7. Low fixed costs.
8. Low funding requirements.
9. Financeable (if not sufficient resources, the project needs to be able to attract finance).
10. Identifiable risks that can be monitored and mitigated.
11. Managerial skills that can be leveraged.
12. Scalability.

Many starting entrepreneurs mistakenly believe that their ground breaking business idea will be equally understood and accepted by customers, and that innovation is so superior that it will sell itself [10, pp.23-32], [9, pp.564-581]. From the idea to its realization and then the acceptance and purchase of the product or service, however, there is a long way to go. A survey in which 120 starting businessmen have shared their experiences has provided insights into the most common business start-up errors, potential customer objections, and appropriate sales models for start-ups [12]. More than half of the surveyed companies designed the final product before receiving customer feedback. The development of a complete product without confrontation and customer feedback was lengthy and yet did not deliver the expected results. Customer notions and requirements must be taken into account from the first day of product development. Although some entrepreneurs began selling early, their enthusiasm and complacency prevented them from accepting criticism and making the changes demanded by customers. Successful entrepreneurs recommended patiently listening to customers, explaining what they wanted to do, offering them their ideas and constantly verifying them. Block et al. [3, pp.1-32] write that the conditions at the birth of a business tend to affect companies over very long periods. The consequences of initial strategic decisions are usually very lasting.

Objectives, research sample and methods of research

The main objective of the research is to deepen the knowledge on business ideas of start-ups, because based on the studied literature, a business idea that is not matched with market need is one of the main causes of start-up failure. The partial objectives are:

1. Know the content of the start-up business idea.

2. To know the causes of origin of the start-up idea.
3. Know the originality of the start-up business idea.

The research sample contains thirty start-ups operating in Slovakia. The original sample size was almost a hundred start-ups and was a follow-up to research in 2015, 2016 and 2017 [16], but it had to be reduced due to incomplete data and the reluctance of start-uppers to devote their time to collaborating with researchers. The common feature of the researched start-ups is approximately the same age and location in large cities with developed infrastructure and start-up scene.

The qualitative research of the business idea is based on induction (deducing a general conclusion based on individual knowledge) and exploring unique and non-recurring phenomena. The advantage of a non-numerical solution is high validity, detailed and in-depth knowledge, but it is associated with a disadvantage that may be of low reliability, because the result may not be generalizable. The qualitative data are studied to find some structure and regularity.

Research results

Nearly half of the start-ups are active in the ICT industry. The rest of the star-ups operate in very diverse industries, e.g. industrial production, agriculture, other research, production of machinery, services, arts, entertainment, recreation and others.

The business ideas of the studied start-ups are implemented as software applications (66%) and products (33%), which are predominantly material in nature. They can be divided into the following groups:
- collecting, pooling and producing specific information and fact sheets, e.g. support for the compilation of CVs, the search for human resources with specific skills and capabilities for temporary use,
- associating things and various objects with information on their properties, operation and use, e.g. information on the status and location of delivery of ordered goods,
- meeting traditional needs in a simpler, faster and more convenient way, e.g. cashier system for restaurants
- convenient, simple and safe access to objects, buildings, apartments, cars, e.g. e-key,
- promoting and developing ideas from other authors, e.g. providing infrastructure for start-ups,
- an original, usually tangible product that satisfies an existing need in a new or more efficient way, e.g. a highly sensitive personal earthquake warning device or a new version of the old service.

The circumstances and causes of origin of the idea are reflected in the content of the business idea. They are very diverse, but here are a few examples:
- the original idea did not have sufficient business potential and has therefore undergone one or more changes (pivot), e.g. one application carries several similar services; from hot water wells to deep wells for oil and gas; from photovoltaics to e-mobility,
- the need for extensive and readily available information, e.g. on operated technology, machinery, devices,
- demand for specialists, e.g. for HRM

- establishing a sense of security, e.g. face and character identification, earthquake warning,
- non-transparent situation in some environment, e.g. real estate market, logistics of orders and deliveries of ready meals,
- personal experience of the entrepreneur, e.g. physical or mental discomfort; urgent lack of something, e.g. information, HR,
- the original services are lengthy, inconvenient, complicated, unautomated and not linked to other services or information systems.

The originality of the business idea is predominantly implicit, officially unconfirmed, and considerably fictitious (90% of the sample). The founders' arguments in favour of the originality of the idea are diverse and based mainly on personal experience, guesses and imperfect knowledge of the competition:
- recognition of an industry and entrepreneurial experience which suggests that the start-up under examination is the only company in the industry,
- the first company of its kind in a certain area, e.g. Central Europe,
- national originality, the only company in country,
- the originality of the idea is small, but the idea is implemented quickly and consistently,
- considerable interest of foreign clients,
- very low price, even 100 times lower than the competitive price,
- technological complexity, but without patent,
- great adaptation to customer needs,
- there are several competitors, but the start-up has strong and loyal customers,
- the product is similar to competing products but just is different,
- the product is complex and complicated or hybrid.

Explicit, officially confirmed originality, patent protection occurred only in 10 % of start-ups in the examined sample. The degree of originality of the whole set of business ideas measured on a five-point scale is 3.87, thus approaching originality at European level.

The development of the business idea was captured in the following phases: 1 - idea/ concept/research, 2 - product development, 3 - prototype of product /testing, 4 - first earnings, 5 - growing earnings. Most business ideas are placed in phase 4 (first earnings) or 5 (growing earnings), several ideas are in phase 3 (prototype of product /testing), the only idea is in phase 1 (idea and concept). The average development phase measured on a five-point scale is 4.1, hence at the level of the first income. Phases 4 and 5 are characterized by continuous product innovation and improvement, so far independently and without exit. Sales revenues are growing, but not fast, rather slow, so start-ups ordinarily do not attract strong investors. Start-ups are beginning to penetrate foreign markets and the challenge is not only to maintain growth but also to speed up significantly.

The start-up financing cycle is captured in the following phases: 1 – pre-start-up capital (angel phase, idea, no product), 2 - start-up capital (seed phase, product work, prototype made/realized, interest-finding), 3 - capital for initial development and further growth (series A/B phase, 1st, 2nd round, investment in a company that already has customers, generates revenue), 4 - developing capital (3rd round, mezzanine capital), 5 - IPO (public the market). Most business ideas (44 %) are occurred in phase 3 of the investment cycle (capital for initial development and next growth). 13 % of ideas are in phase 1 (pre-starting capital), 30 % of ideas are in phase 2. (start-up capital), 10 % of ideas are in stage 4. (developing capital) and 3 %

of ideas are in phase 5. (IPO). The average phase of the funding cycle, measured on a five point scale, is 2.6, hence between start-up capital and initial and development capital.

About 20% of start-ups are self-financing and want to do business with their own resources only. The rest relies on angel and venture investors who are expected to have business experience, advice and business contacts in addition to money. Start-ups do not receive any bank loans and grants, except for the start-up, which received loans from the SZRB and grants from EU projects. Flemio accelerator provides pre-launch capital of 50,000 € and the venture investor Neulogy grants a normal tranche of 200,000 €. The process of raising external capital is lengthy. Domestic venture capital resources are very limited and institutional capital from foreign VC entrepreneurs and strong VC funds is almost unavailable. Start-ups invest mainly in development of prototypes and marketing. However, marketing is a proxy term for advertising that serves to promote brand, product, and service awareness. One-off investments from VC investors are not enough and must be continuously replenished according to the results of product development. Income from sales is not enough to finance current operations and product development, so most start-ups need external investment support.

Discussion

Information and communication technologies dominate the business idea. Software applications are easier to "manufacture" and sell as tangible products, they do not need a manufacturing base and many collaborators, and are in line with technology trends, lifestyles and the spirit of the times. The causes of business idea are trivial, e.g. something is missing or not working, some process is lengthy or inconvenient, but an imaginative entrepreneur sees more sophisticated content in some lack, imperfection or absence e.g. lack of specific information or confusing and disordered information. The conditions of origin of the idea are mainly personal experience and sensitive observation of the surroundings, but is it enough for an original business idea? Often declared exceptional originality of the business idea is a fundamental problem of the researched start-ups. Unconfirmed assumptions, impressions, and various subjective arguments prevail, which later cause failure and disappointment. The notion of self-exceptionality is usually the result of a lack of knowledge of the idea/topic and of ignoring real competition on a wider European and global scale.

It takes too long to develop a business idea. Successful start-ups in the world will show their viability within three years of being established and then scale or exit. In almost five years of existence, the examined start-ups have reached the beginning of phase 4 of the business idea development (4.1: first income) and show no signs of exponential growth. In the funding cycle, they arrived between phases 2 and 3 (2.6: between start-up capital and capital for initial development and further growth). There is a clear difference in the progress of the business and investment cycle. The founders are missing a large amount of several million euros and investors do not trust the underdeveloped product, which is not confirmed by market interest.

Conclusion

Entrepreneurship is becoming a form of self-realization of creative and ambitious people to whom traditional employment relationship does not bring personal and professional satisfaction and the desired standard of living. Such people not only contribute to social and economic progress, but also create jobs for themselves and their co-workers. Entrepreneurship then becomes an expertise and qualification of its kind and one of its key themes is the emergence and development of a business idea. However, the degree of recognisability of the process of creating and developing a business idea is questionable. It is a creative process which excessive algorithm would lead to unification of results and thus to a counterproductive solution. Research should focus more on recognizing the favourable conditions that stimulate original and unconventional solutions and on distinguishing "signal-to-noise" [15], hence the ability to perceive principal trends and not slip down into banalities and false opportunities. It is essential for an entrepreneur not to succumb a captivation by the technical and functional excellence of his idea and to distinguish technical advance from commercial applications. The subject of research should therefore also be cognition of how the technical solution is designing, adjusting, adapting for the needs of business use and implementation. It is envisaged that parameters and criteria could be identified that measure the business potential of the idea and be adapted to the specific development phase of the idea and its content.

References

1. Blank, S., Why the lean start-up changes everything. *Harvard Business Review*, **91,** 5, pp.63-72. (2013)
2. Blank, S., Dorf, B., *The start-up owner's* manual. The step-by-step guide for building a great company. K&S Ranch Publishing Division (2012)
3. Block, J. H., Kohn, K., Miller, D. and Ullrich, K., Necessity Entrepreneurship and Competitive Strategy. *IZA Discussion Papers*, No. 8219, Bonn: Institute for the Study of Labour (IZA) (2014)
4. Blumberg, B. F. and Pfann, G. A., Roads Leading to Self-Employment: Comparing Transgenerational Entrepreneurs and Self-Made Start-Ups. *Entrepreneurship Theory and Practice*, **40**, 2, First published: 06 March 2016 (2016)
5. Burns, P., *New venture creation. A framework for entrepreneurial start-ups.* Palgrave MacMillan (2014)
6. Davila, A., Foster, G., He, X. and Shimizu, C., The rise and fall of start-ups: Creation and destruction of revenue and jobs by young companies. *Australian Journal of Management*, **40**, 1, pp. 6-35. (2015)
7. Foss, N. J. and Saebi, T., Fifteen Years of Research on Business Model Innovation: How Far Have We Come, and Where Should We Go? *Journal of Management*, **43**, 1, pp.200-227 (2017)
8. Fritsch, M. and Wyrwich, M., Regional knowledge, entrepreneurial culture, and innovative start-ups over time and space - an empirical investigation. *Small Business Economics*, **51**, pp.337-353, (2018)

9. Hyytinen, A., Pajarinen, M., and Rouvinen, P., Does innovativeness reduce start-up survival rates? *Journal of Business Venturing*, **30**, 4, pp.564-581 (2015)
10. Kopera, S., Wsendybył-Skulska, E., Cebulak, J., and Grabowski, S., Interdisciplinarity in tech start-ups development - case study of 'UniStartApp' project. *Foundations of Management*, **10**, (2018)
11. Klačmer Čalopa, M., Horvat, J. and Lalić, M., Analysis of financing sources for start-up companies. *Management*, **19**, 2, pp.19-44 (2014)
12. Onyemah, V., Rivera Pesquera, M., and Ali, A., What entrepreneurs get wrong. *Harvard Business Review*, **91**, 5, pp.74-79, (2013)
13. Pettersen, I. B., Aarstad, J., Høvig, Ø. S. and Tobiassen, A. E., Business incubation and the network resources of start-ups. *Journal of Innovation and Entrepreneurship,* **5,** 7, pp.1-17 (2016)
14. Ries, E. *The lean start-up*. New York: Crown Business (2011)
15. Silver, N., *Signál a šum*. Praha – Litomyšl: Nakladatelství Paseka (2014)
16. Slávik, Š., Hagarová, R., Ljudvigová, I., and Zagoršek, B., Business model and team as preconditions of a start-up viability, Entrepreneurship and Sustainability Issues. **6,** 3, (2019)
17. Thiel, P. *Zero to one*. Notes on start-ups, or how to build the future. New York: Crown Business (2014)

Controlling in Manufacturing Enterprises Operating in the Territory of the Slovak Republic

Vladislav Sochanič[1,*], *Paulína* Paprskárová[1]

[1]Faculty of Business Management, University of Economics in Bratislava, Faculty of Business Management, Department of Management, Dolnozemská Cesta 1/b, 852 35 Bratislava, Slovak Republic

Abstract. In recent years, control has become increasingly important than the last management function. The growing interest in this function is also due to the economic crisis, which has led many companies to be on the verge of bankruptcy or to force their business to cease completely. Companies that wanted to survive in a turbulent business environment were forced to streamline processes as well as redistribute their resources more efficiently. The main aim of the paper is to find out the importance of control perceived by employees and managers and the introduced changes in control in manufacturing companies in the Slovak Republic.

1 Introduction

In today's business environment, businesses should be constantly improving, working to increase competitive advantage and adapting constantly to market and customer needs. The aim of the company is to achieve specific results in a specific period and maximize the current value of the company. In order for managers to take the right measures to help them achieve their business goals, they need to know the individual processes in the business and their parts. Control focuses on identifying individual parts of business processes, their analysis and performance measurement and subsequent comparison of measured and planned values. If the process does not deliver planned results, the manager takes corrective action. Through continuous control, the company can continuously improve and streamline the company's business processes. However, the right measure must be found between exaggerated and ineffective control and lack of control or its absence. Understanding current changes and trends in internal control can help businesses adapt to the business environment and gain or maintain their competitive advantage.

2 Theoretical background

Manufacturing is a constantly evolving industry that needs to be sufficiently competitive and adaptable to market changes. For this reason, production and management

* Corresponding author: vladislav.sochanic@euba.sk, paulina.paprskarova@euba.sk

have to be seen as a means of making production more efficient and more flexible for the market and conditions in it.

Production management deals with complex management not only of production, but also pre-production and post-production stages or logistics itself. According to Leščišin et al. (2008) it is necessary to put the focus on the management of the whole production system, which should direct all tangible and intangible processes in the company - development, design and other pre-production activities, production processes and post-production services.

Thanks to control and controlling, it is possible to improve individual business processes and make their results more effective, which can be a major competitive advantage of the company. For example, ISO concepts, quality management or Total Quality Management (TQM) and Kaizen are also based on continual improvement and improvement of currently applied processes, which reduces costs, increases efficiency or brings greater benefits to the process. As Merchant (1998) states, there may be significant problems in checking if an incorrect control deals with wrong objects. Managers must be able to recognize these problems and take corrective action.

Therefore, top management's view of control should include a greater dispersion of factors, according to Horovitz and Webb (2015):

- the degree of detail found in the inspection,
- frequency of control,
- degree of control formalization,
- the degree of centralization of control,
- degree of sophistication of control,
- emphasis on control and its outputs towards top management.

The management levels have the above-mentioned factors at different levels of managers. For example, top managers generally do not carry out detailed employee checks at line level. Bates et al. (2005) define top management as a relatively small group of people who is responsible for general management tasks and performs strategic planning for the enterprise, takes responsibility for the direction of business and sets long-term goals and tasks for the enterprise. It constantly monitors the business environment and sets the vision and future direction of company. Middle management has the task of managing specific functional areas, often a larger group of people than in top management. The middle management implements the decisions of the top management and is responsible for the mid-term planning of the company. It transforms strategies from top management to sub-goals and action plans for lower levels of management. It monitors the business environment that affects the functional area. Managers of first line typically control small segments in a specific department or function. They are focused on the fulfillment of daily tasks and short-term goals assigned to higher levels of management.

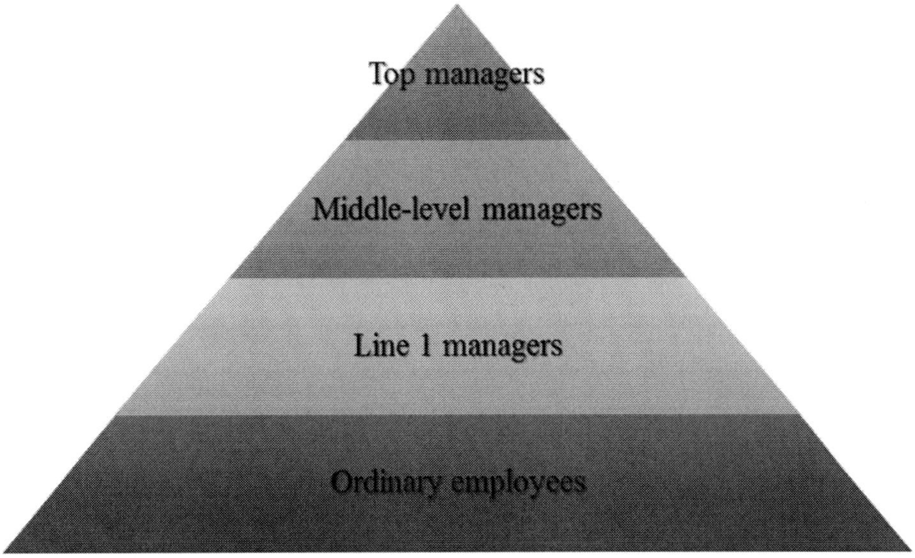

Fig. 1. Levels of management by Robbins and Coulter
Source: *ROBBINS S. P., COULTER M. (2012). Management. p. 6*

3 Materials and methods

The main aim of the paper is to find out the importance of control perceived by employees and managers and the introduced changes in control in manufacturing companies in the Slovak Republic.

Scientific methods such as analysis, synthesis, induction, deduction, abstraction, comparison and generalization were used.

The survey data were collected between autumn 2018 and spring 2019 using a questionnaire created in google forms. The questions in the questionnaire were formulated on the basis of previous research and expert interviews. The rating scale was set from 1 to 6 to avoid mean values. Overall research was not only focused on business units operating in manufacturing sectors, but also on businesses operating in other industries. Altogether 462 questionnaires were collected, of which 70 companies are active in the manufacturing sector. For the purposes of this article, only selected research enterprises operating in the manufacturing sectors have been analysed. The object of the investigation is business units operating in manufacturing industries as defined by NACE operating in the territory of the Slovak Republic. The subject of investigation is control, its importance and intensity, as well as changes in control.

The research sample contains 70 respondents, the first question concerned the level of management of respondents' job positions in selected enterprises, which is also divided in Figure 1. The respondents are:

- 42.85% top managers,
- 21.42% mid-level managers
- 12.85% 1st line managers
- 22.88% non-managerial workers (economist, controller or accountant).

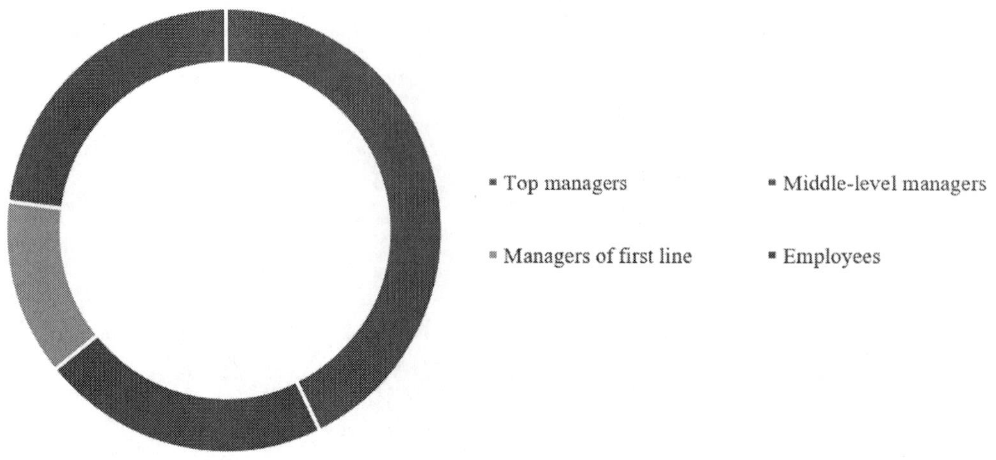

Fig. 2. Levels of managers
Source: *Own work*

The average number of years which managers have worked in individual positions in chosen businesses was 12.43 years for top managers, 9.64 years for middle managers, 11.66 years for first-line managers and 12.33 years for non-managers. The average number of years managers have worked at each level of management was found mainly in addition to qualitative answers, as the overview of business processes and control is usually given to higher managers who work in the company or in a given position for a longer period of time.

Generally, the manufacturing enterprises had the legal form of a limited liability company (53 enterprises), the second most frequently represented legal form was a joint-stock company (13 enterprises) and the rest of the enterprises were a cooperative and self-employed persons. Figure 2 shows respondents in regions of Slovakia. Most of them were from Trenčín, Trnava and Bratislava regions. The Košice Region was the least represented. Data were taken from portal finstat.sk.

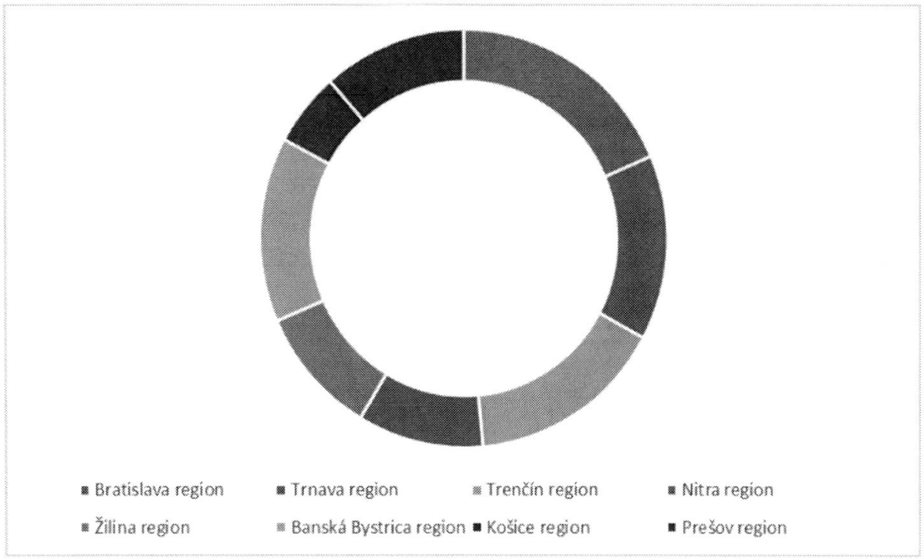

Fig. 3. Respondents by region
Source: *Own work*

4 Research and discussion

The identification questions were followed by a question on the importance that respondents attach to control. The rating scale was determined from 1 (not important) to 6 (very important). The average assigned value was 5.34 (arithmetic mean: μ = 5.34285714286; absolute deviation: 55.2; average deviation: 0.788571428571 ; variance: 0.968163265306; standard deviation σ = 0.983952877584). Figure 4 shows the average values assigned by respondents at each management level.

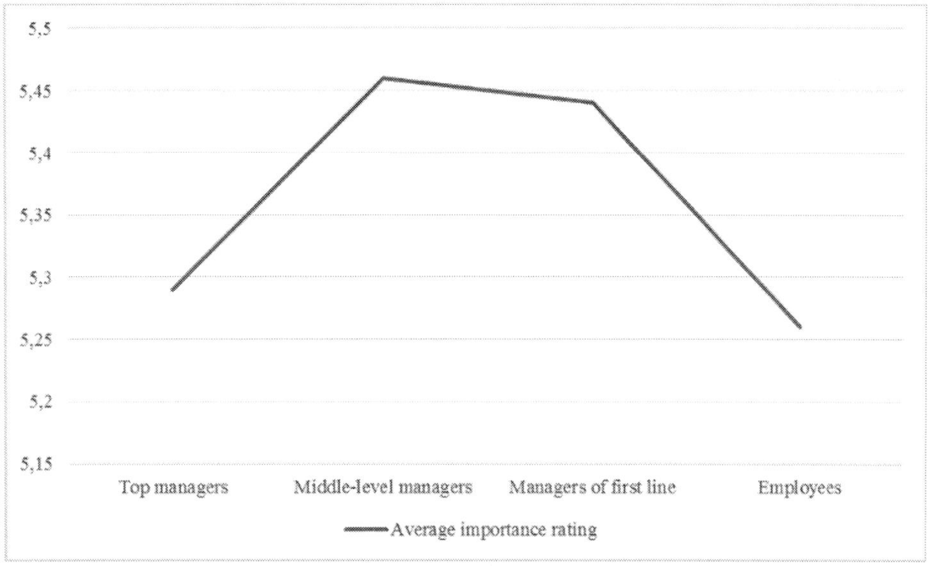

Fig. 4. Average values added to importance of control by managers and non-managers
Source: *Own work*

After assigning a score on the rating scale, another question was aimed at justifying the respondent's attitude. This was therefore a justification given to the importance of control. The lowest value (minimum: 2) was reported in two cases by the business director and the financial director. In one case, it was justified that properly set standards and procedures do not require extensive controls. In the second case, it was a company in extinction, where both internal and external controls were almost negligible compared to controls in previous stages of the company's life cycle. Maximum values (maximum: 6) were reported by up to 42 respondents. Most frequently, respondents stated:

- eliminate or minimize errors in selected business processes,
- preventing errors and mistakes,
- prevention of penalties and fines,
- increase efficiency or quality of selected business processes,
- awareness and truthfulness of information obtained through control,
- development of the company and economic prosperity of the company.

On average, scores of 3 and 4 justified that the company has stable employees who do not require a high degree of control, a low number of employees in the company, or claiming that excessive control also carries negatives, so it does not give it excessive importance.

Subsequently, we investigated whether the importance that managers and employees attach to control also reflected the introduction of changes in selected manufacturing companies and what changes they were and how often they reviewed the effectiveness of the inspection process and the need to introduce changes therein.

Up to 11 respondents stated that they review the review process itself and re-evaluate its benefits as a rule at shorter intervals than once a month. It should be noted, however, that in these cases it was a control of parts of the control process, not the overall control process. At the same time, all respondents had the power to change the review process and introduce changes. A further 12 respondents said they reviewed the review process once a month and two of the respondents had the power to change only some parts of the review process. The rest of the respondents (10) could interfere with the entire control process. Once a quarter, up to 17 respondents review the effectiveness of the review process, of which 7 have the right to change only some parts of the review process, and 10 respondents can introduce changes to all parts of the review process. 21 respondents review the review process once a year, of which 12 can only introduce changes to some parts of the review process, 6 respondents can change all parts of the review process, and 3 have no authority to change anything in the review process. At more than one year intervals, the review process is reviewed by 9 respondents, while only 3 of them can introduce changes to the review process and other respondents have no authority to do so.

Another question was aimed at identifying specific changes that have been made to the review process over the past year. Of the respondents, 19 reported that no changes in their control processes had been implemented in their company over the past year. The rest of the respondents (51) reported that they had made several changes to the review process over the past year. A frequent reason was the introduction of changes due to legislative changes, changes in standards or the introduction of changes due to the increasing intensity of external control (financial administration, other audit bodies), which identified shortcomings in current business processes and their control. At the same time, some changes in the inspection, in particular the intensity of the inspection or the change of the inspected object, were introduced to prevent fines and penalties. Another common reason for introducing changes was technological changes in production and automation. For this reason, the review processes were either re-created or designed, or changes were incorporated into the current review process. The last of the reasons were employees and their turnover or lack of qualifications. Respondents reported that there is a lack of skilled

https://doi.org/10.1051/shsconf/20208301064

labor in the current labor market, often shifting between employers. For this reason, it is necessary to constantly train new employees and constantly intensify their monitoring in order to maintain the quality of business processes at the required level.

5 Conclusion

Managers and employees attach great importance to auditing in general. This is to a certain extent also due to the high quality requirements of the companies operating in the manufacturing sector. Around three quarters of the enterprises surveyed have introduced changes in the review process in the last year. The main reasons were legislative changes, increasing control intensity from external institutions, changes in technology and robotization.

The specifics of the current state of the labor market and the high demand for qualified employees in manufacturing industries in the Slovak Republic are causing changes in control processes in manufacturing companies. Managers are forced to increase the intensity of control of such employees in order to maintain or increase the quality of individual production processes in companies.

References

1. B. Bates et al. *Business Management: fresh perspectives.* (2005)
2. J.H. Horowitz, P. Webb. *Top Management Control in Europe.* (1980)
3. M. Leščišin et al. *Manažement výroby.* (2008)
4. K.A. Merchant. *Modern management control systems: Text and cases.* (1998)
5. S. P. Robbins, M. Coulter. *Management,* (2003)

Place and Role of Personnel Audit in Slovak companies in the Industrial Revolution Stage 4.0 and 5.0

Miroslava Szarková[1,1]

[1]Faculty of Business Management, Department of Management, University of Economics in Bratislava, Dolnozemská cesta 1/b,852 35 Bratislava, Slovak republic

Abstract. The need to effectively use the intellectual and physical potential of human resources increased especially in the phase of the fourth industrial revolution 4.0 and the onset of the fifth industrial revolution 5.0. At these stages the quality of human resources as well as their effective management create a basis for the very existence of the company and at the same time represent a "hatchery" of employees for the future development and competitiveness of the company. In these developmental stages sophisticated information on human resources, their management and leadership as well as on possible reserves in personnel processes gain great importance. This information can be provided to corporate management by a personnel audit. Based on the results of primary and secondary research, the aim of the paper is to give an overview of the use of personnel audit in companies operating in the Slovak Republic. Keywords: personnel audit, human resources, human resources management, intellectual capital.

1 Introduction

Personnel audit is not only a highly actual but also quite complex topic and is not always viewed by corporate management in terms of perception of its need and usefulness, but rather perceived as a method whose results and findings force them to make changes in human resources and management, sometimes even changes that are unpleasant for them [1]. However, as recent developments show, often also necessary. They also force corporate management to learn to react/make decisions in the present with respect to the future of the company. Applying this "new" approach to corporate management means in all areas of company management to abandon established, often experience-based management methods [2, 3] and to learn to use sophisticated procedures and methods.

The use of personnel audits by company management is the result of developments in the labor market that have arisen as a result of the onset of the Industrial Revolution 4.0. It changed not only the entrepreneurial and business environment, but also the perception of personnel work, the positions and roles of employees in the work process and forced a change in the approach to their management. The field of personnel work, especially the perception of employees/staff of the company, was for a long time in theory and practice

[1] Corresponding author: miroslava.szarkova@euba.sk

understood only as "complementary activity", which contained the basic maintenance of administrative processes in the personnel area. This understanding of personnel work was also fostered by the economic crisis, which multiplied the workforce on the labor market and multiplied the illusion of perception of the workforce through its number and not through differences, uniqueness and quality. However, the onset of the Fourth Industrial Revolution has shown that the administrative processes used so far in personnel work have been inefficient and ineffective in terms of the established company needs. For time-compliant decisions, corporate management needed sophisticated information about the workforce, its quality, mental, physical and performance potential [4] and the human resources simply did not have such information. This, as well as other moments related to the massive onset of information and communication technologies and the deepening globalization processes, created the conditions for the emergence of a personnel audit [5,6].

Establishment and content definition of personnel audit reflected the need to obtain information on human resources in the company and on the application of basic legislative and administrative procedures in personnel work, personnel administration and finally in personnel management [7,8]. This basic characteristic is reflected in all definitions of personnel audit, which they define as a comprehensive tool both for monitoring and obtaining relevant information on compliance with legislative procedures in labor relations [9] and activities related to employment of people, with compulsory care of employees and with recruitment, storage and updating of personnel information and documents, as well as the provision of information to company directors and relevant institutions [10, 11, 12]. Its objective is to objectively describe the current state of society and to determine whether specific areas of personnel practice are sufficient and whether they comply with legal standards [13] and to identify and describe deviations that have arisen in this area [14-17 and others].

2 Subject of the research and characteristics of the examined sample

The subject of the submitted study is the personnel audit, its place and functions in the process of human resources creation and their effective management. The basis of the research is the main assumption that personnel managers, personnel departments, human resources departments, corporate management in companies operating in the Slovak Republic do not perceive personnel audit, its tasks and importance for improving human resources work and their management in the full range of currently offers them to streamline personnel work and personnel activities. Therefore, the main objective of secondary research was to systematize current knowledge about personnel audit and to find out in primary research whether companies operating in the Slovak Republic carry out personnel audit and whether they use the information obtained in it to improve the human resources of the company and their management.

The examined sample consisted of personnel managers and managers from small, medium and large companies with 100% Slovak capital participation operating in the SR, personnel managers and employees from the group of small and medium companies with 100% foreign capital participation operating in the SR and personnel managers and managers from a group of small and medium-sized companies with combined ownership. 445 companies were asked and the basic criterion for their inclusion in the next stage of the research was to carry out a personnel audit in the company or "other" forms or ways of obtaining data on human resources for the needs of the company personnel policy.

Based on the knowledge and information obtained in secondary research, we have found that "many managers cannot accurately identify a personnel audit", they cannot distinguish it from other human resource practices... (for example, the cost approach to human resources applied by as part of the accounting audit, also personnel controlling as pointed out by M. Dolenko, 1990 [14], R.M. Grant, 2008 [18] and others, we decided to conduct a pilot survey in order to clean up a sample of respondents who only believed that they performed the personnel audit they actually performed other personnel activities. For this purpose, a questionnaire was created, which was distributed electronically, in print form by post and also through third, contact persons in order to achieve the highest return. This objective was ultimately achieved.

The sample consisted of 160 companies, which were further subdivided according to the set criteria: size, number of employees, subject/sphere of business. One HR manager/personnel professional and two managers replied to the questionnaire from each company, giving a total of 480 questionnaires. These questionnaires were the main source of data to examine the identified issue. Due to the way the questionnaires were distributed, their 100% return was ensured. Standard quantitative and qualitative methods were used to collect, process, evaluate and interpret the data.

3 Results of the Research and Discussion

The aim of the primary research was to find out whether companies operating in the Slovak Republic use personnel audit to identify reserves in the area of human resources and their management. Within the partial goals, we also focused on the characteristics of companies operating in the Slovak Republic, which performed the personnel audit to obtain information on human resources, on the characteristics of companies operating in the Slovak Republic, which performed the personnel audit for information on human resources management and characteristics of companies operating in the Slovak Republic, which carry out personnel audit focused on managers managing and leading human resources in the direction of meeting the strategic objectives of the company, also referred to in the literature as managerial personnel audit [19, 20].

The results obtained through the application of these methods confirmed our assumption that within companies operating in the Slovak Republic the personnel audit is carried out mainly by medium-sized and large companies, companies with combined ownership participation, while the vast majority of companies proceed to personnel audit only or due to the employee complaints or threats of control by public authorities and institutions. Only a small number of companies (2, 1%) proceeded to conduct a personnel audit based on internal, in-house or corporate management needs. In most of the companies examined, a partial internal personnel audit was carried out, focusing on individual personnel activities, which in their entirety reflected the individual procedures applied in the area of human resources management. Most of them were occupational health and safety audits, which corresponded to the above.

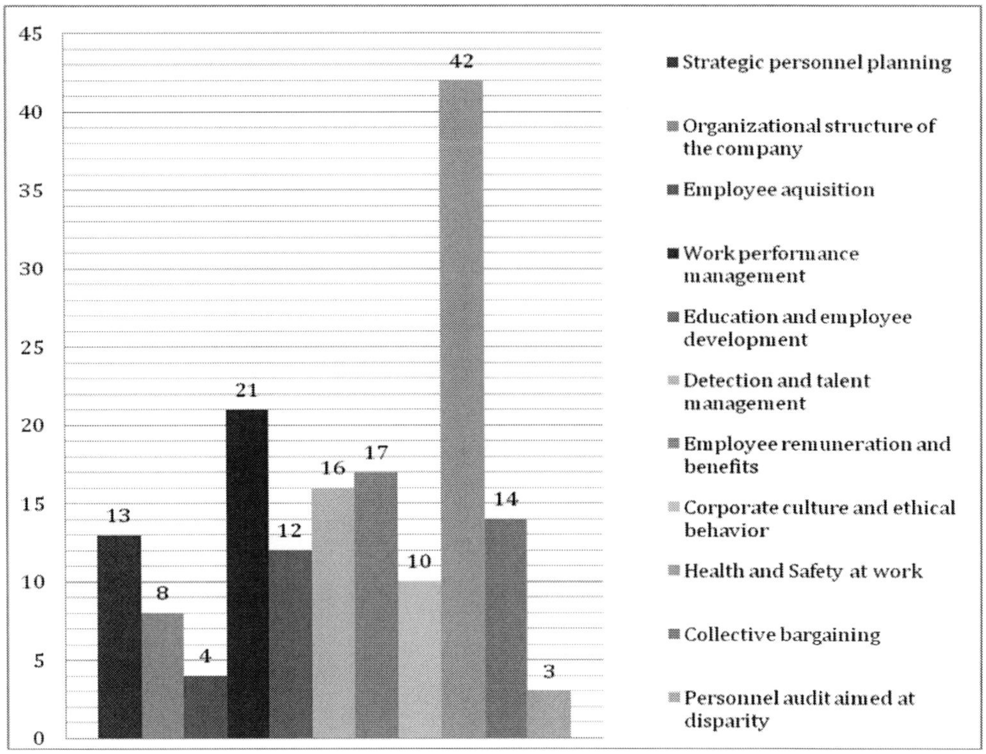

Fig. 1. Personnel activities monitored by personnel audit in examined companies (Source: own elaboration)

Human resources management companies performed partial audits mostly sporadically, which logically provided only a partial view of the company's problems and did not allow corporate management to gain a comprehensive view of the company as a whole. This moment was also reflected in a certain attitude of the company management towards the implementation of the obtained data into the personnel work in the company, respectively streamlining management processes in the company in accordance with the set standards. A more in-depth analysis of this issue provided information that the management of personnel and company managers persisted in learned work stereotypes and working habits, despite being found to be in violation of established standards and elements of effective human management based on findings in a personnel audit resources, respectively intellectual capital, including its individual components. This secondary finding, although very interesting and signaling one of the serious problems in human resources management, was obtained from the content analysis of responses obtained by the method of structured/controlled interviews of some of the examined respondents who did not represent the whole sample and in light of this, we consider this to be only a signal that indicates the direction of future research in the area. It was also a marginal, but in our opinion very interesting finding, that the companies examined conducted a personnel audit in the area of human resources management without a prior analysis of its implementation needs and without defining its objectives - these were defined by the auditor. In most cases, the auditor came from the company's internal environment. The purpose and reasons for conducting a personnel audit are interpreted in Figure 2.

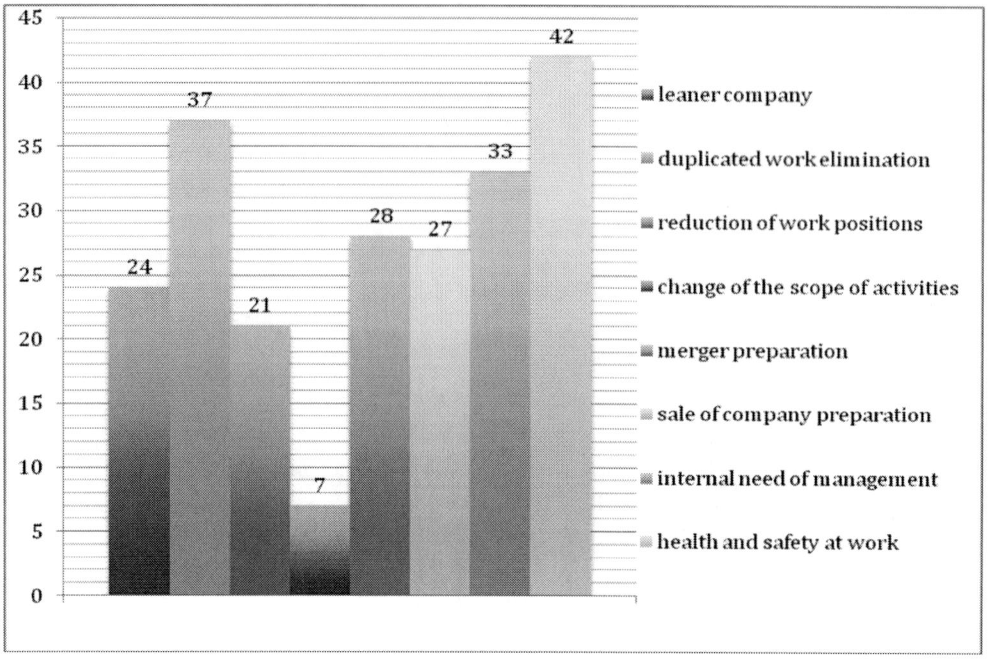

Fig. 2. Purposes/reasons/needs to perform a personnel audit in the company (Source: own elaboration)

The number of companies examined that carried out a human resources audit was less numerous. The data obtained in this area document that the personnel audit focused mainly on compliance to administrative procedures, legal standards and ethical principles resulting from established corporate standards and legislative provisions. Personnel audits focused on human resources were performed mainly by demanded companies that did not have personnel departments or those in which personnel, especially administrative activities, were delegated to human resources managers. The obtained findings entitle us to state that the companies that did not have personnel departments, resp. in which personnel managers have delegated their competencies to managers, have shown a higher number of errors and weaknesses in administrative and legal procedures in the area of human resources than companies with functional personnel departments.

4 Conclusion

Personnel audit, its place and function in the process of the deepening industrial revolution 4.0 and the upcoming revolution 5.0 are still growing. This is due to the fact that, when applied correctly, personnel audit can provide relevant and sophisticated information and knowledge necessary for strategic human resources management, which is becoming increasingly characteristic of the current phase of personnel work and personnel policy of companies. On the one hand, based on the knowledge gained from the personnel audit, the personnel department and the company management can eliminate incorrect procedures and remove the reserves in personnel work, both in the area of selection and training of employees and in the area of their management. The results obtained in the primary research document that companies operating in the Slovak Republic still do not sufficiently use personnel audit to obtain relevant information on personnel work in the company on the quality of the company's intellectual capital or on the reserves of management and leadership in the work process. However, the current trend suggests that personnel audit, which is an independent, objective, critical review of adherence to standard personnel practices in the company, collects data on people, processes, organizational structure and business principles, sets a benchmark for accurately determining the true performance of the human resources department to identify weaknesses in performance and efficiency and detect discrepancies in compliance with legal provisions, administrative procedures, ethical standards, business goals and personnel policy, will be one of the most important personnel work tools in the Industrial Revolution 5.0 stage. In this context, the personnel audit, which enables the company to obtain comprehensive information about the human resources of the company, its management as an essential component of its intangible assets, is undoubtedly a modern and comprehensive tool of personnel work.

The paper was created within the framework of the project VEGA 01/0309/18.

References

1. S. Nutley. Beyond systems: HRM audits in the public sector. Human Resource Management Journal, 2000, 10.2: 21-38. ISSN (2000)
2. S. Hlásna. Prognostické metódy v personálnom marketingu. In Personálny marketing a personálny manažment v organizácii: recenzovaný zborník vedeckých prác. Bratislava: EKONÓM. ISBN 978-80-225-3843-5, p. 17-21 (2013)
3. M. Rešovský, D. Pastor. Porovnanie metód priameho ohodnocovania nehmotných aktív. In: *Ekonomické rozhľady*, vol. 43, no. 4, p. 467 - 480. ISSN 0323-262X (2014)
4. M. Barteková, H. Majduchová 2019. Performance Potential and Development of Creative Industry: The Case of Slovakia. In 33rd International-Business-Information-Management-Association (IBIMA) Conference (Granada, Spain, 5895-5902, 2019)
5. M.F. Olalla, A.S. Castillo. Human Resources Audit. In: International Advances in Economic Research. 2002, vol. 8, no. 1, p. 58 (2002)
6. R.K.Yadav, N. Dabhade. Human Resource Planning and Audit–a Case Study of HEG Limited. Human Resource Management. 4: 5 (2014)
7. M. Szarková, M. Andrejčák, N. Matkovčíková. Personnel audit as a function of personnel marketing and personnel management. Brno: Tribun EU, 2014. ISBN 978-80-263-0809-6 (2014)
8. M. Szarková. Psychológia pre manažérov. Praha: Wolters Kluwer, 2016. 260 s. ISBN978-80-7552-175-0 (2016)
9. N. Matkovčíková. Implementation of EU OSH Standards in companies operating in Slovakia. In Ekonomické rozhľady : vedecký časopis Ekonomickej univerzity v Bratislave. - Bratislava : Ekonomická univerzita v Bratislave, 2015. ISSN 0323-262X, 2015, vol. 44, no. 3, p. 314-322. VEGA 1/0662/15 (2015)
10. G.T. Milkovich, J.W. Boudreau. Řízení lidských zdrojů. ISBN 80-85623-29-3 (1993)
11. R.L Brummet, E.G. Flamhotz, W.C. Pyle. Human Resource Measurement: A challenge for accountants. The accounting Review, no.42, 1968.
12. P. Durai. Human Resource Management Dorling Kindersley (India) Pvt. Limited: Edition (2010).

13. A. Clardy, A. The legal framework of human resources development (Part I). Human Resource Development Review, 2(1), p.26-53 (2004)
14. M. Dolenko. Auditing Human Resource Management. Florida: Institute of Internal Auditors Research Foundation (1990)
15. E.G Flamholz, Y. Randle. Future directions of Human Resource Management, Los Angeles, CA, University California p.369-90 (1996)
16. J.J. Phillips, J. J. Accountability in human resource management. Routledge, (1996)
17. R.L. Brady. Managing An HR Department of One. Old Saybrook: Business & Legal Reports, inc. 2006. p. 47. ISBN 1-55645-184-9 (2006)
18. R.M Grant. Contemporary Strategic Analysis. Blackwell Publishing, 2008. p. 138. ISBN 978-1-4051-6309-5 (2008)
19. M. Vetráková, G. Bočincová. Human Resources Management. Banská Bystrica: Univerzita Mateja Bela v Banskej Bystrici. ISBN 978-80-557-0489-0 (2013)
20. J. Kintler. Eliminácia rizík spojených s investovaním do ľudského kapitálu podniku In: Personálny marketing a personálny manažment v organizácii. Bratislava: EKONÓM. ISBN 978-80-225-3843-5 (2013)
21. IAS 38 – The International Accounting Standard for Intangible Assets is a standard designed to recognize and measure intangible assets in the financial statements of companies.

Management of business processes using stochastic simulation

Miroslav Špaček[1], *Petr* Půlpán[2]and *Zdeněk* Kopecký[3]

[1]Vysoká škola ekonomická v Praze, Nám. W. Churchilla 4, 13700 Praha 3, Czech Republic

[2]WAK System, s.r.o, Petržílkova 2654/21, 15800 Praha 5, Czech Republic

[3]Vysoká škola ekonomická v Praze, Nám. W. Churchilla 4, 13700 Praha 3, Czech Republic

Abstract. The paper deals with risks and uncertainties that are related to business processes. The risks and uncertainties (risk factors) can significantly influence organizational performance, value creation and long term-sustainability. Managers are often puzzled with making decisions in the environment of breakthrough discontinuities. The main outcome of the paper is the presentation of purposefully developed software tool that uses stochastic simulation of risks and uncertainties of business processes to support making optimum managerial decisions. This software is able to simulate any process risk factors that are describable by a mathematical model. The model was tested on the examples of production and investment cases. The research applied Business Process Modelling Notation Approach (BPMN) as well as mathematical and stochastic modelling. In addition, contextual interviews with company managers and risk specialists were used to get unbiased, reliable, and professional information on determination and analysis of process risk factors. The research was also supported by the evidence provided by the case study that illustrates software based decision-making process relating to opting for technology variant.

1 Introduction

Business and entrepreneurship in the environment of Industry 4.0 is tied with multiple risks and uncertainties. These phenomena often roofed by the term risk factors place a burden on future company development. The business is impacted by great many of risk factors (price and sales uncertainty, unpredictability of investment expenditure, competition behaviour, shortening product life cycle etc.). These risk factors can vary both continuously and discontinuously. Individual risk factors act with varying intensity and, in addition, may be mutually statistically dependent. In addition, there are pending externalities which can mediate negative multiplication effects on company prosperity. These externalities usually refer to climatic and weather effects (e.g. extending and repeating dry periods etc.), migration crisis, population aging, various forms of political instability etc. Reliable managerial decisions that would include the influence and interdependence of these factors are thus made with difficulties. It is apparent that managers cannot still cope with high degree of economic uncertainty and very frequently incline to intuitive decisions that are often faulty. This fact should be considered the main problem which is necessary to be solved by the application of advanced tools of risk management. In spite of these problems it is imperative for the company management to create the value for both customers and shareholders. Capability to manage companies in periods of breakthrough business discontinuities remains highly valuated managerial competence. The paper submitted reacts to this situation and by means of the application of SW supported risk analysis tool offers to company managers logic and rational way-out of this problem. The main outcome of the research is the software tool that is able to identify and analyse key business risks, determine the impact of risk factors on main characteristics of company performance and contribute to search for optimum decisions which are supportive to the stability of entrepreneurship. Especially simulation Monte Carlo that has been used since the War II attracted a great deal of popularity. This approach is frequently used in the financial sector where it is applied for various economic analysis [10].

2 Methods used in the research

Research team used Business Process Modelling Notation (BPMN) approach that is proven, quite flexible and sufficiently illustrative. By means of BPMN the basic work consisting in key company process description was carried out.

Stochastic simulation comes out of identification of the most important risk factors. Risk factors were identified either by sensitivity analysis or by expert opinion. The former consisted in sensitivity testing of output variable (usually net profit, cash flow or Net Present Value) against key input variables (risk factors) like unit price, sales volume, investment expenditures or the rate of production capacity utilization. Risk factor identification was accomplished either by creativity methods like brainstorming, brainwriting, Crawford Slip or Delphi [1] or contextual interviews with experts who were pretty well familiar with company business processes. Contextual interviews were conducted in a form of semi-structured interviews that were aimed at following points: (i) processes that are essential for company business, (ii) the approaches to the assessment of process risks (iii) risk measurement and (iv) mitigation provisions that company actually applies.

Eventually, case study research was applied for the confirmation of the model. Qualitative case study is a thorough, holistic description and the analysis of constrained phenomenon that may be program, institution, person, process, or social unit. Case study is a frequently used research technique that enables from qualitative point of view to explore problems and related circumstances Yin [15] perceives case study as contemporary phenomenon which exists in the framework of a real life, especially when the frontiers of the phenomenon are not quite clear. Researchers possess only limited control both over the phenomenon and the context. Case study is then an empirical query that examines specific case by means of purposefully formulated questions "How" and "Why?" Merriam [6] considers case study specific, complex, and functional thing, basically integrated system, that processes certain problematics within specific social constrains.

3 Theoretical aspects of process modelling and simulations

3.1 Process modelling

Business Process Modelling (BPM) represents the set of technologies and standards that are used in design, execution, administration, and monitoring of business processes [4]. BPM facilitates understanding the organization's work, comprehends the process in details and then uses technological support for improvements to human activities [14]. The aim of BPM is to develop a model that mirrors the organization functionality of existing or newly and established business processes and can be regarded as a predecessor to business process simulation that is carried out by using simulation tools or software. The core to BPM is the process modelling languages. Within a variety of BPM languages so called Business Process Modelling Notation (BPMN) plays an important role. This approach that was developed in 2001 represents visual design layer for a transactional workflow system. It involves the set of principles and rules that are oriented to the visualisation of company processes. Over past two decades BPMN has become an acknowledged management standard for graphic notation that displays consequence of steps in a business process. It simultaneously offers users comprehensive business process modelling capabilities in the unified environment. BPMN is composed of five basic categories that are [11]:

- flow objects,
- connecting objects,
- pools and swim lanes,
- data objects, artifacts (i.e. a product created or modified by the enactment of a process element).

Other components that are closely related to BPMN are processes, collaborations, and choreographies. Once a business process is described by using BPMN, a modeller can use a process simulation approach to discover its optimized form [8]. In course of time this approach attracted the attention of many reputable companies (Oracle, Intel, SAP etc.) that are now using BPMN.

3.2 Process simulation

Simulations are a vital component in developing smart manufacturing systems, predicting the behaviour of the manufacturing operations to support production planning, scheduling, and maintenance decision within manufacturing environment [2]. Simulation modelling involves model development which imitates real-world operations, statistical analyses, and their performance with the aim to improve efficiency and effectiveness [9]. In the past the companies were quite satisfied with the "what if" method that provided them with sufficient treatment of uncertain future by means of the development of definite number of strategic scenarios. Nowadays the situation became more complex when number of risk factors dramatically increased, these factors are under permanent change and they are mutually statistically dependent. This situation triggered the attempt to apply process simulation that would include excessive number of risk factors as well as their unexpected behaviour into business planning. Process simulation becomes the method of choice when it is impossible or too expensive to get data, the system to be analysed is too complex, analytical solution to the problem is immensely difficult, and it is impossible or too costly to validate mathematical experiment [12]. Similarly the simulation is applicable when input data are encumbered by quantifiable uncertainties, output value must represent the most probable values of input data, calculated uncertainty of output value must exactly reflect input data uncertainties, calculated uncertainty of output value must be an exact measure of model validity. The nature of simulation approach is a simulation of great many scenarios, number of which may range from ten to hundred thousand [13]. Method is based on choosing pseudorandom numbers. Input values (like any financial figure) are expressed by means of their probability distribution. Output value (e.g. net profit, cash flow, profit margin, Net Present Value - NPV) is obtained in a form of probabilistic distribution (histogram). Successful application of simulation approach is contingent upon viable mathematical model that describes in a mathematical logic the processes in question. Mathematical (sometimes perceived as financial model) represents backbone of simulation approach. Mathematical model describes stochastic dependence between dependent and independent variables as snown by equation 1.

$$Y = f(X_1, X_2, \dots X_n) \tag{1}$$

where parameter Y is output (dependent) variable and parameters X_1, X_2,.....X_n are input (independent) variables.

This model clearly states which output criteria will be observed. Next step is risk factors determination by sensitivity analysis or expert assessment. Proper assignment of specific type of probability distribution to any risk factor is necessary. It is possible to use historic data which may be approximated to fit with pre-defined probabilistic distribution. Oftentimes historic data are not available at all or available at lesser extent. If this is the case there is no chance to deduce probability distribution from them. In this respect experts′ opinions and statements which are mostly grounded on subjective perception and professional intuition are considered. In the extreme they may be even based on the belief in very occurrence of mass (statistical) phenomenon. Despite apparent on-exactness this approach is fully substantiated since it very often exploits long-term experience and knowledge of experts about specific problematics. Moreover, it very often represents the only applicable approach to the determination of probability distribution of risk factors. It is also possible to transfer data from another cases (cross sectional data).

Very important step rests in a proper determination of correlation between pairs of risk factors. By means of statistic dependence of risk factors we understand their mutual interconnection and conditionality. Variables in the model are usually dependent each other. In such a case the value of one input variable is at least partially determined by the value of another input variable. The least probable alternatives are thus excluded. Mutual correlation is of high importance and substantially influences output parameter distribution. Proper correlation setting is achieved by the incorporation of correlation matrix which enables the expert to enter correlation coefficient for any pair of risk factors.

To avoid high complexity of the system, setting correlation coefficient may be simplified and determined on an expert basis as pointed out in Table 1.

Table 1 Setting correlation coefficient

Positive deterministic dependency	High degree positive correlation	Low degree positive correlation	No correlation	Low degree negative correlation	High degree negative correlation	Negative deterministic dependency
1	0,75	0,5	0	-0,5	-0,75	-1

Finally, it is inevitable to set the number of runs (50-100 thous.) that is conditional for smooth and successful simulation. The more runs are carried out the higher preciseness is achieved. When using low number of runs the distribution curve is jagged instead of being smooth. Process simulation is a powerful tool that simplifies and facilitates managerial decision-making. Gudcemir and Hasan [3] proposed an integrated decision support system that combines simulation modelling and multi-criteria decision making. They managed to successfully optimize shop dispatching rules while using three renowned performance-oriented criteria and one customer-oriented criterion: (i) mean flow time, (ii) percentage of tardy orders and (iii) make span and (iv) mean percentage deviation from the customer expectations. Process simulations are also occasionally used for the determination of cash flows that are tied with a high degree of risk. Kazakova [5] developed a system-dynamic model for predicting the financial flows of an enterprise taking into account the risk and uncertainty of interaction. Unfortunately, the model was subjected to testing on experimental level only.

4 Software support for facilitating managerial decisions

The aim of the project was to create a SW-supported managerial decision-making tool that will support and facilitate strategic business decisions. The model was built on the grounds of Osterwalder´s concept of Business Model Canvas (BMC) which was deliberately adapted to the purposes of the project [7]. The main adjustments consisted in the simplification of BMC that rested in exclusive reckoning on elements that could be subjected to clear mathematical description. The elements that were not describable by a mathematical model were excluded from the model. Therefore, the original BMC was partly simplified and adapted to purposes of the project. Fig. 1 illustrates the design of interface of the software tool that represents BMC. Is also shows how was BMC model adapted to one specific testing case (tractors production).

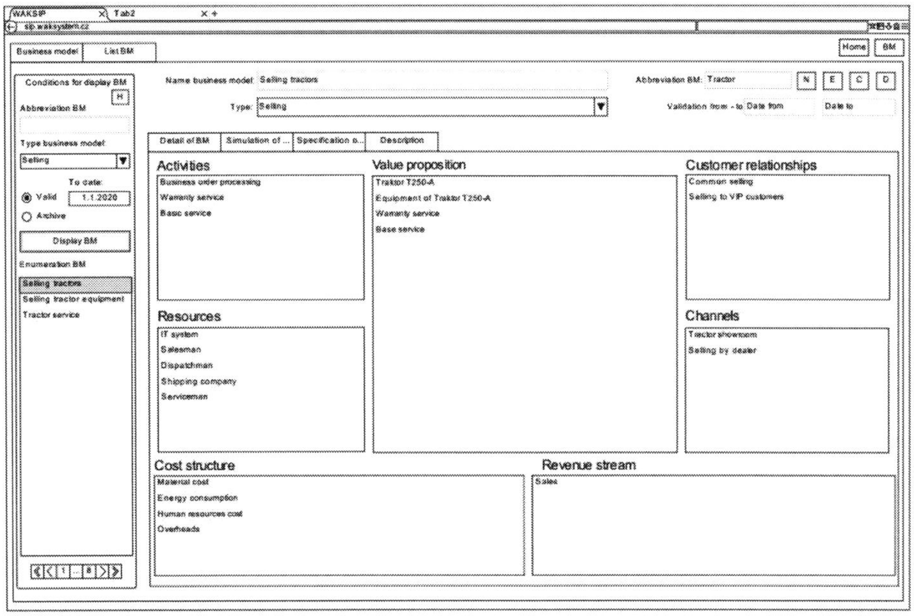

Fig. 1 Adapted Business Model Canvas (BMC)

The tool takes into account business uncertainties by replacing discrete input (independent) variables with their statistical distribution. By this way the possible variability of this variable were addressed. The tool, using a simple correlation matrix, considers the possible degree of cross-correlation of input variables. The tool allows the expert to model output variables (profit, cash flow, return on investment parameters, value creation parameters, etc.) for the considered business variants and thus support the reliability of managerial decisions. The tool allows to consider key risk factors, to link the initial assumptions with the expected results and to propose appropriate management measures. Fig. 2. shows cases of using a software tool from the perspective of a simulation modeller.

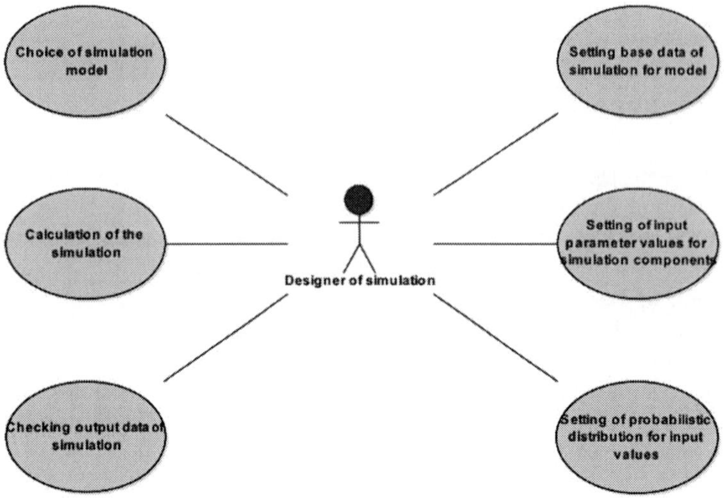

Fig. 2 Cases of using software tool from the perspective of simulation modeller

Design pattern that illustrates general functionality and the interaction among components of the model is shown in Fig. 1. It also illustrates basic software tool libraries integrating BMC elements, BPMN language classes, and simulation model classes.

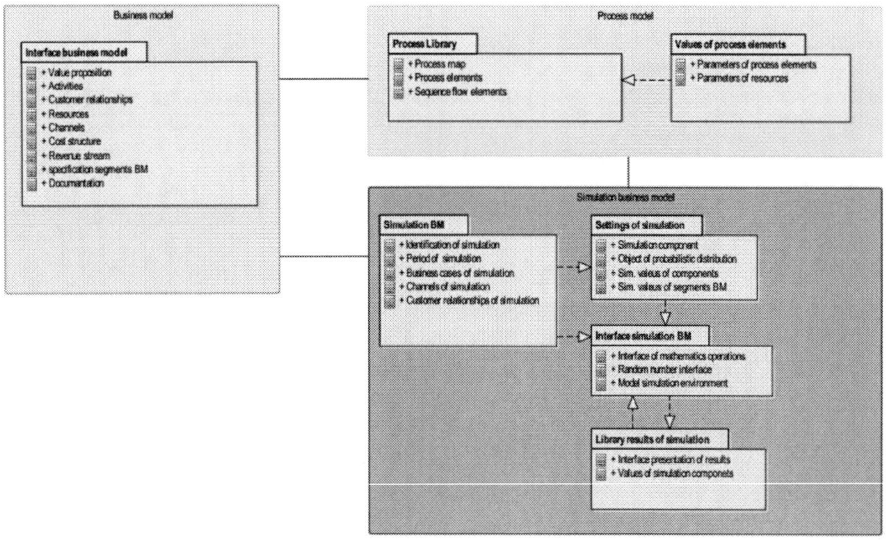

Fig. 1 Design pattern of the simulation model

4.1 Case Study

Chemical company intends to implement technology of the production of Corey lactone which is a versatile chemical compound frequently used in the pharmaceutical production There were several ways how to commercially produce this compound. After a thorough technological research and testing various possibilities in a pilot plant company's R&D put two possible technologies on a short list. The first technology used dicyclopentadiene (DCPD) as a starting material. This technology process consisted of lower number of reaction steps, but the technology was more demanding to operate. Moreover, this technology was highly parameter sensitive. The second technology came of norbornadiene (NB) that is easily available carbohydrate. This technology process was composed of higher number of reaction steps that were quite robust and showed lower sensitivity to the fluctuation of technology parameters. Either technology was considered the candidate for the implementation. Both technologies were technically feasible and met requested safety and environmental standards. The parameters that appeared to be decisive were production costs and reproducibility of the technology.

Key economic data were determined as key input variables (risk factors). The set of risk factors involved material costs, energy costs, personnel costs, capital expenditure etc. Since this product was purposefully produced for company captive production and there was no intention to commercialize this product. The present value of costs (PV) was thus determined to be key output variable. Key input variables were assigned probability distribution that fitted in with triangular distribution. Mathematical model was developed to be in consonance with the formula for the calculation of Present Value (PV) (equation 1):

$$PV(costs) = \frac{CF_1}{(1+k)^1} + \frac{CF_2}{(1+k)^2} + \ldots\ldots + \frac{CF_n}{(1+k)^n} = \sum_{i=1}^{n} \frac{CF_i}{(1+k)^i} \qquad (1)$$

where
CFi - cash flow in i-year
k - discount rate (Weighed Average of Capital Costs – WACC)

Parameters of simulation:

- Financial model of the project was calculated in nominal prices.
- Supposed project lifetime was 10 years.
- The number of risk factors was 117 (58 for NB technology and 59 for DCPD technology).
- Discount rate was calculated as Weighed Average of Capital Costs (WACC)
- The number of simulation steps was in both cases determined as 50000.

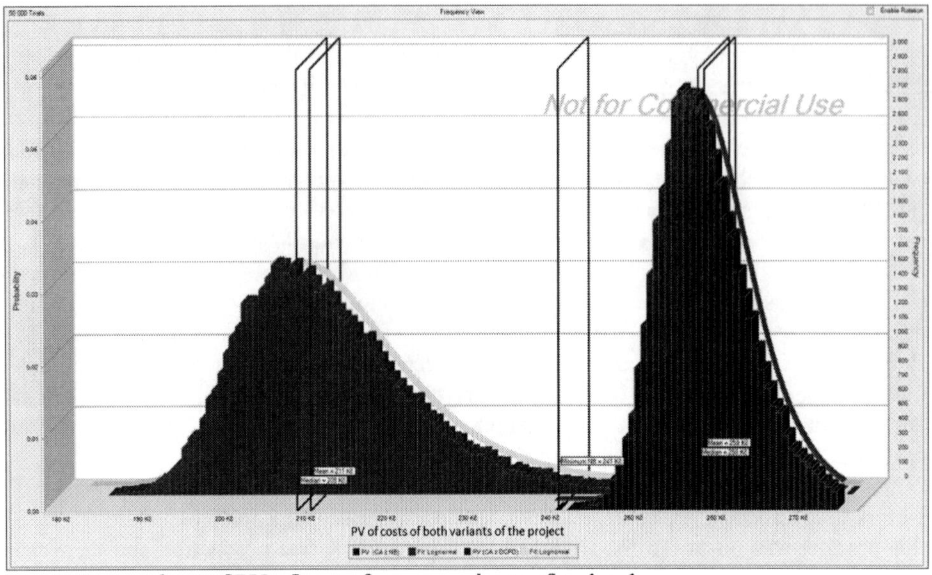

Fig. 2 Comparison of PV of costs for two variants of technology

The interpretation of the results is quite illustrative. Table 2 shows the most probable value of costs PV for both variants (modus) as well as the variance for both variants. It is apparent that DCPD technology operates more economically than NB one, but the former is riskier than the latter. Decision on the option for either type of technology may be also supported by the first rule of stochastic dominance. For the differentiation between technology variants of Corey lactone production from NB or DCPD the first rule of stochastic dominance can be used. In view of the fact that it deals with PV of costs („the lower, the better “) it is necessary to interpret the picture contrarily to routine criteria of revenue type. On the basis of the comparison of statistic distribution it can be concluded that technology variant DCPD dominates technology variant NB since the value of distribution function for variant DCPD is for almost any value of costs criterion more advantageous than for variant NB. On the basis of analyses performed it was recommended to owners to implement the project in technology variant DCPD which offers the best value of economic effectiveness parameters. This decision was made with the notion that risks of this technology will be in course of time and continuing optimization in decline. Moreover, probability that DCPD technology shows worse economic characteristics than NB one is 2,5% only.

Table 2 Comparison of moduses for both technologies

Technology variant	PV of costs (mil. CZK) (modus)	Variance (CZK)
Production of the product from DCPD	206	116
Production of the product from NB	257	33

5 Results and discussion

Software based decision-making tool provided reliable results when being tested on production, investment and service cases. Simulation of key performance characteristics like net profit, operational cash flow or Net Present Value offered a valuable contribution to decision-making process. The project faces some limitation that rest in fact that some processes cannot be easily described by a mathematical model. The advantage of the solution is its openness and adaptability to specific company need. Research team intends this tool to help managers make important decisions upon risk and uncertainty and simultaneously minimize failures of these decisions. The software was purposefully simplified so as not to deter managers from its use. Typically, triangular distribution of input variables (risk factors) was chosen as a preferred alternative because of its easiness to determine the lowest, the highest and the most probable value of the parameter. Since the simulation proceeds very quickly it is recommendable to use 50 000 – 100 000 simulation runs to obtain more reliable results.

Conclusion

Repeated failures made during decision-making processes on complex problematic that requires multicriterial approach persuasively showed that opting for intuitive decision-making approach was untenable. The managers are forced to make key decision under risk and uncertainty in shortening time. Elaboration of the software which was intended for the facilitation of decision-making processes can be considered approach of choice. Software tool enables simulation of any business processes which are describable by a mathematical model. Basically, it is applicable in routine situations where the reality is transformable to mathematical model like decision on purchase, closure analysis of a production unit, decision on outsourcing, assessment of variants of a business plan etc.

Mastering the software in question requires basic training in statistical methods and risk assessment methodology. Interpretation of the output values that are expressed in terms of probabilistic distribution also necessitates some probabilistic and statistical skills. Managers got accustomed to deterministic view of reality which prevents them from multifaceted perception of dynamic business systems. Increasing number of publications dealing with simulation approach showed that there was scarcity of any competitive managerial tools that would be conducive to the solution of complex managerial problems. Research team is in close contact with the experts in practice to get important feedback to this solution. The software based decision-making support shows enormous potential of this approach that can be even increased by the interconnection with artificial intelligence algorithms in the future.

Acknowledgements

This work was subsidized by the Ministry of Industry of the Czech Republic, program TRIO, Tools for the innovation and optimization of company processes performance within the framework of INDUSTRY 4.0 (grant number FV30110).

References

1. K. Al Badi, Discrete event simulation and pharmacy process re-engineering. *International Journal of Health Care Quality* **32,** (2) 398-411 (2019)
2. P. Goodall, R. Sharpe, A. West, A data-drive simulation support remanufacturing operations. *Computers in Industry*, **105**, 48-60 (2019)
3. H. Gucdemir, S. Hasan, Integrating simulation modelling and multi criteria decision making for customer focused scheduling in job shops. Simulation Modelling Practice and Theory **88**, 17-31 (2018)
4. M. Havey, *Essential Business Process Modelling,* O'Reilly Media, Inc., Sebastopol, CA (2005)
5. H. Kazakova, I. Zayarky, M. Medvedev, Forecasting of financial flows in business systems taking into account the risk and uncertainty of interaction. *International Conference on Numerical Analysis and Applied mathematics* (ICNAAM-2018), Rhodes, Greece, AIP Conference proceedings, **2116,** Article No. 430021 (2019)
6. S. B. Merriam, *Qualitative Research and Case Study Applications in Education: Revised and Expanded from Case Study Research in Education.* 2nd edition, San Francisco, CA: Jossey-Bass (2007)
7. Osterwalder, A.; Pigneur, Y. *Business Model Generation.* Wiley & Sons, Hoboken, New Jersey (2010)
8. J. L. Pereira, A. P. Freitas, Towards a characterization of BMP tools ´simulation support: the case of BPMN process models. *International Journal for Quality Research* **13,** (4) 783-796 (2019)
9. V. Schindlerová, I. Šajdlerová, Use of dynamic Simulation to Reduce Handling Complexity in the Manufacturing Process. *Advances in Science and Technology Research Journal* **14**, 1 81-88 (2020)
10. S. A. Silva, P.H.C. de Abreu, F. R. De Amorim, Application of Monte Carlo simulation for analysis of costs and economics risks in a banking agency. *IEEE Latin America Transactions* **17,** (3) 409-417 (2019)
11. B. Silver, *BPMN Method and Style,* 2nd Edition, with BPMN Implementer's Guide. Aptos, CA, USA: Cody-Cassidy Press, 2009.
12. R. Y. Rubinstein, D. P. Kroese, *Simulation and the Monte Carlo Method*, 2nd Edition, Wiley-Interscience (2007)
13. M. Špaček, Pravděpodobnostní přístupy k analýze rizik investičních projektů a jejich využití v praxi. NAVA - Nakladatelská a vydavatelská agentura, Plzeň (2015)
14. D. Tbaishat, Process architecture development using Riva and ARIS: comparative study, *Business Process Management Journal* **24,** (3) 837-858 (2018)
15. R. K. Yin, *Case Study Research: Design and Methods.* SAGE Publications, Inc; 5th edition (2014)

Basic economic indicators of book publishing

Miroslav Tóth[1*], and *Alena* Tóthová[2]

[1]University of Economics in Bratislava, Faculty of Business Management, Department of Business Economy, Dolnozemská cesta 1/b, 852 35 Bratislava, Slovak Republic
[2] University of Economics in Bratislava, Faculty of Business Management, Department of Business Finance, Dolnozemská cesta 1/b, 852 35 Bratislava, Slovak Republic

Abstract. The publishing activity and the publishing of books are part of the creative industry. A peculiarity of the book publishing industry is that it combines art, culture, business and technology. Although it creates less than 1% of job opportunities in Slovakia, it is linked to another 1% of co-workers. The importance of publishing books is also that it contributes to the employment of women and youth. In our paper we explore the book publishing industry. We find out the development of the number of enterprises, the development of the number of employees, basic economic indicators such as cash income, income, costs, profit, profitability. We follow these indicators across the book publishing industry. We examine the effectiveness of business activity by recalculating the above indicators per employee.

1 Introduction

Book publishers are representative of non-periodical printing. Book publishing houses are propagators of artistic, scientific and social ideas. The invention of an innovative printing in the 15th century marks the beginning of a massive spread of new ideas in Europe. Currently, the Internet is helping to spread such new mass information. The book remains popular. We see this also in the fact that in recent years we have seen a recovery in the Slovak book market accompanied by a slight increase in book sales. The general knowledge is the diversification of the reader's interest and a slight shift towards books reflecting social events. It is a gratifying news that there is a growing interest in books for children and young people. For the year 2018, a total of 11442 non-periodical titles were published, which is an increase compared to 2017. Of the total number of non-periodical titles, 9315 books and brochures are 2127 titles. The total number of copies of published publications is 15 408 354 pieces. As a translation from another language, 2634 non-periodical publications were published, which is a decrease compared to 2017. The most frequent were translations from English language 1650 titles, German language 287 titles and French language 172 titles. For this reason, we want to research the basic data about book publishers.

Book publishers bring books to the reader. The book is not an ordinary commodity. The book is an instrument of education, it cultivates us, makes us better. The book, reading,

* Corresponding author: miroslav.toth@euba.sk

improves our creativity. Reading is the process of thinking. Reading books support our analytical and critical thinking. That's why publishers are important to our lives. Countries that develop creative industries, innovate will become leaders of the global economy.

2 Methods

The economics of the arts approach in which economic analysis is used to analyze the arts is the first of those established approaches. The other established approach is the art and commerce approach in which the relative position of the arts in society and in particular its relation to the commercial and industrial world is analyzed. [1] In recent years there has been increased emphasis, in both academic and political arenas, on the economic importance of the creative industries (CIs). As a result, policymakers are increasingly looking to the CIs as a source of potential growth. [2] Cultural consumption is associated with positive externalities. Those who read books, go to theatre or attend music concerts, do not only increase their own utility, but generate also positive effects for the society, for example, in the form of improved education and literacy. [3] Time spent in leisure activities, including the arts, culture and sport, can provide a range of quality of life benefits from positive effects on well-being, to self-development and social and cultural connectedness. [4, 5] Creativity—possibly expressed according to different patterns—can play a mediating role, explaining the local capability to exploit cultural heritage for economic purposes. [6, 7] The extant definition of the creative industries is based on an industrial classification that proceeds in terms of the creative nature of inputs and the intellectual property nature of outputs. [8, 9]

The subject of investigation are publishers of non-periodical printing operating in the Slovak Republic. They are practically publishers that publish book publications and brochures. The time period is 2015 to 2018.

The first aim of the research is to find out the development of the number of publishers on the book market in the Slovak Republic. In the reference year we do research:

- the number of active publishing houses,
- the number of publishers that did not or did not interrupt publishing, and
- the number of publishing houses that have ceased their activities.

The second aim is to research the basic economic indicators (for the year concerned): income, revenue, costs, profit or loss, profitability of costs.

The third aim is to convert basic economic indicators per employee.

From the results of the calculations we want to find out the development of the above mentioned indicators, to look for the cause of positive or negative development.

Publishing companies belong, by size, to the group of small and medium-sized enterprises.

The area under study is, according to the statistical classification of economic activities (SK NACE Rev. 2), class 58.11.0 Book publishing, non-periodical publications. This class includes the publishing of books in printed, electronic (CD, electronic display, etc.) or audio form or on the Internet. This class includes e.g. publishing of books, brochures, leaflets and similar publications, including publishing dictionaries and encyclopaedias, publishing atlases, maps and graphic works, all in physically printed form. These also include the publication of non-periodical publications on CD-ROM, such as the publication of sound recordings of books, publishing encyclopedias, etc. [10, 11, 12]

The subject of the statistical survey is the publication of non-periodical publications (mainly printed publications in book form), which were through statistical survey through the National Educational Center. The Ministry of Culture of the Slovak Republic pursuant to Act No. 540/2001 Coll. on State Statistics, as amended, carries out long-term state statistical surveys in the field of culture every year. Statistical reporting units for non-periodical publications are publishers operating on the Slovak market, but also state institutions, municipalities, non-governmental organizations and individuals who issued non-periodical publications in the relevant year applied for an International Standard Book Number (ISBN) agency in the Slovak Republic and it is the Slovak National Library. Since 2015, two economic modules have been included in the statistical survey. In these modules, we find out basic information on incomes and expenditures of reporting agents, as well as information on employment and persons who provide the given cultural area. Since 2016, the number of employees has been calculated as the average registered number of employees. [13, 14]

A non-periodical publication is a special-purpose printed matter, a reproduction of a literary, photographic, scientific or artistic work, usually issued as a one-off publication, even in parts, irrespective of the form and method of processing, publication or production. The largest group of non-periodical publications are books and brochures.

A publisher of a non-periodical publication is a legal entity or a natural person who fulfills the conditions under special regulations (Commercial Code, Trade Licensing Act) and has paid the costs for their publication or production. A publisher or editor or publisher is an institution or individual who has the intellectual responsibility for publishing a publication. Sometimes a distinction is made between publisher and publisher. The publisher prepares and finances the publication, but the actual publication is carried out by the publisher. The difference is that in the case of publishing the work through the publishing house usually does not bear any cost of publishing the author, but the entire investment is on the shoulders of the publisher (editing, proofreading, graphics, wrapping, printing, distribution, marketing and other activities). The remuneration for the author is usually a share of the profits from the work as agreed with the publisher. Author is the individual who created the work. The author and creator of a work of art that is the result of the author's own creative intellectual activity. According to the law, the author is the owner and has the right to his work which arises at the moment when it is expressed in a form perceptible by the senses, regardless of its form, content, quality, purpose or form of expression. In summary: the publisher is the publisher; the publisher is the institution involved in the publication of books or other documents.

3 Results and Discussions

The first task of the research is to develop the number of non-periodical print publishers. Overall, the number of enterprises increased in the period under review. The highest increase was in 2018, when the number of enterprises increased by 296 business units compared to the previous year. At the same time, we can see that the number of active enterprises increased by 223 business units compared to last year. Despite the fact that the period of 2017 and 2018 was favourable for publishing years (growth of sales), its business was suspended or interrupted year-on-year by up to 68 business units. Five businesses have ended their business. The reason for this is that in 2016 and 2017, businesses in this sector suffered a loss of 22 and 15 mil. €.

In the sample of 2018, there were a total of 1,969 enterprises. Of this number, 84.26% were in active business, 14.68% were not publishing or had discontinued activities. 1.07% ended publishing activities. We compared these data with a four-year average. We found out that even in this period the shares of individual indicators are practically the same.

The development of the number of publishing houses was mainly influenced by the economic results during the period under review. Income decreased by 23 mil. €. In 2017 there was an increase in income from sales of books, but still lagged behind Income from 2015 by 2 million. €. Until 2018 was extraordinarily successful, sales increased by 24% compared to the previous year, the increase in Income 22 mil. €. The revenue indicator developed positively.

Table 1. Number of publishers providing non - periodical publications (book publishing).

Indicator / Year	2014	2015	2016	2017	2018	2018	Average	Share
Number of enterprises	1 635	1 660	1 673	1 673	1 969	100,00%	1 722	100,00%
Year - on - year change	--	25	13	0	296	--	--	--
Actively pursued publishing activities	1 392	1 393	1 417	1 436	1 659	84,26%	1 459	84,75%
Year - on - year change	--	1	24	19	223	--	--	--
Did not/ interrupt publishing activities	230	246	244	221	289	14,68%	246	14,29%
Year - on - year change	--	16	-2	-23	68	--	--	--
Ended publishing	13	21	17	16	21	1,07%	18	1,02%
Year - on - year change	--	8	-4	-1	5	--	--	--

Source: own calculations according to KULT 4-01 Ročný výkaz o neperiodických publikáciách 2014, 2015, 2016, 2017, 2018.

Compared to 2015, it increased by 24 22 mil. €, no increase by 50.28%. Growth rate in 2018 was 18.68%, but the absolute increase revenue was 23.6 22 million. €. The period of 2017 and 2018 was favourable for book publishers, as the positive effects of previous years were reflected. This effect was an increase in book sales. In 2015, 5,640,646 copies were sold, in 2016 it was almost one million more. Adverse developments occurred in 2017, with only 5,449,383 book copies sold. The following year was favourable, sales increased by 35.68%, meaning that 7,393,201 books were sold in 2018.

Table 2. Basic economic indicators – book publishing.

Indicator / Year	2015	2016	2017	2018	Index 2018/2015
Income (€)	93 499 164	70 903 767	91 710 354	113 718 735	--
Index R_n/R_{n-1}	1,0000	0,7583	1,2934	1,2400	1,2163
Revenue (€)	48 463 041	49 216 652	72 830 651	86 432 949	--
Index R_n/R_{n-1}	1,0000	1,0156	1,4798	1,1868	1,7835
Cost (€)	70 838 987	63 842 667	58 164 644	74 349 683	--
Index R_n/R_{n-1}	1,0000	0,9012	0,9111	1,2783	1,0496
Profit, Loss (€)	-22 375 946	-14 626 015	14 666 007	12 083 266	--
Index R_n/R_{n-1}	1,0000	0,6536	-1,0027	0,8239	-0,5400
Year - on - year change (€)	--	7 749 931	29 292 022	-2 582 741	--
Profitability of costs	-31,59%	-22,91%	25,21%	16,25%	--

Source: own calculations according to KULT 4-01 Ročný výkaz o neperiodických publikáciách 2015, 2016, 2017, 2018.

Based on the above data, we have quantified the industry's profit and cost-effectiveness. The years 2015 and 2017 are loss-making and the profitability of the industry is also negative. The big reversal occurs only in 2017, the year-on-year change in profit increased by 29 mil. €. This also means cost-effectiveness of 25.21%. In 2015 and 2016, a loss of 39 mil. €. Despite the favourable years 2017 and 2018, profit total 26.7 mil. €, the deficit is still almost 13 mil. €.

In the next part of the thesis we focus on economic indicators calculated per employee. An employee is a full-time or part-time employee. The number of employees was calculated as the average registered number of employees. In the monitored years we can see that the number of employees increased. In 2016 it was a 13% increase, a year later a 10% increase. In 2014, there was an absolute increase of 501 employees, an increase of 22% over the previous year, but a year-on-year increase of 53%. Income per employee in 2015 was at a high level, 52 205 €. By the following year, the number of employees increased by 235 persons, but at the same time the income per employee indicator decreased by € 17,208. In the following years the number of employees increased. This had a negative impact on the indicator of income per employee, as in the next two years it reached only 80% of the level of 2015. Similarly, the indicator of income per employee developed similarly. However, this indicator increased by approximately 30% in 2017 2018 compared to the previous period. It increased by 16.7% over the four-year period. This increase was despite the almost 53% increase in the number of workers. Significantly to the improvement of the indicator was helped by the fact that revenue in five years increased by up to 78%. The cost development of one employee has generally been favourable. Over the four years it has fallen by almost 12,000 euros, a relative decrease of 31%.

Table 3 Basic economic indicators per employee – book publishing.

Indicator / Year	2015	2016	2017	2018	Index 2018/2015
Employees	1 791	2 026	2 236	2 737	--
Index R_n/R_{n-1}	1,0000	1,1312	1,1037	1,2241	1,5282
Year - on - year change (p.)	--	235	210	501	--
Income (€) / Employee	52 205	34 997	41 015	41 549	--
Index R_n/R_{n-1}	1,0000	0,6704	1,1720	1,0130	0,7959
Year - on - year change (€)	--	-17 208	6 018	533	--
Revenue (€) / Employee	27 059	24 293	32 572	31 579	
Index R_n/R_{n-1}	1,0000	0,8978	1,3408	0,9695	1,1670
Year - on - year change (€)	--	-2 767	8 279	-992	--
Cost (€) / Employee	39 553	31 512	26 013	27 165	
Index R_n/R_{n-1}	1,0000	0,7967	0,8255	1,0443	0,6868
Year - on - year change (€)	--	-8 041	-5 499	1 152	--
Profit, Loss (€) / Employee	-12 494	-7 219	6 559	4 415	--
Index R_n/R_{n-1}	1,0000	0,5778	-0,9086	0,6731	-0,3534
Year - on - year change (€)	--	5 274	13 778	-2 144	--
Profitability of costs / Employee	-31,59%	-22,91%	25,21%	16,25%	--

Source: own calculations according to KULT 4-01 Ročný výkaz o neperiodických publikáciách 2015, 2016, 2017, 2018.

The development of the level of profit per employee copies the development of total profit. In 2015 and 2016, an average loss of 12 resp. 7 thousand euro per employee. This also means that the profitability of costs per employee was negative (-31.59% in 2015 and -22.91% in 2016).

The economic results of the publisher of non-periodical printing (books) are influenced mainly by purchasing demand (sufficient free household money for cultural products). The demand for books is dependent on the business cycle and changes in the prices of goods and services. An important factor is the offer of bestsellers for different age groups of the population, the offer of popular educational publications. The demand for books is also influenced by social networks, film and television production. The book remains aesthetic goods, collectors' items and social links.

Acknowledgement
This paper is the partial result of the scientific grants
VEGA n. 1/0340/19 "The Entrepreneurial Dimension of Creative industries in the Context of Innovation and Smart Growth" (share 50 %) and
VEGA n. 1/0240/20 „Financial Aspects of Sustainable Business - Enterprise Succession Solution for Small and Medium-sized Enterprises" (share 50%).

References

1. E. Dekker, Two approaches to study the value of art and culture, and the emergence of a third. *Journal of Cultural Economics* **39**, 309–326 (2015) https://doi.org/10.1007/s10824-014-9237-y
2. M. S. Sigurdardottir, M. Candi, Growth strategies in creative industries. *Creativity and Innovation Management* **28**, 4, 421-575 (2019) https://doi.org/10.1111/caim.12334
3. K. J. Borowiecki, T. Navarrete, Fiscal and economic aspects of book consumption in the European Union. *Journal of Cultural Economics* **42**, 309–339 (2018) https://doi.org/10.1007/s10824-017-9307-z
4. D. Wheatley, C. Bickerton, Measuring changes in subjective well-being from engagement in the arts, culture and sport. *Journal of Cultural Economics* **43,** 421–442 (2019). https://doi.org/10.1007/s10824-019-09342-7
5. D. Wheatley, C. Bickerton, Subjective well-being and engagement in arts, culture and sport. *Journal of Cultural Economics* **41,** 23–45 (2017) https://doi.org/10.1007/s10824-016-9270-0
6. S. Cerisola, Creativity and local economic development: The role of synergy among different talents. *Papers in Regional Science* **97**, 2, 199 – 209 (2018)
7. S. Cerisola, A new perspective on the cultural heritage–development nexus: the role of creativity. *Journal of Cultural Economics* **43**, 21–56 (2019) https://doi.org/10.1007/s10824-018-9328-2
8. E. Munro, Building soft skills in the creative economy: Creative intermediaries, business support and the 'soft skills gap". *Poetics* **64**, 14-25 (2017) https://doi.org/10.1016/j.poetic.2017.07.002
9. J. Potts, S. Cunningham, J. Hartley, et al., Social network markets: a new definition of the creative industries. *Journal of Cultural Economics* **32**, 167–185 (2008) https://doi.org/10.1007/s10824-008-9066-y
10. www.financnasprava.sk/_img/pfsedit/Dokumenty_PFS/Podnikatelia/Clo_obchodny_tovar/EORI/StatistickaKlasifikaciaEkonomickychCinnosti.pdf [accessed 27.3.2020]
11. Zákon č. 289/2013 o povinných výtlačkoch periodických publikácií, neperiodických publikácií a rozmnožením audiovizuálnych diel
12. Zákon č. 185/2015 autorský zákon

13. Národné osvetové centrum (2019).Správa o štátnom štatistickom zisťovaní v oblasti kultúry 2018. Bratislava, NOC Oddelenie vzdelávania, výskumu a štatistiky december 2019. http://www.culture.gov.sk/ministerstvo/statistika-kultury-1a8.html [accessed 27.3.2020]
14. KULT 4-01 Ročný výkaz o neperiodických publikáciách. http://www.culture.gov.sk /ministerstvo/statistika-kultury-1a8.html [accessed 27.3.2020]

Successes and Failures of Family Business in Slovakia

Alena Tóthová[1*], and *Miroslav* Tóth[2]

[1]University of Economics in Bratislava, Faculty of Business Management, Department of Business Finance, Dolnozemská cesta 1/b, 852 35 Bratislava, Slovak Republic
[2]University of Economics in Bratislava, Faculty of Business Management, Department of Business Finance, Dolnozemská cesta 1/b, 852 35 Bratislava, Slovak Republic

Abstract. Family businesses are part of the world's economies, with an estimated 70-90% share of total performance. The history of family business in Slovakia is not continuous, so we can examine its level from the last 30 years of experience at most. The paper is focused on research of financial results of Slovak family enterprises, their confrontation with other types of enterprises, searching for trends in the level of their financial indicators and finally on the dependence between results and environment. In the research were used several methods, from which we can name mainly analysis, synthesis, comparison, mathematical-statistical methods. The obtained outputs are intended to identify strengths and weaknesses of family business in Slovakia.

1 Introduction

The tradition of family business in Czechoslovakia was violently discontinued during socialism, but after 1989, the family business of family members started to slowly start again. The beginnings were difficult because most Slovaks did not have enough professional knowledge or the necessary skills for doing business. The isolation of Slovakia has caused the loss of an overview of the level and forms of business in the global world. Despite these obstacles, family businesses have gradually become an important component of the Slovak economy over the past thirty years. The following data briefly characterizes the importance of Slovak family businesses.

1.1 Current state of family business in Slovakia

An important source of up-to-date data on Slovak family businesses is the studies of SBA - Slovak Business Agency, an agency focused on supporting small and medium-sized enterprises. Interesting knowledge is also brought by repeatedly conducted surveys of consulting company KPMG International with a global scope, directly focused on family business. Trustworthy facts about family businesses are complemented by surveys by PwC (PricewaterhouseCoopers). Outputs of the above companies declare that about one-eighth

* Corresponding author: alena.tothova@euba.sk,

of family businesses in Slovakia are medium-sized enterprises, approximately 15% are small enterprises and up to 70% are micro-enterprises. Much of the family business is already established on the market, as 34% of respondents surveyed have a history of more than 20 years and another 32% are over 10 years old. In terms of the preferred form of family business in Slovakia, the prevalence of the use of limited liability companies, which is seen by up to 73% of family businesses, can be seen. More than half of family businesses operate by owning only one family member. The importance of family businesses in the European context is clearly stated by the European Commission's estimates of a 60% share of family businesses in all European businesses. Family businesses themselves are aware of their strength. Most think that they provide stability to the wider economy. The characteristics of family businesses, their strengths and weaknesses, succession problems, profitability, competitive struggle, the struggle for skilled labour, innovation and technological progress, the challenges of penetrating international markets, the imperfections of legislation on this type of business and many other open In its scientific work, questions are addressed not only by a significant part of the professional public [1-12], but also by many representatives of practice.

Presented state of family business in Slovakia, combined with the state of family business in the European context, further extend the knowledge stemming from the research of financial indicators of family businesses. The range of financial indicators is very broad and therefore the assessment of the status of family businesses from a financial point of view can be very diverse. The following is a brief illustration of what the financial indicators on family businesses reveal.

1.2 Financial Indicators of Slovak Family Enterprises

Financial indicators provide an indication of the level of success / failure of family businesses. By examining the financial indicators, it is possible to identify the financial condition of family businesses, its development trends, the complexity and nature of interrelationships between the results of family businesses, the dependence of financial results and the degree of severity of the dependence.

For research of the mentioned areas we chose family enterprises engaged in industrial production, namely the production of grape wine (SK NACE 11020). Industrial production is the backbone of the Slovak economy, which is also reflected in the selection of a sample of family businesses for research. The production of grape wine in Slovakia has been internationally successful in recent years, it has the potential of growth, so it is important to deal with the financial condition of family businesses in the given industry. The sample of family businesses examined is linked by several characteristics. These are family businesses with the same activity, employing 10 to 19 employees, with a turnover of up to € 0.8 million.

Table 1. GDP of Slovakia quantified by expenditure method.

Indicator	2018	2017	2016	2015	2014	2013	2012	2011	2010	2009
(1) GDP (1=2+3+4-5)	89 721	84 517	81 038	79 758	76 256	74 355	73 484	71 214	68 093	64 096
(2) Final consumption expenditure	66 882	63 206	59 969	57 960	55 998	54 630	54 237	52 735	52 048	51 062
(3) Gross fixed capital formation	21 022	19 461	18 679	19 346	16 584	15 558	15 192	17 977	16 256	13 141
(4) Export	86 213	80 381	75 955	73 396	69 788	69 608	66 897	60 543	52 647	43 608
(5) Import	84 396	78 531	73 565	70 943	66 114	65 441	62 841	60 041	52 859	43 716

The Slovak economy and the environment of grape production in Slovakia are represented by the indicators shown in Tables 1 and 2, taken from the SO SR. Statistical office of the Slovak Republic

The data in Table 1 show the uneven trend in Slovakia's GDP growth. The main reasons for GDP growth have changed over time. In most cases, the main driver of GDP was investment, then exports and sometimes domestic consumption. The contribution of grape wine production to the Slovak economy is shown in Table 2.

Table 2. Selected financial indicators of beverage production.

Indicator	2018	2017	2016	2015	2014	2013	2012	2011	2010	2009
GVA	80 495	75 786	72 948	71 777	68 843	67 265	66 775	64 088	61 659	57 994
GVA 11	213	193	193	188	209	195	206	202	221	232
Share 11	0,26	0,25	0,26	0,26	0,30	0,29	0,31	0,32	0,36	0,40

Explanatory notes: 11 - Division beverage industry, GVA - gross value added

The trend of beverage production has a different dynamics than the GDP of Slovakia. In the selected time series you can see a repeated alternation of the decline with the increase in beverage production. Since 2016, the situation in the production of beverages is already more favourable. There was no drop in the performance of the division. The last row of the table contains data from the confrontation of beverages production with the overall output of Slovakia. The following presentation of the research relates solely to the financial condition of the beverage industry and to the activities of family businesses in this division. First of all, we have to look at the statistically processed financial results of the beverage companies with a turnover of up to € 0.8 million (these are the companies with the lowest turnover).

Table 3. Financial indicators of beverage enterprises.

Indicator	2018	2017	2016	2015	2014	2013	2012	2011	2010	2009
L3	1,56	1,75	1,37	1,19	1,11	1,22	1,31	1,15	1,19	1,11
DOZ	0,00	0,36	2,53	6,39	12,12	28,18	23,75	12,20	20,37	11,41
DIP	39,87	43,53	55,41	63,45	84,22	83,37	98,88	63,72	89,24	62,78
DIKPzOS	14,13	15,00	25,99	26,03	31,31	32,53	39,85	22,03	25,54	25,61
DSZ	257,39	418,64	380,77	493,02	543,19	457,29	488,35	465,56	479,90	503,07
DSKZzOS	22,88	18,93	24,66	29,52	33,09	42,85	57,76	30,43	33,36	41,51
T / A	0,56	0,36	0,45	0,35	0,32	0,28	0,22	0,27	0,20	0,26
ZA	58,98	51,66	57,80	61,66	67,44	63,65	63,33	63,76	57,52	61,20
DZA	0,00	0,00	0,00	0,00	0,00	0,00	0,00	0,00	0,00	0,02
TZ	35,43		77,34	59,75	130,14	160,81	61,60			55,08
ROE	0,00	-0,73	-1,92	-0,87	-0,12	-0,03	0,00	-0,44	-1,45	0,00
ROA	0,00	-0,36	-0,67	-0,28	0,00	0,00	0,00	-0,10	-0,48	0,00
ROS	0,05	-7,35	-5,52	-3,31	-2,26	-4,70	-0,72	-14,71	-9,58	0,39
PH/T	15,09	12,96	15,14	20,51	21,39	13,71	15,39	15,96	13,32	14,04
NVH/T	0,99	-3,49	-3,78	1,18	0,25	-2,47	-0,13	-14,27	-6,18	1,01

Explanatory notes: L3 total liquidity ratios, DOZ inventory turnover, DIP debtor days ratio, DIKPzOS maturity of short-term trade receivables, DSZ creditor days ratio, DSKZzOS maturity of short-term trade payables, T/A asset turnover, ZA debt ratio, DZA long-terms debt ratio, TZ flow indebtedness,

ROE return on equity, ROA return on assets, ROS return on sales, PH added value, T sales, NVH newly created value.

Table 3 shows the median values of sectoral enterprises, irrespective of the form of business, taken from the Cribis database, the Universal Register [13].

Smaller sector enterprises are usually overcapitalized and more pronounced in the last three years. Stocks are getting faster and faster, and debt collection is shortened. The financial equilibrium helps them to maintain a longer maturity of liabilities than collection of receivables. Longer maturities than direct debits are also maintained for short-term components. Smaller beverage businesses are gradually less indebted over the ten-year period due to better profitability. They never exceeded the 70% debt ratio. The debts of these enterprises are in most cases short-term, but the possibility of repaying all debts is weak. Repayment of debts would take several decades. The median profitability of beverage businesses is low. Quite often they even achieve negative profitability. This result is a tax on labour intensiveness and fixed assets. The normal financial situation of the sectoral enterprises surveyed can generally be assessed favourably, as the results did not show any serious shortcomings and the trend of their development is positive. The assessment of the financial performance of the smallest beverage business raises the question: How does family business thrive in such an environment? The basis for the answer to the question is contained in Table 4. The quantification of the data shown in Table 4 was preceded by the search for family-owned grape holders, with a staff of 10 to 19, with a turnover of up to € 0.8 million from FinStat [14].

Table 4. Financial indicators of family enterprises sector.

Indicator	2018	2017	2016	2015	2014	2013	2012	2011	2010	2009
L3	0,86	0,88	0,97	1,03	0,97	1,56	1,82	1,71	2,11	1,58
DOZ	487,93	427,94	338,18	329,06	299,51	329,10	326,76	351,28	365,37	496,94
DIP	45,78	42,18	37,46	34,91	45,71	72,97	75,83	60,68	46,81	90,75
DIKPzOS	29,15	36,73	29,70	34,25	41,73	36,56	41,61	49,18	45,48	56,79
DSZ	199,54	140,74	184,97	343,54	266,72	225,45	133,85	208,08	143,25	345,85
DSKZzOS	43,39	57,41	39,87	35,02	29,90	52,85	27,60	40,64	11,81	16,85
T / A	0,30	0,34	0,33	0,32	0,39	0,28	0,40	0,36	0,47	0,32
ZA	45,49	45,67	50,47	50,59	49,23	46,81	34,29	25,15	38,38	47,56
DZA	5,85	8,01	9,32	11,25	4,65	9,10	0,05	0,03	2,90	6,49
Flow indebtedness	9,84	3,36	6,91	2,73	4,03	3,10	4,46	3,03	1,08	2,58
ROE	0,01	0,48	0,31	1,85	1,23	0,44	2,44	0,74	1,20	1,47
ROA	0,01	0,35	0,15	0,45	0,58	-0,04	0,30	0,20	0,36	0,04
ROS	0,02	0,02	0,00	1,32	0,01	-0,19	0,60	0,01	0,14	0,01
PH/T	36,98	38,49	41,49	28,53	24,33	41,55	28,59	31,92	33,45	36,08
NVH/T	-30,65	-15,82	-12,78	-6,59	-8,48	-7,01	-2,27	-7,76	-1,51	-1,52

After compiling the set of family businesses, it was necessary to calculate their financial indicators, in accordance with the methodology of the Cribis database, the Universal Register. The method of calculating the financial indicators of family businesses was conditioned by the research intentions. The aim of the presented research is to compare the financial results of family enterprises with other comparable enterprises in the industry, to verify their interrelationships, development trends, to recognize the complexity and nature of the interrelationships between family enterprises, Finally, we proceeded to quantify the median values of family business indicators. The median values represent whole sets of enterprises and can serve the desired confrontation.

Family grape holdings cannot be permanently overcapitalised. The last three years have shown a discrepancy between current assets and current liabilities. The mismatch undermines the financial stability of the family businesses under review. Removal of stocks has been protracted, and this period has been increasing more and more over the last two years. The scales of financing the selected set of enterprises are partly offset by shorter collection of total receivables compared to the maturity of all liabilities. Significantly longer maturity of liabilities increases the scope for cash inflow into the company. It is assumed that an adequate amount of cash flows into the enterprise during a long maturity period. The supplier-customer relations of family businesses are not so favourably consolidated for them. Only roughly in the last half of the time series will they collect cash more quickly than they use to pay short-term liabilities to suppliers. Immediately after the global crisis and for about the next three years, family businesses were not able to raise funds from their customers in due time. The low turnover of assets is given by the nature of the activity, its demands on the asset. The indebtedness of family businesses has not changed very much over the last ten years and almost does not exceed 50% at all. The debt portfolio also includes long-term debt, which, given the difficulty of maintaining the stability of financing, can be considered a positive situation.

Table 5. Cohen correlation coefficient interpretation.

ρ_{xy}	Interpretation	ρ_{xy}	Interpretation
0,0-0,1	trivial correlation	0,5-0,7	big
0,1-0,3	small	0,7-0,9	very big
0,3-0,5	middle	0,9-1,0	almost perfect

The strong correlations between the financial results of family businesses and the results of other sectoral enterprises are defined by the data in Tables 6 and 7.

Table 6. Interrelations between indicators of comparison sets of enterprises.

Indicators		Sectoral indicators						
		L3	DOZ	DIP	DIKPzOS	DSZ	DSKZzOS	T / A
Family business	L3		0,788	0,694			0,658	-0,827
	DOZ			-0,635	-0,660			
	DIP						0,761	
	DIKPzOS					0,640		-0,744
	DSKZzOS	0,627						
	Flow indebtedness					-0,832		0,878
	ROE			0,649	0,674	0,705	0,704	-0,644
	NVH/T	-0,711	0,681	0,751	0,699	0,891	0,654	-0,922

Family businesses are able to handle debt repayment at a reasonable level, except for 2018. Although the profitability of family businesses producing grapes is low, it was only negative in 2013. Profitability affects the nature of the business and its intensity on production factors.

The relationships between the financial results of the surveyed enterprise files can be verified through the MS Excel statistical function, the CORREL function. It has a formula: = correl (array1; array2). It aims to evaluate the intensity of the dependency between the independent variable "x" and the dependent variable "y". Individual indicators of industrial enterprises represent an independent variable "x" and indicators of family enterprises a dependent variable "y". The results of the individual measurements are in the range <-1; 1>. Based on their position, we can only focus on those that show strong links, whether positive or negative (Table 5). We set the limit of strong dependence at the point „0.6“, resp. "-0.6".

Table 7. Interrelations between indicators of comparison sets of enterprises.

Indicators		Sectoral indicators							
		ZA	DZA	Flow indebtedness	ROE	ROA	ROS	PH/T	NVH/T
Family business	DOZ		0,607	-0,603					
	DIP		0,662		0,625	0,626			
	DIKPzOS		0,673						
	ZA						0,605		0,745
	ROA							0,644	
	PH/T							-0,761	

Numerous strong interrelationships are not confirmed between sectoral results and family-owned grape holdings We can state that the most strongly interrelated indicators are NVH / T, ROE and L3. Nearly perfect negative binding is found exclusively in NVH / T. It seems that on the Slovak beverage market in smaller turnovers, the family businesses producing grapes do not have enough room for realization. In contrast, NVH's increasing share of sales for beverage-producing businesses with different forms of business reduces the share of the grape-producing family. Probably everything is related to competition in the beverage market. If other businesses in the turnover zone of € 0.8 million tear off most of the common cake, the family businesses will remain smaller. The correlations in Tables 6 and 7 are predominantly positive. Among the indicators of the comparative sets of enterprises, mainly a large link was quantified, a little less very large and only in one case cited almost perfect.

Research on the financial condition of family-run grape holdings with lower turnover can be complemented by conclusions from the verification of relations with the macro environment and / or sector / internal environment. Specifically, these are relations with GDP components and typical sectoral indicators, as shown in Table 8.

Table 8. Relationships between family business indicators and external / internal environment.

Indicators	Final consumption expenditure	Gross fixed capital formation	Export	Import	Sectoral PP	Sectoral T
L3	-0,816	-0,672	-0,768	-0,777		
DOZ					-0,659	
DIP	-0,607	-0,821	-0,635	-0,672		
DIKPzOS	-0,813	-0,756	-0,917	-0,907		
DSZ						-0,685
DSKZzOS	0,624		0,752	0,743	-0,737	
Flow indebtedness	0,766		0,690	0,701		
NVH/T	-0,924	-0,801	-0,797	-0,839		

Explanatory notes: Sectoral PP - Sectoral labour productivity, Sectoral T - Sectoral sales

The internal environment in Table 8 is represented by the indicators of sectoral labour productivity and sectoral revenues. Other indicators (final consumption expenditure, gross capital formation, exports, imports) represent the macro environment of Slovakia. The strong impact of the internal environment on the family businesses examined was only confirmed by the turnover of stocks and the repayment of liabilities. Better utilization of the workforce within the sector benefits the grape family business from faster stock picking and short-term trade payables. Increasing labour productivity is reflected in business revenues. Revenue from implementation is due to the removal of production and a larger volume of sales can be better used to pay off liabilities. The impact of the macro environment on the selected family business is somewhat clearer and more numerous. The relationships with the macro-environment are mostly negative. Increasing domestic consumption, increasing investment and exports have demonstrably shortened the collection of all receivables, including trade receivables, to family businesses, but also worsens overall liquidity. The better-named macroeconomic indicators press strongly to extend the maturities of short-term liabilities to of all debts. Paradoxically, more performance in the economy reduces the share of newly created value of family businesses in sales. What is it given? Nature of activity and competition. Increasing demand hides increasing demand, which all business entities want to use. But the business being monitored is only slightly profitable, almost to the edge, which does not allow businesses to create enough of their own resources. Businesses must therefore seek foreign resources in order to make the most of the growing demand. At the same time, foreign sources are being favoured by favourable conditions on financial markets, materialized in attractive conditions for obtaining foreign sources. Low profitability and more debt must automatically aim to extend the debt settlement period. Better times in the country's economy encourage the optimistic mood of economic agents, and as a result, it may be expected to provide or tolerate a longer reimbursement period by suppliers for higher purchases. What family families producing grape can certainly not be pleased to find is that, despite optimism, rising household spending, increasing investment in long-term assets and more intensive exports, they fail to change the share of newly created value in sales to such an extent as to improve their profitability. Now is the time to draw attention to the competitive struggle between businesses with less turnover in grape production. It turns out that increasing domestic consumption is not necessarily directed towards the production of the most vulnerable family businesses, investment in fixed assets usually reduces operating costs, but also increases the cost of wear and tear. The cost savings are not likely to be so high as to sufficiently exceed labour requirements and the cost of wear and tear on fixed assets.

2 Conclusion

The facts presented in this paper clearly demonstrate the importance of family businesses not only for the Slovak economy, but also for the world economy. Family business has its specifics, which brings certain advantages but also disadvantages. Surveys and studies by reputable companies discuss in detail the rules for the functioning of family businesses, their financial decision-making, mentioning unused opportunities and the most important challenges for the future. The knowledge base of the surveys is connected with the results of the examination of financial indicators of a selected sample of Slovak family businesses. The paper deals with the evaluation of financial condition of family businesses, its development trends, mutual relations between the results of their external and internal environment. Presented data and method of their processing are an example of analytical work with parameters of family businesses.

Acknowledgement
The article is a partial result of research within the VEGA projects n. 1/0240/20 „Financial Aspects of Sustainable Business - Enterprise Succession Solution for Small and Medium-sized Enterprises“ share 50% and
VEGA MŠ SR n. 1/0708/20 "Socio-economic determinants of sustainable consumption and production in terms of impact on performance and competitiveness of enterprises" share 50 %.

References

1. B. Maury, Family Ownership and Firm Performance: Emperical Evidence from Western European Corporations. *Journal of Corporate Finance* **12,** pp. 321-341. ISSN 0929-1199. (2006).
2. R. C Anderson, D. M. Reeb, Founding- family ownership and firm performance: Evidence from the S&P 500. *The Journal of Finance* **58**, Issue 3, pp. 301-1328. ISSN 1540-6261. (2003)
3. K. Birdi et al., The impact of human resource and operational management practices on company productivity: a longitudinal study. *Personnel Psychology* **61**, Issue 3, pp. 467-501. ISSN 1744-6570. F. De Lillo, F. Cecconi, G. Lacorata, A. Vulpiani, Sedimentation speed of inertial particles in laminar and turbulent flows. *EPL*, **84**, 314-381 (2008)
4. B. Leavy, Outsourcing strategy and a learning dilemma. Production and Inventory *Management Journal* **4**, 50-54 (1996).
5. O. Hart, The Costs and Benefits of Ownership: A Theory of Vertical and Lateral Integration. *Journal of Political Economy* **94**, 4, 691-719 (1986).
6. C. Lynch, Logistics Outsoucing – A Management Guide. Council of Logistics Management, Oak Brook (2000).
7. KPMG, Enterprise European family business barometer Succession top of mind as business families eye their legacy 2019 Eighth edition home.kpmg/familybusiness www.europeanfamilybusinesses.eu (2020)
8. E. Arrigo, Global Sourcing in Fast Fashion Retailers: Sourcing Locations and Sustainability Considerations. *Sustainability* **12**, 508 (2020).
9. M. Aoki, Horizontal vs. Vertical Information Structure of the Firm. *American Economic Review* **76**, 5, 971-983 (1986).
10. M. Pankowska, Information Technology Outsourcing Chain: Literature Review and Implications for Development of Distributed Coordination. *Sustainability* **11**, 1460 (2020).
11. A. M. Abubakr, A. T. Abbas, T. Italo, M. S. Soliman, M. Luqman, H. Hegab, Sustainable and Smart Manufacturing: An Integrated Approach. *Sustainability* **12**, 2280 (2020).
12. M. W. Rutherford, D. F. Kuratko, D. T. Holt, Examining the Link Between "Familiness" and Performance: Can the F-PEC Untangle the Family Business Theory Jungle?. *Entrepreneurship Theory and Practice* **32**, pp. 1089–1109. ISSN 1540-6520, (2008)
13. Cribis, Universal register. https://www2.cribis.sk/ExtendedSearch.aspx
14. Finstat, Database of companies and organizations. https://finstat.sk/databaza-firiem-organizacii

Specifics of Creative Business in Architecture

Miroslav Uhliar[*]

University of Economics in Bratislava, Faculty of Business Management, Department of Business Economy, Dolnozemská cesta 1/b, 852 35 Bratislava, Slovak Republic

Abstract. Architecture is one of the sectors which belong to the cultural and creative industry. It is characterized by a high level of individual creativity and creativeness. The main subjects in the architectural sector are mostly individuals or legal entities with a very small number of employees. Architecture in general is closely connected to other sectors, like construction building, and the nature of the architectural job positions requires the employees to have a wide variety of skills apart from formal education. The leading professionals in this sector are architects, landscape architects and engineers who often carry out their activities in the form of a free enterprise. As to the remuneration, the sector of architecture is considered above-average.

1 Introduction

The importance of creativity in business has been gradually increasing since the beginning of the 21st century. Creativity has become one of the essential driving powers in economics. As a result, the significance of different sectors of the creative industry has been growing as well. Creative industry represents the core of the creative economics and it consists of multiple sectors, which are based on the creativity and creativeness of individuals, their abilities, talents, and it has the potential to build and increase the number of job positions. These sectors tend to have significant differences and each of them has its own business specifics. Architecture is considered one of these creative sectors.

The objective of this article is to define architecture as one of the sectors of the creative industry, to define its subjects, make an overview of their numbers and other statistical features within the environment of the Slovak economics and to highlight the specifics of their business.

In the study, fundamental research methods were used. The data were gathered from personal interviews and statistical data were obtained mainly from Eurostat and Slovak Statistical Office. They were processed and visualized in MS Excel. Current knowledge base from available sources was analyzed and the frameworks of multiple organizations to the cultural and creative industry were compared.

[*] Corresponding author: miroslav.uhliar@euba.sk

2 Creativity, Creative Industry and Its Characteristics

From the psychologic or sociologic point of view, creativity is understood to be an ability. From the psychological point of view, it is "an ability based on intellectual processes which lead to new ideas, solutions, concepts, artistic forms, theories or products which are considered unique and unusual." [1] From the sociologic point of view, it is "an ability to create or find new, intriguing solutions for both old and recently discovered problems." [2]

It is important to define the indicators of creativity, i.e. indicators, which enable to judge the level of creativity/creativeness in business subjects, and thus it is possible to specify the different sectors which belong to the area of the creative industry. It is quite complicated to set up the adequate indicators because there are no general definitions of terms which are related to the creative industry – e.g. what the creative or cultural industries are. [3] The creative economy includes scientific, research, artistic, and other activities falling under the protection of intellectual property rights. Creative industries are seen as a source of localized innovation. [4]

The issues of the creative economics and creative industry have been tackled by a number of authors who have been observing various factors related to the creativity in their studies – the creativity factor [5], the factor of creative classes and the factor of the career [6], the factor of information and communication technologies [7], the factor of the communicativeness of a cultural product [8], the social factors and networking [9]. Florida's approach, based on the evaluation of the creative environment, may be the most discussed one. It concentrates on the essence of the creative economics, which lies in the individual, and it mostly tries to define the creative class. It recognizes three manifestations of the creativity, which, according to him, are the following: artistic creativity (cultural creativity), economic creativity (entrepreneurial spirit) and technological creativity (inventions).

Certain perspectives have been established based on the ideas of the afore-mentioned authors and these have brought new ways and documents to define the sectors of the cultural and creative industries. The most renowned ones are the following – the KEA study [10], the UNESCO methodology [11] and the ESSnet-Culture methodology [12]. The methodologies and the understanding of the structure of the different creative sectors' classes are dissimilar in these approaches.

The British Department for Digital, Culture, Media and Sport (DCMS) defined the creative industry as a set of those sectors which derive from the individual creativity and creativeness, skills and talents, and which have the potential to create wealth and working positions through the creation and use of the intellectual property. At the same time, it recognizes close economical relationships with other sectors, e.g. tourism, museums, sports. [13]

The creative industry can be perceived as a complex of sectors which depend on the use of the results of the creative intellectual activities from the cultural, artistic and other creative sectors. It is based on business activities, which derive from the individual creativity/creativeness, talent and skill. These activities have the potential to make capital and job positions through the use of the intellectual property. [14]

As there are two main groups of sectors, the terms cultural and creative industry have been distinguished both worldwide and in Slovakia. The European Union has been using the definition framework from the KEA study, where the whole area is divided into the cultural and creative sectors. In Slovakia, the cultural and creative industries have been defined according to the latter methodology of the ESSnet-Culture. The NACE codes of the cultural and creative industries have been set up as well, thus fulfilling the requirement for comparison within the Eurostat.

"Doing business in these sectors means to have creative ideas and to realize them commercially with the objective of gaining profits. However, the profit itself is not the driving power; it is the creativity and the possibility to build something, the self-fulfillment or the ability to pursue one's creative interests. It is a mix of business and creative aspects." [15] Creativity supports the establishment of innovative processes in the business entity. Such innovation process in the creative industry is characterized by close cooperation with the client, networking and partnerships with other subjects (e.g. universities, innovation incubators, cooperation with competitors, etc.) and it requires advanced skills for cooperation, team building, conflict and problem solving. [16]

The following table shows the list of different sectors which belong to the cultural and creative industries, organized by various approaches and classifications of certain organizations, arranged chronologically.

Table 1. Classification of the cultural and creative industries according to different organizations.

UNESCO (1986)	**WIPO (2003)**	**Americans for the Arts (2005)**	**KEA (2007)**	**UNCTAD (2008)**	**DCMS (2013)**	**Concept of the Slovak Republic (2014)**
Social-cultural activities						
Environment and Nature						
Sport and Games						
Cultural Heritage		Museums and Collections	Cultural Heritage	Cultural Heritage		Cultural and Natural Heritage
Press and Literature	Press and Literature	Publishing House	Books and Press	Publishing House and Printed Media	Publishing House	Literature and Books Market
Music and Dramatic Art	Music, Theatrical Production, Opera	Dramatic Art	Dramatic Art	Dramatic Art	Dramatic Art and Entertainment Industry	Theatre
			Music		Recording Studios and Music Publishing	Music
Visual Arts (incl. arts and crafts)	Visual and Graphics	Visual Arts and Photography	Visual Arts	Visual Arts		Crafts
Photography	Photography				Photography	Arts Market
						Visual Arts
Film	Film and Video	Film	Film and Video	Audiovisual Media	Film, Video and TV Programs	Film and Video
Broadcasting and Television	Broadcasting and Television	Broadcasting and Television	Television and Broadcasting		Programs and Broadcasting	Communication Media
	Advertising Services		Advertising	Functional Creations: Design, New	Advertising	Advertisement Industry
					PR and Communication	Information-Communicational

UNESCO (1986)	WIPO (2003)	Americans for the Arts (2005)	KEA (2007)	UNCTAD (2008)	DCMS (2013)	Concept of the Slovak Republic (2014)
				Media (software, digital content and games, creative services like architecture and advertising, digital services and holiday services)		Technologies
					Translation and Interpretation	Hardware and Related Services
		Design	Design		Specialized Design	Design
			Architecture		Architecture	Architecture
	Software and Databases		Video Games		Software Publishing	Multimedia Industry
					Computer Programming	Software
					Computer Consulting	Entertainment Industry
	Collective Management of Copyrights	Artistic Schools and Services			Cultural Education	Memory Institutions
						Fashion Industry
						Supporting Services

Source: Author's own adaptation based on [17] and [14]

Notes: abbreviations:

UNESCO - United Nations Educational, Scientific and Cultural Organization, headquarters: Paris, France, est.: 1945
WIPO - World Intellectual Property Organization, headquarters: Geneva, Switzerland, est.: 1967
Americans for the Arts – headquarters: Washington, USA, est.: 1960
KEA – KEA European Affairs, est.: 1999
UNCTAD - United Nations Conference on Trade and Development, headquarters: Geneva, Switzerland, est.: 1964
DCMS - Department for Digital, Culture, Media and Sport, headquarters: London, UK, est.: 1997
Concept of the Slovak Republic – Východiská stratégie rozvoja kreatívneho priemyslu v Slovenskej republike (Base of the Development Strategy for the Creative Industry in the Slovak Republic), approved in 2014

3 Results and Discussions

3.1 Sector of Architecture – Part of the Creative Industry and its Position in Slovakia

Architecture is a complex and creative industry. Various concepts and definitions of the creative industry place it into the sector of the cultural and creative industry. KEA from 2007 defines this sector as an independent class – Architecture; according to the UNCTAD study it belongs to the class Functional Creations (which includes design and new media like software, digital content and games, creative services such as architecture and advertising, digital and holiday services). It also has its independent place in the study from the DCMS and in the Concept of the Slovak Republic from 2014. These results reflect the fact that the business subjects from the architecture sector present a high rate of human creativity, which is included in the creation of the added value – this is divided into an economic and symbolic value, such as image and reputation.

Architecture mostly means providing a certain service. It is the kind of individual or corporate business activity which requires a certificate of professional competence. According to the SK NACE classification, it belongs to the class 71.11 – Architectural Activities, which includes two types of architectural counseling: Design and Development of a Building Project, Projecting of Towns and Villages and Landscape Architecture. In Slovakia, business in this sector is regulated, among other things, by the Act No. 138/1992 Coll. – Act by the Slovak National Council on Authorized Architects and Authorized Construction Engineers. [18] This Act limits the activities of the authorized persons followingly:

- **architects:**
  a) realization of preliminary pre-project activities,
  b) realization of complex project activities,
  c) processing of landscape-planning templates and landscape-planning documentation,
  d) project management,
  e) graphics and modeling of the architectural work of art,
  f) providing related technical services, mainly architectural counselling,
  g) elaboration of expert reports, assessments and references.

- **landscape architects:**
  a) realization of pre-project activities, mainly for the processing of landscape-planning templates and landscape-planning surveys and analyses,
  b) cooperation on the creation of the landscape-planning documentation within the landscape formation,
  c) realization of complex project activities,
  d) incorporation of constructions into the environment with the objective to minimize any negative impacts,
  e) realization of professional author's supervision over the creation of the works of art based on official approved documents,
  f) providing related services, mainly professional counselling,
  g) graphic output and modeling of landscape and garden architecture works of art, including the elaboration of specifications and technical documentation
  h) elaboration of expert reports, assessments and references.

- **engineers:**

a) realization of complex architectural and engineering services and related technical counselling or

b) providing services from the following categories:

1. engineer for the construction of surface structures,
2. engineer for the construction of engineering structures,
3. engineer for the structure statics,
4. engineer for technical, technological and energetic equipment of the structures.

In Slovakia, the structure of the creative architectural sector is mostly made by studios, individual architects and designer offices. Architects from linked sectors (designing, real estate, sales of interiors, etc.) also belong here, together with professions like restorers and persons with exclusive expert qualification for the realization of ancient monuments' research.

In terms of the development of the overall creative industry, Slovakia can be assessed as stagnant in the period between the years 2010 and 2015. The slight growth in the number of enterprises in the creative industries and the labor productivity of their employees can be viewed positively. [19]

The following table shows the number of subjects in the architectural sector in 2013-2017 in Slovakia and its neighboring countries.

Table 2. Number of subjects in the sector of Architecture and its development in selected EU countries

Country	**2013**	**2014**	**2015**	**2016**	**2017**
Slovakia	10112	10502	10986	11721	12298
Czech Republic	32334	32713	33360	34229	34942
Hungary	19706	21132	21809	22764	24009
Austria	15578	15805	15889	15947	16125
Poland	53070	55849	54429	57489	58102

Source: Eurostat, Author's own adaptation

The number of subjects in the creative sector of architecture has been increasing constantly since 2013 not only in Slovakia, but also in its neighboring countries. According to the last data from 2019, although including only corporations from the given sector, the number of legal entities was almost 1500, with mostly the form of limited liability companies (97.53 %), rarely joint-stock companies (1.53 %). The structure of the architectural sector significantly favors natural persons who make business through free enterprise within this sector.

The following graph shows the number of employed people and the development of the employment rate in the sector of architecture in Slovakia and the neighboring countries. In 2017, there were 21 579 people working in the sector of architecture, which represented approximately 0.9% of the total employment. It may appear as a small number at the first sight, but it is only one part of the creative industry.

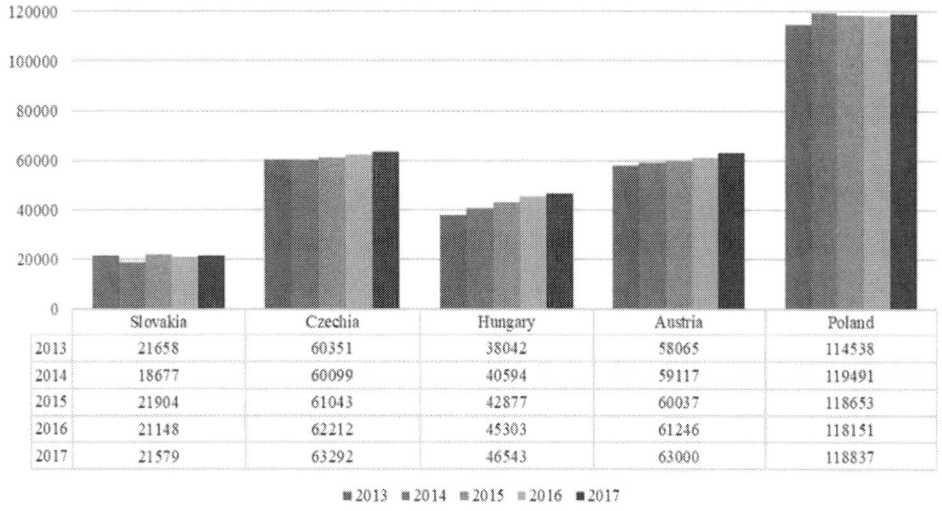

	Slovakia	Czechia	Hungary	Austria	Poland
2013	21658	60351	38042	58065	114538
2014	18677	60099	40594	59117	119491
2015	21904	61043	42877	60037	118653
2016	21148	62212	45303	61246	118151
2017	21579	63292	46543	63000	118837

Figure 1. Development of the Number of People Employed in Architecture in Selected EU Countries

Source: Eurostat, Author's own adaptation

The people employed in the sector of architecture have mostly gained university degrees. This fact, together with the extent of their knowledge and experience, impacts the remuneration within the sector. The table below shows orientational salaries for selected work positions.

Table 3. Selected work positions of the creative sector of architecture and the range of gross salaries, Slovakia, 2019

Work Position	**Salary Range (Min – Max; Mid-Point)**
Architect	826 – 1767 EUR; 1296,5 EUR
Interior Architect	672 – 1645 EUR; 1158,5 EUR
Landscape Architect	580 – 1610 EUR; 1095,0 EUR

Source: www.platy.sk, Author's own adaptation

The average gross monthly salary in Slovakia, according to the Statistical Office, was 1 063 euros in 2019. It increased from 1 013 euros in 2018 and 954 euros back in 2017. If we compare the general average salary in Slovakia to the salaries of selected work positions from the architectural sector (mentioned in the table above), we can declare that the salary conditions in this creative sector are above-average.

3.2 Business in the Sector of Architecture and its Specifics

Business and the basic relationships, which influence the subjects of the creative architectural sector, are determined by the relationship triangle between the investor, architect and supplier. This creative sector also needs to be perceived as a part of a complex interconnected industry of architecture – engineering – construction building. Business in the sector of architecture is directly dependent on the current development in construction building, on the general situation on the real estate market, on the purchase power of the citizens, on the scope and level of public investment projects and on the capital cumulation as well. One of the factors, which significantly impact the market, is the state of the public space, its vision and the method of its development. High sensibility to economic recession is one of the specific features of this business, as recession damages a high number of investment and development projects, which leads to a reduction of many architectural services. As this sector is considered creative, there is a high amount of innovations, a wide use of software solutions and modern technologies. Application of virtual and extended reality is one of the current trends.

The architectural profession is well remunerated, and it is a well-organized expert profession with highly skilled labor. The business subjects of architecture create value through processes, which require them to have wider knowledge than their clients, either in the sense of education or practical experience. This is the reason why there is high dependence on people, their creativity, quality, knowledge and experience, because they are the main factor of value creation and they make important interactions with the clients. According to these specifics, there is a presumption that anyone with the required education and certificate of expertise can do business in this sector on their own behalf. As a result, there are mostly individual entrepreneurs – natural persons or small corporations (especially limited liability companies) doing business in this sector. If we analyze the number of employees and the number of business subjects in the creative architectural sector, we can see that the number of employees per one business entity is very small, as it is shown in the following graph.

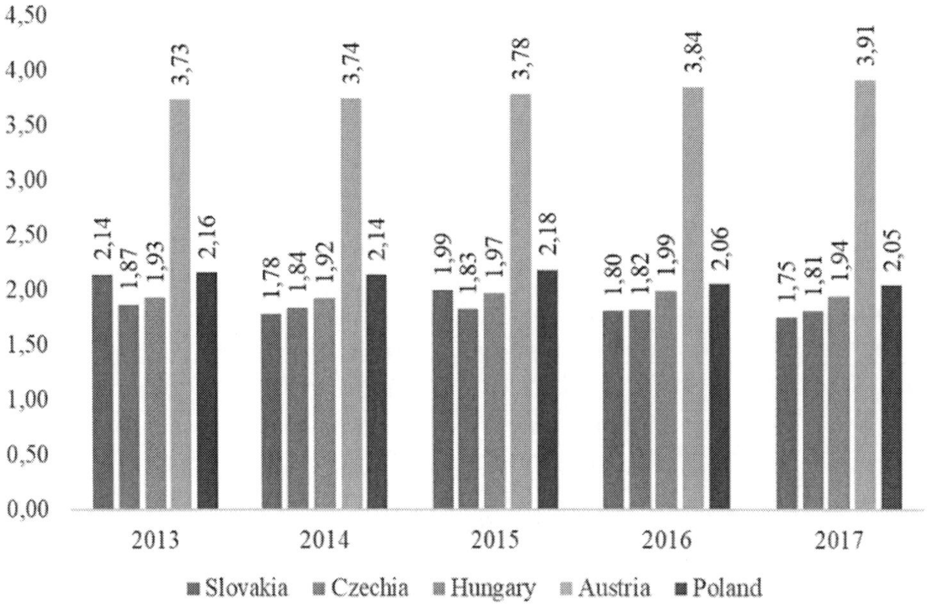

Figure 2. Number of Employees per Business Subjects in Architecture in Selected EU Countries

Source: data from Eurostat and Finstat, Author's own adaptation

The average number of employees per legal entity in the sector of architecture is very similar in countries like Slovakia, Czech Republic, Hungary and Poland. It is usually 2 employees per entity. Within Slovakia's neighboring countries, only Austria has shown a major deviation – the business subjects in architecture are larger and reach the average number of 4 employees.

The fact that the subjects in architecture are mostly individuals and small companies with a low number of employees stresses the need for wider specialization and more universal focus of the employees in this sector. Apart from a certain level of expertise, it is necessary for them to have practical experience, to be able to work with modern digital technologies, to master the system of project management, to be skilled at communication and psychology and, at the same time, they need to understand the related legislative, pricing and business economy.

The future of the architecture as a part of the creative sector in Slovakia depends mainly on the subjects themselves and their ability to respond to current economic, technologic and social trends in the world. Rapidly developing areas show that the most important thing is the capability of invention of the subjects themselves, their ability to cooperate and to organize internally. It is important that the state and public policies also help in this process. [20]

4 Conclusion

The importance of creativity as a motive force has been increasing. This fact made it necessary to define the creative industry and its sectors. Architecture, as an independent creative sector, was first defined in the KEA concept in 2007; it has an equally strong position in more recent approaches. According to the Concept of the Slovak Republic from 2014, it is a strong component of the creative sector as well. This sector is mostly represented by individual architects, studios and design offices.

The number of subjects on the architectural sector has increased during the past years. The same can be stated about the number of employees in this sector. In Slovakia, the average number of employees per one architectural business subject is two – generally, the employees have reached high formal education and they also show a high level of expertise. This fact impacts the remuneration, which is considered above-average in Slovakia – the salaries of the architectural experts overcome the monthly national average.

Business in this sector has its characteristic features and these are determined by the relationship triangle between the investor, architect and supplier. The sector is also significantly influenced by the current state of the national economics and the phase of the economic cycle; it depends on the development of the construction industry, the real estate market and the purchase power of the citizens. It greatly reacts to the economic recession, as it slows down the developers' investment projects. There is a high amount of innovations applied in architecture; a wide use of software solutions and modern technologies is also a typical feature. Application of virtual and extended reality is one of the current trends, accompanied by the use of cloud services. All these facts, together with the development of the IT and the human society in general, will continue to push further increase of the creativity and creativeness in the sector of architecture and its products and services.

This article is a partial output of the VEGA MŠ SR 1/0340/19 project – "Podnikateľský rozmer subjektov kreatívneho priemyslu v kontexte inovácií a inteligentného rastu" to the extent of 100%.

References

1. P. Hartl, H. Hartlová. Psychologický slovník. [Psychological Dictionary.]. 1st ed. Praha: Portál. ISBN 80-7178-303-X. (2000).

2. J. Linhart, M. Petrusek, A. Vodáková, H. Maříková. *Velký sociologický slovník*. Praha: Karolinum. ISBN 80-7184-310-5. (1996).

3. M. Kačírková. *Kreativita v kontexte znalostnej spoločnosti*. Bratislava: Ekonomický ústav SAV. ISSN 1337-5598 (2017).

4. R. Boix, F. Capone, L. D. Propris, Comparing creative industries in Europe. *European Urban and Regional Studies* **23**, 935-940 (2016)

5. D. Throsby. *Economics and Culture*. Cambridge University Press. Cambridge. ISBN 0-521-58639-9. (2001)

6. R. Florida. *The Rise of the Creative Class: And How It's Transforming Work, Leisure and Everyday Life*. New York: Basic Books, 2002. ISBN 978-0-455-02993-8. (2002).

7. J. Hartley. The New Economy, Creativity and Consumption. *International Journal of Cultural Studies*, **7** (1) (2004).

8. D. Hesmondhalgh. *The Cultural Industries*. London: SAGE Publications, ISBN 978-1-4129-0808-5. (2007).

9. J. Potts, S. Cunningham, J. Hartley, P. Ormerod. Social network markets: A new definition of the creative industries. *Journal of Cultural Economics*, **32**, č. 3, s. 167 – 185. (2008)

10. KEA European affairs. The Economy of Culture in Europe. (2006). http://ec.europa.eu/culture/library/studies/cultural-economy_en.pdf, [accessed 3.16.2020]

11. UNESCO. The UNESCO Framework for Cultural Statistics. Statistical Commission and Economic Commission for Europe. Conference of European Statisticians. Third Joint meeting on Cultural Statistics, 17-20 March 1986. CES/AC/44/11. 13 February 1986. (1986)

12. EUROPEAN COMMISSION. ESSnet-Culture 2012. (2012) http://ec.europa.eu/culture/library/reports/ess-net-report_en.pdf, [accessed 3.16.2020]

13. UK GOVERNMENT. Creative Industries Mapping Documents 2011 (2011). https://assets.publishing.service.gov.uk/government/uploads/system/uploads/attachment_data/file/183544/2001part1-foreword2001.pdf, [accessed 3.16.2020]

14. MK SR a MH SR. Stratégia rozvoja kreatívneho priemyslu v Slovenskej republike. (2014). http://www.culture.gov.sk/extdoc/5963/strat%E9gia, [accessed 3.16.2020]

15. HKU Utrecht Scholl of the Arts. *The Entrepreneurial Dimension of the Cultural and Creative Industries*, Utrecht. ISBN/EAN: 978-90-817243-1-9. (2010).

16. H. Majdúchová and M. Kmety Barteková. Innovations in the Creative Industry Entities. Globalization and its Socio-Economic Consequences 2019: Sustainability in the Global-Knowledge Economy, pp. 1-8. ISSN 2261-2424 (2019).

17. C. Jones, M. Lorenzen, J. Sapsed. *The Oxford Handbook of Creative Industries*. New York: Oxford University Press. ISBN 978-0199603510 (2018).

18. Act No. 138/1992 Coll. – Act of the Slovak National Council on Authorized Architects and Authorized Construction Engineers. (1992)

19. D. Rybárová. Creative Industry as a Key Creative Component of the Slovak Economy. Globalization and its Socio-Economic Consequences 2019: Sustainability in the Global-Knowledge Economy, pp. 1-8. ISSN 2261-2424 (2019).

20. S. Šagátová. The Impact of Globalization on the Development of Creative Industries Businesses. Globalization and its Socio-Economic Consequences 2019: Sustainability in the Global-Knowledge Economy, pp. 1-8. ISSN 2261-2424 (2019).

Environmental indicators

Katarina Vavrova[1,*]

[1]University of Economics in Bratislava, Faculty of Business Management, Dolnozemská 1, 85235 Bratislava, Slovak Republic

Abstract. Environmental indicators currently are considered to be very important in achieving environmental policy objectives. The aim of the paper is to present the most important information on environmental indicators in the European Union. The environmental taxes are measured using indicators, reports and sets of indicators. The evaluation is based on the methodology of the Organization for Economic Cooperation and Development (OECD) and the European Environment Agency (EEA). The methodology for assessing indicators examines the causal link between the current state of environmental taxes and current environmental trends. The effective introduction of environmental taxes based on an environmental strategy can ultimately lead to many desirable effects in European Union countries. Achieving environmental excellence should be a priority objective for both countries and companies, as the development and successful implementation of policies and environmental strategies is a key element of companies around the world.

1 INTRODUCTION

Today's society is characterized by an immediate benefit of business and does not reflect on the threat to the future with respect to the environment. Countries are also trying to tackle the unfavorable situation through taxes. Environmental taxes are key instruments for achieving sustainability in the economy by increasing the prices of environmentally harmful goods or the cost of production inputs. Developed countries, such as the European Union, have been drawing attention to this area of the economy more and more recently.

Growing concerns about climate change have put environmental issues at the forefront of the economic agenda in many European countries. Global warming is one of the most important challenges facing the world [1]. Taxes, fees, tradable permits and other economic instruments play an important role in achieving cost-effective greenhouse gas emission control. Their potential scale and revenues increase the much broader implications of economic and fiscal policy. European countries introduced carbon taxes in the 1990s, although the proposal for a European energy tax was ultimately unsuccessful. Recently, the focus has been on emissions trading. Several tax measures have been introduced in the EU, mainly with environmental objectives. They included national environmental taxes along with the existence of environmental tax measures. The increasing use of environmental taxes, emissions trading and other economic instruments has been partly conditional on the

* Corresponding author: katarina.vavrova@euba.sk

recognition of the limitations of conventional environmental regulation. They can initiate the necessary changes in the economy, resource use, behavior and universal access to nature [2]. The relationship between environmental knowledge, attitudes and behavior is complex. Moreover, electricity consumption in the six Gulf Cooperation Council countries has increased rapidly over the last decades as a result of rapid population growth and relatively rapid economic growth [3]. Energy is an integral part of our daily life in society in a market economy. We need it for heating, cooling, lighting and movement; this is important for the functioning of our offices, workplaces and the whole economy [4]. This is one of the reasons why the Commission EU has proposed its Energy Union strategy.

2 Data and methods

Indicators are measurable quantities providing information on the development and trends of phenomena and processes, in quantitative and valitative terms. The environmental indicator significantly helps in planning, setting policy objectives, including monitoring their implementation, and in developing follow-up measures and instruments to achieve them in various policy and strategy documents. It is also a comprehensive source of information on the state and development of the environment and related aspects for the general public [5]. It also constitutes an important basis for the subsequent drafting of the relevant types of reports. The Slovak Environmental Agency processes and regularly evaluates various sets of indicators, namely [6]:

- Key indicators.
- Sector indicators.
- Sustainable Development (TUR) indicators.
- Green growth indicators, **environmental indicators.**
- Resource efficiency indicators.

The stochastic indicator consists of two curves: the% K curve and the% D curve. The Mathematical calculation is apparent from the attached formulas:

$$\%K = 100 \times [(C - Lx) \div (Hx - Lx)] \qquad (1)$$

Where C = the last closing price (close) at the relevant time period (timeframe); Lx = the lowest point (low) for a selected time period; and Hx = the highest point (high) for the selected time period.

$$\%D = 100 \times [Hn \div Ln] \qquad (2)$$

where Hn = n-day sum (C-Lx); Ln = n-day amount (Hx-Lx) (For daily graph).

The results of these two mathematical expressions are lines that oscillate between 0 and 100. The K(%) curve is usually indicated by a solid line and a slow D(%) curve by a dashed or dotted line. The curve K shows (expressed as a percentage), which is the cost of closing toward the x-day (or period) range. Row D is nothing but the n-day floating version of K. The results of these two mathematical expressions are two lines that oscillate between 0 and 100. The curve shows K (expressed as a percentage), which is the closing price toward the x-day range (or period). Indicators are measurable quantities that provide information on developments in quantitative and qualitative terms. They should be evaluated on the basis of a set of sectoral indicators, green indicators, resource indicators. Sectoral indicators are a means of assessing progress in implementing sectoral policies in relation to the environment. Based on the analysis of indicators and regular evaluation of the European Environment Agency (EEA) [7], the evaluation options of the Organization for Economic Development and Economic Cooperation (OECD) and of the European Union Office of Indicators (Eurostat) were evaluated. Flexible and liquid markets can enable supply and demand to be more efficiently adjusted, reducing production costs and therefore prices. Such situations should also trigger bilateral contract prices in the most developed markets [8]. Gradually, these separate national wholesale markets are merging with neighboring markets, which together with several interconnections of the transmission network create more efficient markets. Prices are influenced by various factors, including fuel mix, cross-border connections, market connections, market concentration of suppliers and weather conditions. Energy efficiency and weather also affect prices, demand of consumers and industry business, Figure 1.

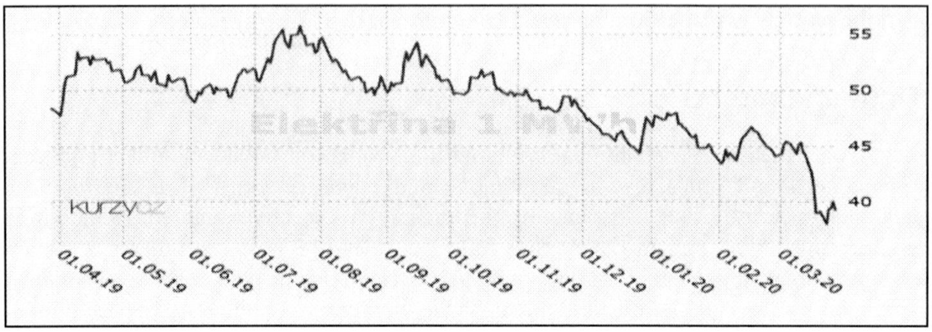

Fig.1. Figure of electricity price development in years 2019-2020. Source: Eurostat

Why are we monitoring the development of electricity prices on the wholesale market? Econometric analysis suggests that a 1% increase in the share of fossil fuels (coal, natural gas and oil) in the energy mix leads to an increase in wholesale electricity prices of EUR 0.2 - 1.3 / MWh depending on the regional market, an increase in several markets The low marginal cost of solar and wind electricity reduces wholesale prices. The econometric analysis also suggests that each percentage increase in the share of renewable energy on average reduces the wholesale electricity price by EUR 0.4 / MWh in the EU; the actual reduction depends on the regional market and the fuel source is replaced by renewable sources. In Northwest Europe, the Baltic and Central and Eastern Europe, the impact of renewable energy sources is even greater (0.6-0.8 EUR / MWh) Figure 2.

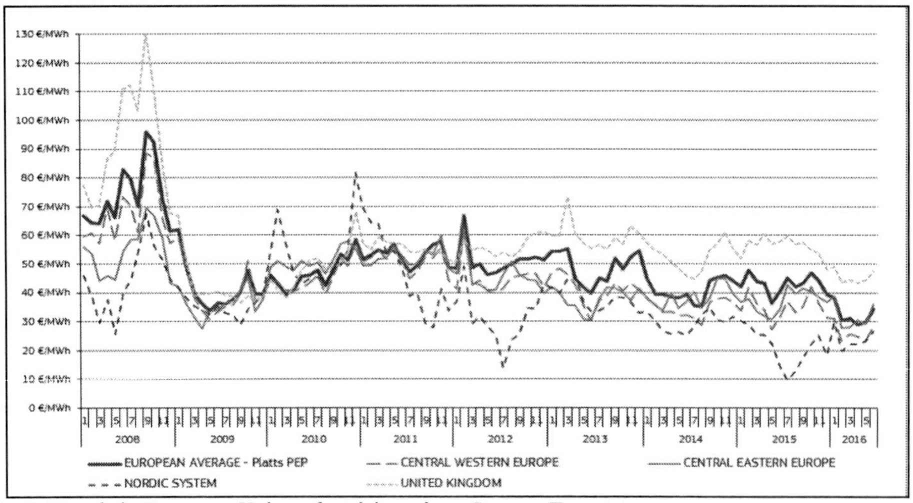

Fig. 2. Trends in European Union, electricity prices. Source: Eurostat

2.1 Environmental indicators

Indicators are measurable quantities providing information on the development and trends of phenomena and processes, in quantitative and qualitative terms. Regarding the indicators, we state that the fundamental aim of the research is to identify key environmental indicators that credibly reflect environmental policy. The basic starting point for their identification was the results of the correlation analysis, where the individual environmental indicators obtained from the data of individual sectors of the sample were used as variables. Based on the analysis of indicators developed and regularly evaluated by the European Environment Agency (EEA), the Organization for Economic Development and Economic Cooperation (OECD), the European Community Statistical Office (EUROSTAT). Environmental taxes are one of the most important means of the European Union's environmental policy. The evolution of total revenue from this tax category can be seen in Figure 3, where the revenue is shown across the EU over a 10-year period. From graph no. 3 and after calculating the evolution index over time, it is clear, that over the last ten years the total EU revenue from environmental taxes has increased by 19.67% between 2008 and 2018. In 2018, total revenues from environmental taxes amounted to 381, 381. 85 Million Eura, which represents an increase in revenues of 59, 891 Million Eura compared to 2007. The data show that the overall revenue from environmental taxes in the EU is increasing [9]. The justification for revenue growth stems from rising environmental tax rates. The exception to this upward trend is the years 2007 and 2009, during which Europe went through the global financial crisis [10].

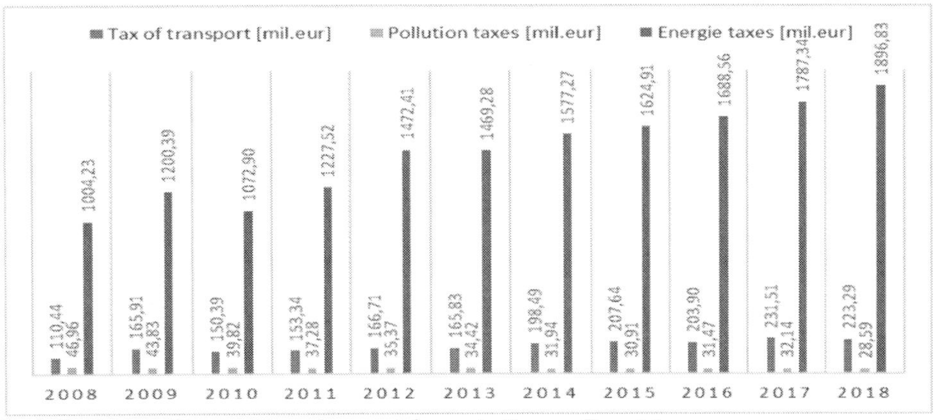

Fig. 3. Evolution of taxes with environmental aspects. Source: Eurostat

From figure no. 4 it is clear, how the Percentage of Total Revenues from Taxes and Social contributions has developed in years 2009-2018 in the seven countries of the European Union. For the numerator, the OECD Secretariat uses revenue figures that are submitted annually by correspondents from national Ministries of Finance, Tax Administrations or National Statistics Offices [11]. Although provisional figures for most countries become available with a lag of about six months, finished data become available with a lag of around one and a half years. Final revenue data for 2017 were received during the period May-August 2019. In 33 OECD countries the reporting year coincides with the calendar year. This has been sought to be achieved through flexible recruitment practices designed to induct quality applicants, a focus on training and development, and the development of systems that capture and allow the repetition of best practice. As they say, globalization is also one of the most important realities [12]. The modern trend for tax agencies is for more highly skilled officers, with merit-based promotion, and performance management and development systems [13].

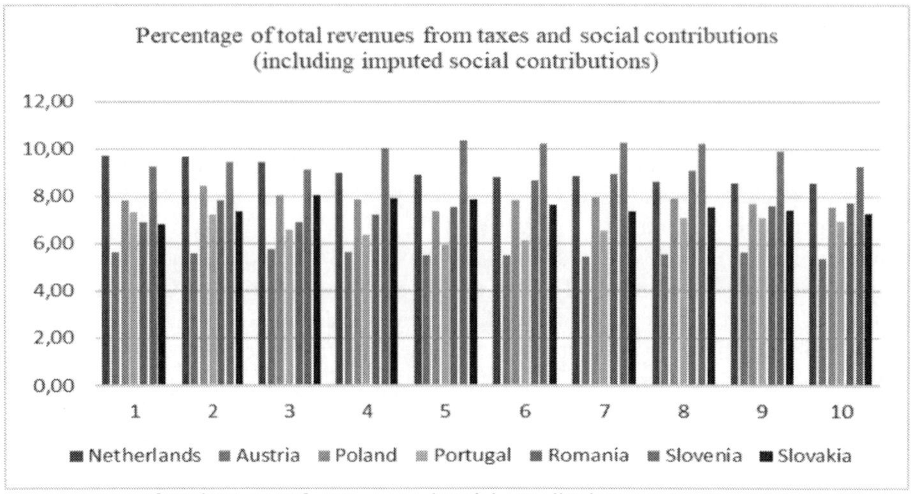

Fig. 4. Percentage of total revenues from taxes and social contributions. Source: Eurostat

The figure 5 is obvious, that in 2018, in OECD countries had an increase in their tax-to-GDP ratio relative to 2017. In all of these countries, GDP growth was positive, although to a lesser degree than tax revenue growth. Of the 15 OECD countries that experienced a decline in their tax-to-GDP ratio in 2018, thirteen had higher levels of tax revenues in nominal terms than the preceding year, but the increase in nominal tax revenues was less than the increase in nominal GDP levels. Two countries (the United States and Israel) had positive nominal GDP growth and negative tax revenue growth; but no countries experienced declines in nominal GDP (Figure 5). The figure 5 is obvious, that in 2018, the governments in the EU collected environmental tax revenue of EUR 324.6 billion. The value represents 2.4 % of the EU gross domestic product (GDP) and 6.0 % of the EU total government revenue from taxes and social contributions.

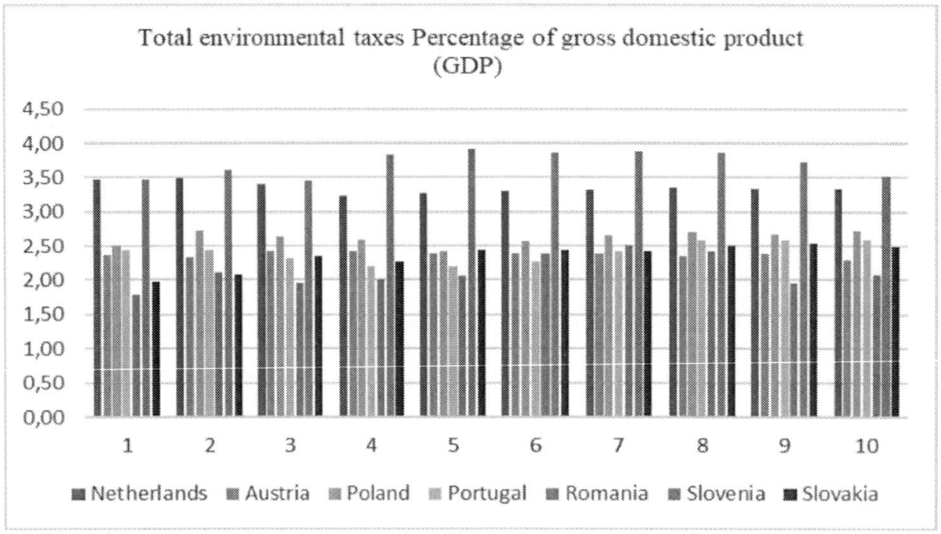

Fig. 5. Total environmental taxes – percentage of gross domestic product. Source: Eurostat

Figure 6 is obvious, still, the evolution of the environmental tax revenue relative to GDP and TSC seems to have changed its pattern in 2008, and again in 2016. After five years of consecutive decrease, it started to increase in 2009 (presumably due to the economic recession and drop in both nominal GDP and government revenue, as a result of the financial crisis) and it remained relatively stable thereafter. Starting from 2016, the ratios have been slowly falling again.

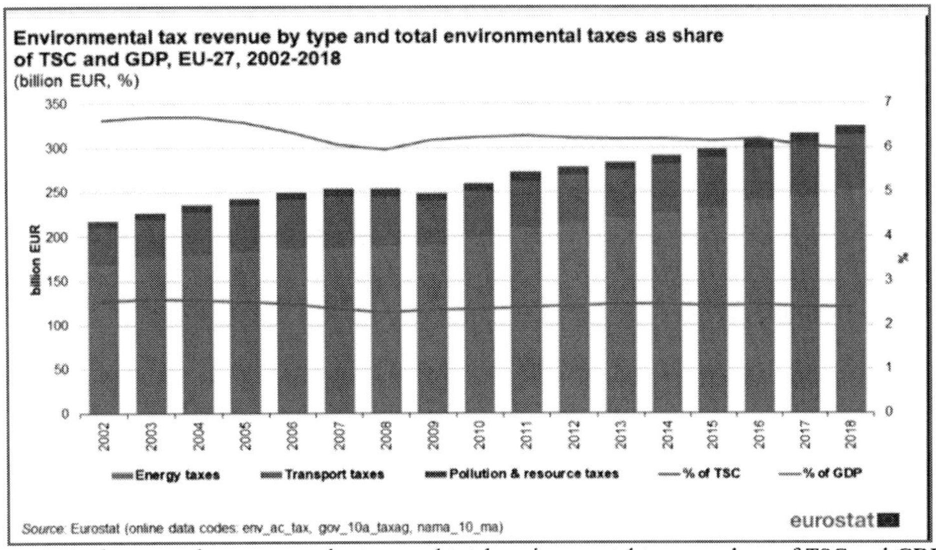

Fig. 6. Environmental tax revenue by type and total environmental taxes as share of TSC and GDP, EU-27, 2002-2018. Source: Eurostat

3 Results and Discussions

Figure 7 is obvious, that economic instruments for pollution control and natural resource management are an important part of environmental policy in the EU Member States [14]. The range of instruments that are available includes, among others, environmental taxes, fees and charges, tradable permits, deposit-refund systems and subsidies. Environmental taxes are used to influence the behaviour of economic operators, whether producers or consumers. The EU has progressively favoured these instruments because they provide a flexible and cost-effective means for reinforcing the polluter-pays principle and for reaching environmental policy objectives. The use of economic tools for the benefit of the environment is promoted in the EU Environment Action Programme to 2020, the EU sustainable development goals and the Europe 2020 strategy .
Environmental indicators, together with indicators supporting Europe 2020, continue a series of Eurostat's major publications supporting Europe 2020 by monitoring progress towards the objectives defined under the three mutually reinforcing priorities of smart, sustainable and inclusive growth. The analysis is based on the Europe 2020 headline indicators chosen to monitor progress towards the objectives of the strategy. Further breakdowns targeting specific sub-groups of society or economy are also used to deepen the analysis and provide a more detailed picture. The data come mainly from official statistics produced by the European Statistical System and disseminated by Eurostat.

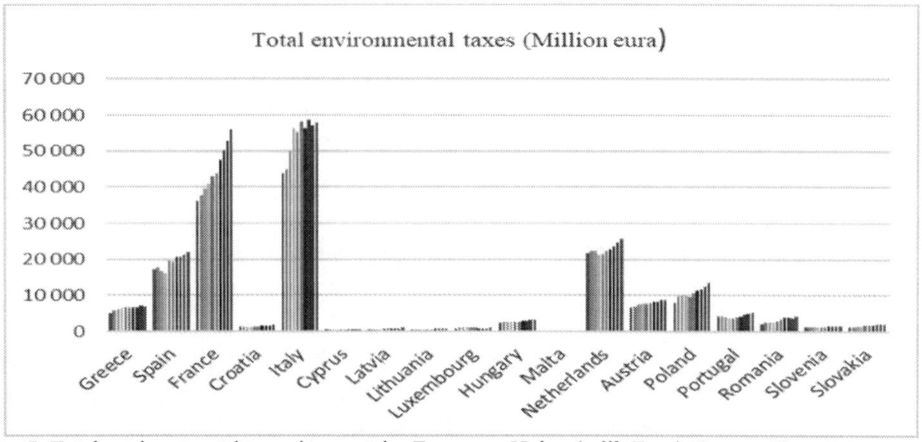

Fig. 7. Total environmental taxes in countries European Union (mill. Eura). Source: Eurostat

4 Conclusion

In the context of environmental indicators, needs considered in energy. Primary energy production is any extraction of energy products in a usable form from natural sources [15]. This occurs either in the exploitation of natural resources (for example, in coal mines, oil fields, hydropower plants) or in the production of biofuels. For example, primary energy production in the EU-28 was 771 million tonnes of oil equivalent (peak) and 757 million peaks in 2016 (from available data) in 2015, while world production reached 13.79 billion peaks in 2015. Members of the G20 include China, the United States and Russia. These countries recorded higher levels of production than in the EU-28. The increased awareness of the need to combat environmental pollution and preserve natural resources has led to an increase in the supply and demand of environmental goods and services, in other words, products to prevent, measure, control, limit, minimise or correct environmental damage and resource depletion. Statistics on environmental protection expenditure enable to identify and measure society's response to environmental concerns and how it is financed. Environmental protection expenditure accounts (EPEA) quantify the
resources devoted to environmental protection by resident economic units. The results of the 2018 data collection on environmental protection expenditure accounts (EPEA) provided by European Union (EU) Member States with reporting obligation (Cyprus was granted a derogation) and selected non-EU countries [16]. It provides information on the EU's expenditure on prevention, reduction and elimination of pollution or any other degradation of the environment and covers the total spending by a country (i.e. by its households, businesses and government) on environmental protection services, e.g. waste and wastewater management, protection of biodiversity, as well as protection of soil, research and development, education and training.

Acknowledgment
The presented working paper is the output of the scientific grants VEGA n. 1/0007/19 "Allocation of assets in the environment of low interest rates in financial and non-financial corporations in Slovakia" in the range of 100%.

References

1. N. Suwirman, Environmental Altruistic Behavioral. *International Journal of Environmental Research and Development* **4**, 1 (2014)
2. F. Aidyn, Secondary school students' perceptions towards global warming: a phenomenographic analysis. *American Journal of Nursing Science,* Essays **5**, 2328 (2017)
3. J.M. Hines, H.R. Hungerford, A.N. Tomera, Analysis and Synthesis of Research on Responsible Environmental Behavior: A Meta-Analysis. *The Journal of Environmental Education* **18**, 1 (2010)
4. J. Squalli, Electricity consumption and economic growth: bounds and causality analyses of OPEC members. *Energy Economics Journal* **29**, 1192 (2007)
5. C. Fischer, W.H. Parry, W.A. Pizer, Instrument choice for environmental protection when technological innovation is endogenous. *Journal of Environmental Economics and Management* **45**, 523 (2003)
6. Energy, transport and environment statistics 2019 edition, *The Statistical books, Eurostat, Printed by Imprimeries Bietlot Freres in Belgium*, European Union, (2019).
7. I. Ioannou, G. Serafeim, What drives corporate social performance, International evidence from social, environmental and governance scores. *Journal of International Business Studies* **43**, 834 (2012)
8. P. Deng, A. Delios, M.P. Pemg, A geographic relational perspective on the internationalization of emerging market firms. *Journal of International Business Studies* **51**, 50 (2020)
9. G. Tomas, M. Hult, M.A. Gonzalez-Perez, K. Lagerstrom, The theoretical evolution and use of the Uppsala Model of internationalization in the international business ecosystem. *Journal of International Business Studies* **51**, 38 (2020)
10. M.A. Marinov, S.T. Marinova, J.A. Larimo, T. Lepovsky, Characteristics and Future Trends. *Internetional Business and Emerging Economy Firms* **1**, 1 (2020)
11. J. Kwan Kim, R. Mudambi, An ecosystem-based analysis of design innovation infringements : South Korea and China in the global tire industry. *Journal of International Business policy* **3**, 38 (2020)
12. M. Minárová, D. Malá, M. Sedliačiková, Emotional Intelligence of Manageres. 4th Conference on Business. *Economics and Management Journal* (WCBEM). *Ephesus, Turkey* **124**, 1119 (2015)
13. J. H. Rogers, C. Scotti, J. H. Wright, Evaluating asset-market effects of unconventional monetary policy: a cross-country comparison. *Journal Economic Policy* **29**, 749 (2014)
14. J. Dobrovič, A. Korauš, L. Dančišinová, Sustainable economic development of Slovakia: Factors determining optimal tax collection. *Journal of Security and Sustainability Issuess, International Entrepreneurial perspectives and Innovative Outcomes* **5**, 533 (2016)
15. D. Youová, Environmental taxation. *Brussels,* [electronic source]. Available on the Internet: [cit. 2018-02-02]
16. K. Mc Cullagh, O. Tambou, S. Bourton, (Eds.), National Adaptations of the GDPR, *Collection Open Access Book, Blogdroiteuropeen*, Luxembourg, 1- 130 (2019)

Factors influencing local competition intensity

Branislav Zagorsek*

University of Economics in Bratislava, Department of Management, 852 35 Bratislava, Slovak republic

Abstract. This paper investigates the factors influencing the intensity of local competition. The research results may be of interest to companies if deciding whether to enter a market or by defining a new market. Understanding which factors impact the competition the most also helps to address these factors when dealing with competitors and other subjects. The studied factors were selected based on Porter's model of five competitive forces. The data were obtained from the World Economic Forum database. To analyze the data, the statistics software PSPP and the programming language Python with statistical package Statsmodels were used. The results were considered to be statistically significant if they were on significance level $p \leq 0.05$. The finding is that higher competition intensity is positively influenced by the number of local suppliers, foreign competition, intellectual property protection, and negatively by spending on research and development. This research may serve as a guideline for companies. If a company is entering a new market, the results presented in this paper may help to assess the intensity level of local competition by identifying the most influential factors.

1 Introduction

The globalization of the economy is represented by foreign trade, the flow of capital, and the spread of technologies, along with market integration [1]. It affects companies on the national as well as on the local scale by intensifying competition. Among other issues, local businesses do not only compete with their direct neighbors but also with businesses on the other end of the world. Information technologies are another factor that contributes to this intensification of competition. Thanks to the widespread and easily accessible information the consumers are more informed about the products and the product providers. It is essential to understand these mechanisms and factors influencing competitiveness so that the companies can adjust their business models and activities accordingly.

* branislav.zagorsek@euba.sk

In this paper, the factors that impact the intensity of local competition were examined. The objective is to identify factors that are influencing this intensity. We want to find out whether there are factors that companies can focus on to deal with competition more effectively.

2 Literature review

Alexander [2] by showing how territorial location can shape the sustainability of the product, depending on whether adequate waste processing facilities are available and how related practices are regulated, created a basis for the argument that the presence of local suppliers might influence the business. Ma [3] argues that competitive advantage is not defined clearly enough and that competitive advantage and performance are rather two different constructs. Dollinger [4] wrote that sustainable competitive advantage arises when the enterprise creates and captures value and protects it against erosion by competitors. Dollinger also describes the four conditions affecting higher competition intensity. These conditions include numerous equally balanced competitors, slow growth in the industry, high fixed costs and a commodity-type product. Porter [5] has identified five competitive forces to describe the competition and subsequently adjust the strategy. These five competitive forces include supplier power, buyer power; the threat of substitution; the threat of new entry and competitive rivalry. Malecki and Oinas [6] claim that the level of innovativeness and competitiveness of firms everywhere depends not only on the degree to which they are embedded within local networks of suppliers but also on the presence of links to external markets. Jankelova et al. [7] claim that by optimizing the organizational structure, the company could overcome economic insecurity more easily during the times of the crisis. To keep the company competitive, Rocha and Abreu [8] believe it is important to create conditions for flexible governance regimes. This would help to reduce the power of the large retailers in setting prices and production standards across the value chain, immensely reducing the financial pressures on producers to continuously cut costs, and creating the possibility to invest in a technological and organizational upgrade. Hamel [9] found that a firm with no ambition beyond the investment avoidance and substitution of its partner's competitiveness for its own lack of competitiveness may be perfectly content not to learn from its partner what is likely to undermine the competitiveness and independence. Krugman [10] claims that the company is not competitive unless it can maintain its position on the market, so if it does not increase its performance, it will have to leave the market. Collins and Troilo [11] made an argument for public-private partnerships for innovation as these may increase the competitiveness of small and medium-sized companies. The positive relation between competition and higher economic growth or economic progress was argued by Dutz and Hayri [12] and Porter [13]. Meanwhile, Ireland and Webb [14] and Dyer and Singh [15] associated spending on research and development with a possible competitive advantage. Dosi et al.[16] argue that market shares at the country level are mainly influenced by technology, while the cost advantage or disadvantage does not have a significant impact. Siriphattrasophon [17] claims that businesses that have more international activities are more competitive than others. Bharati and Chaudhury [18] claim that although small and medium-sized enterprises are not generally technology-demanding, they are the innovators themselves as they produce 13 to 14 times more patents per employee per year than large enterprises. Slavik and

Zagorsek [19] claim that in order for the business model to be set as the entrepreneurs want, it must be matched with strategic factors. If the company has some idea of its competitive behavior, it must have a strategy that will allow to exercise it. The challenge of competitiveness (Gibb) [20] is one of finding new ways of converting the learning skills needs for SMEs and the stakeholders into the training demands of the future. According to Quack et al. [21], competences and capabilities of a company can influence its ability to adapt to changing business environment and pressure from global competition with consequent effects for their own economic performance, the performance of their sector, and that of their home nation-state.

3 Methods and Objectives

This paper aims to identify the factors that influence the intensity of local competition the most. Analyzed data were obtained from the World Economic Forum database. The analyzed variables were selected on the basis of Porter's five competitive forces model [5], which identifies five groups of factors affecting the competition: supplier power, buyer power; the threat of substitution; the threat of new entry and competitive rivalry. Based on this model, a search for variables in the database that represent individual groups of factors was conducted. The Global Competitiveness Index dataset that was used contains data for 137 countries of the world, where each county was presented by one data point per variable representing the mean value. The dataset used in this research was acquired between 2017-2018. In general, the survey data could have values from 1 to 7, with 1 being the lowest and 7 being the highest outcome. The variables number of days and the number of tasks required before starting a business are not on a scale, but they are the real values in each country. Questions and variables are described to greater detail in the Global Competitiveness Report 2017-2018[22]. The database was imported into Python and analyzed using the Numpy, Pandas, Matplotlib, and Statsmodels packages. The data were subjected to graphical analysis and inductive statistics. The package used for linear regression is further described by Seabold and Perktold [23]. When analyzing the data, the methods of descriptive statistics and graphical analysis were used to describe the data. Consequently, multiple linear regression, ANOVA, Durbin-Watson test and Jarque-Bera test were used. The method of linear regression was utilized as a tool for comparing the effects of individual exogenous variables on an endogenous variable. Statistically significant were considered results with $p \leq 0.05$.

4 Results

Our research focuses on the intensity of local competition. In order to account for the impact of competition from abroad, the intensity of foreign competition was used as the control variable. Indicators of barriers to market entry (number of days to start a business; the number of procedures to start a business) and production process sophistication represent the threat of new entry and the threat of substitution. These factors make it more difficult for potential competitors to enter the market. They represent barriers that must be overcome by new businesses at a time of uncertainty before their products can be sold or barriers that must be surpassed to acquire customers. Supplier power (local supplier quality; local supplier quantity) and buyer power (buyer sophistication; domestic market size) indicators represent the negotiating power based on the quantity and quality or sophistication of these subjects. The individual variables are described in Table 1.

Table 1. Variable description

Variable	N	Avg	St. d.	Med	Min	Max
Intensity of local competition	137	5.0	0.6	5.1	2.8	6.2
Intensity of foreign competition	137	4.5	0.8	4.5	2.6	6.4
Intellectual property protection	137	4.3	1.0	4.1	2.0	6.6
Number of days to start a business	137	19.4	26.2	11	0.5	230
Number of procedures to start a business	137	6.8	3.1	6	1	20
Buyer sophistication	137	3.5	0.8	3.4	1.8	5.3
Degree of customer orientation	137	4.7	0.7	4.6	2.4	6.2
Domestic market size	137	3.7	1.2	3.7	1	7
Ease of access to loans	137	3.9	0.8	3.9	1.5	5.7
Quality of overall infrastructure	137	4.1	1.1	4.1	1.5	6.7
Production process sophistication	137	4.0	1.1	3.8	2.0	6.5
Company spending on R&D	137	3.6	0.9	3.3	1.9	6.1
Local supplier quality	137	4.4	0.7	4.4	2.4	6.3
Local supplier quantity	137	4.5	0.6	4.6	2.4	6.2

To get the final model, the non-significant variables were removed with respect to adjusted R-squared with the goal of creating a relevant model of the intensity of local competition. The relevant model means that the model explains as much of local competition intensity as possible and consists of only or almost only significant variables. After adjustments, the model displayed in Table 2 emerged. By eliminating non-significant variables, the model improved in the goodness of fit. The model of the intensity of local competition with adjusted $R^2 = 0.746$ consists of the intensity of foreign competition, local supplier quantity, spending on research and development, intellectual property protection, and degree of customer orientation. In this model, the quantity of local suppliers influenced the intensity of local competition the most, followed by the negative impact of company spending on research and development and the positive impact of intensity of foreign competition, the degree of customer orientation and intellectual property protection.

Unlike, in theory, the research found that a larger number of suppliers is associated with higher competition intensity. This may be caused by the nature of the local market where the reason for a large number of suppliers might be given by a number of competing companies whom the suppliers provide the products and so the competition is more intense. This research also found that investing more in research and development is associated with the lower intensity of local competition. As a part of natural explanation research and development is considered to be a potential source of competitive advantage that may create barriers for competitors and thus reduce the intensity of competition. The expected result that the higher intensity of foreign competition will lead to higher intensity of local competition can be explained by the fact that local businesses compete not only with their direct neighbors but also with foreign companies what makes them take into account more subjects making the situation more complex and so the competition more intense.

Table 2. Regression model of the intensity of local competition

	coefficient	standard error	p
Intensity of foreign competition	0.157	0.048	0.001**
Intellectual property protection	0.134	0.058	0.023*
Degree of customer orientation	0.148	0.068	0.032*
Company spending on R&D	-0.161	0.061	0.009**
Local supplier quantity	0.639	0.061	0.000**
R^2	0.755		
Adjusted R^2	0.746		
F (5,131)	80.92**		
Durbin-Watson test	1.840		
Jarque-Bera test	0.226		

*p≤0.05 **p≤0.01

A higher level of intellectual property protection is associated with a higher intensity of local competition. A possible explanation is that the need for proper protection itself arises in a situation that is more likely to be associated with intense competition. In other words, if the protection of the intellectual property is not needed, it is likely that the intensity of competition is not high, because the market is large enough for companies not to worry about their survival. In turn, the positive effect of the degree of customer orientation on the intensity of local competition may be explained as companies that care more for their customers and consider more variables of customer satisfaction. This makes the competition more intense, as competing companies also have to take into account more information.

5 Discussion

The intensity of local competition is a condition with a great impact on business success. This can be seen either as an a priori argument supported by probability, the argument from biology or argument supported by economic research. If there are more subjects selling the same product, there are more options for the buyer from whom to buy the product. This lessens the probability of buying it from one specific company by applying basic probability calculation. If an experiment must result in one of the n outcomes and the outcomes are equally likely, the probability of an outcome is 100/n percent, and so the more companies, the less likely is the success. In biology research, where competition between and within species (Hardin)[24] is studied, the basic principle of competition can be derived as there cannot be coexisting complete competitors. That means increased competition results in increased default. However, economic research further suggests that competition may lead to higher economic growth or progress [12,13]. Therefore, the higher intensity of local competition can be perceived as a threat for emerging companies and companies with prominent weaknesses and as an opportunity for companies with a competitive advantage. This impact of intensity of competitiveness is the reason for the importance of understanding the competitive environment for companies operating in a market or entering one. The results presented in this paper can be interpreted in the following ways. The most interesting findings reveal the nature of the intensity of local competition and factors influencing it. By far, the biggest impact among the studied variables was the impact of local suppliers' quantity. This can be the result of causation or can be used as an indicator. The larger number of suppliers results in stronger negotiating power of buyers. These better conditions may lead to increased incentives to enter the market by other companies. The increased interest leads to a higher number of competitors and so to more intense competition. Other interpretation is from the perspective of the observer using the number of local suppliers as an indicator of competition. In this case, the higher number of companies indicates that there must be a higher number of buyers and so a more intense competition among them. At first sight, the most surprising result is the negative impact of research and development. In this case, the causation is not clear. It could be expected that companies are spending more on research and development because they perceive higher competition intensity. However, the results indicate that spending on research and development might be an important competitive tool creating a possible competitive advantage, as suggested by literature [14,15]. Companies with higher spending on research and development perceive local competition as less intense. This might be the outcome of the competitive advantage gained by the R&D. As the results show in the time of globalisation, the impact of foreign competition is also significant and should not be ignored. Companies entering a local market that might seem vacant have to reconsider this first sight perception and a further look at whether there is a foreign competition that might interfere with their operations. The degree of customer orientation seems to have a straightforward interpretation. The more attention is drawn to customer orientation in the market the more intensity of local competition there is. The explanation would be that if there are more competitors and some of them start to differentiate themselves by intensifying customer orientation, other competitors are also forced to increase their customer orientation. This leads to a higher level of competition. Intellectual property protection indicates the intensity of local competition. If companies compete and the competition arises, the need for intellectual property rights protection also rises, and companies may pressure the government to do so.

6 Conclusion

This paper investigated the factors influencing the intensity of local competition. Examined factors were selected in accordance with Porter's model of the five competitive forces framework, while the data were obtained from the World Economic Forum database. The finding presented in this paper pointed out the factors affecting the most the intensity of local competition. The finding is that a higher number of local suppliers affected the higher intensity of local competition. Furthermore, the intensity of local competition was impacted positively by foreign competition, intellectual property protection, and negatively by spending on research and development. This research may serve as a guideline for companies. If companies are entering a new market, it may help to assess the intensity of local competition by identifying the most influential factors. Furthermore, businesses can use the number of local suppliers and intellectual property protection as indicators of competition intensity. They can also fight the perceived intensity of local competition by investing in research and development.

The limitations of this study are as follow. There is a possibility of generalization of the study as there might be more meaningful influences on an individual basis, that did not show in statistical analysis. There is a possibility of hidden variables that were not considered in the model. The limited control over data collection is given by the use of World Economic Forum databases.

Additionally, an original value was added by this paper with a different point of view. The data used are relevant data selected from a database usually used in macroeconomics. This allowed us to study a problem from a different perspective, which is a valuable scientific approach, observing one phenomenon using different research channels. In future research, it should be considered worthwhile to understand further the mechanisms of factors influencing the intensity of local competition identified in this paper.

Acknowledgment

This paper is a part of VEGA n.1/0017/20.

References

1. G. Shangquan, Economic globalization: trends, risks and risk prevention. *Economic & Social Affairs, CDP Background Paper 1,* (2000)
2. R. Alexander, Sustainability in global production networks–Introducing the notion of extended supplier networks. *Competition & Change* **22**, 255-273. (2018)
3. H. Ma, Competitive advantage and firm performance. *Competitiveness Review: An International Business Journal* **10**, 15-32. (2000)
4. M.J. Dollinger, *Entrepreneurship: Strategies and resources.* (Marsh Publications, 2008)
5. M.E. Porter, The five competitive forces that shape strategy. *Harvard business review* **86**, 78-93 (2008)
6. E.J. Malecki, P. Oinas, *Making connections: technological learning and regional economic change.* (Ashgate Publishing Company, 1999)
7. N. Jankelova, A. Jankurova, M. Benova, Z. Skorkova, Security of the business organizations as a result of the economic crisis. *Entrepreneurship and Sustainability Issues* **5**, 659-671 (2018)
8. R. Rocha, M. Abreu, Emerging strategies and flexible forms of governance: The dynamics of role exchange in local value chains. *Competition & Change* **22**, 363–382 (2018)
9. G. Hamel, Competition for Competence and Inter-Partner Learning Within International Strategic Alliances. *Strategic Management Journal* **12**, 83-103 (1991)
10. P. Krugman, Competitiveness: A Dangerous Obsession. *Foreign Affairs* **73**, 28-44 (1994)
11. M.J. Collins, M.L. Troilo, National factor effects on firm competitiveness and innovation. *Competitiveness Review* **25**, 392-409 (2015)
12. M.A. Dutz, A. Hayri, *Does more intense competition lead to higher growth?* The World Bank (1999)

13. M.E. Porter, The Adam Smith address: location, clusters, and the 'new' microeconomics of competition. *Business Economics* **33**, 7-13 (1998)
14. D.R. Ireland, J.W. Webb, Strategic entrepreneurship: Creating competitive advantage through streams of innovation. *Business horizons* **50**, 49-59 (2007)
15. J.H. Dyer, H. Singh, The relational view: Cooperative strategy and sources of interorganizational competitive advantage. *Academy of management review* **23**, 660-679 (1998)
16. G. Dosi, M. Grazzi, D. Moschella, Technology and costs in international competitiveness: from countries and sectors to firms. *Research Policy* **44**, 1795-1814 (2015)
17. S. Siriphattrasophon, Multi-level factors affecting firm competitiveness in ASEAN region of small and medium enterprises of Thailand. *Kasetsart Journal of Social Sciences* (2017)
18. P. Bharati, A. Chaudhury, *SMEs and competitiveness: The role of information systems* (2009)
19. S. Slavik, B. Zagorsek, How can strategy influence the internal nature of business model? *Journal of applied economic sciences* **12**, 1080-1090 (2017)
20. A.A. Gibb, Small firms' training and competitiveness. Building upon the small business as a learning organization. *International small business journal* **15**, 13-29 (1997)
21. S. Quack, G. Morgan, R. Whitley, *National capitalisms, global competition, and economic performance.* (John Benjamins Publishing, 2000)
22. World Economic Forum, The Global Competitiveness Index dataset 2007-2017, available at: www3.weforum.org/docs/GCR2017-2018/GCI_Dataset_2007-2017.xlsx (2018)
23. S. Seabold, J. Perktold, Statsmodels: Econometric and statistical modeling with python. *Proceedings of the 9th Python in Science Conference* **57**, 61 (Austin, SciPy society, 2010).
24. G. Hardin, The competitive exclusion principle. *Science* **131**, 3409, 1292-1297 (1960)

AUTHOR INDEX